T0183073

Lecture Notes in Computer Science   9124

Commenced Publication in 1973
Founding and Former Series Editors:
Gerhard Goos, Juris Hartmanis, and Jan van Leeuwen

More information about this series at http://www.springer.com/series/7412

Marzena Kryszkiewicz · Sanghamitra Bandyopadhyay
Henryk Rybinski · Sankar K. Pal (Eds.)

Pattern Recognition and Machine Intelligence

6th International Conference, PReMI 2015
Warsaw, Poland, June 30 – July 3, 2015
Proceedings

 Springer

Editors

Marzena Kryszkiewicz
Institute of Computer Science
Warsaw University of Technology
Warsaw
Poland

Henryk Rybinski
Institute of Computer Science
Warsaw University of Technology
Warsaw
Poland

Sanghamitra Bandyopadhyay
Machine Intelligence Unit
Indian Statistical Institute
Kolkata, West Bengal
India

Sankar K. Pal
Indian Statistical Institute
Kolkata, West Bengal
India

ISSN 0302-9743   ISSN 1611-3349   (electronic)
Lecture Notes in Computer Science
ISBN 978-3-319-19940-5   ISBN 978-3-319-19941-2   (eBook)
DOI 10.1007/978-3-319-19941-2

Library of Congress Control Number: 2015940415

LNCS Sublibrary: SL6 – Image Processing, Computer Vision, Pattern Recognition, and Graphics

Springer Cham Heidelberg New York Dordrecht London

Springer International Publishing AG Switzerland is part of Springer Science+Business Media
(www.springer.com)

Preface

This volume contains the papers selected for presentation at the 6th International Conference on Pattern Recognition and Machine Intelligence (PReMI 2015), which was held at Warsaw University of Technology, Warsaw, Poland, during June 30 – July 3, 2015.

PReMI is a conference series that started in 2005. Held every two years, PReMI provides an international forum for exchanging scientific, research, and technological achievements in pattern recognition, machine intelligence, and related fields. In particular, major areas selected for PReMI 2015 include pattern recognition, machine intelligence, image processing, retrieval and tracking, data mining techniques for large-scale data, fuzzy computing, rough sets, bioinformatics, and applications of artificial intelligence. In addition, two special sessions were organized; namely, Special Session on Data Mining Techniques for Large-Scale Data and Special Session on Scalability of Rough Set Methods. Four plenary keynote talks and two tutorials were delivered. The PReMI 2015 conference was accompanied by the Industrial Session on Machine Intelligence and Big Data in the Industry.

PReMI 2015 received 90 submissions that were carefully reviewed by three or more Program Committee members or external reviewers. Papers submitted to special sessions were subject to the same reviewing procedure as those submitted to regular sessions. After a rigorous reviewing process, 54 papers were accepted for presentation at the conference and publication in the PReMI 2015 proceedings volume. This volume also contains one invited paper and three extended abstracts by the plenary keynote speakers.

It is truly a pleasure to thank all those people who helped this volume to come into being and to turn PReMI 2015 into a successful and exciting event. In particular, we would like to express our appreciation for the work of the PReMI 2015 Program Committee members and external reviewers who helped assure the high standards of accepted papers. We would like to thank all the authors of PReMI 2015, without whose high-quality contributions it would not have been possible to organize the conference. We are grateful to the organizers of the PReMI 2015 special sessions: Julian Szymański and Marcin Błachnik (Special Session on Data Mining Techniques for Large-Scale Data) as well as Jarosław Stepaniuk (Special Session on Scalability of Rough Set Methods). We would also like to express our appreciation to the Organizing Committee chairs (Robert Bembenik and Łukasz Skonieczny) for their involvement in all the organizational matters related to PReMI 2015 as well as the creation and maintenance of the conference website. We are grateful to Bożenna Skalska and Joanna Konczak for their administrative work.

Special thanks go to Andrzej Skowron who was *spiritus movens* for this conference to take place in Warsaw and become a successful scientific event in Warsaw. Furthermore, we want to thank the Industrial Session chairs (Piotr Gawrysiak and Dominik Ryżko), and we also gratefully acknowledge the generous help of the remaining PReMI

2015 chairs – Piotr Andruszkiewicz, Grzegorz Protaziuk, Santanu Chaudhury, Sergei Kuznetsov, as well as of the Steering Committee members – Malay K. Kundu and Andrzej Skowron. We wish to express our thanks to George Karypis, Sankar K. Pal, Roman Słowiński, and Xin Yao for accepting to be plenary speakers at PReMI 2015. We also thank the PReMI 2015 tutorial speakers – Gerald Schaefer and Santanu Chaudhury.

Our thanks are due to Alfred Hofmann of Springer for his continuous support and to Anna Kramer and Christine Reiss for their work on the proceedings.

We believe that the proceedings of PReMI 2015 will be a valuable source of reference for your ongoing and future research activities.

June – July 2015

Marzena Kryszkiewicz
Sanghamitra Bandyopadhyay
Henryk Rybinski
Sankar K. Pal

Organization

PReMI 2015 was organized by the Institute of Computer Science, Warsaw University of Technology, Warsaw, Poland, in cooperation with Machine Intelligence Unit, Indian Statistical Institute, Kolkata, India, and Center for Soft Computing Research, Indian Statistical Institute, Kolkata, India.

PReMI 2015 Conference Committee

Honorary Chair

Sankar K. Pal   Indian Statistical Institute, Kolkata, India

Steering Committee

Malay K. Kundu   Indian Statistical Institute, Kolkata, India
Andrzej Skowron   University of Warsaw, Poland

General Chair

Henryk Rybinski   Warsaw University of Technology, Poland

Program Chairs

Sanghamitra   Indian Statistical Institute, Kolkata, India
  Bandyopadhyay
Marzena Kryszkiewicz   Warsaw University of Technology, Poland

Organizing Chairs

Robert Bembenik   Warsaw University of Technology, Poland
Łukasz Skonieczny   Warsaw University of Technology, Poland

Special Session and Tutorial Chairs

Piotr Andruszkiewicz   Warsaw University of Technology, Poland
Grzegorz Protaziuk   Warsaw University of Technology, Poland

Industrial Session Chairs

Piotr Gawrysiak   Warsaw University of Technology, Poland
Dominik Ryżko   Warsaw University of Technology, Poland

International Liaison Chairs

Santanu Chaudhury   Indian Institute of Technology, Delhi, India
Sergei O. Kuznetsov   Higher School of Economics, Moscow, Russia

Program Committee

Cesare Alippi	Politecnico di Milano, Italy
Piotr Andruszkiewicz	Warsaw University of Technology, Poland
Annalisa Appice	Università degli Studi di Bari, Italy
R. Venkatesh Babu	Indian Institute of Science, Bangalore, India
Sanghamitra Bandyopadhyay	Indian Statistical Institute, Kolkata, India
Sivaji Bandyopadhyay	Jadavpur University, Kolkata, India
Minakshi Banerjee	RCC Institute of Information Technology, Kolkata, India
Jan G. Bazan	University of Rzeszow, Poland
Robert Bembenik	Warsaw University of Technology, Poland
Sambhu Nath Biswas	Indian Statistical Institute, Kolkata, India
Marcin Błachnik	Silesian University of Technology, Katowice, Poland
Ilona Bluemke	Warsaw University of Technology, Poland
Michelangelo Ceci	Università degli Studi di Bari, Italy
Basabi Chakraborty	Iwate Prefectural University, Japan
Goutam Chakraborty	Iwate Prefectural University, Japan
Mihir Chakraborty	CSCR, Indian Statistical Institute, Kolkata, India
Amitava Chatterjee	Jadavpur University, Kolkata, India
Sung-Bae Cho	Yonsei University, Korea
Ananda Shankar Chowdhury	Jadavpur University, Kolkata, India
Paweł Cichosz	Warsaw University of Technology, Poland
Davide Ciucci	University of Milano-Bicocca, Italy
Andrzej Czyżewski	Gdansk University of Technology, Poland
Amit Das	Bengal Engineering and Science University, Shibpur, India
Partha Pratim Das	India Institute of Technology, Kharagpur, India
Sukanta Das	Bengal Engineering and Science University, Shibpur, India
Rajat De	Indian Statistical Institute, Kolkata, India
Andries Engelbrecht	University of Pretoria, South Africa
Paolo Gamba	University of Pavia, Italy
Tomasz Gambin	Warsaw University of Technology, Poland
Niloy Ganguly	Indian Institute of Technology, Kharagpur, India
Bernhard Ganter	Dresden University of Technology, Germany
Piotr Gawkowski	Warsaw University of Technology, Poland
Ashish Ghosh	Indian Statistical Institute, Kolkata, India
Kuntal Ghosh	Indian Statistical Institute, Kolkata, India
Susmita Ghosh	Jadavpur University, Kolkata, India
Jarek Gryz	York University, Toronto, Canada
Jerzy Grzymała-Busse	University of Kansas, Kansas, USA
Francisco Herrera	University of Granada, Spain

External Reviewers

Alcalde, Cristina
Artiemjew, Piotr
Banerjee, Abhirup
Banerjee, Sanghita
Benítez Caballero, María José
Betliński, Paweł
Chu, Henry
De, Arijit
Dembski, Jerzy
Demidova, Elena
Diaz, Elizabeth
Gamba, Paolo
Gambin, Tomasz
García-Osorio, César
Gawkowski, Piotr
Guerra, Francesco
Kożuszek, Rajmund

Markkassery, Sreejith
Martincic-Ipsic, Sanda
Meina, Michal
Mondal, Ajoy
Nayak, Losiana
Pal, Jayanta Kumar
Patra, Braja Gopal
Paul, Sushmita
Pio, Gianvito
Roy, Rahul
Roy, Shaswati
Rupino Da Cunha, Paulo
Terziyan, Vagan
Trillo Lado, Raquel
Trzciński, Tomasz
Świeboda, Wojciech

Invited Talks

Granular Mining and Rough-Fuzzy Computing: Data to Knowledge and Big Data Issues

Sankar K. Pal

Machine Intelligence Unit
and
Center for Soft Computing Research
Indian Statistical Institute
Kolkata 700108, India
http://www.isical.ac.in/sankar

Extended Abstract

Pattern recognition and data mining in the framework of machine intelligence are explained. The role of rough sets in uncertainty handling and granular computing is highlighted. Relevance of its integration with fuzzy sets to result in a stronger paradigm for uncertainty handling is explained. Generalized rough sets, rough-fuzzy entropy, different f-information measures, and fuzzy granular social network (FGSN) model are described. FGSN handles the uncertainty arising from vaguely defined closeness or relations of the actors (nodes). Various measures towards this are stated.

Rough-fuzzy image entropy takes care of the fuzziness in boundary regions as well as the rough resemblance among nearby pixels and gray levels. Rough-fuzzy case generation with variable reduced dimension is useful for mining data sets with large dimension and size. Fuzzy granular model of social networks provides a generic platform for its analysis. Fuzzy-rough communities, detected thereby, are more significant when the degree of overlapping between communities increases. f-information measures quantify well the mutual information in efficient feature selection, and the conditional information in measuring the goodness of community structures in network mining. These characteristics are demonstrated for tasks like video tracking, social network analysis and gene/microRNA selection. The role of different kinds of granules is illustrated as well as the concepts of *fuzzy granular* computing and granular *fuzzy computing*.

The talk concludes mentioning their relevance in handling Big data, the challenging issues and the future directions of research.

References

1. Pal, S.K., Mitra, P.: Case generation using rough sets with fuzzy representation. IEEE Trans. Knowl. Data Eng. **16**(3), 292–300 (2004)
2. Sen, D., Pal, S.K.: Generalized rough sets, entropy and image ambiguity measures. IEEE Trans. Syst. Man Cyberns. Part B **39**(1), 117–128 (2009)
3. Maji, P., Pal, S.K.: Feature selection using f-Information measures in fuzzy approximation spaces. IEEE Trans. Knowl. Data Eng. **22**(6), 854–867 (2010)

4. Pal, S.K., Meher, S.K., Dutta, S.: Class-dependent rough-fuzzy granular space, dispersion index and classification. Pattern Recogn. **45**(7), 2690–2707 (2012)
5. Pal, S.K.: Granular mining and rough-fuzzy pattern recognition: a way to natural computation (feature article). IEEE Intell. Inform. Bull. **13**(1), 3–13 (2012)
6. Kundu, S., Pal, S.K.: FGSN: fuzzy granular social networks - model and applications. Inf. Sci. **314**, 100–117 (2015). doi:10.1016/j.ins.2015.03.065
7. Kundu, S., Pal, S.K.: Fuzzy-rough community in social networks. Pattern Recogn. Lett. doi:10.1016/j.patrec.2015.02.005

Constructive Learning of Preferences with Robust Ordinal Regression

Roman Słowiński

Institute of Computing Science, Poznań University of Technology, 60-965 Poznań,
and Systems Research Institute, Polish Academy of Sciences,
01-447 Warsaw, Poland
roman.slowinski@cs.put.poznan.pl

Extended Abstract

The talk is devoted to preference learning in Multiple Criteria Decision Aiding. It is well known that the dominance relation established in the set of alternatives (also called actions, objects, solutions) is the only objective information that comes from a formulation of a multiple criteria decision problem (ordinal classification, or ranking, or choice, with multiobjective optimization being a particular instance). While dominance relation permits to eliminate many irrelevant (i.e., dominated) alternatives, it does not compare completely all of them, resulting in a situation where many alternatives remain incomparable. This situation may be addressed by taking into account preferences of a Decision Maker (DM). Therefore, all decision-aiding methods require some preference information elicited from a DM or a group of DMs. This information is used to build more or less explicit preference model, which is then applied on a non-dominated set of alternatives to arrive at a recommendation (assignment of alternatives to decision classes, or ranking of alternatives from the best to the worst, or the best choice) presented to the DM. In practical decision aiding, the process composed of preference elicitation, preference modeling, and DM's analysis of a recommendation, loops until the DM accepts the recommendation or decides to change the problem setting. Such an interactive process is called constructive preference learning.

I will focus on processing DM's preference information concerning multiple criteria ranking and choice problems. This information has the form of pairwise comparisons of selected alternatives. Research indicates that such preference elicitation requires less cognitive effort from the DM than direct assessment of preference model parameters (like criteria weights or trade-offs between conflicting criteria). I will describe how to construct from this input information a preference model that reconstructs the pairwise comparisons provided by the DM. In general, construction of such a model follows logical induction, typical for learning from examples in AI. In case of utility function preference models, this induction translates into ordinal regression. I will show inductive construction techniques for two kinds of preference models: a set of utility (value) functions, and a set of "*if. . ., then. . .*" monotonic decision rules. An important feature of these construction techniques is identification of all instances of the preference model that are compatible with the input preference information – this permits to draw robust conclusions regarding DM's preferences when any of these models is applied on the considered set of alternatives. These techniques are called Robust Ordinal Regression and Dominance-based Rough Set Approach.

I will also show how these induction techniques, and their corresponding models, can be embedded into an interactive procedure of multiobjective optimization, particularly, in Evolutionary Multiobjective Optimization (EMO), guiding the search towards the most preferred region of the Pareto-front.

References

1. Branke, J., Greco, S., Słowiński, R., Zielniewicz, P.: Learning value functions in interactive evolutionary multiobjective optimization. IEEE Trans. Evol. Comput. **19**(1), 88–102 (2015)
2. Corrente, S., Greco, S., Kadziński, M., Słowiński, R.: Robust ordinal regression in preference learning and ranking. Mach. Learn. **93**, 381–422 (2013)
3. Figueira, J., Greco, S., Słowiński, R.: Building a set of additive value functions representing a reference preorder and intensities of preference: GRIP method. Eur. J. Oper. Res. **195**, 460–486 (2009)
4. Szeląg, M., Greco, S., Słowiński, R.: Variable consistency dominance-based rough set approach to preference learning in multicriteria ranking. Inf. Sci. **277**, 525–552 (2014)
5. Słowiński, R., Greco, S., Matarazzo, B.: Rough-set-based decision support. In: Burke, E.K., Kendall, G. (eds.) Search Methodologies: Introductory Tutorials in Optimization and Decision Support Techniques, Chap. 19, 2nd edn., pp. 557–609. Springer, New York (2014)

Ensemble Approaches in Learning

Xin Yao

CERCIA, School of Computer Science
University of Birmingham, Birmingham B15 2TT, UK
http://www.cs.bham.ac.uk/~xin

Extended Abstract

Designing a monolithic system for a large and complex learning task is hard. Divide-and-conquer is a common strategy in tackling such large and complex problems [1,2]. Ensembles can be regarded an automatic approach towards automatic divide-and-conquer [3,4]. Many ensemble methods, including boosting [5], bagging [6], negative correlation [4], etc., have been used in machine learning and data mining for many years. This talk will describe three research topics in ensemble learning, i.e., multi-objective learning [7,8], online learning with concept drift [9,10], and multi-class imbalance learning [11,12]. Given the important role of diversity in ensemble methods [13,14], some discussions and analyses will be given to gain a better understanding of how and when diversity may help ensemble learning.

Multi-objective learning might first sound strange because the sole objective of learning should be to maximise the generalisation ability of learned models. However, in ensemble learning, we are interested in finding an ensemble of accurate and diverse individual learner. The accuracy and diversity naturally become two objectives. While one could aggregate these two objectives into a single function using a weighted sum, it is often very difficult in practice to tune the weights appropriately. An alternative to aggregating two objectives into one is to use a multi-objective optimisation algorithm to learn the two objectives simultaneously [7]. This is advantageous because a multi-objective optimisation algorithm will find a set of non-dominated solutions, rather than just a single solution, which can naturally be used as individual learners in an ensemble. In a sense, multi-objective ensemble learning natural solves the problem of determining what individuals should be in an ensemble. Once we have the multi-objective learning framework, it is very straightforward to add additional objectives, e.g., by adding an additional regularisation objective [8].

In the real world, data streams are very difficult to deal with, especially their underlying data distributions change, e.g., with concept drifts. What was correct long time ago may not be correct anymore, and what was incorrect might become correct now. It is always a huge challenge to deal with such concept drifts in online learning of data streams. Ensemble methods have long been used in online learning with concept drifts. However, there were few analyses of different roles that the diversity plays during online learning. A detailed analysis of its roles [9] can actually provide insight into different roles of the diversity at different stages of online learning. Such insight can be exploited to develop new online ensemble learning algorithms [10], which have been shown to perform well with or without and the presence of concept drift.

In classification, we are often faced with the situation where the numbers of training examples from different classes are very imbalanced. For example, in fault diagnosis or fraud detection, the majority of available examples are normal cases, the examples for faults or frauds are minority. In such cases, applying off-the-shelf machine learning algorithms may not lead to an appropriate learning outcome because of the bias toward the majority class. Cost sensitive learning can be used for class imbalance learning *if* we know the cost matrix of making different errors. Unfortunately, costs are very difficult to define for many real-world problems. In this case, two alternative approaches have often been followed. One is sampling, i.e., by manipulating the training data, either over-sampling of the minority class or under-sampling of the majority class or both. The other alternative is to design specific algorithms for class imbalance. Ensemble methods have often been used. However, few studies exist on the analysis of why ensembles would be a good choice and what role(s) the diversity plays. An in-depth investigation of the impact of diversity on single class performance in multiple class imbalance learning can shed some light on the challenging problem and potential solutions [11]. One future possibility is to embed the strength of DyS [12], a single MLP learner for multi-class imbalance learning, into an ensemble.

Recently, online class imbalance learning has attracted more attentions from researchers [15]. This is more than just the combination of online learning and class imbalance learning because it is impossible to know in advance which class is a majority and which is a minority. In fact, whether a class is a minority or majority depends on time. A learning algorithm has to learn that. The learning algorithm also has to detect concept drifts and react accordingly. There are still many unsolved problems on this new research topic [15].

References

1. Darwen, P.J., Yao, X.: Speciation as automatic categorical modularization. IEEE Trans. Evol. Comput. **1**(2), 101–108 (1997)
2. Khare, V., Yao, X., Sendhoff, B.: Multi-network evolutionary systems and automatic problem decomposition. Int. J. Gen. Syst. **35**(3), 259–274 (2006)
3. Yao, X., Liu, Y.: Making use of population information in evolutionary artificial neural networks. IEEE Trans. Syst. Man Cybern. Part B: Cybern. **28**(3), 417–425 (1998)
4. Liu, Y., Yao, X.: Ensemble learning via negative correlation. Neural Netw. **12**(10), 1399–1404 (1999)
5. Schapire, R.E.: The strength of weak learnability. Mach. Learn. **5**(2), 197–227 (1990)
6. Breiman, L.: Bagging predictors. Mach. Learn. **24**(2), 123–140 (1996)
7. Chandra, A., Yao, X.: Ensemble learning using multi-objective evolutionary algorithms. J. Math. Model. Algorithms **5**(4), 417–445 (2006)
8. Chen, H., Yao, X.: Multiobjective neural network ensembles based on regularized negative correlation learning. IEEE Trans. Knowl. Data Eng. **22**(12), 1738–1751 (2010)
9. Minku, L.L., White, A., Yao, X.: The impact of diversity on on-line ensemble learning in the presence of concept drift. IEEE Trans. Knowl. Data Eng. **22**(5), 730–742 (2010)
10. Minku, L.L., Yao, X.: DDD: a new ensemble approach for dealing with concept drift. IEEE Trans. Knowl. Data Eng. **24**(4), 619–633 (2012)

11. Wang, S., Yao, X.: Multi-class imbalance problems: analysis and potential solutions. IEEE Trans. Syst. Man Cybern. Part B **42**(4), 1119–1130 (2012)
12. Lin, M., Tang, K., Yao, X.: A dynamic sampling approach to training neural networks for multi-class imbalance classification. IEEE Trans. Neural Netw. Learn.Syst. **24**(4), 647–660 (2013)
13. Tang, E.K., Suganthan, P.N., Yao, X.: An analysis of diversity measures. Mach. Learn. **65**, 247–271 (2006)
14. Brown, G., Wyatt, J.L., Harris, R., Yao, X.: Diversity creation methods: a survey and categorisation. Inf. Fusion **6**(1), 5–20 (2005)

Contents

Image Retrieval

Image Tracking

Pattern Recognition

Rough Sets

Bioinformatics

Applications of Artificial Intelligence

Invited Paper

Recent Advances in Recommender Systems and Future Directions

Xia Ning[1] and George Karypis[2(✉)]

[1] Department of Computer and Information Science, IUPUI, Indianapolis,
IN 46202, USA
xning@cs.iupui.edu
[2] Department of Computer Science and Engineering, University of Minnesota,
Twin Cities, MN 55455, USA
karypis@cs.umn.edu

Abstract. This article presents an overview of recent methodological advances in developing nearest-neighbor-based recommender systems that have substantially improved their performance. The key components in these methods are: (i) the use of statistical learning to estimate from the data the desired user-user and item-item similarity matrices, (ii) the use of lower-dimensional representations to handle issues associated with data sparsity, (iii) the combination of neighborhood and latent space models, and (iv) the direct incorporation of auxiliary information during model estimation. The article will also provide illustrative examples for these methods in the context of item-item nearest-neighbor methods for rating prediction and Top-N recommendation. In addition, the article will present an overview of exciting new application areas of recommender systems along with the challenges and opportunities associated with them.

1 Introduction

Recommender systems [1] are designed to identify the items that a user will like or find useful based on the user's prior preferences and activities. These systems have become ubiquitous and are an essential tool for information filtering and (e-)commerce [2]. Over the years, collaborative filtering (CF) [3], which derives these recommendations by leveraging past activities of groups of users, has emerged as the most prominent approach in recommender systems. Among the multitude of CF methods that have been developed, user- and item-based nearest-neighbor approaches are the simplest to understand and are easy to extend to capture different user behavioral models and types of available information. However, in their classical forms [3–8], the performance of these methods is worse than that of latent-space based approaches [9–14].

In this article, we present an overview of recent methodological advances in developing nearest-neighbor-based CF methods for recommender systems that have substantially improved their performance. In specific, we overview the methods that (i) use statistical learning to estimate from the data the desired

© Springer International Publishing Switzerland 2015
M. Kryszkiewicz et al. (Eds.): PReMI 2015, LNCS 9124, pp. 3–9, 2015.
DOI: 10.1007/978-3-319-19941-2_1

user-user and item-item similarity matrices, (ii) use lower-dimensional representations to handle issues associated with sparsity, (iii) combine neighborhood and latent space models, and (iv) directly incorporate auxiliary information during model estimation. We provide illustrative examples for these methods in the context of item-item nearest-neighbor methods for rating prediction and Top-N recommendation. We also briefly discuss the reasons as to why such methods achieve superior performance and derive insights from there for further development. In addition, we present an overview of exciting new application areas of recommender systems along with the challenges and opportunities associated with them.

2   Review of Previous Research

In the conventional nearest-neighbor-based CF methods [3–8, 15–17], the user-item ratings stored in the system are directly used to predict ratings or preferences for a user on certain items. This has been done in two ways known as *user-based* recommendation and *item-based* recommendation. In user-based recommendation methods such as those used in GroupLens [6], Bellcore video [15] and Ringo [17], a set of *nearest user neighbors* for a target user is first identified as the users who have most similar preference patterns as the target user over a set of common items. Then the preferences from such neighboring users on a certain item are leveraged to produce a recommendation score of that item to the target user. In item-based approaches [3,5,7], on the other hand, a set of *nearest item neighbors* for a certain item is first identified as those that have been preferred in a most similar fashion as the item of interest by a set of common users. Then the recommendation score of the item to a user is generated by incorporating the user's preferences on the neighboring items.

The fact that the conventional nearest-neighbor-based CF methods work well in practice is largely due to that the available information of user-item preferences is typically very sparse, but such CF methods can capture and utilize the most important signals among the sparse data using simplistic and non-parametric approaches. Nearest-neighbor-based CF methods are intuitive, easy in computation, and very scalable to large e-commerce datasets and thus suitable for really applications. Although there have been numerous other recommendation methods developed over the years, particularly the latent-space-based methods [9–14], which involve more complicated modeling, demand much more computational resources and could achieve better recommendation performance, nearest-neighbor-based CF methods still remain as a strong baseline particularly when the trade-off between computational costs and performance is a major consideration.

3   CF from Data-Driven Nearest Item Neighbors: SLIM

Conventionally, the item-item similarities used in CF methods are calculated using a pre-defined similarity function, typically cosine similarity, correlation coefficient or their variations. A drawback of using pre-defined similarity functions is that they cannot adapt to different datasets and therefore may lead to poor neighborhood structures and thus sub-optimal recommendation results. A recent advance is to derive the similarity matrices from data rather than use any pre-defined similarity functions. A representative neighborhood-learning recommendation method is the Sparse LInear Methods (SLIM) [18]. In SLIM, the recommendation score $\tilde{r}_{ui}$ for a user u on an item i is predicted as a sparse aggregation of existing ratings in a user's profile, that is,

$$\tilde{r}_{ui} = \mathbf{r}_u \mathbf{w}_i^\mathsf{T},$$

where $\mathbf{r}_u$ is the user u's rating profile over items and $\mathbf{w}_i^\mathsf{T}$ is a sparse row vector of item similarities with respect to item i. The non-zero entries in $\mathbf{w}_i^\mathsf{T}$ correspond to the nearest items neighbors of item i. The item neighborhood matrix $W = [\mathbf{w}_1, \mathbf{w}_2, \cdots, \mathbf{w}_n]$ is learned by minimizing the reconstruction error of the user-item data R using item-based CF with item neighbors represented in W. In specific, the optimization problem is formulated as follows,

$$\underset{W}{\text{minimize}} \quad \frac{1}{2}\|R - RW\|_F^2 + \frac{\beta}{2}\|W\|_F^2 + \lambda\|W\|_1$$
$$\text{subject to} \quad W \geq 0$$
$$\text{diag}(W) = 0,$$

where both the non-negativity constraint and the ℓ_1 regularization on W enforce a sparse and positive neighborhood for each item. Extensive experiments as in [18] demonstrate that SLIM outperforms the state-of-the-art latent-space-based methods in terms of recommendation performance. Meanwhile, SLIM is scalable to large datasets, which makes SLIM much more applicable in real applications. The success of SLIM validates CF as a fundamental framework for recommendation problems, and meanwhile demonstrates the advantage of data-driven item neighborhoods over the conventional hand-crafted similarity metrics in real problems.

4 CF from Factorized Item Similarities: FISM

A remaining issue for SLIM is that when the use-item data is very sparse, it is difficult to well estimate W. The data sparsity issue has substantially challenged almost all the CF based recommendation methods, while latent-space-based (LS) methods provide an appropriate remedy that consequently inspires the combination of CF and LS. The Factorized Item Similarity Method (FISM) [19] represents a recent effort along this line. In FISM, the recommendation score $\tilde{r}_{ui}$ for a user u on an item i is calculated from an aggregation of the items that have been rated by u and that are also similar to item i, where the item-item similarity between two items i and j is factorized and calculated as a dot product of two latent item factors $\mathbf{p}_j$ and $\mathbf{q}_i$. In specific, $\tilde{r}_{ui}$ is calculated as follows,

$$\tilde{r}_{ui} = b_u + b_i + (n_u^+)^{-\alpha} \sum_{j \in \mathcal{R}_u^+} \mathbf{p}_j \mathbf{q}_i^\mathsf{T}, \tag{1}$$

where $\mathcal{R}_u^+$ is the set of items that have been rated by user u, $n_u^+ = |\mathcal{R}_u^+|$, and b_u and b_i are user and item bias, respectively. The learning of $\mathbf{p}_j$ and $\mathbf{q}_i$ can be done by minimizing the reconstruction error or by minimizing the ranking divergence using Eq. 1 on the training data. The experiments in [19] show that when the user-item data is sparse, FISM outperforms SLIM in the recommendation performance. FISM provides a general framework that combines neighborhood-based CF and LS-based factorization of data-driven item-item similarities so as to effectively handle the data sparsity issues and achieve good recommendation performance.

5   CF from User-Specific Feature-Based Similarities: UFSM

In addition to leveraging advanced modeling and learning techniques as in FISM to accommodate for data sparsity, an alternative is to leverage additional information sources. The increasing amount of auxiliary information associated with the items in E-commerce applications has provided a very rich source of information that, once properly exploited and incorporated, can significantly improve the performance of the conventional CF methods. Thus, a recent trend is to incorporate auxiliary information to improve nearest-neighbor-based CF methods [20–22]. For example, in the User-specific Feature-based Similarity Models (UFSM) [20], the recommendation score $\tilde{r}_{ui}$ for a user u on an item i is calculated as the aggregation of multiple user-specific item-item similarities (i.e., l different similarity functions $\text{sim}(i, j)$), that is,

$$\tilde{r}_{ui} = \sum_{j \in \mathcal{R}_u^+} \sum_{d=1}^{l} m_{u,d}\, \text{sim}_d(i, j),$$

where $\text{sim}_d(i, j)$ is the d-th similarity between item i and item j, and it is estimated from the feature vectors $\boldsymbol{f}_i$ and $\boldsymbol{f}_j$ of items i and j, respectively, as follows,

$$\text{sim}_d(i, j) = \boldsymbol{w}_d(\boldsymbol{f}_i \odot \boldsymbol{f}_j)^{\mathsf{T}},$$

The Feature-based factorized Bilinear Similarity Model (FBSM) proposed in [21] extends UFSM by modeling the item-item similarity $\text{sim}(i, j)$ as a bilinear function of their features, that is,

$$\text{sim}(i, j) = \boldsymbol{f}_i^{\mathsf{T}} W \boldsymbol{f}_j$$

where W is the weight matrix which captures correlation among item features, and it is further factorized as follows so as to deal with data sparsity issues during learning,

$$W = D + V^{\mathsf{T}} V,$$

where D is a diagonal matrix and V is low-rank.

UFSM and FBSM calculate item-item similarities only from item features. This characteristics enables them to conduct cold-start recommendations for new items when there is no existing rating information for the new items. As demonstrated in [20] and [21], the performance of UFSM and FBSM for cold-start recommendations is superior to that of the state-of-the-art methods.

A different way to leverage auxiliary information is to use such information to bias the learning of an existing CF method. For example, SLIM is extended to incorporate item features in such a way [22]. In the collective SLIM method (cSLIM), it is imposed that the item-item similarities calculated from user-item information and the item-item similarities calculated from item features are identical, while in the relaxed collective SLIM (RCSLIM), the item-item similarities calculated from both aspects are close. In these methods, the item features are used to bias the learning of item neighbors so that the neighbor structures conform to and also encode the information from item features. It is demonstrated in [22] that when the user-item information is sparse, item features can play an important role for CF methods that use such information to achieve good recommendation performance.

6   Future Directions on Nearest-Neighbor-Based CF

There have existed other methods that have substantially improved conventional CF methods. Such methods include the ones that can capture high-order relations among item similarities [23], the methods that learn and utilize non-linear relations among items [24], etc. However, to make CF methods fully personalized, highly scalable and sufficiently robust against data sparsity and meanwhile produce high-quality recommendations, significant efforts from recommender system communities have been continuously dedicated. It has been recognized [25] that items may fall into clusters and thus item-item similarities may have local structures that may be sufficiently different from other local structures and from global structures, which leads to potential future research that discovers and incorporates local item neighbors into conventional CF methods. Fast and scalable learning algorithms are demanded for such methods once non-linear similarity structures are involved. On the other hand, dynamic components (e.g., user preferences change over time) have become ubiquitous among recommender systems, which may result in dynamically evolving user/item neighborhood structures. Such evolvement may exhibit interesting signals from which novel knowledge can be derived and used to predict future user preference/needs and make recommendations correspondingly (e.g., to recommend TV shows, to recommend courses). Another interesting research topic would be to develop scalable and efficient methods that can effectively incorporate heterogeneous auxiliary information from various static/dynamic sources in a systematical way into CF methods.

References

1. Ricci, F., Rokach, L., Shapira, B., Kantor, P.B. (eds.): Recommender Systems Handbook. Springer, New York (2011)

2. Schafer, J.B., Konstan, J., Riedl, J.: Recommender systems in e-commerce. In: Proceedings of the 1st ACM Conference on Electronic Commerce, EC 1999, pp. 158–166. ACM, New York (1999)
3. Sarwar, B., Karypis, G., Konstan, J., Reidl, J.: Item-based collaborative filtering recommendation algorithms. In: WWW 2001, Proceedings of the 10th International Conference on World Wide Web, pp. 285–295. ACM, New York, NY, USA (2001)
4. Delgado, J., Ishii, N.: Memory-based weighted majority prediction for recommender systems. In: Proceedings of the ACM SIGIR 1999 Workshop on Recommender Systems (1999)
5. Deshpande, M., Karypis, G.: Item-based top-N recommendation algorithms. ACM Trans. Inf. Syst. **22**, 143–177 (2004)
6. Konstan, J.A., Miller, B.N., Maltz, D., Herlocker, J.L., Gordon, L.R., Riedl, J.: GroupLens: applying collaborative filtering to usenet news. Commun. ACM **40**, 77–87 (1997)
7. Linden, G., Smith, B., York, J.: Amazon.com recommendations: item-to-item collaborative filtering. IEEE Internet Comput. **7**, 76–80 (2003)
8. Nakamura, A., Abe, N.: Collaborative filtering using weighted majority prediction algorithms. In: ICML 1998, Proceedings of the 15th International Conference on Machine Learning, pp. 395–403. Morgan Kaufmann Publishers Inc., San Francisco, CA, USA (1998)
9. Hu, Y., Koren, Y., Volinsky, C.: Collaborative filtering for implicit feedback datasets. In: Proceedings of the 2008 Eighth IEEE International Conference on Data Mining, pp. 263–272. IEEE Computer Society, Washington, DC, USA (2008)
10. Koren, Y.: Factorization meets the neighborhood: a multifaceted collaborative filtering model. In: Proceeding of the 14th ACM SIGKDD International Conference on Knowledge Discovery and Data Mining, KDD 2008, pp. 426–434. ACM, New York, NY, USA (2008)
11. Pan, R., Zhou, Y., Cao, B., Liu, N.N., Lukose, R., Scholz, M., Yang, Q.: One-class collaborative filtering. In: Proceedings of the 2008 Eighth IEEE International Conference on Data Mining, pp. 502–511. IEEE Computer Society, Washington, DC, USA (2008)
12. Rennie, J.D.M., Srebro, N.: Fast maximum margin matrix factorization for collaborative prediction. In: Proceedings of the 22nd International Conference on Machine learning, ICML 2005, pp. 713–719. ACM, New York, NY, USA (2005)
13. Sindhwani, V., Bucak, S.S., Hu, J., Mojsilovic, A.: One-class matrix completion with low-density factorizations. In: Proceedings of the 2010 IEEE International Conference on Data Mining, ICDM 2010, pp. 1055–1060, IEEE Computer Society, Washington, DC, USA (2010)
14. Weimer, M., Karatzoglou, A., Smola, A.: Improving maximum margin matrix factorization. In: Daelemans, W., Goethals, B., Morik, K. (eds.) ECML PKDD 2008, Part I. LNCS (LNAI), vol. 5211, p. 14. Springer, Heidelberg (2008)
15. Hill, W., Stead, L., Rosenstein, M., Furnas, G.: Recommending and evaluating choices in a virtual community of use. In: CHI 1995, Proceedings of the SIGCHI Conference on Human Factors in Computing Systems, pp. 194–201. ACM Press/Addison-Wesley Publishing Co., New York, NY, USA (1995)
16. Resnick, P., Iacovou, N., Suchak, M., Bergstrom, P., Riedl, J.: GroupLens: an open architecture for collaborative filtering of netnews. In: CSCW 1994, Proceedings of the 1994 ACM Conference on Computer Supported Cooperative Work, pp. 175–186. ACM, New York, NY, USA (1994)

17. Shardanand, U., Maes, P.: Social information filtering: algorithms for automating "word of mouth". In: CHI 1995, Proceedings of the SIGCHI Conference on Human factors in Computing Systems, pp. 210–217. ACM Press/Addison-Wesley Publishing Co., New York, NY, USA (1995)
18. Ning, X., Karypis, G.: Slim: sparse linear methods for top-n recommender systems. In: Proceedings of 11th IEEE International Conference on Data Mining. pp. 497–506 (2011)
19. Kabbur, S., Ning, X., Karypis, G.: FISM: factored item similarity models for top-n recommender systems. In: Proceedings of the 19th ACM SIGKDD International Conference on Knowledge Discovery and Data Mining, KDD 2013, pp. 659–667. ACM, New York, NY, USA (2013)
20. Elbadrawy, A., Karypis, G.: User-specific feature-based similarity models for top-n recommendation of new items. ACM Trans. Intell. Syst. Technol. 6(3) (2015)
21. Sharma, M., Zhou, J., Hu, J., Karypis, G.: Feature-based factorized bilinear similarity model for cold-start top-n item recommendation (2015)
22. Ning, X., Karypis, G.: Sparse linear models with side-information for top-n recommender systems. In: RecSys 2012 (2012)
23. Christakopoulou, E., Karypis, G.: HOSLIM: higher-order sparse linear method for top-N recommender systems. In: Tseng, V.S., Ho, T.B., Zhou, Z.-H., Chen, A.L.P., Kao, H.-Y. (eds.) PAKDD 2014, Part II. LNCS, vol. 8444, pp. 38–49. Springer, Switzerland (2014)
24. Kabbur, S., Karypis, G.: Nlmf: Nonlinear matrix factorization methods for top-n recommender systems. In: 2014 IEEE International Conference on Data Mining Workshop (ICDMW), pp. 167–174 (2014)
25. Lee, J., Bengio, S., Kim, S., Lebanon, G., Singer, Y.: Local collaborative ranking. In: Proceedings of the 23rd International Conference on World Wide Web, WWW 2014, pp. 85–96. ACM, New York, NY, USA (2014)

Foundations of Machine Learning

On the Number of Rules and Conditions in Mining Data with Attribute-Concept Values and "Do Not Care" Conditions

Patrick G. Clark[1] and Jerzy W. Grzymala-Busse[1,2]([✉])

[1] Department of Electrical Engineering and Computer Science,
University of Kansas, Lawrence, KS 66045, USA
patrick.g.clark@gmail.com
[2] Department of Expert Systems and Artificial Intelligence,
University of Information Technology and Management, 35-225 Rzeszow, Poland
jerzy@ku.edu

Abstract. In this paper we discuss two interpretations of missing attribute values: attribute-concept values and "do not care" conditions. Experiments were conducted on eight kinds of data sets, using three types of probabilistic approximations: singleton, subset and concept. Rules were induced by the MLEM2 rule induction system. Our main objective was to test which interpretation of missing attribute values provides simpler rule sets in terms of the number of rules and the total number of conditions. Our main result is that experimental evidence exists showing rule sets induced from data sets with attribute-concept values are simpler than the rule sets induced from "do not care" conditions.

1 Introduction

The most fundamental ideas of rough set theory are lower and upper approximations. In this paper we study probabilistic approximations. A probabilistic approximation, associated with a probability α, is a generalization of the standard approximation. For $\alpha = 1$, the probabilistic approximation becomes the lower approximation; for very small positive α, it becomes the upper approximation. Research on theoretical properties of probabilistic approximations started from [16] and then continued in many papers, see, e.g., [15–17,19–21].

Incomplete data sets may be analyzed using global approximations such as singleton, subset and concept [8–10]. Probabilistic approximations for incomplete data sets and based on an arbitrary binary relation were introduced in [12]. The first experimental results using probabilistic approximations were published in [1].

For our experiments we used eight incomplete data sets with two types of missing attribute values: attribute-concept values [11] and "do not care" conditions [4,13,18]. Additionally, in our experiments we used three types of probabilistic approximations: singleton, subset and concept.

In [3], the results indicate that rule set performance in terms of error rate is not significantly different for both missing attribute value interpretations. As a

© Springer International Publishing Switzerland 2015
M. Kryszkiewicz et al. (Eds.): PReMI 2015, LNCS 9124, pp. 13–22, 2015.
DOI: 10.1007/978-3-319-19941-2_2

result, given two rule sets with the same error rate, the more desirable would be the least complex, both for comprehension and computation performance. Therefore, the main objective of this paper is research on the complexity of rule sets induced from data sets with attribute-concept values and "do not care" conditions. Complexity is defined in terms of the number of rules and the number of rule conditions, with larger numbers indicating greater complexity.

Initially, the total number of rules and conditions in rule sets induced from incomplete data sets with attribute-concept values and "do not care" conditions were studied in [2]. However, in [2] only one type of probabilistic approximations was considered (concept) while in this paper we consider three types of probabilistic approximations (singleton, subset and concept). Additionally, in [2] only three values of α were discussed (0.001, 0.5 and 11.0) while in this paper we consider eleven values of α (0.001, 0.1, 0.2,..., 1.0).

Note that there are dramatic differences in complexity of rule sets induced from data sets with attribute-concept values and "do not care" conditions. For example, for the *bankruptcy* data set and concept approximation with $\alpha = 1.0$, the rule set induced from this data set in which missing attribute values were interpreted as attribute-concept values has four rules with seven conditions, while the rule set induced from the same data set in which missing attribute values were interpreted as "do not care" conditions has 13 rules with 31 conditions. The error rate, measured by ten-fold cross validation for the data set with attribute-concept values is 24.24 %, while the error rate for the same data set with "do not care" conditions is 37.88 %.

Our main result is that the simpler rule sets are induced from data sets in which missing attribute values are interpreted as attribute-concept values.

Our secondary objective was to identify the probabilistic approximation (singleton, subset or concept) that is associated with the lowest rule complexity. Our conclusion is that there is weak evidence that the best probabilistic approximation is subset.

2   Incomplete Data

We assume that the input data sets are presented in the form of a *decision table*. Rows of the decision table represent *cases*, while columns are labeled by *variables*. The set of all cases will be denoted by U. Independent variables are called *attributes* and a dependent variable is called a *decision* and is denoted by d. The set of all attributes will be denoted by A. The value for a case x and an attribute a will be denoted by $a(x)$.

In this paper we distinguish between two interpretations of missing attribute values: attribute-concept values and "do not care" conditions. *Attribute-concept values*, denoted by "$-$", indicate that the missing attribute value may be replaced by any of the values that have been specified for that attribute in a given concept. For example, if a patient is sick with flu, and if for other such patients the value of temperature is high or very-high, then we will replace the missing attribute values of temperature by values high and very-high, for details see [11].

"Do not care" conditions, denoted by "*", mean that the original attribute values are irrelevant, so we may replace them by any attribute value, for details see [4,13,18].

One of the most important ideas of rough set theory [14] is an indiscernibility relation, defined for complete data sets. Let B be a nonempty subset of A. The indiscernibility relation $R(B)$ is a relation on U defined for $x, y \in U$ as follows:

$$(x, y) \in R(B) \text{ if and only if } \forall a \in B \ (a(x) = a(y)).$$

The indiscernibility relation $R(B)$ is an equivalence relation. Equivalence classes of $R(B)$ are called *elementary sets* of B and are denoted by $[x]_B$. A subset of U is called *B-definable* if it is a union of elementary sets of B.

The set X of all cases defined by the same value of the decision d is called a *concept*. The largest B-definable set contained in X is called the *B-lower approximation* of X, denoted by $appr_B(X)$, and defined as follows

$$\cup\{[x]_B \mid [x]_B \subseteq X\},$$

while the smallest B-definable set containing X, denoted by $\overline{appr}_B(X)$ is called the *B-upper approximation* of X, and is defined as follows

$$\cup\{[x]_B \mid [x]_B \cap X \neq \emptyset\}.$$

For a variable a and its value v, (a, v) is called a variable-value pair. A *block* of (a, v), denoted by $[(a, v)]$, is the set $\{x \in U \mid a(x) = v\}$ [5].

For incomplete decision tables the definition of a block of an attribute-value pair is modified in the following way.

– If for an attribute a there exists a case x such that the corresponding value is an attribute-concept value, i.e., $a(x) = \ -$, then the corresponding case x should be included in blocks $[(a, v)]$ for all specified values $v \in V(x, a)$ of attribute a, where $V(x, a)$ is defined as follows

$$\{a(y) \mid a(y) \text{ is specified, } y \in U, \ d(y) = d(x)\},$$

– If for an attribute a there exists a case x such that $a(x) = \ *$, i.e., the corresponding value is a "do not care" condition, then the case x should not be included in any blocks $[(a, v)]$ for all values v of attribute a.

For a case $x \in U$ and $B \subseteq A$, the *characteristic set* $K_B(x)$ is defined as the intersection of the sets $K(x, a)$, for all $a \in B$, where the set $K(x, a)$ is defined in the following way:

– If $a(x)$ is specified, then $K(x, a)$ is the block $[(a, a(x))]$ of attribute a and its value $a(x)$,
– If $a(x) = \ -$, then the corresponding set $K(x, a)$ is equal to the union of all blocks of attribute-value pairs (a, v), where $v \in V(x, a)$ if $V(x, a)$ is nonempty. If $V(x, a)$ is empty, $K(x, a) = U$,
– If $a(x) = *$ then the set $K(x, a) = U$, where U is the set of all cases.

3   Probabilistic Approximations

For incomplete data sets we may define approximations in many different ways [8]. For the lack of space, we are going to define only probabilistic approximations.

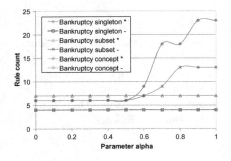

Fig. 1. Size of the rule set for the *Bankruptcy* data set

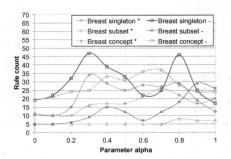

Fig. 2. Size of the rule set for the *Breast cancer* data set

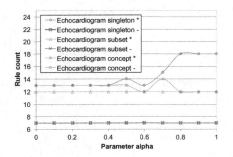

Fig. 3. Size of the rule set for the *Echocardiogram* data set

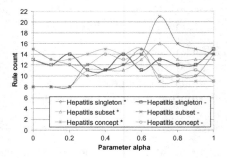

Fig. 4. Size of the rule set for the *Hepatitis* data set

A B-singleton probabilistic approximation of X with the threshold α, $0 < \alpha \leq 1$, denoted by $appr_{\alpha,B}^{singleton}(X)$, is defined as follows

$$\{x \mid x \in U,\ Pr(X \mid K_B(x)) \geq \alpha\},$$

where $Pr(X \mid K_B(x)) = \frac{|X \cap K_B(x)|}{|K_B(x)|}$ is the conditional probability of X given $K_B(x)$ and $|Y|$ denotes the cardinality of set Y.

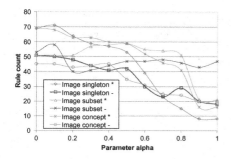

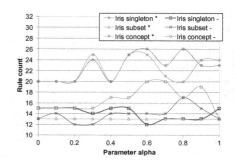

Fig. 5. Size of the rule set for the *Image segmentation* data set

Fig. 6. Size of the rule set for the *Iris* data set

A B-subset probabilistic approximation of the set X with the threshold α, $0 < \alpha \leq 1$, denoted by $appr^{subset}_{\alpha,B}(X)$, is defined as follows

$$\cup\{K_B(x) \mid x \in U,\ Pr(X \mid K_B(x)) \geq \alpha\}.$$

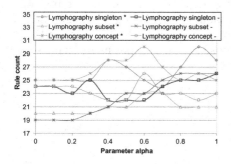

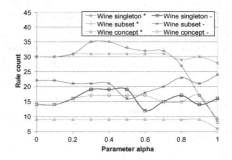

Fig. 7. Size of the rule set for the *Lymphography* data set

Fig. 8. Size of the rule set for the *Wine recognition* data set

A B-concept probabilistic approximation of the set X with the threshold α, $0 < \alpha \leq 1$, denoted by $appr^{concept}_{\alpha,B}(X)$, is defined as follows

$$\cup\{K_B(x) \mid x \in X,\ Pr(X \mid K_B(x)) \geq \alpha\}.$$

For simplicity, the *A-singleton probabilistic approximation* will be called a *singleton probabilistic approximation*, *A-subset probabilistic approximation* will be called a *subset probabilistic approximation*, and *A-concept probabilistic approximation* will be called a *concept probabilistic approximation*.

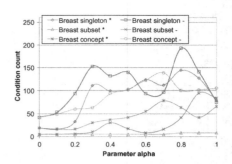
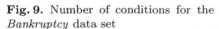

Fig. 9. Number of conditions for the *Bankruptcy* data set

Fig. 10. Number of conditions for the *Breast cancer* data set

4 Experiments

Our experiments were conducted on eight data sets that are available from the University of California at Irvine *Machine Learning Repository*. For every data set a template was created by replacing randomly 35 % of existing specified attribute values by *attribute-concept values*. The same template was used for constructing a corresponding data set with *"do not care" conditions*, by replacing "−"s by "*"s. For two data sets, *bankruptcy* and *iris*, replacing more than 35 % of existing specified values by missing attribute values resulted in cases where all attribute values were missing. Hence we used for our experiments data sets with exactly 35 % missing attribute values.

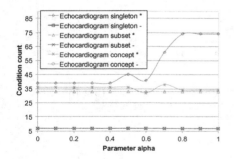
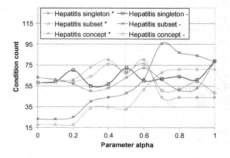

Fig. 11. Number of conditions for the *Echocardiogram* data set

Fig. 12. Number of conditions for the *Hepatitis* data set

In our experiments, for any data set with given type of missing attribute values a rule set was induced using three types of probabilistic approximations: singleton, subset and concept, resulting in 24 combinations. For every such combination, rule sets induced from a data set with attribute-concept values and the corresponding data set with "do not care" conditions were induced, for all eleven values of the parameter α, $\alpha = 0.001$, 01, 0.2,..., 1.0. Both the total number of rules and the total number of conditions in the rule set were compared using the Wilcoxon matched-pairs signed rank test with a 5 % level of significance, two-tailed test.

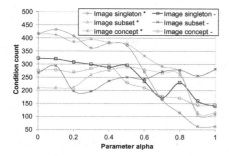

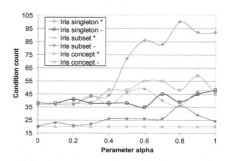

Fig. 13. Number of conditions for the *Image segmentation* data set

Fig. 14. Number of conditions for the *Iris* data set

In our experiments, we used the MLEM2 rule induction algorithm of the Learning from Examples using Rough Sets (LERS) data mining system [1,6,7]. Results of our experiments are presented in Figs. 1, 2, 3, 4, 5, 6, 7, 8, 9, 10, 11, 12, 13, 14, 15 and 16.

The total number of rules was smaller for attribute-concept values than for "do not care" conditions for 13 combinations: for the *bankruptcy* and *echocardiogram* data sets with all three types of probabilistic approximations, for the *image* data set with concept probabilistic approximations, and for the *iris* and *lymphography* data sets for singleton and concept probabilistic approximations. On the other hand, the total number of rules was smaller for "do not care" conditions than for attribute-concept values for five combinations: for the *breast cancer* data set with singleton, subset and concept approximations and for the *hepatitis* and *wine recognition* data sets with subset probabilistic approximations. For the remaining six combinations of the data set and probabilistic approximation type the difference between the number of rules induced from the attribute-concept values and "do not care" conditions was statistically insignificant.

Similarly, for the same 24 combinations we compared the total number of conditions in rule sets. For 13 combinations the total number of conditions was smaller for data sets with attribute-concept values than for "do not care" conditions: for the *bankruptcy* and *echocardiogram* data sets with all three types of probabilistic approximations, for the *image* data set and concept probabilistic approximations and for the *iris* and *lymphography* data sets with singleton and subset probabilistic approximations and for the and *wine recognition* data set with singleton and subset approximations. However, for 5 combinations the

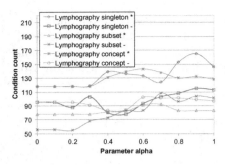
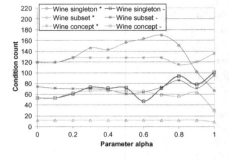

Fig. 15. Number of conditions for the *Lymphography* data set

Fig. 16. Number of conditions for the *Wine recognition* data set

total number of conditions was smaller for "do not care" conditions than for attribute-concept values: for the *breast cancer* data set with all three types of probabilistic approximations, for the *hepatitis* data set with subset probabilistic approximations and for the *wine recognition* data set with concept approximations.

We may conclude that there is some evidence to support the idea that rule sets induced from data sets with attribute-concept values are simpler than rule sets induced from data sets with "do not care" conditions.

Our secondary objective was to select a type of probabilistic approximation that should be used for induction the simplest rules. Results of our experiments were divided into four groups, based on the type of the missing attribute values (attribute-concept values and "do not care" conditions) and whether the number of rules or the total number of conditions was used as a criterion of quality. Within each group we had 24 combinations (eight data sets and three types of probabilistic approximations). The Friedman multiple comparison rank sum test was applied, with 5 % significance level.

In our first group, where attribute-concept values were concerned with the number of rules, in one combination, associated with the *breast cancer* data set, the subset probabilistic approximations were better than the singleton probabilistic approximations and for another combination (for the *iris* data set) the subset probabilistic approximations were better than the concept probabilistic approximations. For the *wine recognition* data set, in two combinations, the concept probabilistic approximations were better than the remaining two probabilistic approximations. For the remaining 20 combinations results were statistically inconclusive.

For a group associated with "do not care" conditions and the number of rules, for nine combinations the subset approximations were better than other probabilistic approximations (for the *breast cancer*, *iris*, *lymphography* and *wine recognition* the subset probabilistic approximations were better than the remaining two probabilistic approximations and for the *echocardiogram* data set the subset probabilistic approximations were better than the singleton probabilistic approximations). For the 15 other combinations the results were statistically inconclusive.

For the remaining two groups, both associated with the total number of conditions, the results were similar. In four combinations of attribute-concept values, the subset approximations were the best. For the remaining 15 combinations of attribute-concept values, the results were statistically inconclusive. For nine combinations of "do not care" conditions, the subset probabilistic approximations were the best. In the remaining 15 combinations of "do not care" conditions, the results were inconclusive. In summary, there is weak evidence that the subset probabilistic approximations are the best to be used for inducing the simplest rule sets.

5 Conclusions

As follows from our experiments, there is evidence that the rule set size is smaller for the attribute-concept interpretation of missing attribute values than for the "do not care" condition interpretation. The total number of conditions in rule sets is also smaller for attribute-concept interpretation of missing attribute values than for "do not care" condition interpretation. Thus we may claim attribute-concept values are better than "do not care" conditions as an interpretation of a missing attribute value in terms of rule complexity.

Furthermore, all three kinds of probabilistic approximations (singleton, subset and concept) do not differ significantly with respect to the complexity of induced rule sets. However, there exists some weak evidence that the subset probabilistic approximations are better than the remaining two: singleton and concept.

References

1. Clark, P.G., Grzymala-Busse, J.W.: Experiments on probabilistic approximations. In: Proceedings of the 2011 IEEE International Conference on Granular Computing, pp. 144–149 (2011)
2. Clark, P.G., Grzymala-Busse, J.W.: Complexity of rule sets induced from incomplete data sets with attribute-concept values and and "do not care" conditions. In: Proceedings of the Third International Conference on Data Management Technologies and Applications, pp. 56–63 (2014)
3. Clark, P.G., Grzymala-Busse, J.W.: Mining incomplete data with attribute-concept values and "do not care" conditions. In: Polycarpou, M., de Carvalho, A.C.P.L.F., Pan, J.-S., Woźniak, M., Quintian, H., Corchado, E. (eds.) HAIS 2014. LNCS, vol. 8480, pp. 156–167. Springer, Heidelberg (2014)
4. Grzymala-Busse, J.W.: On the unknown attribute values in learning from examples. In: Raś, Zbigniew W., Zemankova, M. (eds.) ISMIS 1991. LNCS, vol. 542, pp. 368–377. Springer, Heidelberg (1991)
5. Grzymala-Busse, J.W.: LERS—a system for learning from examples based on rough sets. In: Slowinski, R. (ed.) Intelligent Decision Support. Handbook of Applications and Advances of the Rough Set Theory, pp. 3–18. Kluwer Academic Publishers, Dordrecht (1992)

6. Grzymala-Busse, J.W.: A new version of the rule induction system LERS. Fundamenta Informaticae **31**, 27–39 (1997)
7. Grzymala-Busse, J.W.: MLEM2: a new algorithm for rule induction from imperfect data. In: Proceedings of the 9th International Conference on Information Processing and Management of Uncertainty in Knowledge-Based Systems, pp. 243–250 (2002)
8. Grzymala-Busse, J.W.: Rough set strategies to data with missing attribute values. In: Notes of the Workshop on Foundations and New Directions of Data Mining, in conjunction with the Third International Conference on Data Mining, pp. 56–63 (2003)
9. Grzymała-Busse, J.W.: Characteristic relations for incomplete data: a generalization of the indiscernibility relation. In: Tsumoto, S., Słowiński, R., Komorowski, J., Grzymała-Busse, J.W. (eds.) RSCTC 2004. LNCS (LNAI), vol. 3066, pp. 244–253. Springer, Heidelberg (2004)
10. Grzymala-Busse, J.W.: Data with missing attribute values: generalization of indiscernibility relation and rule induction. Trans. Rough Sets **1**, 78–95 (2004)
11. Grzymala-Busse, J.W.: Three approaches to missing attribute values—a rough set perspective. In: Proceedings of the Workshop on Foundation of Data Mining, in conjunction with the Fourth IEEE International Conference on Data Mining, pp. 55–62 (2004)
12. Grzymała-Busse, J.W.: Generalized parameterized approximations. In: Yao, J.T., Ramanna, S., Wang, G., Suraj, Z. (eds.) RSKT 2011. LNCS, vol. 6954, pp. 136–145. Springer, Heidelberg (2011)
13. Kryszkiewicz, M.: Rough set approach to incomplete information systems. In: Proceedings of the Second Annual Joint Conference on Information Sciences, pp. 194–197 (1995)
14. Pawlak, Z.: Rough sets. Int. J. Comput. Inf. Sci. **11**, 341–356 (1982)
15. Pawlak, Z., Skowron, A.: Rough sets: some extensions. Inf. Sci. **177**, 28–40 (2007)
16. Pawlak, Z., Wong, S.K.M., Ziarko, W.: Rough sets: probabilistic versus deterministic approach. Int. J. Man Mach. Stud. **29**, 81–95 (1988)
17. Ślęzak, D., Ziarko, W.: The investigation of the bayesian rough set model. Int. J. Approximate Reasoning **40**, 81–91 (2005)
18. Stefanowski, J., Tsoukias, A.: On the extension of rough sets under incomplete information. In: Zhong, N., Skowron, A., Ohsuga, S. (eds.) RSFDGrC 1999. LNCS (LNAI), vol. 1711, pp. 73–82. Springer, Heidelberg (1999)
19. Yao, Y.Y.: Probabilistic rough set approximations. Int. J. Approximate Reasoning **49**, 255–271 (2008)
20. Yao, Y.Y., Wong, S.K.M.: A decision theoretic framework for approximate concepts. Int. J. Man Mach. Stud. **37**, 793–809 (1992)
21. Ziarko, W.: Probabilistic approach to rough sets. Int. J. Approximate Reasoning **49**, 272–284 (2008)

Simplifying Contextual Structures

Ivo Düntsch[1]($\boxtimes$) and Günther Gediga[2]

[1] Brock University, St. Catharines, ON L2S 3A1, Canada
duentsch@brocku.ca
[2] Department of Psychology, Institut IV, Universität Münster, Fliednerstr. 21,
Münster, Germany
gediga@uni-muenster.de

Abstract. We present a method to simplify a formal context while retaining much of its information content. Although simple, our ICRA approach offers an effective way to reduce the complexity of a concept lattice and/or a knowledge space by changing only little information in comparison to a competing model which uses fuzzy K-Means clustering.

1 Introduction

A very simple data structure is a triple $\mathfrak{C} = \langle U, V, R \rangle$ where R is a binary relation between elements of U and elements of V which is sometimes called a *formal context* [6,19]. From this, various data models can be obtained, one of the more popular ones being the *concept lattice* obtained from $\mathfrak{C}$ introduced by Wille [19]. With each concept a line diagram can be associated which depicts the concept lattice in a consolidated way. For lack of space we shall not describe this further; for details we invite the reader to consult, for example, [20] or [6].

As a context $\mathfrak{C}$ grows large, the construction of the concept lattice is costly and it is difficult to interpret the structure and its associated line diagram. Therefore, various techniques have been proposed to simplify a formal context $\mathfrak{C}$ or its associated concept lattice such as stability indices [1,11,14,15] which only consider only part of the concept lattice, simplification using fuzzy K-Means clustering (FKM) [13] or object similarity [2], or selection of relevant concepts in the presence of noisy data [11]. All these techniques can be subsumed under one of the following strategies:

1. Omit attributes (or objects), or
2. Merge attributes (or objects) which are similar according to some criterion, or
3. Remove concepts with low index values.

In each case, the adjacency matrix of R is changed. However, reducing the matrix does not guarantee that the associated concept lattice will be reduced as well, see Example 3 of [12]. In this paper we propose a simple algorithm to simplify a concept which does not increase the size of its associated concept lattice.

The ordering of authors is alphabetical and equal authorship is implied.
The author gratefully acknowledges support by the Natural Sciences and Engineering Research Council of Canada.

© Springer International Publishing Switzerland 2015
M. Kryszkiewicz et al. (Eds.): PReMI 2015, LNCS 9124, pp. 23–32, 2015.
DOI: 10.1007/978-3-319-19941-2_3

2   Notation and Definitions

Throughout we suppose that $U = \{p_1, \ldots, p_n\}$ is a finite set of objects (such as problems) and $V = \{s_1, \ldots, s_k\}$ is a finite set of attributes (such as skills). $R \subseteq U \times V$ is a binary relation between elements of U and elements of V. For each $p \in U$ we set $R(u) \stackrel{\mathrm{df}}{=} \{s \in V : pRs\}$, and $\mathscr{R} \stackrel{\mathrm{df}}{=} \{R(u) : u \in U\}$. The identity relation on U is denoted by $1'_U$. The relational converse of R is denoted by $R^{\smallsmile}$, and $-R$ is the complement of R in $U \times V$. The set $\mathscr{R}$ is partially ordered by $\subseteq$. The *adjacency matrix of R* has rows labeled by the elements of U, and columns labeled with the elements of V. An entry $\langle u, v \rangle$ is 1 if and only if $u_i R s_j$, otherwise, the entry in this cell is left empty. A formal context $\langle U, V, R \rangle$ gives rise to several set operators frequently used in modal logics: Let $X, X' \subseteq U$ and define

$$\langle R \rangle(X) = \{b \in V : (\exists a \in X)aRb\} = \{b \in B : R^{\smallsmile}(b) \cap X \neq \emptyset\}, \qquad \text{(Possibility)}$$

$$[R](X) = \{b \in V : (\forall a \in U)[aRb \Rightarrow a \in X]\} = \{b \in B : R^{\smallsmile}(b) \subseteq X\}, \qquad \text{(Necessity)}$$

$$[[R]](X) = \{b \in V : (\forall a \in U))[a \in X \Rightarrow aRb]\} = \{b \in B : X \subseteq R^{\smallsmile}(b)\}. \qquad \text{(Sufficiency)}$$

The mappings $\langle R \rangle$ and $[[R]]$ are, respectively, the existential (disjunctive) and universal (conjunctive) extensions of the assignment $x \mapsto R(x)$ to subsets of U, since it follows immediately from the definitions that for all $x \in U, X \subseteq U$,

$$\langle R \rangle(\{x\}) = [[R]](\{x\}) = R(x), \qquad (1)$$

$$\langle R \rangle(X) = \bigcup_{x \in X} R(x), [[R]](X) = \bigcap_{x \in X} R(x). \qquad (2)$$

The operators $[[R]]$ and $[R]$, as well as $\langle R \rangle$, are related since

$$b \in [[R]](X) \Longleftrightarrow X \subseteq R^{\smallsmile}(b) \Longleftrightarrow U - R^{\smallsmile}(b) \subseteq U \setminus X \Longleftrightarrow b \in [-R](U \setminus X). \qquad (3)$$

For unexplained notation and concepts in lattice theory we refer the reader to [8].

3   Data Models Based on Modal Operators

Suppose we have a formal context $\mathfrak{C} = \langle U, V, R \rangle$ which we regard as "raw data". The image sets $R(x)$ are our basic constructs. As a first approach to a data model based on $\langle U, V, R \rangle$, which, in our view, is a structural representation of raw data, we define a quasiorder $\preceq$ on U by setting $x \preceq y$ if and only if $R(x) \subseteq R(y)$. We also define the *incomparability relation* by

$$x \# y \stackrel{\mathrm{df}}{\Longleftrightarrow} (x \npreceq y) \text{ and } (y \npreceq x). \qquad (4)$$

From this starting point, several more involved data models can be developed. One of the better known models are those based on the sufficiency operators $[[R]]$ ("intent") and $[[R^{\smallsmile}]]$ ("extent"): For each $X \subseteq U$, $[[R]](X)$ is the set of all attributes common to all elements of X, and for $Y \subseteq V$, $[[R^\vee]](Y)$ is the set of all objects which possess all attributes in Y. A pair $\langle [[R^\vee]][[R]](X), [[R]](X) \rangle$ is called a *formal concept*. The set of all formal concepts can be made into a lattice which can be drawn as a consolidated line diagram [19] as in Fig. 1[1]. Each node of the diagram represents a formal concept, and for each object x, $R(x)$ is the set of all attributes above the node labelled x (we interpret "above" and "below" as reflexive relations). In the line diagram of R, $x \preceq y$ if and only if x and y label the same node or the node labelled by y is below the node labelled by x.

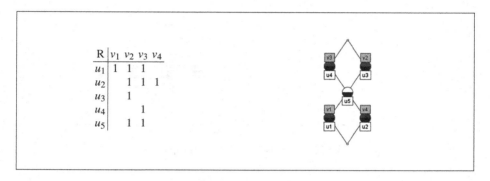

Fig. 1. A context and its line diagram

A data model which in some sense competes with concept lattices are the *knowledge spaces* introduced in [4]. These are set systems closed under union and can be related to the modal operator $\langle R \rangle$ which is called the *span operator* in [3]. It was shown in [7] that the models arising from $[[R]]$ and $\langle R \rangle$ have the same expressive power and are useful in situations different from those where conjunctive assignments such as the (DINA) model [9,10,16] and the rule space model [18] are employed.

Taking $\{ R(x) : x \in U \}$ as a starting point, the set of spans and the set of intent go into different directions. It follows from (1) and (2) that $\mathscr{K}_R \stackrel{\mathrm{df}}{=} \{ \langle R \rangle(X) : X \subseteq U \}$ is the $\cup$ – semilattice generated by $\{ R(x) : x \in U \}$, and $\mathscr{I}_R \stackrel{\mathrm{df}}{=} \{ [[R]](X) : X \subseteq U \}$ is the $\cap$ – semilattice generated by $\{ R(x) : x \in U \}$. For $X \subseteq U$, $[[R]]$ is the set of all attributes lying above all objects in X, and $\langle R \rangle(\{x\})$ is the set of all attributes <u>not</u> upwards reachable from object x in the line diagram of $-R$.

4 Reducing the Complexity

The simplest way to change the adjacency matrix is to change one bit at a time, according to a given criterion. The question arises which criterion we shall

---

[1] The diagrams were drawn by the ConExp package [21].

use. If $\preceq$ is a linear quasi order – i.e. if any two objects of U are comparable – then $\mathscr{K}_R$ and $\mathscr{I}_R$ coincide and are equal to $\langle \mathscr{K}_R, \subseteq \rangle$ (possibly with added $\emptyset$ or V); nothing is gained by going from the simple model $\langle |C, \preceq \rangle$ to one of the more involved ones. At the other extreme, if no two different elements of U are comparable with respect to $\#$, then the representations obtained from $\mathfrak{C}$ very strongly depend on the modal operator used and may widely differ. Consider the simple relation depicted in Fig. 2. There, $\mathscr{I}_R$ consists of the singletons $\{v_i\}$ and the empty set, while $\mathscr{K}_R$ is the set of all nonempty subsets of V. If we consider the complement of $-R$, then situation is reversed, see Fig. 3.

R	v_1	v_2	v_3	v_4
u_1	1			
u_2		1		
u_3			1	
u_4				1

Fig. 2. $\# = U^2 \setminus 1'_U$, 1st example

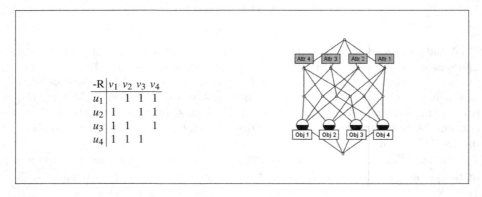

-R	v_1	v_2	v_3	v_4
u_1		1	1	1
u_2	1		1	1
u_3	1	1		1
u_4	1	1	1	

Fig. 3. $\# = U^2 \setminus 1'_U$, 2nd example

Therefore, if the incomparability relation is large, choosing one operator over the other may not provide a meaningful interpretation, and it may not be the wisest choice at the outset to prefer one over the other. Keeping in mind the problem/skill situation, we suggest the *relative incomparability* of objects as a measure of context complexity which we aim to reduce: If $\mathfrak{C} = \langle U, V, R \rangle$ is a formal context and $u \in U$, then we let

$$\texttt{incomp}(u) \stackrel{\mathrm{df}}{=} \{v \in U : u \# v\}, \quad \texttt{incomp}(\mathfrak{C}) \stackrel{\mathrm{df}}{=} \frac{|\{\langle u, v \rangle : u \# v\}|}{n^2 - n},$$

where $n = |U|$. Now, incomp($\mathfrak{C}$) $= 0$ if and only if $\preceq$ is a linear quasiorder, and incomp($\mathfrak{C}$) $= 1$ if no two different elements are $\preceq$ – comparable. The measure of success is the reduction of incomp($\mathfrak{C}$) relative to the number of bit changes.

Our *InComparablity Reduction Analysis* algorithm (ICRA)[2] is based on a simple steepest descent method: We consider objects u for which $|$incomp(u)$|$ is maximal and then invert a bit – i.e. an entry in the adjacency matrix of the relation under consideration – for which the drop of the number of overall incomparable pairs is maximal. This will increase the comparability of objects with respect to $\preceq$ or, equivalently, of sets $R(x)$ without increasing the number of intents, respectively, knowledge states. Indeed, in most cases we have looked at, the complexity of the concept lattice was significantly reduced. If one bit is inverted, so that the resulting relation is R' and $x \preceq_{R'} y$, then there will be a path from y to x in the line diagram of R' as well, so that the new representation is closer to the data as represented by R.

The basic concept is that we assume some of the data to be faulty, but we do not know which entries. More concretely, we assume that some (or all) incomparabilities are caused by faulty data. In this sense, our proposed procedure is a trade – off measure.

The stop criterion is a predetermined relative value of incomparable pairs, i.e. a value for incomp($\mathfrak{C}$), where $\mathfrak{C}$ is the current context, or no more complexity reduction is possible. As a rule of thumb we suggest to require that 50 % of

```
noEntry:=FALSE.
pout:=p                                          ▷ Initialize stop criterion to 0 ≤ p ≤ 1.
Unmark all object-attribute-pairs.                ▷ No pair changed yet.
repeat
    Find the set OBJ of objects belonging to unmarked pairs for which incomp(u) is maximal.
    if incomp(𝔠) ≤ pout then                      ▷ Goal reached
        NoEntry:=TRUE
    else
        Using OBJ find the object-attribute-pairs, which maximally reduce the in comparability,
        when inverting one bit of the matrix under consideration.
        if no reduction is achieved for any of these then
            NoEntry:=TRUE
        else
            Invert the entry of one of the maximal object-attribute-pairs and use the new rela-
tion.
            Mark the chosen object-attribute-pair.
            Replace 𝔠 with the revised context.
        end if
    end if
until NoEntry=TRUE.
```

Fig. 4. Pseudocode of the algorithm

---

[2] The algorithm is implemented in R [17] and the source code is available at http:// roughsets.net/FCred.R.

pairs with different components should be comparable (*Median InComparablity Reduction Analysis*). An overview of the pseudocode the ICRA algorithm is shown in Fig. 4.

5   Experiments

Even though our procedure is simple, it compares well with other simplification measures. As a case in point we shall consider the reduction using fuzzy K-Means clustering (FKM) proposed in [13]. This method is based on partitioning a set of vectors into k fuzzy clusters, specifying to what degree a vector belongs to the cluster centre. Owing to lack of space we cannot explain their method in detail and refer the reader to [13]. The context $\mathfrak{C}$ of their first example relates documents with keywords and it is shown in Fig. 5 along with its context lattice. The relative incomparability of $\mathfrak{C}$ is 94 %.

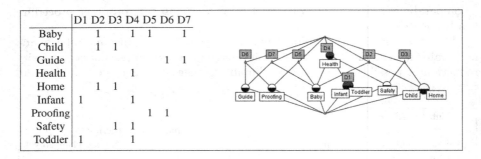

	D1	D2	D3	D4	D5	D6	D7
Baby	1			1	1		1
Child	1	1					
Guide					1	1	
Health			1				
Home	1	1					
Infant	1			1			
Proofing					1	1	
Safety		1	1				
Toddler	1			1			

Fig. 5. Example from [13], p. 2699

After applying FKM based clustering with $k = 2$, the columns D1 – D2 are identified and the entry $\langle T_i, D1' - -D4 \rangle$ of the resulting adjacency matrix is $\max\{\langle T_i, D1 \rangle, \ldots, \langle T_i, D4 \rangle\}$. The simplified context $\mathfrak{C}_1$ and its concept lattice are shown in Fig. 6.

To achieve the FKM result $\mathfrak{C}_1$ from $\mathfrak{C}$ requires to change 15 bits for a relative incomparability of 49 %; this includes the effort to identify columns. In comparison, our algorithm needs only 4 bits for a 50 % incomparability, and 9 bits for 0 % incomparability. The resulting context along with its line diagram is shown in Fig. 7. It has the same number of concepts as the concept lattice obtained from FKM (9), and the same number of edges (14).

In classification tasks, there is often a trade – off between the (relative) number of correctly classified objects and, for example, the (relative) cost of obtaining the classification or the clarity of a pictorial representation. In some instances, this may be expressed as the amount of errors we are prepared to allow to achieve another aim. A case in point are curves based on receiver operating characteristics (ROC), where the sensitivity (benefit) of a binary classifier is

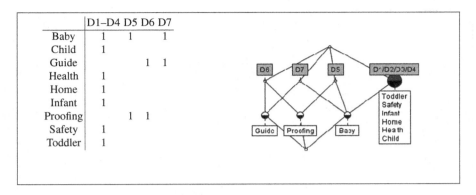

	D1–D4	D5	D6	D7
Baby	1	1		1
Child	1			
Guide			1	1
Health	1			
Home	1			
Infant	1			
Proofing			1	1
Safety	1			
Toddler	1			

Fig. 6. Example from [13], p. 2699, reduced

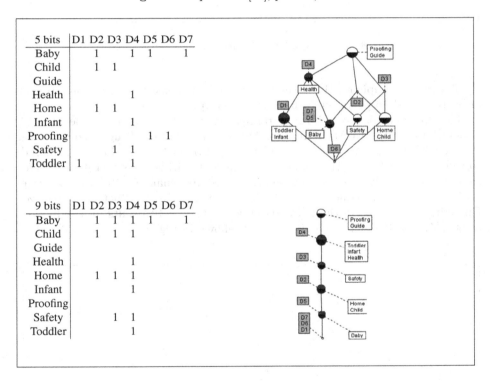

5 bits	D1	D2	D3	D4	D5	D6	D7
Baby	1		1	1		1	
Child	1	1					
Guide							
Health			1				
Home	1	1					
Infant			1				
Proofing					1	1	
Safety		1	1				
Toddler	1		1				

9 bits	D1	D2	D3	D4	D5	D6	D7
Baby	1	1	1	1		1	
Child	1	1	1				
Guide							
Health			1				
Home	1	1	1				
Infant			1				
Proofing							
Safety		1	1				
Toddler			1				

Fig. 7. Reduction of Example 1 from [13] using ICRA

plotted as a function of its FP rate (cost), see [5] for an overview. We can plot the relative incomparability as a function of the number of bits changed to achieve it, see the graph in Fig. 8. If we interpret (in-)comparability as sensitivity and the number of changed bits as cost to retrieve the original data, this can be interpreted as a ROC curve.

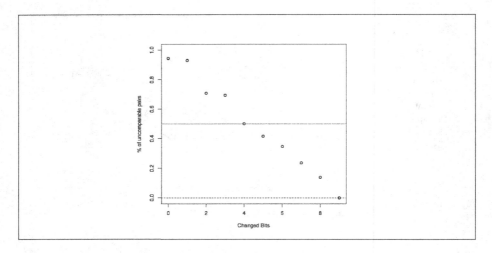

Fig. 8. Reducing relative incomparability with ICRA

The next example for [13] investigates a dataset consisting of various species of bacteria and 16 phenotypic characters, shown in Table 1.

For this context $\mathfrak{C}$, the incomparability incomp($\mathfrak{C}$) turns out to be 81 %. $\mathfrak{C}$ is reduced with the FKM method for $k = 5$ and $k = 9$, resulting in contexts $\mathfrak{C}_5$ and $\mathfrak{C}_9$ with incomp($\mathfrak{C}_5$) = 34.5 % and incomp($\mathfrak{C}_9$) = 64.7 %. 40 bits are required to reduce $\mathfrak{C}$ to C_5, and the reduction to $\mathfrak{C}_9$ with 64.7 % incomparability needs changing 11 bits. In contrast, our algorithm requires changing 19 bits to achieve an incomparability reduction to 34.6 %, and 8 bits for a reduction to 66.1 %. Changing 11 bits (as in the FKM reduction with k = 9) results in a reduction to 60.2 %. The ICRA reducibility graph is shown in Fig. 9.

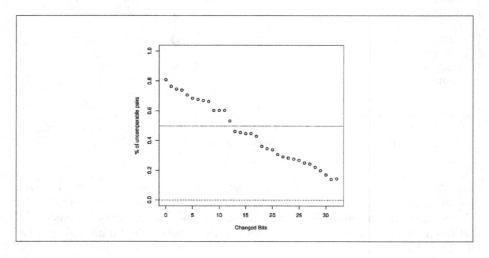

Fig. 9. Reducing relative incomparability of the bacterial dataset with ICRA

Table 1. Bacterial dataset from [13]

	H2S	MAN	LYS	IND	ORN	CIT	URE	ONP	VPT	INO	LIP	PHE	MAL	ADO	ARA	RHA
ecoli1	0	1	1	1	0	0	0	1	0	0	0	0	0	0	1	1
ecoli2	0	1	0	1	1	0	0	1	0	0	0	0	0	0	1	0
ecoli3	1	1	0	1	1	0	0	1	0	0	0	0	0	0	1	1
styphi1	0	1	1	0	0	0	0	0	0	0	0	0	0	0	1	0
styphi2	0	1	1	0	0	0	0	0	0	0	0	0	0	0	0	0
styphi3	1	1	1	0	0	0	0	0	0	0	0	0	0	0	1	0
kpneu1	0	1	1	1	0	1	1	1	1	1	0	0	0	1	1	1
kpneu2	0	1	1	1	0	1	1	1	1	1	0	0	1	0	1	1
kpneu3	0	1	1	1	0	1	1	1	1	1	0	0	1	1	1	1
kpneu4	0	1	1	1	0	1	1	1	0	1	0	0	1	1	1	1
kpneu5	0	1	1	1	0	1	0	1	1	1	0	0	1	1	1	1
pvul1	1	0	0	1	0	1	1	0	0	0	0	1	0	0	0	0
pvul2	1	0	0	1	0	0	0	0	0	0	0	1	0	0	0	0
pvul3	1	0	0	1	0	0	1	0	0	0	0	1	0	0	0	0
pmor1	0	0	1	1	1	0	1	0	0	0	0	1	0	0	0	0
pmor2	0	0	0	1	1	0	0	0	0	0	0	1	0	0	0	0
smar	0	1	1	0	1	1	0	1	1	0	1	0	0	0	0	0

6   Conclusion and Outlook

We have introduced a simple algorithm ICRA to simplify a formal context, the success criterion of which is a prescribed reduction of incomparable pairs. As a rule of thumb, we propose a relative frequency of incomparable pairs of objects of 50 %. This seems a fair compromise between closeness to the data on the one hand, and the additional structure introduced by the chosen model on the other. We have compared the success of our algorithm with several examples of [13] and have found that fewer bits are needed than FKM to obtain similar incomparability ratios. Furthermore, the FKM algorithm requires much more effort and additional model assumptions so that its cost/benefit ratio is much smaller than for the median comparability algorithm. Furthermore, it is not clear which k should used for the reduction.

In the available space, only an indication of the impact of the median comparability algorithm could be given. Further work will include investigation of the powers and limitations of the ICRA algorithm using both theoretical and practical analysis. In particular, we shall consider its effects on implication sets and association rules.

Acknowledgement. We thank the referees for careful reading and constructive comments.

References

1. Buzmakov, A., Kuznetsov, S.O., Napoli, A.: Scalable estimates of concept stability. In: Glodeanu, C.V., Kaytoue, M., Sacarea, C. (eds.) ICFCA 2014. LNCS, vol. 8478, pp. 157–172. Springer, Heidelberg (2014)
2. Dias, S.M., Vieira, N.J.: Reducing the size of concept lattices: the JBOS approach. In: Proceedings CLA, pp. 80–91 (2010)

3. Düntsch, I., Gediga, G.: Approximation operators in qualitative data analysis. In: de Swart, H., Orłowska, E., Schmidt, G., Roubens, M. (eds.) Theory and Applications of Relational Structures as Knowledge Instruments. LNCS, vol. 2929, pp. 214–230. Springer, Heidelberg (2003)

4. Falmagne, J.C., Koppen, M., Villano, M., Doignon, J.P., Johannesen, J.: Introduction to knowledge spaces: how to build, test and search them. Psychol. Rev. **97**(2), 201–224 (1990)

5. Fawcett, T.: An introduction to ROC analysis. Pattern Recog. Lett. **27**, 861–874 (2006)

6. Ganter, B., Wille, R.: Formal Concept Analysis: Mathematical Foundations. Springer, Berlin (1999)

7. Gediga, G., Düntsch, I.: Skill set analysis in knowledge structures. Br. J. Math. Stat. Psychol. **55**, 361–384 (2002). http://www.cosc.brocku.ca/duentsch/archive/skills2.pdf

8. Grätzer, G.: General Lattice Theory, 2nd edn. Birkhäuser, Basel (2000)

9. Haertel, E.H.: Using restricted latent class models to map the skill structure of achievement items. J. Educ. Meas. **26**, 301–324 (1989)

10. Junker, B.W., Sijtsma, K.: Cognitive assessment models with few assumptions, and connections with nonparametric item response theory. Appl. Psychol. Meas. **25**, 258–272 (2001)

11. Klimushkin, M., Obiedkov, S., Roth, C.: Approaches to the selection of relevant concepts in the case of noisy data. In: Kwuida, L., Sertkaya, B. (eds.) ICFCA 2010. LNCS, vol. 5986, pp. 255–266. Springer, Heidelberg (2010)

12. Krupka, M.: On complexity reduction of concept lattices: three counterexamples. Inf. Retr. **15**(2), 151–156 (2012). http://dx.doi.org/10.1007/s10791-011-9175-7

13. Kumar, C.A., Srinivas, S.B.: Concept lattice reduction using fuzzy K-means clustering. Exp. Syst. Appl. **37**, 2696–2704 (2010)

14. Kuznetsov, S.O., Obiedkov, S., Roth, C.: Reducing the representation complexity of lattice-based taxonomies. In: Priss, U., Polovina, S., Hill, R. (eds.) ICCS 2007. LNCS (LNAI), vol. 4604, pp. 241–254. Springer, Heidelberg (2007)

15. Kuznetsov, S.: On stability of a formal concept. Ann. Math. Artif. Intel. **49**(1–4), 101–115 (2007). http://dx.doi.org/10.1007/s10472-007-9053-6

16. Macready, G.B., Dayton, C.M.: The use of probabilistic models in the assessment of mastery. J. Edu. Stat. **2**, 99–120 (1977)

17. R Core Team, R.: A Language and Environment for Statistical Computing. R Foundation for Statistical Computing, Vienna, Austria (2014). http://www.R-project.org/

18. Tatsuoka, K.K.: Rule space: an approach for dealing with misconceptions based on item response theory. J. Edu. Meas. **20**(4), 345–354 (1983)

19. Wille, R.: Restructuring lattice theory: an approach based on hierarchies of concepts. In: Rival, I. (ed.) Ordered Sets, pp. 445–470. NATO Advanced Studies Institute, Reidel, Dordrecht (1982)

20. Wolff, K.E.: A first course in formal concept analysis - how to understand line diagrams. In: Faulbaum, F. (ed.) Softstat '93: Advances in Statistical Software 4, pp. 429–438. Stuttgart, Fischer (1993)

21. Yevtushenko, S.: The concept explorer (2000). retrieved 24 December 2011. http://conexp.sourceforge.net/index.html

Towards a Robust Scale Invariant Feature Correspondence

Shady Y. El-Mashad$^{(\boxtimes)}$ and Amin Shoukry

Computer Science and Engineering Department, Egypt-Japan University
for Science and Technology (E-JUST), Alexandria, Egypt
{shady.elmashad,amin.shoukry}@ejust.edu.eg

Abstract. In this paper, we introduce an improved scale invariant feature correspondence algorithm which depends on the Similarity-Topology Matching algorithm. It pays attention not only to the similarity between features but also to the spatial layout of every matched feature and its neighbours. The features are represented as an undirected graph where every node represents a local feature and every edge represents adjacency between them. The topology of the resulting graph can be considered as a robust global feature of the represented object. The matching process is modeled as a graph matching problem; which in turn is formulated as a variation of the quadratic assignment problem. The Similarity-Topology Matching algorithm achieves superior performance in almost all the experiments except when the image has been exposed to scaling deformations. An amendment has been done to the algorithm in order to cope with this limitation. In this work, we depend not only on the distance between the two interest points but also on the scale at which the interest points are detected to decide the neighbourhood relations between every pair of features. A set of challenging experiments conducted using 50 images (contain repeated structure) representing 5 objects from COIL-100 data-set with extra synthetic deformations reveal that the modified version of the Similarity-Topology Matching algorithm has better performance. It is considered more robust especially under the scale deformations.

Keywords: Features matching · Features extraction · Topological Relations · Graph matching · Performance evaluation

1 Introduction

Image matching or in other words, comparing images in order to obtain a measure of their similarity, is an important computer vision task. It is involved in many different applications, such as object detection and recognition, image classification, content based image retrieval, video data mining, image stitching, stereo vision, and 3D object modeling. A general solution for identifying similarities between objects and scenes within a database of images is still a faraway goal. There are a lot of challenges to overcome such as viewpoint or lighting

© Springer International Publishing Switzerland 2015
M. Kryszkiewicz et al. (Eds.): PReMI 2015, LNCS 9124, pp. 33–43, 2015.
DOI: 10.1007/978-3-319-19941-2_4

variations, deformations, and partial occlusions that may exist across different examples.

Furthermore, image matching as well as many other vision applications rely on representing images with sparse number of distinct keypoints. A real challenge is to efficiently detect and describe keypoints, with robust representations invariant against scale, rotation, view point change, noise, as well as combinations of them [1].

Keypoint detection and matching pipeline has three distinct stages which are feature detection, feature description and feature matching. In the feature detection stage, every pixel in the image is checked to see if there is a unique feature at this pixel or not. Subsequently, during the feature description stage, each region (patch) around the selected keypoints is described with a robust and invariant descriptor which can be used to match against other descriptors. Finally, at the feature matching stage, an efficient search for prospective matching descriptors in other images is made [2].

In the context of matching, a lot of studies have been used to evaluate interest point detectors as in [3,4]. On the other hand, little efficient work has been done on the evaluation of local descriptors. K. Mikolajczyk and C. Schmid [5], proposed and compared different feature detectors and descriptors as well as different matching approaches in their study. Although this work proposed an exhaustive evaluation of feature descriptors, it is still unclear which descriptors are more appropriate in general and how their performance depends on the interest point detector.

D.G. Lowe [6], proposed a new matching technique using distinctive invariant features for object recognition. Interest points are matched independently via a fast nearest-neighbour algorithm to the whole set of interest points extracted from the database images. Therefore, a Hough transform to identify clusters belonging to a single object has been applied. Finally, least-squares solution for consistent pose parameters has been used for the verification.

Another technique to find correspondences is RANSAC. The most beneficial side of RANSAC is the ability of jointly estimating the largest set of mutual compatible correspondences between two views. Zhang and Kosecka [7] demonstrate the shortcomings of RANSAC when dealing with images containing repetitive structures. The failure of RANSAC in these cases is due to the fact that similarity measure is used to find matching based only on feature descriptor and, with repetitive structures, the chosen descriptors can change dramatically. Therefore, the nearest neighbor strategy is not an appropriate solution.

There are two levels to measure the images similarity which are patch and image levels. In the patch level, the distance between any two patches is measured based on their descriptors. In the image level, the overall similarity between any two images is calculated which in most cases contain many patches.

The Minkowski-type metric has been used to measure the distance between patches in most of researches. The Minkowski metric is defined as in (1):

$$D(X,Y) = (\sum_{i=1}^{P} |X_i - Y_i|^r)^{1/r} \tag{1}$$

when r = 2, it is the Euclidean distance (L2 distance), and it is the Manhattan distance (L1 distance) when r = 1 [8].

In the approach proposed in the present paper both local and global features are considered simultaneously. We try to retain the locality of the features advantages in addition to preserving the overall layout of the objects. The similarity between the local features has been used in conjunction with the topological relations between them as a global feature of the object.

In this paper, the approach presented in [9,10] is modified to be scale invariant. In addition, intensive experiments are conducted mainly focused on images with different resolutions as the objective of the modified algorithm to be more scale invariant. The images contain a duplication of the same object which reflects the scope of work (dealing with repeated structure).

This paper is organized as follows: The proposed scale invariant feature correspondence algorithm is introduced in Sect. 2. Section 3, presents the conducted experiments to evaluate the performance of the modified matching approach. Finally, the conclusions of this work and the recommendations for future work are presented in Sects. 4 and 5, respectively.

2   Proposed Matching Approach

Conventional matching approaches reduce the matching problem to a metric problem. Therefore, the choice of a metric is substantial for the matching of local features. Most approaches depend mainly on finding the minimum distance between features (descriptors) in feature space as shown in (2), where D_{ij} is the distance measure between feature i from the first image and feature j from the second image. X_{ij} is a matching indicator between feature i and feature j, i.e. $X_{ij} = 1$ if feature i in the 1st image is mapped to feature j in the 2nd image and $X_{ij} = 0$ otherwise. Note that $X_{ij} \in \{0, 1\}$.

$$Min\ F = \sum_{\forall i,j} D_{ij}\ X_{ij} \qquad (2)$$

Limitations: The similarity measure between features deals with each feature individually rather than a group of features. Consequently, the minimum distance between features can be misleading in some cases and as a result the performance of the algorithm deteriorates. In other words, the minimum distance criterion has no objection for a feature to be wrongly matched as long as it successfully achieves the minimum distance objective.

2.1   Similarity-Topology Matching Algorithm

In [9], a new matching algorithm called "Similarity-Topology Matching" has been proposed. This algorithm pays attention not only to the similarity between features but also to the spatial layout of every matched feature and its neighbors. A new term, describing the neighbourhood/ topological relations between every pair of features has been added $\alpha \sum_{\forall i,j,k,l} X_{ij}\ X_{kl}\ P_{ij,kl}$. In addition, another term has been added to relax the constraints $\beta \left(Min(m,n) - \sum_{\forall_{i,j}} X_{ij}\right)$ as shown below in (3).

$$Min\ F = \sum_{\forall_{i,j}} D_{ij}\ X_{ij} + \alpha \sum_{\forall_{i,j,k,l}} X_{ij}\ X_{kl}\ P_{ij,kl} + \beta \left(Min(m,n) - \sum_{\forall_{i,j}} X_{ij}\right) \quad (3)$$

Subject to:

$$\sum_{j=1}^{n} X_{ij} \leq 1 \qquad\qquad (a)$$

$$\sum_{i=1}^{m} X_{ij} \leq 1 \qquad\qquad (b)$$

The second term in (3) represents a penalty term over all pairs of features. $P_{ij,kl}$ is called a penalty matrix. It is used to penalize matching pairs of features i and k in one image with corresponding pairs j and l in the other image if they have different topologies. It is binary and of $(m \times n, m \times n)$ dimension; where m, n are the number of features in the first and the second images respectively. $P_{ij,kl} = 1$ if the features j, l in the second image have different topology when compared to features i, k in the first image. Accordingly, the penalty matrix is calculated by applying the XOR logical operation to the adjacency matrices(AM1, AM2) of the two images as in (4). In XOR, the output is true whenever both inputs are different from each other. For example, if one input is true and the other is false. The output is false whenever both inputs are similar to each other, i.e., both inputs are true or false.

$$P(i,j,k,l) = XOR\left(AM1(i,k),\ AM2(j,l)\right) \quad (4)$$

(α) is called a topology coefficient. It indicates how much the matching algorithm depends on the topology between images. In the experiments, (α) is chosen in a range from 0 to 0.1. The topology coefficient is effective and has a great impact when the interest points are similar to each other. On the contrary, it has almost no impact when the difference of similarities between the interest points is high. (β) is called a threshold coefficient. It indicates how much the matching algorithm depends on the features matching threshold. In the experiments, (β) is chosen in a range from 0 to 0.5. These parameters are determined by cross validations.

Constraints Interpretation: Constraint (a): There exists at most one $'1'$ in every column of x. Constraint (b): There exists at most one $'1'$ in every row of x. The two constraints ensure that every feature in the first image should match to at most one feature in the second image.

2.2   Scale Invariant Similarity-Topology Matching Algorithm

Analysis and Modification. An analysis is done to determine why the algorithm isn't accurate enough in case of the scaling deformation. It is noticed that the adjacency matrix (AM) of an image is constructed using the neighbourhood idea. In other words, if the distance between any two interest points in the same image is less than a threshold then they are called neighbours to each other. Consequently, the neighbourhood relation between each two interest points depends only on the distance between them, which is not valid specially when dealing with different scales.

The two interest points in Fig. 1 are the same. The algorithm considers them as neighbours to each other in the left image but not in the right image which is counter intuitive, as they are neighbours in both cases.

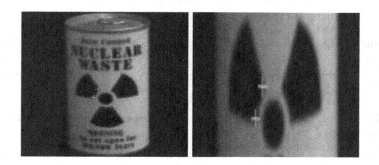

Fig. 1. Scaling problem example

An amendment is done to the algorithm in order to cope with this limitation. The modification makes the Neighbourhood Relation (NR) depend not only on the distance between the two interest points as in the Similarity-Topology but also on the scales at which the two interest points are detected. Hence, the Neighbourhood Relation (NR) between two interest points i and k in an image is defined as shown in (5).

$$NR = \frac{Distance\ between\ two\ interest\ points}{Average\ scale\ of\ the\ two\ interest\ points}$$

$$= \frac{d_{ik}}{Avg(\sigma_i, \sigma_k)} \tag{5}$$

Accordingly, the adjacency matrix is modified and calculated as in (6):

$$AM(i,k) = \left\{ \begin{matrix} 1 & if & NR < Threshold \\ 0 & & otherwise \end{matrix} \right\} \tag{6}$$

where d_{ik} is the Euclidean distance between interest points i and k in the same image spatial domain. σ_i and σ_k are the scales at which the interest points i and k are detected respectively.

Scale Invariant Similarity-Topology Matching Algorithm. Algorithm (1) gives a summary of the modified version of the "Similarity-Topology Matching" approach. This new algorithm achieves superior performance in almost all the experiments specially when the images are exposed to scaling deformations.

Algorithm 1. Scale Invariant Similarity-Topology Matching

Input: A pair of images, topology coefficient (α), and threshold coefficient (β).
1. For every image:
    (a) Detect interest points (select strongest n);
    (b) Extract a descriptor for every interest point;
    (c) Construct adjacency matrix using equations (5) and (6);
2. For every feature (descriptor) in the 1st image: compute the distance between it and all the features in the 2nd image using the Euclidean distance in feature space;
3. Penalize pairs of matched features taking into consideration their adjacency relations;

$$P(i,j,k,l) = XOR(AM1(i,k), AM2(j,l))$$

4. Solve the optimization problem using (7) (features similarity and topological constraints);

Output: List of features correspondences.

This investigated problem has a quadratic-objective function which is subject to linear constraints. It is called a binary (0-1) Quadratic Programming problem. Consequently, the objective function formulated in (3) can be rewritten as in (7):

$$Min\ F = \sum_{\forall i,j} X_{ij}(D_{ij} - \beta) + \alpha \sum_{\forall i,j,k,l} X_{ij}\,X_{kl}\,P_{ij,kl} \qquad (7)$$

This optimization problem is solved using IBM ILOG CPLEX Optimization Studio (usually called just CPLEX for simplicity) which is an optimization software package.

3   Experimental Results

3.1   Data-Set

Columbia Object Image Library (COIL-100) has been used in the experiments [11]. COIL-100 is a database of color images which has 7200 images of 100 different objects (72 images per object). These collections of objects have a wide diversity of complex geometric and reflectance characteristics. Consequently, it is the most suitable data-set which can be helpful in the proof of concept of the proposed feature correspondence approach. Figure 2, depicts 10 objects from the Coil-100 data-set.

Fig. 2. Examples of objects from the COIL-100 data-set used for the evaluation

The Challenge. Fifty images representing five objects of the aforementioned data-set are chosen to perform the experiments. These objects with extra synthetic deformations such as rotation, scaling, partial occlusion and heavy noise are used for this purpose. In addition, a duplication of the same object is put in the same image with deformations, but one as a whole and one as parts to make the matching more challenging and to test the principle goal of the new matching strategy. In this case, a feature in the first image has almost two similar features in the second image. Figure 3, shows an example to illustrate the idea. The feature in the first image (left) has two similar features in the second image (right). This raises a question, which one should be matched. This challenge demonstrates the idea of the proposed approach, that rely on the similarity as well as the topological relations between the features as shown in the experiments in the next subsection.

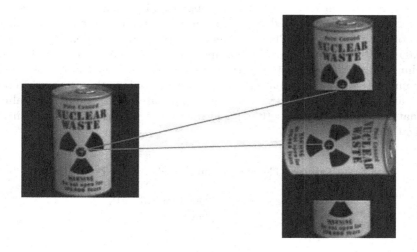

Fig. 3. An illustrative example of the duplication of the same feature

3.2   Experiments

Three different experiments are conducted to test the modification introduced in the "Similarity-Topology Matching" algorithm to make it scale invariant. All of these tests are done on images having different resolutions. The first test is done between a pair of images with different scales only. The second test is done between a pair of images with different scales as well as a duplication of the same object as parts in the second image. The last test is done like the second experiment but with extra deformations such as rotation and view point changes. These tests are ranged in difficulty from easiest to hardest as shown Table 2.

Features Detection and extraction: the interest points are detected and extracted using SURF (Speeded Up Robust Features) [12]. We demonstrate in [10] that SURF algorithm can be used prior to the proposed matching approach to get more robust feature correspondence.

Evaluation criterion: For each pair of images, every interest point in image 1 is compared to all interest points in image 2 according to their descriptors. The detection Rate and the False Positive Rate (FPR) are calculated in order to evaluate the performance. The detection rate R is defined as the ratio between the number of correct matches and the number of all possible matches (number of correspondence). The target is to maximize the detection rate and to minimize the false positive rate.

$$R = \frac{\text{Number of correct matches}}{\text{Number of possible matches within the full instance}}$$

The experiments have been done using three state-of-the-art strategies which are Threshold, Nearest Neighbour (NN) and Nearest Neighbour Distance Ratio (NNDR) in addition to the Similarity-Topology Matching as well as its modified version. In the proposed algorithm and its modified version as well, the values of the topology penalty coefficient and the threshold penalty coefficient are 0.05 and 0.3 respectively. The modified version of the "Similarity-Topology Matching" algorithm has better performance. It is considered more robust specially under the scale deformations. As shown in Table 1, the Modified-Version of the algorithm not only has higher detection rate (0.65), but also it almost eliminates the false matches (0.01) which is more important specially in the localization problem.

Table 1. The experimental results summary

Matching strategy	Detection rate	FPR
NNDR	0.40	0.04
NN	0.48	0.13
Threshold	0.55	0.28
Similarity-topology	0.46	0.08
Modified-version	0.65	0.01

Table 2. Scale invariant feature correspondence examples

NN	NNDR	Threshold	proposed	Modified

4   Conclusions

In this paper, an improved scale invariant feature correspondence algorithm which depends on the "Similarity-Topology Matching" algorithm has been introduced. In this approach, both local and global features are considered simultaneously and a set of control parameters is employed to tune the performance by adjusting the significance of global vs. local features. The major contribution of this research is depending not only on the distance between the two interest points but also on the scale at which the interest points are detected to decide the neighbourhood relations between every pair of features. Three different tests focusing on scaling deformations have been conducted. From the experimental results, it is noticed that the number of correctly matched features is increased.

In conclusion, the modified version of the "Similarity-Topology Matching" algorithm has superior performance specially when the images have been exposed to scaling deformations.

5 Future Work

After the proof of concept of the aforementioned approach has been verified, a lot of work remains to be done in order to generalize the local features matching approach and achieve high degree of robustness and computational efficiency. First, a preprocessing step is required to automatically evaluate the parameters values (alpha, beta). Second, an optimization of the algorithm to be more computationally efficient should be made without any loss in the algorithm accuracy as this algorithm may be used in real-time applications. Finally, applying this approach in a particular robot application such as mobile robot localization. The proposed approach can be used in conjunction with other approach [13] which depends on wifi-signals to determine the location of a mobile robot (such as KheperaIII) in indoor limited areas.

Acknowledgments. This research has been supported by the Ministry of Higher Education (MoHE) of Egypt through a Ph.D. fellowship. Our sincere thanks to Egypt-Japan University for Science and Technology (E-JUST) for guidance and support. I wish to express an extended appreciation to Prof. Mohamed Hussein for his fruitful discussions and helpful suggestions.

References

1. Zitova, B., Flusser, J.: Image registration methods: a survey. Image Vision Comput. **21**(11), 977–1000 (2003)
2. Szeliski, R.: Computer Vision: Algorithms and Applications. Springer, London (2011)
3. Mikolajczyk, K., Schmid, C.: An affine invariant interest point detector. In: Heyden, A., Sparr, G., Nielsen, M., Johansen, P. (eds.) ECCV 2002, Part I. LNCS, vol. 2350, pp. 128–142. Springer, Heidelberg (2002)
4. Mikolajczyk, K., Tuytelaars, T., Schmid, C., Zisserman, A., Matas, J., Schaffalitzky, F., Kadir, T., Van Gool, L.: A comparison of affine region detectors. Int. J. Comput. Vision **65**(1–2), 43–72 (2005)
5. Mikolajczyk, K., Schmid, C.: A performance evaluation of local descriptors. IEEE Trans. Pattern Anal. Mach. Intell. **27**(10), 1615–1630 (2005)
6. Lowe, D.G.: Distinctive image features from scale-invariant keypoints. Int. J. Comput. Vision **60**(2), 91–110 (2004)
7. Zhang, W., Kosecka, J.: Generalized RANSAC framework for relaxed correspondence problems. In: Third International Symposium on 3D Data Processing, Visualization, and Transmission, pp. 854–860. IEEE (2006)
8. Liu, Y., Zhang, D., Lu, G., Ma, W.Y.: A survey of content-based image retrieval with high-level semantics. Pattern Recogn. **40**(1), 262–282 (2007)

9. El-Mashad, S.Y., Shoukry, A.: A more robust feature correspondence for more accurate image recognition. In: 2014 International Conference on Computer and Robot Vision (CRV). IEEE (2014)
10. El-Mashad, S.Y., Shoukry, A.: Evaluating the robustness of feature correspondence using different feature extractors. In: 2014 International Conference on Methods and Models in Automation and Robotics. IEEE (2014)
11. Nayar, S., Nene, S., Murase, H.: Columbia object image library (coil 100). Technical Report CUCS-006-96, Department of Computer Science, Columbia University (1996)
12. Bay, H., Tuytelaars, T., Van Gool, L.: SURF: speeded up robust features. In: Leonardis, A., Bischof, H., Pinz, A. (eds.) ECCV 2006, Part I. LNCS, vol. 3951, pp. 404–417. Springer, Heidelberg (2006)
13. Elbasiony, R., Gomaa, W.: WiFi localization for mobile robots based on random forests and GPLVM. In: 2014 13th International Conference on Machine Learning and Applications (ICMLA), pp. 225–230, December 2014

A Comparison of Two Approaches to Discretization: Multiple Scanning and C4.5

Jerzy W. Grzymala-Busse[1,2]([✉]) and Teresa Mroczek[2]

[1] Department of Electrical Engineering and Computer Science, University of Kansas, Lawrence, KS 66045, USA
jerzy@ku.edu
[2] Department of Expert Systems and Artificial Intelligence, University of Information Technology and Management, 35-225 Rzeszow, Poland
tmroczek@wsiz.rzeszow.pl

Abstract. In a Multiple Scanning discretization technique the entire attribute set is scanned many times. During every scan, the best cut-point is selected for all attributes. The main objective of this paper is to compare the quality of two setups: the Multiple Scanning discretization technique combined with the C4.5 classification system and the internal discretization technique of C4.5. Our results show that the Multiple Scanning discretization technique is significantly better than the internal discretization used in C4.5 in terms of an error rate computed by ten-fold cross validation (two-tailed test, 5 % level of significance). Additionally, the Multiple Scanning discretization technique is significantly better than a variant of discretization based on conditional entropy introduced by Fayyad and Irani called Dominant Attribute. At the same time, decision trees generated from data discretized by Multiple Scanning are significantly simpler from decision trees generated directly by C4.5 from the same data sets.

1 Introduction

Mining numerical data sets requires an additional step called *discretization*. Discretization is a process of transforming numerical values into intervals.

For a numerical attribute a with an interval $[i, j]$ as a range, a partition of the range into k intervals

$$\{[i_0, i_1), [i_1, i_2), ..., [i_{k-2}, i_{k-1}), [i_{k-1}, i_k]\},$$

where $i_0 = i$, $i_k = j$, and $i_l < i_{l+1}$ for $l = 0, 1, ..., k-1$, defines a discretization of a. The numbers i_1, i_2,..., i_{k-1} are called *cut-points*.

A new discretization technique, called *Multiple Scanning*, introduced in [11,12], was very successful when combined with rule induction and a classification system of LERS (Learning from Examples based on Rough Sets) [9]. The novelty of this paper is a comparison of the C4.5 classification system applied to data discretized using Multiple Scanning with C4.5 applied directly to the original data sets with numeric attributes. Additionally, we compare the Multiple

© Springer International Publishing Switzerland 2015
M. Kryszkiewicz et al. (Eds.): PReMI 2015, LNCS 9124, pp. 44–53, 2015.
DOI: 10.1007/978-3-319-19941-2_5

Scanning discretization technique with a variant of the well-known discretization based on conditional entropy introduced by Fayyad and Irani [7,8] and called *Dominant Attribute* [11,12].

In Multiple Scanning, during every scan, the entire attribute set is analyzed. For all attributes the best cutpoint is selected. At the end of a scan, some subtables that still need discretization are created. The entire attribute set of any subtable is scanned again, and the best corresponding cutpoints are selected. The process continues until the stopping condition is satisfied or the required number of scans is reached. If the required number of scans is reached and the stopping condition is not satisfied, discretization is completed by Dominant Attribute, in which first the best attribute is selected, then for this attribute, the best cutpoint, again using conditional entropy, is selected. This process continues recursively until the same stopping criterion is satisfied. Multiple Scanning ends up with an attempt to reduce the number of intervals called merging. Since Multiple Scanning uses Dominant Attribute as the last resort, if we skip scanning, or equivalently set the required number of scans to zero, discretization is reduced to Dominant Attribute. Thus we may include a comparison of Multiple Scanning with Dominant Attribute. Typically, in Multiple Scanning the required number of scans should be set to some small number. In our experiments, for all data sets, after six scans the error rate computed using ten-fold cross validation was constant, because new intervals created in consecutive scans were merged together during the last step of discretization. The stopping criterion used in this paper is based on rough set theory.

The main objective of this paper is to compare the quality of two setups: the Multiple Scanning discretization technique combined with the C4.5 classification system and the internal discretization technique of C4.5. For 12 numerical data sets two sets of experiments were conducted: first the C4.5 system of tree induction was used to compute an error rate using ten-fold cross validation, then the same data sets were discretized using Multiple Scanning and for such discretized data sets the same C4.5 system was used to establish an error rate. Thus we may compare two discretization techniques: Multiple Scanning with the internal discretization of C4.5.

Our results show that the Multiple Scanning discretization technique is significantly better than the internal discretization used in C4.5 or the Dominant Attribute discretization in terms of an error rate computed by ten-fold cross validation (two-tailed test, 5 % level of significance). Additionally, decision trees generated from data discretized by Multiple Scanning are significantly simpler than decision trees generated directly by C4.5 from the same data sets.

2   Entropy Based Discretization

Discretization based on conditional entropy of the concept given the attribute is considered to be one of the most successful discretization techniques [2–8,10, 11,13–15,19,20].

An example of a data set with numerical attributes is presented in Table 1. In this table all cases are described by variables called *attributes* and one variable called a *decision*. The set of all attributes is denoted by A. The decision is denoted by d. The set of all cases is denoted by U. In Table 1 the attributes are *Max_Speed* and *Number_of_Seats* while the decision is *Price*. Additionally, $U = \{1, 2, 3, 4, 5, 6, 7\}$. For a subset S of the set U of all cases, an entropy of a variable v (attribute or decision) with values $v_1, v_2, ..., v_n$ is defined by the following formula

$$H_S(v) = -\sum_{i=1}^{n} p(v_i) \cdot \log p(v_i),$$

where $p(v_i)$ is a probability (relative frequency) of value v_i in the set S, $i = 0, 1, ..., n$. All logarithms in this paper are binary.

Table 1. An example of a data set with numerical attributes

Case	Attributes		Decision Price
	Max_Speed	Number_of_Seats	
1	280	2	very-high
2	220	4	small
3	180	5	small
4	220	5	medium
5	220	2	high
6	280	4	medium
7	180	4	small

A conditional entropy of the decision d given an attribute a is

$$H_S(d|a) = -\sum_{j=1}^{m} p(a_j) \cdot \sum_{i=1}^{n} p(d_i|a_j) \cdot log\ p(d_i|a_j),$$

where $a_1, a_2, ..., a_m$ are all values of a and $d_1, d_2, ..., d_n$ are all values of d, all values are restricted to S. There are two fundamental criteria of quality based on entropy. The first is an *information gain* associated with an attribute a and defined by

$$I_S(a) = H_S(d) - H_S(d|a)$$

the second is *information gain ratio*, for simplicity called *gain ratio*, defined by

$$G_S(a) = \frac{I_S(a)}{H_S(a)}.$$

Both criteria were introduced by J.R. Quinlan, see, e.g., [18] and used for decision tree generation.

Let a be an attribute and q be a cutpoint that splits the set S into two subsets, S_1 and S_2. The conditional entropy $H_S(d|q)$ is defined as follows

$$\frac{|S_1|}{|U|} H_{S_1}(a) + \frac{|S_2|}{|U|} H_{S_2}(a),$$

where $|X|$ denotes the cardinality of the set X. The cut-point q for which the conditional entropy $H_S(d|q)$ has the smallest value is selected as the best cut-point. The corresponding information gain is the largest.

2.1 Stopping Criterion for Discretization

A stopping criterion of the process of discretization, described in this paper, is the *level of consistency* [3], based on *rough set theory* [16,17]. For any subset B of the set A of all attributes, an *indiscernibility* relation $IND(B)$ is defined, for any $x, y \in U$, in the following way

$$(x, y) \in IND(B) \text{ if and only if } a(x) = a(y) \text{ for any } a \in B,$$

where $a(x)$ denotes the value of the attribute $a \in A$ for the case $x \in U$. The relation $IND(B)$ is an equivalence relation. The equivalence classes of $IND(B)$ are denoted by $[x]_B$ and are called *B-elementary sets*. Any finite union of B-elementary sets is *B-definable*.

A partition on U constructed from all B-elementary sets of $IND(B)$ is denoted by B^*. $\{d\}$-elementary sets are called *concepts*, where d is a decision. For example, for Table 1, if $B = \{Max\_Speed\}$, $B^* = \{\{1, 6\}, \{2, 4, 5\}, \{3, 7\}\}$ and $\{d\}^* = \{\{1\}, \{2, 3, 7\}, \{4, 6\}, \{5\}\}$. In general, arbitrary $X \in \{d\}^*$ is not B-definable. For example, the concept $\{2, 3, 7\}$ is not B-definable. However, any $X \in \{d\}^*$ may be approximated by a *B-lower approximation* of X, denoted by $\underline{B}X$ and defined as follows

$$\{x \mid x \in U, [x]_B \subseteq X\}$$

and by *B-upper approximation* of X, denoted by $\overline{B}X$ and defined as follows

$$\{x \mid x \in U, [x]_B \cap X \neq \emptyset\}.$$

In our example, $\underline{B}\{2, 3, 7\} = \{3, 7\}$ and $\overline{B}\{2, 3, 7\} = \{2, 3, 4, 5, 7\}$.

The B-lower approximation of X is the greatest B-definable set contained in X. The B-upper approximation of X is the least B-definable set containing X. A *level of consistency* [3], denoted by $L(A)$, is defined as follows

$$L(A) = \frac{\sum_{X \in \{d\}^*} |\underline{A}X|}{|U|}.$$

Practically, the requested level of consistency for discretization is 1.0, i.e., we want the discretized data set to be *consistent*. For example, for Table 1, the level of consistency $L(A)$ is equal to 1.0, since $\{A\}^* = \{\{1\}, \{2\}, \{3\}, \{4\}, \{5\}, \{6\}, \{7\}\}$ and, for any X from $\{Price\}^* = \{\{1\}, \{2, 3, 7\}, \{4, 6\}, \{5\}\}\}$, we have $\underline{A}X = X$. Additionally, $L(B) \approx 0.286$.

Table 2. Partially discretized Table 1 using Multiple Scanning

Case	Attributes		Decision Price
	Max_Speed	Number_of_Seats	
1	[280, 280]	[2, 4)	very-high
2	[180, 280)	[4, 5]	small
3	[180, 280)	[4, 5]	small
4	[180, 280)	[4, 5]	medium
5	[180, 280)	[2, 4)	high
6	[280, 280]	[4, 5]	medium
7	[180, 280)	[4, 5]	small

2.2 Multiple Scanning Strategy

This discretization technique needs some parameter denoted by t and called the total number of scans. In Multiple Scanning algorithm,

- for the entire set A of attributes the best cutpoint is computed for each attribute $a \in A$, based on minimum of conditional entropy $H_U(d|a)$, a new discretized attribute set is A^D, and the set U is partitioned into a partition $(A^D)^*$,
- if the number t of scans is not reached, the next scan is conducted: we need to scan the entire set of partially discretized attributes again, for each attribute we need only one cutpoint, the best cutpoint for each block $X \in (A^D)^*$ is computed, the best cutpoint, among all such blocks is selected,
- if the requested number t of scans is reached and the data set needs more discretization, the Dominant Attribute technique is used for remaining subtables,
- the algorithm stops when $L(A^D) = 1$, where A^D is the discretized set of attributes.

We illustrate this technique by scanning Table 1 once, i.e., $t = 1$. First we are searching for the best cut-point for both attributes, *Max_Speed* and *Number_of_Seats*. For the attribute *Max_Speed* there exist two potential cutpoints: 220 and 280 with three potential intervals: [180, 220), [220, 280) and [280, 280]. The corresponding conditional entropies are

$$H_{Max\_Speed}(220, U) = \frac{5}{7}((-\frac{1}{5} \cdot \log \frac{1}{5})(3) + (-\frac{2}{5} \cdot \log \frac{2}{5})) + \frac{2}{7}(0) \approx 1.373,$$

$$H_{Max\_Speed}(280, U) = \frac{5}{7}((-\frac{1}{5} \cdot \log \frac{1}{5})(2) + (-\frac{3}{5} \cdot \log \frac{3}{5})) + \frac{2}{7}(1) \approx 1.251.$$

The better cutpoint is 280. Similarly, there are three potential cutpoints for the attribute *Number_of_Seats*: 4 and 5, with three potential intervals: [2, 4), [4, 5) and [5, 5]. The corresponding conditional entropies are

$$H_{Number\_of\_Seats}(4, U) = \frac{5}{7}((-\frac{3}{5} \cdot \log \frac{3}{5}) + (-\frac{2}{5} \cdot \log \frac{2}{5})) + \frac{2}{7}(1) \approx 0.979,$$

$$H_{Number\_of\_Seats}(5, U) = \frac{5}{7}((-\frac{1}{5} \cdot \log \frac{1}{5})(3) + (-\frac{2}{5} \cdot \log \frac{2}{5})) + \frac{2}{7}(1) \approx 1.229.$$

The better cut-point is 4. Table 1, partially discretized this way, is presented as Table 2.

The level of consistency for Table 2 is 0.429 since $A^* = \{\{1\}, \{2, 3, 4, 7\}, \{5\}, \{6\}\}$, we need to distinguish cases 2, 3, and 7 from the case 4. Therefore we need to use the *Dominant Attribute* technique for a subtable, with four cases, 2, 3, 4 and 7. This data set is presented in Table 3.

Table 3. The remaining data set that still needs discretization

Case	Attributes		Decision Price
	Max_Speed	Number_of_Seats	
2	220	4	small
3	180	5	small
4	220	5	medium
7	180	4	small

3 Experiments

Our experiments were conducted on 12 data sets available on the University of California at Irvine *Machine Learning Repository*, with the exception of *bankruptcy*. The *bankruptcy data* set is a well-known data set used by E.I. Altman to predict a bankruptcy of companies [1].

Both discretization methods, Multiple Scanning and C4.5, were applied to all data sets, with the level of consistency equal to 100 %. For a choice of the best attribute, we used gain ratio.

Table 4 presents results of ten-fold cross validation, using increasing number of scans. Obviously, for any data set, after some fixed number of scans, an error rate is stable (constant). For example, for *Australian* data set, the error rate is 14.93 % for the scan number 4, 5, etc. Thus, any data set from Table 4 is characterized by two error rates: minimal and stable [12]. For a given data set, the smallest error rate from Table 4 is called *minimal* and the last entry in the row that corresponds to the data set is called *stable*. For example, for the *Australian* data set, the minimal error rate is 13.48 % and the stable error rate is 14.93 %. For some data sets (e.g., for *bankruptcy*), minimal and stable error rates are identical.

Table 5 presents the size of decision trees generated from all 12 data sets discretized by Multiple Scanning. In Table 6 error rates are shown for decision trees generated directly by C4.5 and for the decision trees generated by C4.5 from data sets discretized by Multiple Scanning, only the minimal error rates are presented with the corresponding scan numbers. Finally, Table 7 presents tree

Table 4. Error rates for Multiple Scanning

Data set	Error rate for scan number						
	0	1	2	3	4	5	6
Australian	14.49	13.48	13.77	14.93			
Bankruptcy	10.61	3.03	3.03				
Bupa	41.74	29.86	30.43	29.28	29.86		
Connectionist bench	27.89	16.83					
Echocardiogram	27.03	14.86	14.86	24.32	22.97		
Glass	34.11	30.84	28.50	24.77	25.23	27.10	26.64
Image segmentation	13.81	18.10	11.90	12.38			
Iris	5.33	5.33	4.67				
Pima	27.73	25.78	24.09	24.61	25.00	26.17	
Wave	32.81	23.05	26.17	24.80			
Wine recognition	7.87	3.93					
Yeast	56.40	53.84	54.65	51.75	51.75	51.75	

Table 5. Tree size for Multiple Scanning

Data set	Tree size for scan number						
	0	1	2	3	4	5	6
Australian	3	13	26	27			
Bankruptcy	14	3	4				
Bupa	13	10	9	11	20		
Connectionist bench	6	31					
Echocardiogram	13	5	10	7	7		
Glass	126	72	67	58	40	40	40
Image segmentation	16	33	24	24			
Iris	6	4	4				
Pima	73	34	27	44	49	48	
Wave	7	55	94	105			
Wine recognition	8	11					
Yeast	414	276	491	362	458	442	

size for decision trees generated directly by C4.5 and for decision trees generated by C4.5 from data sets discretized by Multiple Scanning.

It is clear from Tables 4–7 that the minimal error rate is never associated with 0 scans, i.e., with a special case of the Multiple Scanning discretization technique: Dominant Attribute. Using the Wilcoxon matched-pairs signed-ranks

Table 6. Error rates for C4.5 and the best results of Multiple Scanning

Data set	C4.5	Multiple Scanning	
	Error rate	Error rate	Scan number
Australian	16.09	13.48	1
Bankruptcy	6.06	3.03	1
Bupa	35.36	29.28	3
Connectionist bench	25.96	16.83	1
Echocardiogram	28.38	14.86	1
Glass	33.18	24.77	3
Image segmentation	12.38	11.90	2
Iris	5.33	4.67	2
Pima	25.13	24.09	2
Wave	26.37	23.05	1
Wine recognition	8.99	3.93	1
Yeast	44.41	51.75	3

Table 7. Tree size for C4.5 and the best results of Multiple Scanning

Data set	C4.5	Multiple Scanning
Australian	63	13
Bankruptcy	3	3
Bupa	51	11
Connectionist bench	35	31
Echocardiogram	9	5
Glass	45	58
Image segmentation	25	24
Iris	9	4
Pima	43	27
Wave	85	55
Wine recognition	9	11
Yeast	371	362

test, we conclude that the following three statements are statistically significant (with the significance level equal to 5 % for a two-tail test):

- the minimal error rate associated with Multiple Scanning is smaller than the error rate associated with Dominant Attribute,
- the minimal error rate associated with Multiple Scanning is smaller than the error rate associated with C4.5,
- the size of decision trees generated from data discretized by Multiple Scanning is smaller than size of decision trees generated directly by C4.5.

4   Conclusions

This paper presents results of experiments in which three different techniques were used for discretization: Multiple Scanning, the internal discretization of C4.5, and Dominant Attribute. All techniques were validated by conducting experiments on 12 data sets with numerical attributes. Our discretization techniques were combined with decision tree generation using the C4.5 system. Results of our experiments show that the Multiple Scanning technique is significantly better than discretization included in C4.5 and that decision trees generated from data discretized by Multiple Scanning are significantly simpler than decision trees generated directly by C4.5 from the same data sets (two-tailed test and 0.05 level of significance). Additionally, the Multiple Scanning discretization technique is significantly better than the Dominant Attribute technique. Thus, we show that there exists a new successful technique for discretization.

References

1. Altman, E.I.: Financial ratios, discriminant analysis and the prediction of corporate bankruptcy. J. Financ. **23**(4), 189–209 (1968)
2. Blajdo, P., Grzymala-Busse, J.W., Hippe, Z.S., Knap, M., Mroczek, T., Piatek, L.: A comparison of six approaches to discretization—a rough set perspective. In: Wang, G., Li, T., Grzymala-Busse, J.W., Miao, D., Skowron, A., Yao, Y. (eds.) RSKT 2008. LNCS (LNAI), vol. 5009, pp. 31–38. Springer, Heidelberg (2008)
3. Chmielewski, M.R., Grzymala-Busse, J.W.: Global discretization of continuous attributes as preprocessing for machine learning. Int. J. Approximate Reasoning **15**(4), 319–331 (1996)
4. Clarke, E.J., Barton, B.A.: Entropy and MDL discretization of continuous variables for bayesian belief networks. Int. J. Intell. Syst. **15**, 61–92 (2000)
5. Elomaa, T., Rousu, J.: General and efficient multisplitting of numerical attributes. Mach. Learn. **36**, 201–244 (1999)
6. Elomaa, T., Rousu, J.: Efficient multisplitting revisited: optima-preserving elimination of partition candidates. Data Min. Knowl. Disc. **8**, 97–126 (2004)
7. Fayyad, U.M., Irani, K.B.: On the handling of continuous-valued attributes in decision tree generation. Mach. Learn. **8**, 87–102 (1992)
8. Fayyad, U.M., Irani, K.B.: Multi-interval discretization of continuous-valued attributes for classification learning. In: Proceedings of the Thirteenth International Conference on Artificial Intelligence, pp. 1022–1027 (1993)
9. Grzymala-Busse, J.W.: A new version of the rule induction system LERS. Fundamenta Informaticae **31**, 27–39 (1997)
10. Grzymala-Busse, J.W.: Discretization of numerical attributes. In: Kloesgen, W., Zytkow, J. (eds.) Handbook of Data Mining and Knowledge Discovery, pp. 218–225. Oxford University Press, New York (2002)
11. Grzymala-Busse, J.W.: A multiple scanning strategy for entropy based discretization. In: Proceedings of the 18th International Symposium on Methodologies for Intelligent Systems, pp. 25–34 (2009)
12. Grzymala-Busse, J.W.: Discretization based on entropy and multiple scanning. Entropy **15**, 1486–1502 (2013)

13. Kerber, R.: Chimerge: discretization of numeric attributes. In: Proceedings of the 10-th National Conference on AI, pp. 123–128 (1992)
14. Kohavi, R., Sahami, M.: Error-based and entropy-based discretization of continuous features. In: Proceedings of the Second International Conference on Knowledge Discovery and Data Mining, pp. 114–119 (1996)
15. Nguyen, H.S., Nguyen, S.H.: Discretization methods in data mining. In: Polkowski, L., Skowron, A. (eds.) Rough Sets in Knowledge Discovery 1: Methodology and Applications, pp. 451–482. Physica-Verlag, Heidelberg (1998)
16. Pawlak, Z.: Rough sets. Int. J. Comput. Inform. Sci. **11**, 341–356 (1982)
17. Pawlak, Z.: Rough Sets. Theoretical Aspects of Reasoning about Data. Kluwer Academic Publishers, Dordrecht (1991)
18. Quinlan, J.R.: C4.5: Programs for Machine Learning. Morgan Kaufmann Publishers, San Mateo (1993)
19. Stefanowski, J.: Handling continuous attributes in discovery of strong decision rules. In: Polkowski, L., Skowron, A. (eds.) RSCTC 1998. LNCS (LNAI), vol. 1424, pp. 394–401. Springer, Heidelberg (1998)
20. Stefanowski, J.: Algorithms of Decision Rule Induction in Data Mining. Poznan University of Technology Press, Poznan (2001)

Hierarchical Agglomerative Method
for Improving NPS

Jieyan Kuang[1], Zbigniew W. Raś [1,2(⊠)], and Albert Daniel[1]

[1] KDD Laboratory, College of Computing and Informatics,
University of North Carolina, Charlotte, NC 28223, USA
{jkuang1,ras}@uncc.edu
[2] Institute of Computer Science, Warsaw University of Technology,
00-665 Warsaw, Poland

Abstract. The paper proposes a new strategy called HAMIS to improve
NPS (Net Promoter Score) of certain companies involved in heavy equipment repair in the US and Canada - we call them clients. HAMIS is
based on the semantic dendrogram built by using agglomerative clustering strategy and semantic distance between clients. More similar is the
knowledge extracted from two clients, more close these clients semantically are to each other. Each company is represented by a dataset which
is built from answers to the questionnaire sent to a number of randomly
chosen customers using services offered by this company. Before knowledge is extracted from these datasets, each one is extended by merging
it with datasets which are close to it in the semantic dendrogram, have
higher NPS, and if classifiers extracted from them have higher FS-score.
Action rules are extracted from these extended datasets and used for providing recommendations to clients of how to improve their businesses.

1 Introduction

Improving companies' performance is an important issue nowadays and Net Promoter Score (NPS) is one of the most popular measure for such purpose [7–9].
Net Promoter Score assumes that customers are categorized into three categories: promoter, passive and detractor, which represent customers' satisfaction,
loyalty and the likelihood of recommending this client in a descending order.

Our dataset involves 34 clients who are located in different areas crossing the
whole United States as well as some parts of Canada. These clients provide similar services to over 25,000 customers. The dataset consists of three categories of
values which are collected from the questionnaire answered by randomly selected
customers during 2011 and 2012. The first and second category in the questionnaire provide information about customers and services they received and the
third category (the key part of the questionnaire) relates to the customers feelings about the services. Here are some examples of questions in these three
categories:

© Springer International Publishing Switzerland 2015
M. Kryszkiewicz et al. (Eds.): PReMI 2015, LNCS 9124, pp. 54–64, 2015.
DOI: 10.1007/978-3-319-19941-2_6

- Information about the customer (name, contact phone number).
- Information about the service (name of the client, invoice amount, type of equipment to be repaired).
- Feeling about the service (how many days were needed to finish the job, was the job completed correctly, are you satisfied with the job, likelihood to refer to friends).

Customers are asked to share their feelings about the service by scoring 0 to 10 for all the asked questions in third category. Higher the score is, more pleased the customer is with the service. Based on the average score of all the collected answers for each customer, over 99 % of customers are divided into three groups: customers falling into interval 9–10 are seen as promoter, into 7–8 as passive, and into 0–6 as detractor. With the determined NPS status in our dataset, the NPS efficiency rating (defined as the percentage of customers labeled promoter minus the percentage of customers labeled detractor) can be computed for each client.

Our ultimate goal is to improve the service of every client, in another word, improve its NPS. The semantic distance (similarity) between clients, which indicates the similarity of clients' knowledge concerning *Promoter*, *Passive* and *Detractor* hidden in datasets, can be computed. Smaller the semantic distance is, more similar the clients are. Using the notion of semantic distance, we build semantic similarity based dendrogram by following agglomerative clustering algorithm in the domain of datasets representing 34 clients. Next, we propose a method called Hierarchical Agglomerative Method for Improving NPS (HAMIS). Besides semantic similarity in HAMIS, NPS efficiency rating is another primary measure we consider before we merge two semantically similar clients. As a matter of fact, the NPS rating of the newly merged dataset will be higher than or at least equal to the dataset which is used for merging and it is with lower NPS rating than the other dataset. So we can expect that by analyzing the merged dataset of two most semantically similar clients we should be able to offer recommendations to the client with lower NPS rating. However, the consistency of data may decrease in the merged dataset so we also evaluate its representing classifier and if the results are satisfactory we merge two datasets in HAMIS.

Action rules mining is a known strategy in the area of data mining and it was firstly proposed by Ras and Wieczorkowska in [6] and investigated further in [2–5, 10]. In early papers, action rules have been constructed from two classification rules $[(\omega \wedge \alpha) \rightarrow \phi]$ and $[(\omega \wedge \beta) \rightarrow \psi]$, where ω is a stable part for both rules. Action rule was defined as the term $[(\omega) \wedge (\alpha \rightarrow \beta)] \Rightarrow (\phi \rightarrow \psi)$, where ω is the description of clients for whom the rule can be applied, $(\alpha \rightarrow \beta)$ shows what changes in values of attributes are required, and $(\phi \rightarrow \psi)$ gives the expected effect of the action. Let us assume that ϕ means *detractors* and ψ means *promoters*. Then, the discovered knowledge shows how values of attributes need to be changed under the situation required by stable part of the rule so the customers classified as detractors will become promoters.

2   Introduction of Semantic Similarity

The concept of semantic similarity between clients was introduced in [1]. Each client was represented by a tree classifier extracted from its extended dataset. More similar are the tree classifiers representing clients, more close they are semantically. Figure 1 shows the hierarchical clustering of 34 clients with respect to their semantic similarity. In the dendrogram we can easily identify groups of clients which are semantically close to each other. If we use tree structure based terminology, every leaf node represents a client as the numbers show. The depth of a node is the length of the path from it to the root. So larger is the depth of the earliest common ancestor of two clients, more semantically similar they are to each other.

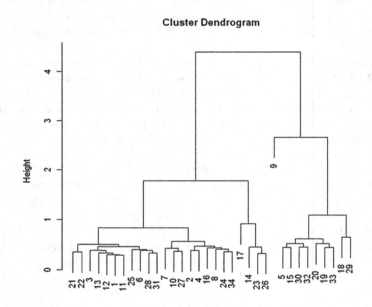

Fig. 1. Hierarchical clustering of 34 clients

3   Hierarchical Agglomerative Method for Improving NPS

HAMIS is enlarging the dataset of any specified client by following a bottom-up path in the hierarchically structured dendrogram based on semantic similarity. In the dendrogram, every leaf node represents a dataset of a corresponding client and every parent node represents the merged dataset of its mergeable children. Therefore, higher the bottom-up path ends in the dendrogram, larger the resulting merged dataset is, namely, more generalized dataset is returned by HAMIS.

The bottom-up path formed during the process links all the successfully merged nodes. Mergeable node means the node that can be used for merging and it is identified by the following criteria:

– It is the most semantically similar node in current situation.
– Its NPS rating is not less than the targeted client.

Given the definition of semantic similarity in the second section, we are capable to quantify the concept of how similar customers from different clients feel about the provided service. Therefore, if the semantic distance between two clients is relatively small, in other words, these clients are semantically close, then we could infer that the customers from these clients think of *Promoter*, *Passive* and *Detractor* in a more similar way, comparing to customers from clients that are further away regarding semantic distance. So it is possible that action rules extracted from the dataset covering all these semantically similar clients are also useful for improving the NPS rating of individual client. Based on the semantic distance retrieved, we clustered all the 34 clients using the agglomerative clustering algorithm and generated a dendrogram as shown in Fig. 1 which provides us with very efficient way to identify the most similar clients. As mentioned previously, each leaf node of the dendrogram stands for each client correspondingly, so the nodes that are semantically closest should be all the leaf nodes on the sibling side. For instance, if the sibling node is a leaf node, then there is only one node available for being the closest. If the sibling node is a parent node, it complicates the situation since the union set of all the leaf nodes under this sibling node should be the most semantically similar, then certainly, all the leaf nodes on the sibling side should be counted in and be checked one by one in a top down sequence following the depth of these nodes.

However, merging a targeted client with a semantically similar client whose NPS rating is lower won't fully match our expectation, since our goal is to improve the target's NPS rating, not conversely. But what we can be certain of is that merging a client with other client whose NPS rating is not lower gives us a dataset with higher or at least the same NPS rating. Let's assume that $NPS[i]$ and $NPS[j]$ are the NPS ratings of two clients i and j. Then

$$NPS[i] = \frac{Num[i, Promoter]}{Num[i, *]} - \frac{Num[i, Detractor]}{Num[i, *]}.$$

By $Num[i, Promoter]$ and $Num[i, Detractor]$ we mean number of *Promoter* and *Detractor* records in dataset of client i respectively, and by $Num[i, *]$ we mean the total number of records in i regardless of the class categories.

Meanwhile, $NPS[j] = \frac{Num[j, Promoter]}{Num[j, *]} - \frac{Num[j, Detractor]}{Num[j, *]}.$

By $Num[j, Promoter]$ and $Num[j, Detractor]$ we mean number of *Promoter* and *Detractor* records in dataset of client j respectively, and by $Num[j, *]$ we mean the total number of records in j regardless of the class categories.

Also we assume that $NPS[j] \geq NPS[i]$ and $NPS[i \cup j]$ is the NPS rating of the union set of client i and j, so we can expect $NPS[i \cup j] \geq NPS[i]$, because if $NPS[j] - NPS[i] =$

$(\frac{Num[j,Promoter]}{Num[j,*]} - \frac{Num[j,Detractor]}{Num[j,*]}) - (\frac{Num[i,Promoter]}{Num[i,*]} - \frac{Num[i,Detractor]}{Num[i,*]}) \geq 0,$

then $NPS[i \cup j] - NPS[i] =$

$(\frac{Num[j,Promoter]+Num[i,Promoter]}{Num[j,*]+Num[i,*]} - \frac{Num[j,Detractor]+Num[i,Detractor]}{Num[j,*]+Num[i,*]}) -$
$(\frac{Num[i,Promoter]}{Num[i,*]} - \frac{Num[i,Detractor]}{Num[i,*]}) \geq 0.$

Thus, we can surely get a joined dataset with non-decreased NPS rating. In addition, continuously keeping track of the quality of classifiers extracted from the merged dataset during the entire procedure is advantaging for achieving the best performance of generalization. Classification results show the quality of datasets for mining action rules and worse classifiers lead to poor confidence of action rules. Accordingly, we must make sure the classifiers are under improvement. To evaluate the classifiers, we use F-score that includes both accuracy and coverage of classification into consideration. As a popular measure of assessing the classification performance, F-score offers us a comprehensive and accurate view on our data.

Therefore, the three criteria mentioned above make up the foundation of algorithm HAMIS which is presented thoroughly in the next section.

3.1   Presentation of HAMIS

Technically speaking. the purpose of the algorithm HAMIS is to keep expanding the targeted client by unionizing it with all the clients satisfying the conditions. Unless the resulting dataset for chosen client can't be expanded any further, the algorithm would be repeatedly executed. And the algorithm returns resulting dataset when it ends. As HAMIS is built on the basis of a dendrogram regarding semantic distance, we describe the procedure using tree structure related terminology. The algorithm is designed as presented in Algorithm 1.

In the procedure of HAMIS, the resulting node is defined as N and it is initialized with the input targeted node N_{target}. Once N has been given, the nodes that are semantically closest to it are retrieved and stored in a list naming N_c. Accordingly, N_c contains all the leaf nodes on the sibling side of current N in the dendrogram and they are the candidates for being mergeable with N. It is apparent that at least one candidate is required to proceed, otherwise, it means the node N has reached the root and there is no more node available for merging. When proceeding, the following part is the main part in HAMIS and it iterates through all the candidates in N_c on the foundation of other two merging criteria mentioned above: NPS rating and F-score. If a candidate $N_c[i]$ does not have lower NPS rating than the targeted node N_{target}, then the candidate is qualified for merging. And the merged result is temporarily stored as N_m. N_m can't become the new resulting node N yet unless its F-score is greater or at least equal to F-score of current N. Thus, if the resulting node N is replaced by the merged result N_m, it suggests the merging process for current candidate succeeds and the new N will be used for next candidate in N_c if there are still

Algorithm 1. Hierarchical Agglomerative Method for Improving NPS

> **procedure** HAMIS(N_{target})
> $N \leftarrow N_{target}$ ▷ N is a node including only the targeted node initially
>   **repeat**
> $N_0 \leftarrow N$
>   retrieve N_c which is a list of candidates $N_c[1], N_c[2], ..., N_c[n]$;
>   **while** $N_c \neq \emptyset$ **do**
>   get next available candidate $N_c[i] \in N_c$ ($i \in \{1, 2, ..., n\}$);
>   **if** $NPS[N_c[i]] \geq NPS[N_{target}]$ **then**
> $N_m \leftarrow \{N_c[i]\} \cup N$
>   **if** $Fs[N_m] \geq F[N]$ **then**
> $N \leftarrow N_m$ ▷ N is replaced by N_m
>   **end if**
>   **end if**
> $N_c \leftarrow N_c \setminus \{N_c[i]\}$ ▷ remove it from the list
>   **end while**
>   **if** $N_0! = N$ **then**
> N climbs to its parent node in upper level;
>   **end if**
>   **until** $N_0 = N$ ▷ We have the final dataset return if no more merging happens.
>   **return** N
> **end procedure**

any. When a candidate fails merging with N, the same resulting node N will be used for another generalization attempt with next available candidate. The main part will not end until all the candidates have been checked. If there are more than one candidate found in N_c, they will be checked in a top down order based on the depth of them in the dendrogram, smaller the depth of a candidate is, earlier the candidate will be checked. So candidates are stored in an ascending order with regards to their depth in dendrogram, saying that for each candidate $N_c[i](i \in \{1, 2, ..., n\})$,

$depth[N_c[i+1]] > depth[N_c[i]]$, where $depth[N_c[i]]$ is the depth of node $N_c[i]$ in dendrogram and $i \in \{1, 2, ..., n-1\}$.

Each candidate is examined in almost same way, while the only difference is a new resulting node iteratively generated by successful merging process. Every time a new node is merged in the resulting node, the newly updated resulting node is replacing the current one. When the main part of the algorithm is finished with the resulting node being updated, it will climb up one level in the dendrogram and become the parent node of previous position. With a new resulting node at a new depth, HAMIS will keep going until the resulting node is not changed after the main part ends or it has reached the root.

4 Experiment

To show the running process of algorithm HAMIS in our domain, we take Client 2 as a target example, and the relevant data used during this procedure are shown

in Table 1. As the semantic similarity based clustering dendrogram is given in
Fig. 1 and the part of it related to our example is shown in Fig. 2, we observe
that node {2} representing Client 2 is labeled in green at the bottom and it is
the initial node. As a sibling node to node {2}, node {4} is the only candidate
in N_c which is most semantically similar to node {2}, and NPS rating of node
{4} shown in Table 1 is higher. In addition, F-score calculated by J48 in WEKA
for merged node {2, 4} is also higher than current resulting node {2}, which are
0.788 comparing to 0.783, hence the merged node {2, 4} successfully replaces {2}
and become the new resulting node. Meanwhile, there are no more unchecked
candidate in N_c, so HAMIS is done with current depth and will continue with
the new resulting node by climbing up to the parent node which is labeled in
blue. At a new position, because the sibling node of current resulting node is not
a leaf node, and leaf nodes {16}, {8}, {24} and {34} on the sibling side should
be included in candidate set as we defined, and they are labeled in blue as well.
According to the depth of each candidate in dendrogram, they will be checked
following the top down sequence which is node {16} first, then node {8} and
{24}, and {34} at the end. Then HAMIS attempts to merge these candidates to
resulting node individually, but it turns out none of them can successfully merge
with {2, 4}. When it comes to node {16}, although its NPS rating is just a little
bit higher than for the targeted node {2}, the F-score of {2, 4, 16} is much lower
than for {2, 4}, so the merging of {16} and {2, 4} fails and the main part goes to
the next one, which is node {8}. The case for node 8 is exactly the same as for
node {16}, so {2, 4} is still the resulting node without being changed and it keeps
going to node {24} and {34}. But neither of them can be merged with {2, 4} due
to either low NPS ratings or lower F-score of joined nodes. Consequently, node
{2, 4} has not been replaced with any new merged node after all the candidates
have been checked, which suggests {2, 4} is the most generalized in our program
for Client 2. Thus, HAMIS ends here and returns {2, 4}.

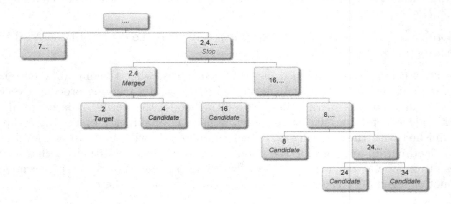

Fig. 2. Example of running HAMIS with Client 2 selected (Color figure online)

Table 1. NPS rating and Fscore of relevant nodes

	$N\{2\}$	$N\{4\}$	$N\{16\}$	$N\{8\}$	$N\{24\}$	$N\{34\}$
NPS rating	0.765	0.803	0.767	0.802	0.724	0.779

	$N\{2\}$	$N\{2,4\}$	$N\{2,4,16\}$	$N\{2,4,8\}$	$N\{2,4,24\}$	$N\{2,4,34\}$
F-score	0.783	0.788	0.776	0.786	NA	0.778

In the next step, we are going to generate action rules for both generalized dataset and original dataset of Client 2. Before the program starts, we need to specify the necessary attributes. Certainly the promoter status should be the decision attributes and the transitions we are interested in are from *Detractor* to *Promoter*. The customers' personal information related attributes should be seen as stable attributes, in our experiment, attributes like customers' name, location and contact number are set as stable attributes. Then the attributes about customers' feeling and comment are selected as attributes which can change (they are flexible), and these are the keys for improving NPS ratings since they will tell us about what actions we should adopt. For example, attributes evaluating if the job is done correctly and the timeframe of technician's arrival are flexible attributes. Based on our personal knowledge about the dataset, we expect that a huge number of action rules will be generated and we only pay attention to the ones with sufficiently high confidence, so we intend to get the action rules with at least 80 % confidence.

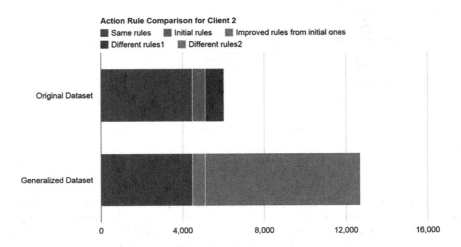

Fig. 3. Action rule comparion for Client 2 (Color figure online)

Figure 3 shows the results of comparing action rules extracted from dataset $\{2,4\}$ to dataset $\{2\}$ alone. In the figure, blue bars display the number of exact same rules with same support and confidence found in both datasets, red bar represents the rules extracted from dataset {2} which are not found the same in dataset {2, 4} but the action sets associated with these rules are contained in the action sets associated with rules extracted from {2, 4} with higher confidence or support, which is marked using orange bar on the bottom. Last but not the least, green bar and pink bar show the unique rules in both action rule sets respectively that don't exist in the other action rule set. Firstly, we can easily see that there are twice as many as rules generated from the expanded dataset. More specifically, we found 12, 715 action rules from the larger dataset while 6, 026 from the original dataset. At the same time, nearly 75 % of action rules from dataset of Client 2 can be found in the set of action rules from the more generalized dataset {2, 4} with same support and confidence. And over 10 % of action rules found in original dataset can be found in the set of action rules from generalized dataset with higher support or confidence. Furthermore, a lot of new action rules have been discovered and over 70 % of the new action rules are with remarkably high confidence.

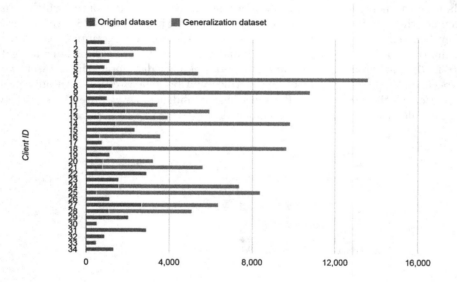

Fig. 4. Performance of HAMIS on 34 clients

In order to get more convincing results, we apply HAMIS to all 34 clients individually and retrieve the generalized datasets. From the results in Fig. 4, we get 18 out of 34 clients who are generalized by HAMIS, and averagely, the generalized dataset for each client is three times as large as the original dataset. The largest generalized dataset is from Client 7 which is far more than other expanded datasets. For comparing action rules, the results vary with the different generalized datasets associated with clients, and the number of action rules generated from those generalized dataset of clients is at least two times bigger than that from a client alone, which is still within our expectation.

5   Conclusion and Further Work

The paper presents HAMIS which is one of the main modules of a hierarchically structured recommender system for improving NPS. We have shown that by expanding datasets assigned to nodes of the dendrogram, recommender systems can give clients more promising suggestion for improving their NPS score. With the hierarchical dendrogram, HAMIS continually enlarges the dataset by following a bottom up path starting from the chosen node, higher the path ends, more generalized the dataset becomes. After applying HAMIS to all 34 clients in our domain, we notice that over half of the datasets can be expanded as shown in Fig. 4. As what we expected, action rules mined from extended datasets are far more promising than the ones from original datasets, no matter in quantity or in quality. Thus, the more generalized (more extended are the datasets) recommender system is built using HAMIS, better recommendations for improving the NPS score of clients can be given.

However, there are still some clients who can't benefit from HAMIS due to the failure of generalization based on semantic similarity. And 90 % of them are caused by the lower NPS score of their most semantically similar clients. For example, Client 5 (in Nevada) is one of the clients which failed in HAMIS, but its geographical neighbor Client 17 (in California) has been expanded with Client 23 (in Mississippi) which can be seen in Fig. 4. Meanwhile, we speculate that knowledge hidden in dataset of Client 5 could be similar to the knowledge hidden in Client 17 to some extent, since customers locating nearby possibly have common thoughts about how they felt about the services and how they would like to be served. Therefore, it is interesting to think that Client 5 can benefit from the advices of Client 17 and even from Client 23, but the problem is how the geographic distance influences the generalization. What's more, the fact that the clients are geographically located close to each other does not necessarily imply they are semantically similar. The most fitting example is Client 24 which is in California and Client 34 which stays at Georgia. They are physically far away, but they are treated as most semantically similar as it is shown in Fig. 1, and Client 34 can't be generalized with Client 24 because of lower NPS rating. So this makes us wonder, maybe we can find another client near Client 34 which can offer some help. When looking into the location of Client 34, we happen to discover another interesting case, for Client 2 (in South Carolina), although it has been generalized using semantic similarity, it is surrounded by several clients with higher NPS ratings. Living such competitive environment, customers here could have stricter requirement for clients and harder to satisfy. With all the concerns above, we still have some space to make progress.

References

1. Kuang, J., Daniel, A., Johnston, J., Raś, Z.W.: Hierarchically Structured Recommender System for Improving NPS of a Company. In: Cornelis, C., Kryszkiewicz, M., Ślęzak, D., Ruiz, E.M., Bello, R., Shang, L. (eds.) RSCTC 2014. LNCS (LNAI), vol. 8536, pp. 347–357. Springer, Heidelberg (2014)

2. Rauch, J., Šimůnek, M.: Action rules and the GUHA method: preliminary considerations and results. In: Rauch, J., Raś, Z.W., Berka, P., Elomaa, T. (eds.) ISMIS 2009. LNCS (LNAI), vol. 5722, pp. 76–87. Springer, Heidelberg (2009)
3. Paul, R., Hoque, A.S.: Mining irregular association rules based on action and non-action type data. In: Proceedings of the Fifth International Conference on Digital Information Management (ICDIM), pp. 63–68 (2010)
4. Qiao, Y., Zhong, K., Wang, H.-A., Li, X.: Developing event-condition-action rules in real time active database. In: Proceedings of the 2007 ACM Symposium on Applied Computing, Seoul, pp. 511–516 (2007)
5. He, Z., Xu, X., Deng, S., Ma, R.: Mining action rules from scratch. Expert Syst. Appl. **29**(3), 691–699 (2005). Elsevier
6. Raś, Z.W., Wieczorkowska, A.A.: Action-Rules: how to increase profit of a company. In: Zighed, D.A., Komorowski, J., Żytkow, J.M. (eds.) PKDD 2000. LNCS (LNAI), vol. 1910, pp. 587–592. Springer, Heidelberg (2000)
7. Reichheld, F.F.: The one number you need to grow. Harvard Bus. Rev. **18**, 46–54 (2003)
8. SATMETRIX, NET Promoter (2014). http://www.satmetrix.com/net-promoter/
9. SATMETRIX: Improving your net promoter scores through strategic account management, White paper (2012)
10. Tzacheva, A., Ras, Z.W.: Association action rules and action paths triggered by meta-actions. In: Proceedings of 2010 IEEE Conference on Granular Computing, Silicon Valley, CA, pp. 772–776. IEEE Computer Society (2010)

A New Linear Discriminant Analysis Method to Address the Over-Reducing Problem

Huan Wan[2(✉)], Gongde Guo[2], Hui Wang[1],
and Xin Wei[2]

[1] School of Computing and Mathematics, University of Ulster at Jordanstown,
Newtownabbey, Ireland, UK
h.wang@ulster.ac.uk
[2] Key Lab of Network Security and Cryptology School of Mathematics and
Computer Science, Fujian Normal University, Fuzhou, P.R. China
huanwan.mail@qq.com

Abstract. *Linear discriminant analysis* (LDA) is an effective and efficient linear dimensionality reduction and feature extraction method. It has been used in a broad range of pattern recognition tasks including face recognition, document recognition and image retrieval. When applied to fewer-class classification tasks (such as binary classification), however, LDA suffers from the *over-reducing* problem – insufficient number of features are extracted for describing the class boundaries. This is due to the fact that LDA results in a fixed number of reduced features, which is one less the number of classes. As a result, the classification performance will suffer, especially when the classification data space has high dimensionality. To cope with the problem we propose a new LDA variant, *orLDA* (i.e., *LDA for over-reducing problem*), which promotes the use of individual data instances instead of summary data alone in generating the transformation matrix. As a result orLDA will obtain a number of features that is independent of the number of classes. Extensive experiments show that orLDA has better performance than the original LDA and two LDA variants – uncorrelated LDA and orthogonal LDA.

Keywords: Dimensionality reduction · Binary classification · Linear discriminant analysis · LDA for over-reducing problem

1 Introduction

Linear discriminant analysis (LDA) is an effective and efficient method for dimensionality reduction (feature extraction). It has been successfully used in many pattern recognition problems such as face recognition [1,2], document recognition [3] and image retrieval [4,5]. It uses within-class scatter matrix S_w to evaluate the compactness within same class, and between-class scatter matrix S_b to evaluate the separation between different classes. The objective of LDA is to find an optimal transformation matrix W which minimizes the within-class scatter matrix S_w and simultaneously maximizes the between-class scatter matrix S_b.

© Springer International Publishing Switzerland 2015
M. Kryszkiewicz et al. (Eds.): PReMI 2015, LNCS 9124, pp. 65–72, 2015.
DOI: 10.1007/978-3-319-19941-2_7

In past two decades various improvements over the origianl LDA have been proposed in order to enhance its performance in different ways, resulting in different LDA variants. These LDA variants can be put into two categories. In the first category, the LDA variants attempt to tackle the singularity problem of within-class scatter matrix (S_w). In LDA, we take the leading eigenvectors of $S_w^{-1}S_b$ as the columns of optimal transformation matrix W. In order to guarantee S_w nonsingular, it requires at least $N + C$ samples [6], where N is data dimension and C is the number of classes. However, in realistic world it does not always happen and it is almost impossible in high-dimensionality space. Therefore, sigularity makes within-class scatter matrix irreversible and we can not use $S_w^{-1}S_b$ to obtain transformation matrix W. In order to address the singularity problem, Li-Fen Chen et al. [1] proposed NLDA, which is short for null space linear discriminant analysis. It is based on a new Fisher's criterion function and calculates the transformation matrix in the null space of the within-class scatter matrix, which avoids the singularity problem implicitly. In [7] regularized linear discriminant analysis (RLDA) is proposed. It gets optimal constant α by heuristic approach and adds α to the diagonal elements of the within-class scatter matrix to overcome the singularity problem. Some new approaches are proposed to solve the singularity problem recently. For example, Alok Sharma et al. [8] proposed a new method to compute the transformation matrix W, which gave a new perspective to NLDA and presented a fast implementation of NLDA using random matrix multiplication with scatter matrices; Alok Sharma et al. [9] proposed an improvement of RLDA, which presented a recursive method to compute the optimal parameter; and Xin Shu et al. [10] proposed LDA with spectral regularization to tackle the singularity problem. Other LDA variants for solving the singularity problem can be found in [11].

In the second category, the LDA variants apply the original LDA in local data space instead of whole data space. For example, Zizhu Fan et al. [12] presented two local linear discriminant analysis (LLDA) approaches: vector-based LLDA (VLLDA) and matrix-based LLDA (MLLDA), which select a proper number of nearest neighbors of a test sample from a training set to capture the local data structure and use the selected nearest neighbors of the test sample to produce the local linear discriminant vectors or matrix. Chao Yao et al. [13] proposed a subset method for improving linear discriminant analysis, which divided the whole set into several subsets and used the original LDA in each subset. There are other LDA variants such as nonparametric discriminant analysis [14], sparse discriminant analysis [15], semi-supervised linear discriminant analysis [16], incremental LDA [17], tensor-based LDA [18], and local tensor discriminant analysis [19].

The original LDA and most of its variants have elegant mathematical properties, one of which being that the dimensionality of the data space can be reduced to at most one less the number of classes. One consequence is that if there are few classes in a data set, e.g., two classes in a binary classification problem, there will be one or only a few features left after the dimensionality reduction, probably insufficient for deciding the class boundaries. This leads to the *over-reducing* problem, meaning that dimensionality reduction is over done.

In this paper we propose changes to the original LDA to address the over-reducing problem. Instead of using only the means of each class and the whole data to evaluate the separation between different classes, our new LDA variant uses a new method to compute the between-class scatter matrix. As a result we get more between-class information and more features (than before) after dimensionality reduction even for binary classification.

The rest of the paper is organized as follows. Section 2 reviews the original LDA and two well known LDA variants – Uncorrelated LDA and Orthogonal LDA. Section 3 presents our orLDA, Sect. 4 presents our experimental results and Sect. 5 concludes the paper.

2 Linear Discriminnant Analysis

2.1 The Original LDA

LDA has been widely used for dimensionality reduction and feature extraction. In the original LDA, the within-class scatter matrix and the between-class scatter matrix are used to measure the class compactness and separability respectively. They are defined as [20]:

$$S_w = \frac{1}{N} \sum_{i=1}^{C} \sum_{j=1}^{n_i} (x_{ij} - \mu_i)(x_{ij} - \mu_i)^T, \tag{1}$$

$$S_b = \frac{1}{N} \sum_{i=1}^{C} n_i (\mu_i - \mu)(\mu_i - \mu)^T, \tag{2}$$

where N denotes the number of data samples, C denotes the number of the classes, n_i denotes the number of samples in class i, μ_i denotes the mean of samples in class i, μ denotes the mean of whole samples, and x_{ij} is the jth sample in class i. The original LDA aims to find a transformation matrix $W_{opt} = [w_1, w_2, ..., w_f]$ that maximizes the Fisher's criterion

$$J(W) = \frac{W^T S_b W}{W^T S_w W} \tag{3}$$

Mathematically, the solution to this problem corresponds to an eigenvalue decomposition of $S_w^{-1} S_b$, taking its leading eigenvectors as the columns of W_{opt}.

From Eq. (2), we can see that LDA uses only the centers of classes and whole data set to compute between-class scatter matrix. This may lose much class-separating information. Because the rank of the between-class matrix is at most $C - 1$, the number of extracted features by LDA is at most $C - 1$. However, it is insufficient to separate the classes well with only $C - 1$ features, especially for binary classification in high-dimensional spaces.

2.2 Uncorrelated LDA and Orthogonal LDA

Uncorrelated LDA (ULDA) and Orthogonal LDA (OLDA) were presented in [21]. In this paper, Jieping Ye proposed a new optimization criterion to obtain the optimal transformation matrix W_{opt}. W_{opt} is defined as: $W_{opt} = X_q M$, where X is a matrix that simultaneously diagonalizes S_b, S_w, $S_t{}^1$, X_q is the matrix consisting of the first q columns of X, and M is an arbitrary nonsingular matrix. When M is the identity matrix, we can get Uncorrelated LDA algorithm and make features in the reduced space uncorrelated; however, if we let X_q=QR be the QR decomposition of X_q and choose M as the inverse of R, we get Orthogonal LDA algorithm and make the discriminant vectors of OLDA orthogonal to each other.

3 Linear Discriminnant Analysis that Avoids Over-Reducing

In this section we present the proposed changes to the original LDA in order to address the over-reducing problem, which are related to how to compute the between-class matrix.

Suppose there are N samples $X_i \in R^n$ for $i = 1, 2, \ldots, N$ from two classes, N_k is the number of samples in class k $(k = 1, 2)$ such that $\sum_{k=1}^{2} N_k = N$, μ_k is the mean of the samples in class k, and x_{kj} is the jth sample in class k. Two scatter matrices, the within-class scatter matrix $(\widetilde{S_w})$ and between-class scatter matrix $(\widetilde{S_b})$ are defined as follows:

$$\widetilde{S_w} = \frac{1}{N} \sum_{k=1}^{2} \sum_{j=1}^{N_k} (x_{kj} - \mu_k)(x_{kj} - \mu_k)^T \tag{4}$$

$$\widetilde{S_b} = \frac{1}{N}(N_1 \sum_{j=1}^{N_1} (x_{1j} - \mu_2)(x_{1j} - \mu_2)^T + N_2 \sum_{j=1}^{N_2}(x_{2j} - \mu_1)(x_{2j} - \mu_1)^T) \tag{5}$$

When the number of classes is two, Eq. (1), the computation of within-class scatter matrix in the original LDA, is the same as Eq. (4). However, Eq. (5), the computation of between-class scatter matrix, is quite different from the original LDA. In Eq. (5), we use every sample in one class to subtract the mean of another class.

It is clear that computing the between-class scatter matrix in this way will capture more between-class information than the original LDA hence we can expect better classification performance. Besides, by Eq. (5), we can get more than 1 feature in binary classification. According to linear algebra and Eqs. (4) and (5), we can obtain $rank(S_b) = min(n, N-2)$ and $rank(S_w) = min(n, N-2)$, where n is the dimensionality of data space and N is the total number of samples.

---

[1] S_t denotes total scatter matrix, which is defined as: $S_t = \frac{1}{N} \sum_{j=1}^{N}(x_j - \mu)(x_j - \mu)^T$, where x_j denotes the jth sample.

Then we can find that the ranks of S_b and S_w depend only on n and N. Therefore, $rank(S_w^{-1}S_b) = min(rank(S_w^{-1}), rank(S_b))$ is not limited by 1 extracted feature. Our optimal transformation matrix W_{opt} maximizes $J(W) = \frac{W^T \widetilde{S_b} W}{W^T \widetilde{S_w} W}$ and we get the eigenvectors corresponding to the top eigenvalues of the eigenequation $\widetilde{S_w}^{-1}\widetilde{S_b}$ as columns of W_{opt}.

4 Experiments

In this section we take *K-Nearest Neighbor* (KNN, K=1) as the classifier and use ten-fold cross-validation to evaluate our method on three face datasets – ORL face database[2], Labeled Faces in the Wild (LFW) [22], and Extended Cohn-Kanade [23]; and one DNA microarray gene expression datasets from Kent Ridge Bio-medical Dataset (KRBD)[3].

The ORL face database consists of a total of 400 images of 40 distinct people. Each people has ten different images and the size of each image is 92*112 pixels, with 256 grey levels per pixel. All the images were taken against a dark homogeneous background with the subjects in an upright, frontal position.

LFW face dataset consists of 13,233 images of 5,749 people, which are organized into 2 views – a development set of 3,200 pairs for building models and choosing features; and a ten-fold cross-validation set of 6,000 pairs for evaluation. The size of each image is 250*250 pixels. All the images are collected from the Internet with large intra-personal variations. There are three versions of the LFW: original, funneled and aligned. In our experiment, we use the aligned version [24].

For the above two face datasets, we do face verification experiment, which is a binary classification problem. The goal of face verification is to decide if two given face images match or not. We use subset of view2 of LFW. We randomly choose 200 matched face pairs and 200 mismatched face pairs from view2 and crop each image to an image of 80 * 150 pixels as in [25]. However, for ORL face dataset, through randomly matching face images, we obtain 80 matched face pairs and 391 mismatched face pairs for face verification. Therefore, we have 400 samples of LFW and 471 samples of ORL. The dimensionality of each sample in LFW and ORL are 24,000 and 20,608, respectively.

Extended Cohn-Kanade dataset (CK+) is a complete dataset for action unit and emotion-specified expression. In this paper, we focus on emotion-specified expressions. There are 593 sequences from 123 subjects which are FACS coded at the peak frame, but only 327 of the 593 sequences have emotion sequences and use the last frame of each sequence to do expression classification. There are seven kinds of emotion expression, including: neutral, anger, contempt, disgust, fear, happy, sadness and surprise. Here, we do positive and negative expression classification experiment and take happy as positive expression and the rest of emotion as negative expression. Therefore, we have 69 positive expression

---

[2] http://www.cl.cam.ac.uk/research/dtg/attarchive/facedatabase.html.
[3] http://datam.i2r.a-star.edu.sg/datasets/krbd/.

samples and 258 negative expression samples and the dimensionality of each sample is 10,000.

Acute Leukemia dataset [26] consists of DNA microarray gene expression data of human acute leukemia data for cancer classification. There are two types of acute leukemia: 47 acute lymphoblastic leukemia (ALL) and 25 acute myeloid leukemia (AML), over 7129 probes from 6817 human genes.

We compare our orLDA with three discriminant dimension reduction methods, which are the original LDA, Uncorrelated LDA (ULDA) and Orthogonal LDA (OLDA) [21]. To guarantee that S_w does not become singular, we use two-stage PCA+LDA [27] – we reduce the data dimensionality by PCA, retaining principal components which explain 95 % of variance, before original LDA and orLDA methods are used.

Experimental results on the four datasets are shown in Table 1. It is clear that our orLDA has better classification performance than the original LDA, ULDA and OLDA on all datasets except Extended Cohn-Kanade. We credit the better performance to the facts that (1) orLDA obtains more between-class information than the other three LDA variants; (2) more than 1 extracted features can better separate two classes.

Table 1. Mean accuracy and standard error of the mean on four datasets

Datasets	ORL	LFW	Extended Cohn-Kanade	Acute Leukemia
Original LDA	0.8536 ± 0.0103	0.5675 ± 0.0190	0.9695 ± 0.0101	0.9857 ± 0.0143
orLDA	0.8832 ± 0.0102	0.625 ± 0.0194	0.9695 ± 0.0101	0.9857 ± 0.0143
ULDA	0.7684 ± 0.0179	0.58 ± 0.0244	0.9757 ± 0.0099	0.9589 ± 0.0299
OLDA	0.7684 ± 0.0179	0.58 ± 0.0244	0.9757 ± 0.0099	0.9589 ± 0.0299

5   Conclusion

In this paper, we propose a new LDA, orLDA, to address the over-reducing problem associated with LDA. orLDA uses a new method to compute between-class scatter matrix, which contains more between-class information and allows extracting more features. Experiments have shown that orLDA outperformed the original LDA, ULDA and OLDA significantly on two face datasets, outperformed ULDA and OLDA on the gene expression dataset. orLDA achieved the same performance as the original LDA on the emotion expression dataset and the gene expression dataset, and underperformed ULDA and OLDA slightly on the emotion expression dataset. It is then reasonable to conclude that the new LDA variant is an improvement over the state of the art.

References

1. Chen, L.-F., Liao, H.-Y.M., Ko, M.-T., Lin, J.-C., Yu, G.-J.: A new LDA-based face recognition system which can solve the small sample size problem. Pattern Recognit. **33**(10), 1713–1726 (2000)
2. Zhao, X., Evans, N., Dugelay, J.: Semi-supervised face recognition with LDA self-training. In: 2011 18th IEEE International Conference on Image Processing (ICIP), pp. 3041–3044. IEEE (2011)
3. He, C.L., Lam, L., Suen, C.Y.: Rejection measurement based on linear discriminant analysis for document recognition. Int. J. Doc. Anal. Recognit. (IJDAR) **14**(3), 263–272 (2011)
4. Swets, D.L., Weng, J.J.: Using discriminant eigenfeatures for image retrieval. IEEE Trans. Pattern Anal. Mach. Intell. **18**(8), 831–836 (1996)
5. He, X., Cai, D., Han, J.: Learning a maximum margin subspace for image retrieval. IEEE Trans. Knowl. Data Eng. **20**(2), 189–201 (2008)
6. Martínez, A.M., Kak, A.C.: PCA versus LDA. IEEE Trans. Pattern Anal. Mach. Intell. **23**(2), 228–233 (2001)
7. Guo, Y., Hastie, T., Tibshirani, R.: Regularized linear discriminant analysis and its application in microarrays. Biostatistics **8**(1), 86–100 (2007)
8. Sharma, A., Paliwal, K.K.: A new perspective to null linear discriminant analysis method and its fast implementation using random matrix multiplication with scatter matrices. Pattern Recognit. **45**(6), 2205–2213 (2012)
9. Sharma, A., Paliwal, K.K., Imoto, S., Miyano, S.: A feature selection method using improved regularized linear discriminant analysis. Mach. Vis. Appl. **25**(3), 775–786 (2014)
10. Shu, X., Lu, H.: Linear discriminant analysis with spectral regularization. Appl. Intel. **40**(4), 724–731 (2014)
11. Ye, J., Ji, S.: Discriminant analysis for dimensionality reduction: An overview of recent developments. In: Boulgouris, N.V., Plataniotis, K.N., Micheli-Tzanakou, E. (eds.) Biometrics: Theory, Methods, and Applications. Wiley-IEEE Press, New York (2010)
12. Fan, Z., Xu, Y., Zhang, D.: Local linear discriminant analysis framework using sample neighbors. IEEE Trans. Neural Netw. **22**(7), 1119–1132 (2011)
13. Yao, C., Lu, Z., Li, J., Xu, Y., Han, J.: A subset method for improving linear discriminant analysis. Neurocomputing **138**, 310–315 (2014)
14. Li, Z., Lin, D., Tang, X.: Nonparametric discriminant analysis for face recognition. IEEE Trans. Pattern Anal. Mach. Intell. **31**(4), 755–761 (2009)
15. Clemmensen, L., Hastie, T., Witten, D., Ersbøll, B.: Sparse discriminant analysis. Technometrics **53**(4), 406–413 (2011)
16. Zhao, M., Zhang, Z., Chow, T.W., Li, B.: A general soft label based linear discriminant analysis for semi-supervised dimensionality reduction. Neural Netw. **55**, 83–97 (2014)
17. Pang, S., Ozawa, S., Kasabov, N.: Incremental linear discriminant analysis for classification of data streams. IEEE Trans. Syst. Man Cybern. Part B: Cybern. **35**(5), 905–914 (2005)
18. Li, M., Yuan, B.: 2D-LDA: a statistical linear discriminant analysis for image matrix. Pattern Recognit. Lett. **26**(5), 527–532 (2005)
19. Nie, F., Xiang, S., Song, Y., Zhang, C.: Extracting the optimal dimensionality for local tensor discriminant analysis. Pattern Recognit. **42**(1), 105–114 (2009)
20. Webb, A.R.: Statistical Pattern Recognition. Wiley, New York (2003)

21. Ye, J.: Characterization of a family of algorithms for generalized discriminant analysis on undersampled problems. J. Mach. Learn. Res. **6**, 483–502 (2005)
22. Huang, G.B., Ramesh, M., Berg, T., Learned-Miller, E.: Labeled faces in the wild: a database for studying face recognition in unconstrained environments. Technical report, Technical Report 07–49, University of Massachusetts, Amherst (2007)
23. Lucey, P., Cohn, J.F., Kanade, T., Saragih, J., Ambadar, Z., Matthews, I.: The extended Cohn-Kanade dataset (CK+): a complete dataset for action unit and emotion-specified expression. In: 2010 IEEE Computer Society Conference on Computer Vision and Pattern Recognition Workshops (CVPRW), pp. 94–101. IEEE (2010)
24. Wolf, L., Hassner, T., Taigman, Y.: Similarity scores based on background samples. In: Zha, H., Taniguchi, R., Maybank, S. (eds.) ACCV 2009, Part II. LNCS, vol. 5995, pp. 88–97. Springer, Heidelberg (2010)
25. Kan, M., Xu, D., Shan, S., Li, W., Chen, X.: Learning prototype hyperplanes for face verification in the wild. IEEE Trans. Image Process. **22**(8), 3310–3316 (2013)
26. Golub, T.R., Slonim, D.K., Tamayo, P., Huard, C., Gaasenbeek, M., Mesirov, J.P., Coller, H., Loh, M.L., Downing, J.R., Caligiuri, M.A., et al.: Molecular classification of cancer: class discovery and class prediction by gene expression monitoring. Science **286**(5439), 531–537 (1999)
27. Belhumeur, P.N., Hespanha, J.P., Kriegman, D.: Eigenfaces vs. fisherfaces: recognition using class specific linear projection. IEEE Trans. Pattern Anal. Mach. Intell. **19**(7), 711–720 (1997)

Image Processing

Procedural Generation of Adjustable Terrain for Application in Computer Games Using 2D Maps

Izabella Antoniuk[(✉)] and Przemysław Rokita

Institute of Computer Science, Warsaw University of Technology,
Nowowiejska 15/19, 00-665 Warsaw, Poland
I.Antoniuk@stud.elka.pw.edu.pl,
P.Rokita@ii.pw.edu.pl

Abstract. This paper describes method for generating 3D terrain for usage in computer games by processing set of 2D maps and employing of user-specified parameters. Most of existing solutions don't allow for modifications during generation process, while introducing any changes usually requires complex activities, or is limited to adjusting input maps. We present our solution that allows not only for easy edition of created terrain, but also verification of its quality at each step of generation.

1 Introduction

Computer games, depending on their genre, can include numerous terrains with different properties and details. Some areas are represented schematically, while others are created with great level of detail, showing even the smallest elements.

Terrain designed for usage in computer games must posses series of different properties as well as allow easy introduction of various changes. Those concern not only terrain shape but also other game elements connected to them such as: story, quests, placement of objects, enemies and other constituents. Because of those requirements most of maps are modelled by hand, usually requiring considerable amount of time to create. Procedural content generation in it most common form is rarely used in those applications, because of low level of control over created content, as well as many difficulties with further edition of obtained objects. However it may provide interesting results.

While considering different terrains and arrangement of occurring elements, it is possible to notice that most of them can be divided by similar features and inclusive properties. Parts that contain distinctive similarities (considering that they are smaller and content inside them is mostly uniform) are much easier to generate and combine. It is also easier to decide precise set of constraints for single tile. That leads to the idea of generating 3D terrain using image processing approach based on simplified 2D maps.

While this concept is not new, existing methods still have a few drawbacks. They allow for creation of suitable and complex terrains, but their final product lacks the capability for further terrain adjustments. Usually procedure allows introducing modifications either by editing input maps (which greatly decreases

© Springer International Publishing Switzerland 2015
M. Kryszkiewicz et al. (Eds.): PReMI 2015, LNCS 9124, pp. 75–84, 2015.
DOI: 10.1007/978-3-319-19941-2_8

level of control over final shape of generated object) or requires conversion of generated terrain to 3D modelling environment (where results of such actions are not always satisfactory and often hard to predict). Another issue is complexity of obtained terrain, which not always meets computer games requirements.

In our approach we propose a method for creating 3D terrains by processing set of simplified maps. This allows for fast and precise generation, easy adjustment and regeneration of both entire terrain and its individual parts. As a test platform for our algorithm we chose Blender 2.73a application (for Blender documentation see [19]).

The rest of the paper is organized as follows. In Sect. 2 we review other works related to our area of research. Section 3 describes initial assumptions for our method. Section 4 contains procedure overview. Section 5 outlines some areas of future work. Finally we conclude our work in Sect. 6.

2   Related Work

Procedural generation for computer games is not a new topic similarly to generating terrain by processing 2D maps or other simplified input. Among existing algorithms some focus only on generating 2D maps from their more basic versions [6], while other generate complex terrains, containing various data and details [17, 18]

One of many problems described in some of the existing works is building entire worlds in real time (which is not an easy case, considering amount of data that needs to be processed) [2]. Other interesting area of research is creating objects from input given by the users in form of pre-processed objects [5]. Some works focus on increasing level of control over generated output by defining terrain with various constraints and parameters such as set of actions that will be performed at generated map [4] or by restricting some of its properties to better fit desired results [13].

Simple and complex algorithms proposed for creation of various virtual worlds, range from basic methods [7] to entire frameworks and methodologies [11, 16]. Among the interesting issues is representation of real world data through sets of height maps [10].

When it comes to generation itself, there are many different ways to realize it. Some solutions use sets of height maps to create complex objects [12], others introduce genetic algorithms, evolving terrain according to user preferences [8], or focus on efficient introduction of various changes to created map(i.e. adding roads) [9].

There are many methods for generation of objects and terrains. For detailed study of procedural content generation algorithms see [1, 3, 14, 15].

3   Initial Assumptions

In this section we present assumptions that led to design of our algorithm in its current form.

While planning our procedure, we assumed that generated terrain must meet certain constraints and have a few significant properties, if it is to be used in computer games.

Most important features concern terrain itself. We assumed that any area must allow for introducing various changes, both during algorithm operations and after it terminates its work. At the same time, any created object should be as close to desired output as possible, providing high level of control over terrain properties.

Another crucial issue is generation process. Designing and modelling of complex 3D objects takes considerable amount of time. We decided to use image processing and base our procedure on 2D maps to simplify the process. In our approach each pixel represents different region inside generated object. Since terrain tiles contain different data, that would be difficult to represent on single image, also increasing its complexity. We decided to use set of maps, where each file contains different information. Currently we consider three type of maps: height map (containing information about basic terrain level, represented in greyscale colour value), dispersion map (describing dispersion range for height value in single region, stored in text file) and terrain map (showing different terrains distribution through the scene, represented as RGB colour value). Each map is independent from others therefore performing changes to one of them does not require updating others (for example, if we want to slightly adjust dispersion range, we don't need to change terrain level or type).

While creating 2D images is easier than modelling 3D environment, very complex maps can still take a lot of time to finish, especially if we consider large maps with great level of detail. Since each terrain can be divided into patches containing elements of the same type (like mountains, plains, forests etc.) we decided to simplify that input as much as possible. Such patches are represented by pixels in our maps and are further processed adding details during algorithm operations and merging terrains according to user-defined properties.

After generation is complete, each of the regions processed from pixels in input maps is represented as different object. Although its internal properties are to some point determined by adjacent regions, it can be further adjusted or regenerated independently, therefore allowing for easier management of terrain data. At the same time, one faulty region doesn't disqualify entire terrain - it can simply be regenerated (or manually adjusted), without influencing other, correct regions.

4   Algorithm Overview

In this section we describe procedure we use for generating 3D maps by processing set of 2D images and using chosen properties of emerging terrain. As a test platform for our algorithm we use newest Blender application (version 2.73a). We chose this 3D modelling environment because it contains complete python interpreter and also allows for easy access to program functionality.

Terrain generation algorithm consists of two main steps: processing terrain maps and building terrain according to input data and predefined properties.

4.1   Processing Terrain Maps

Storing different terrain data in separate files requires precise defining what type of information is taken from which map. In our method we currently consider three types of data, defining our output object.

First map contains data about basic altitude level of region represented by each pixel. Since we assume that RGB values in that map are equal we read only one colour value and represent terrain height with it (Fig. 1. left). This parameter defines basic terrain level, which is a foundation for any further calculations (like obtaining dispersion or performing terrain generation).

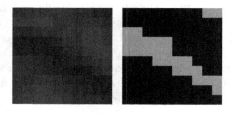

Fig. 1. Input Files: height map (left), and terrain map (right)

Second map, stored in text file, contains values of dispersion for each region. Total range of this parameter is described by two variables - one concerning dispersion value below terrain level, and one for terrain above it (depending on desired results, those values might differ rather significantly).

Final map stored in RGB image (Fig. 1. right), contains description of different terrains occurring in generated object. Each terrain type is assigned a colour, and is then recognized by simply comparing pixel RGB values to those stored in terrain dictionary (each item, apart from terrain ID also contains basic properties for 3D generation algorithms to work with).

4.2   3D Terrain Generation

After information contained in map files is processed and interpreted we obtain few sets of data that are used in next step, which is terrain generation. Our algorithm currently consist of three main parts: creating and placing basic regions, adding details to those regions and finally joining created terrain tiles and reducing visibility of borders between them, as presented in Algorithm 1.

In first step of our procedure we use values obtained by processing image containing height map to place basic grids across the scene, as shown at Fig. 2 (top). Each grid has predefined number of vertices. We also normalize height values for it to better fit terrain characteristic, using user-defined parameter (normalization parameter allows us to adjust height differences between regions, decreasing or increasing tiles dispersion along "z" axis).

Algorithm 1. Terrain Generation

Require: Size, Heights, Dispersions, Terrains, Regions, Normalization
  for all Regions **do**
    CreateGrid(Size)
    GetHeight(Heights, Region)
    PlaceGrid(Height, Normalization)
  end for
  for all Grids **do**
    AssignTerrainType(Region)
    ChooseAlgorithm(TerrainType)
    ReadDispersion(Dispersions, Region)
    GenerateTerrain(Region, Algorithm, Dispersion)
  end for
  for all Grids **do**
    CheckNeighbours()
    AlignBorders(Regions, TerrainType)
    SmoothTransition()
  end for

Although in this step we do not perform any complex operations, obtained results give quite good impression about overall terrain shape. It also provides the possibility to fix any defects that generated terrain can have at this point (like to high dispersion between regions) before main generation process begins.

After we create basic version of our terrain, we then use information from remaining maps, to add details. First we check type of terrain for each existing region (Fig. 2. bottom), since that information decides what kind of algorithm will be used for generation, as well as some of its input parameters. After obtaining required data, we run proper algorithm with given parameters and obtained dispersion range. At this point we gain set of separate regions, where each of them contains part of output terrain, generated according to data from map files. Results of such generation are shown at Fig. 3 (top).

Final step of our procedure connects created regions and blurs border between them. This part of algorithm works on two neighbouring tiles. It first checks placement of border vertices (since grids are generated next to each other, global x and y coordinates for those vertices will be the same) and moves them to new location. Final placement of each vertex is calculated according to set of constraints specified by user (i.e. they can be placed in the middle, or closer to one region location, depending from influence that each terrain will have at calculations outcome). Results of joining tiles are shown at Fig. 3 (middle).

Although at this point created regions are connected, borders between tiles are clearly visible. We use simple procedure to blur those transitions, by defining percentage of terrain that will be aligned (it corresponds directly to number of vertices in each region that will be modified). Subsequently we level that terrain to its border (already connected with neighbouring region), gradually decreasing alignment influence. Final terrain is shown in Fig. 3 (bottom).

Fig. 2. Basic generation: placing regions (top), and assigning terrain type (bottom)

After this final step we obtain terrain made up from set of separate tiles, where each of elements can be processed and modified independently. Since computer games rarely store terrain as single object, and rather load small parts of it when they are needed (i.e. while player travels to another locations or approaches borders between them), such data representation helps to avoid unnecessary operations (like dividing terrain after it was created, to include it in computer game). At the same time we also allow possibility to connect regions into single object (in this operation we consider both joining all regions as well as restricting this action to chosen set of elements).

In its current form (see Fig. 4), our algorithm has both advantages and disadvantages. We still use rather simple procedures for generating terrain inside tiles and blurring borders between them, therefore the amount of available terrains as well as their variety is limited. At the same time terrains we can generate can be easily edited and adjusted because they are editable Bender objects. Unfortunately using Blender application as our test platform greatly limits our performance. Our algorithm can generate terrain (including complex maps, with many different terrain tiles), from relatively simple input, but without inserting any additional detail (like vegetation or buildings). One big advantage is that each step of our procedure can be verified, limiting number of errors occurring in final object. Also due to its modular structure, appending new functionality is an easy task.

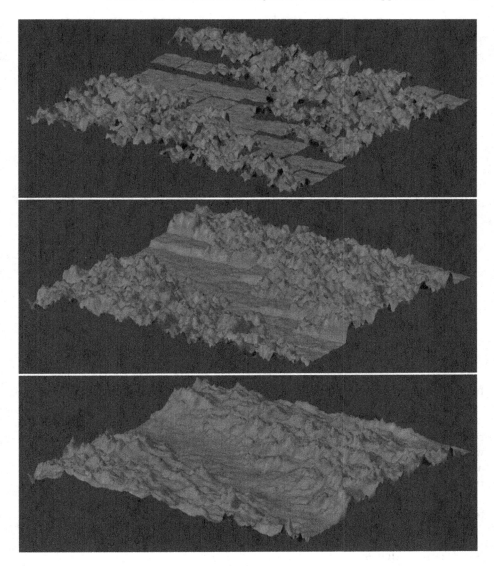

Fig. 3. Terrain generation: generating terrain details in regions (top), joining regions (middle) and blurring borders between regions (bottom)

5   Future Work

Our algorithm in its current form can create rather large set of different terrains and is only limited by generation procedure that we are using for different terrain types. Currently we use diamond-square algorithm for that purpose. In the

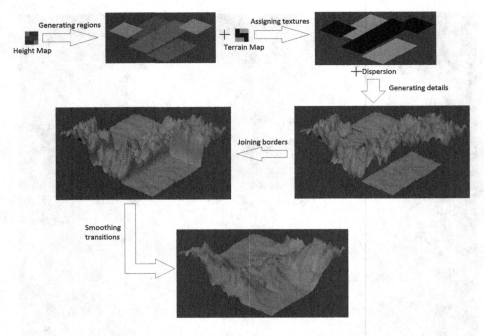

Fig. 4. Algorithm overview

future we would like to add different and more complex procedures for generating terrain.

Another issue is that our joining procedure does not provide satisfactory results, since from time to time, borders between regions are still visible. We would like to work on that problem and try some other solutions, like genetic algorithms, or more complex, mathematical functions than the one we are currently using.

Finally, we would like to add some procedures to further add details to created terrain (like vegetation, rocks, buildings and other).

6   Conclusions

In this work we presented procedure for creating 3D terrain by processing set of schematic maps, and user-defined properties. Terrains obtained during generation process could not only be easily adjusted at different parts of their creation, but also are suitable base for further modifications. At the same time any adjustments can be performed without further actions, either by editing maps, or modifying terrain itself, since our output areas are editable Blender objects.

Our algorithm is at early stage of development and still has some drawbacks, like to clear borders between terrain tiles, or insufficient base of generation algorithms. We plan to address those problems.

Even in this form our procedure creates interesting and playable terrains, that can also be easily used in computer games and further adjusted by designers, providing suitable base for future research.

References

1. Ebert, D.S., Kenton Musgrave, F., Peachey, D., Perlin, K., Worley, S.: Texturing and Modelling: A Procedural Approach, 3rd edn. Morgan Kaufmann Publishers Inc., San Francisco (2002)
2. Greuter, S., Parker, J., Stewart, N., Leach, G.: Real-time procedural generation of 'Pseudo Infinite' cities. In: Proceedings of the 1st International Conference on Computer Graphics and Interactive Techniques in Australasia and South East Asia (2003)
3. Hendrix, M., Mejer, S., van der Velden, J., Iosup, A.: Procedural content generation for games: a survey. Commun. Appl. ACM Trans. Multimed. Comput. **9** (2013)
4. Linden, van der R., Lopes, R., Bidarra, R.: Designing procedurally generated levels. In: Proceedings of IDPv2 - Workshop on Artificial Intelligence in the Game Design Process (2013)
5. Merrell, P., Manocha, D.: Model synthesis: a general procedural modeling algorithm. IEEE Trans. Vis. Comput. Graph. **17**(6), 715–728 (2011)
6. Prachyabrued, M., Roden, T.E., Benton, R.G.: Procedural generation of stylized 2D maps. In: Proceedings of the International Conference on Advances in Computer Entertainment Technology (2007)
7. Prusinkiewicz, P., Hammel, M.: A fractal model of mountain with rivers. In: Proceedings of Graphics Interface, pp. 174–180 (1993)
8. Raffe, W.L., Zambetta, F., Li, X.: Evolving patch-based terrains for use in video games. In: Proceedings of the 13th Annual Conference on Genetic and Evolutionary Computation (2011)
9. Raman, S., Jianmin, Z.: Efficient terrain triangulation and modification algorithms for game applications. Int. J. Comput. Games Technol. **2008**, 5 (2008)
10. Roberts, G., Balakirsky, S., Foufou, S.: 3D reconstruction of rough terrain for USARSim using a height map method. In: Proceedings of the 8th Workshop on Performance Metrics for Intelligent Systems (2008)
11. Roden, T., Parberry, I.: From artistry to automation: a structured methodology for procedural content creation. In: Rauterberg, M. (ed.) ICEC 2004. LNCS, vol. 3166, pp. 151–156. Springer, Heidelberg (2004)
12. Santos, P., Toledo, de R., Gattas, M.: Solid height-maps sets: modelling and visualisation. In: Proceedings of the ACM Symposium on Solid and Physical Modelling (2008)
13. Smelik, R.M., Galka, K., Kraker, de K.J., Kujiper, F., Bidarra, R.: Semantic constraints for procedural generation of virtual worlds. In: Proceedings of the 2nd International Workshop on Procedural Content Generation in Games (2011)
14. Smelik, R.M., Kraker, de K.J., Groenewegen, S.A.: A Survey of procedural methods for terrain modelling. In: Proceedings of the Workshop on 3D Advanced Media in Gaming and Simulation, pp. 25–34 (2009)
15. Smelik, R.M., Tutenel, T., Bidarra, R., Benes, B.: A survey on procedural modelling for virtual worlds. Comput. Graph. Forum (2014)
16. Smelik, R.M., Tutenel, T., Kraker, de K.J., Bidarra, R.: A proposal for a procedural terrain modelling framework. In: Proceedings of the 14th Eurographics Symposium on Virtual Environments (2008)

17. Smelik, R.M., Tutenel, T., Kraker, K.J., Bidarra, R.: Declarative terrain modelling for military training games. Int. J. Comput. Games Technol. **2010**, 11 (2010). Article no. 2
18. Smelik, R.M., Tutenel, T., de Kraker, K.J., Bidarra, R.: A declarative approach to procedural modelling of virtual worlds. Comput. Graph. **35**(2), 352–363 (2010)
19. Blender application web page (2015). http://www.blender.org/. Accessed: 2 February 2015

Fixed Point Learning Based 3D Conversion of 2D Videos

Nidhi Chahal[✉] and Santanu Chaudhury

Indian Institute of Technology, Delhi, India
{nidhi.vce,schaudhury}@gmail.com

Abstract. The depth cues which are also called monocular cues from single still image are more versatile while depth cues of multiple images gives more accurate depth extraction. Machine learning is a promising and new research direction for this type of conversion in today scenario. In our paper, a fast automatic 2D to 3D conversion technique is proposed which utilizes a fixed point learning framework for the accurate estimation of depth maps of query images using model trained from a training database of 2D color and depth images. The depth maps obtained from monocular and motion depth cues of input images/video and ground truth depths are used in training database for the fixed point iteration. The results produces with fixed point model are more accurate and reliable than MRF fusion of both types of depth cues. The stereo pairs are generated then using input video frames and their corresponding depth maps obtained from fixed point learning framework. These stereo pairs are put together to get the final 3D video which can be displayed on any 3DTV and seen using 3D glasses.

Keywords: Monocular · ZPS · Artifacts

1 Introduction

The world of 3D includes the third dimension of depth which can be perceived in form of binocular disparity by human vision. The different views of real world are perceived by both eyes of human as these are located at different positions. From these different views, the brain is able to reconstruct the depth information. A 3DTV display presents two slightly different images of every scene to the individual eyes and thus 3D perception can be realized.

There is great impact of depth map quality on overall 3D output. The better visualization of 3D views demands accurate and denser depth map estimation. In general, depth cues of multiple images gives more accurate depth extraction. And monocular depth cues which are utilized for depth estimation from single still image are more versatile. For the accurate and high quality conversion of 2D images to 3D models, a single solution does not exist. To enhance the accuracy of results, various depth cues should be combined in such a way that more dense and fine depth maps can be obtained. The depth consistency should also

© Springer International Publishing Switzerland 2015
M. Kryszkiewicz et al. (Eds.): PReMI 2015, LNCS 9124, pp. 85–94, 2015.
DOI: 10.1007/978-3-319-19941-2_9

be maintained for the accurate depth estimation. The principle of depth consistency states that if the color values or intensities of the neighboring pixels are similar, these also should have similar depth values. In our paper, machine learning framework is introduced which gives faster and more accurate 2D to 3D conversion. In training phase, the video frames are extracted first from training video and their appearance features are utilized; the depth values of neighboring blocks of a given image block are used as contextual features. Thus, a trained model is obtained which is used in testing phase where new testing video frames are used as input and their depth values are obtained as output using this learning framework. So, a 3D structure of testing video frames is obtained from a training database of video frames and their depths.

The depth extraction is achieved by using monocular depth cue of images/video frames considering single image at a time and motion depth cue using more than one image. The contextual prediction function is used in this model which gives labeling i.e. assigning depth values of image blocks or individual pixels as output while input being both its features and depth values of rest of the image blocks or pixels. Finally, stereo pair generation is achieved by using input images and their corresponding depth maps using image warping technique. And thus, final 3D output images/video is obtained which can be displayed on 3DTV display and 3D output can be viewed using 3D glasses.

2   Related Work

In the last decade, a new family of machine learning approaches are introduced for the depth estimation of different 2D images. In these techniques, training database of color and depth images are used to estimate the depth map of a query color image. In this way, the information is transferred from the structure correlations of the color and depth images in the database to the query color image and its 3D structure is obtained. Some 3D conversion systems have used monocular depth cues only for the depth extraction from single still image which has their own limitations. Most of the prior work have focused on obtaining the better quality of estimated depth maps, but at the expense of high computational cost algorithms.

A depth estimated in [1] from a single color image using MRF framework and image parsing processing. The machine learning approach has also been adopted in [2] that exchanges transference of labels by directly depth map data. The semantic labels and a higher complex supervised model is incorporated in [3] to achieve more accurate depth maps. Konard in [4] has not used image registration step to reduce the computational burden of previous approaches and used a matching based search framework based on HOG features to find similar structured images. This approach has less computational cost than previous approaches, but for many practical applications, it is still too high. The depth maps are combined using MRF model which incorporates contextual constraints into fusion model described in [7], but depth learning is not done in this method. MRF and CRF are often limited to capturing less neighborhood interactions due to heavy computational burden in training and testing stages and thus, their modeling capabilities are also limited.

In our paper, fixed point learning framework provides depth learning from training database of color images, local depth values from focus and motion cues and their ground truth depth values. The images features and depth values from focus and motion cues are used as training data. And we are using ground truth depth values which we are obtained after manual labeling of input images because ground truth depth maps are available for limited data sets. These ground truth depths have been used as training labels which results to fixed point iteration. The learned fixed point function captures rich contextual/structural information and is easy to train and test and much faster also which balances the performance and learning time. For the testing images, reliable depth maps can be obtained using this trained model. The proposed learning framework provides automatic and fast 2D to 3D conversion method which converges at very less number of iterations with good accuracy in results. Fixed point model [8] is a simple and effective solution to the structured labeling problem. We compared the results with MRF fusion as described in [7] of depth cues. Depth maps from focus and motion are fused using MRF approach. But the depths obtained from fixed point model have higher accuracy than MRF fused depths. The results are also obtained with other fusion methods like weighted averaging, least squares, maximizing approach, window technique. All the fusion depths have less accuracy as compared to fixed point model results.

3   Depth Estimation

The HVS (Human Visual System) exploits a set of visual depth cues to perceive 3D scenes. The extraction of depth information from single view is known as monocular cue. There are various monocular depth cues as shading, texture gradient, accommodation, linear perspective and others. The motion depth cue provides depth extraction from multiple images. Both methods are used in the paper to utilize the benefits of both depth cues. The two important depth cues for depth extraction are focus/defocus and motion parallax on the basis of their relative importance in the human brain. The importance depends on the distance (JND-Just noticeable difference) from the observer. A depth cue with larger JND means that it is harder to be detected in the human vision system.

3.1   Depth Extraction from Monocular Depth Cue

The defocusing factor is used as monocular depth cue as discussed above. The principle that stuff which is more blurry is further away is used in depth estimation from focus depth cue. The method used here recovers depth from a single defocused image captured by uncalibrated conventional camera as described in [5]. From the blur amount at each edge location, a sparse depth map is estimated. Then, propagate the depth estimates to the whole image and obtain full depth map. The diameter of CoC characterizes the amount of defocus and can be written as

$$c = \frac{|d - d_f|}{d} \cdot \frac{f_0^2}{N(d_f - f_0)} \tag{1}$$

where d_f is the focus distance, d is the object distance f_0 and N are the focal length and the stop number of the camera respectively.

3.2   Depth Extraction Using Motion Parallax

The other depth cue used is depth from motion. There are two ways to calculate depth from motion parallax and these are motion blur and optical flow. The drawback of depth extraction using motion blur is the cost due to the multiple of high quality cameras, image processing programming and a big hardware background. In our experiments, optical flow is used for depth extraction and it is calculated between two consecutive frames taken from the scene as described in [6]. The length of the optical flow vectors will be inverse proportional to the distance of the projected point. The optical flow can also be used to recover depth from motion by using:

$$Z = v_c \cdot \frac{D}{V} \tag{2}$$

where v_c is the velocity of camera,
D is the distance of the point on image plane from focus of expansion,
V is the amplitude of flow and
Z is the depth of the point in the scene projected.

4   Fixed Point Learning Based 2D to 3D Conversion

There are infinite number of possible solutions for recovering 3D geometry from single 2D projection. The problem can be solved using learning based methods in vision. Initially a set of images and their corresponding depth maps are gathered in supervised learning. Then suitable features are extracted from the images. Based on the features and the ground truth depth maps, learning is done using learning algorithms. The depths of new images are predicted from this learned algorithm. The local as well as global features are used in a supervised learning algorithm which predicts depth map as a function of image. If we take examples of monocular depth cues, local information such as variation in texture and color of a patch can give some information about its depth, these are insufficient to determine depth accurately and thus global properties have to be used. For example, just by looking at a blue patch, it is difficult to tell whether this patch is of a sky or a part of a blue object. Due to these difficulties, one needs to look at both the local and global properties of an image to determine depth.

The entire image is initially divided into small rectangular patches which are arranged in a uniform grid. And a single depth value for each patch is estimated. Absolute depth features are used to determine absolute depth of a patch which captures local feature processing. To capture additional global features, the features used to predict the depth of a particular patch are computed from that patch as well as the neighboring patches which is repeated at each of the multiple

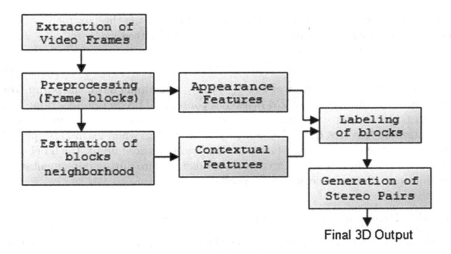

Fig. 1. 2D video to 3D output block diagram

scales, so that the feature vector of a patch includes features of its immediate neighbors at large spatial scale, its far neighbors at a larger spatial scale, and again its very far neighbors at an even larger spatial scale. The fixed point model approach is an effective and simple solution to the problem of structured labeling in image processing. A fixed point function is obtained with the labeling of the structure being both the output and the input. The overall fixed point function is a vector form of contextual prediction function of the nodes. The contextual prediction function gives labels of an individual node as output while inputs are both its features and labeling of rest of the nodes or neighboring nodes. The learned function captures rich contextual/structural information and is easy to train and test.

In image processing, a structured input is an image of all pixels and the structured outputs are the corresponding labels of these pixels. The structured input as shown in Fig. 3a, can be represented as a graph, denoted as $\chi = (\mathbf{I}, \mathbf{J})$. The block $b_i \in \mathbf{I}$ represents i^{th} block with its features denoted as f_i. And $\mathbf{J}$ represents the neighborhood of each block. M denotes the number of neighbors a block can have in its neighborhood specification. That is, M number of blocks specifies the neighborhood N_i of block b_i in four directions: left, right, top and bottom. The labeling of the neighborhood of b_i is denoted as $\mathbf{q}\,N_i$ where $\mathbf{q}$ denotes labeling of all the blocks in χ. The labels for all blocks can be represented as $\mathbf{p} = (p_i : 1...\phi)$, where $p_i \in \phi$ and ϕ is the label space. For each block b_i, a context prediction function Ψ takes both b_i's appearance features f_i and contextual features $\mathbf{q}N_i$ as inputs. And this prediction function predicts the labeling $p_i \in \phi$ for the blocks of image. The contextual prediction function defined in [8] can be represented as follows:

$$q_i = \psi(f_i, \mathbf{q}N_i; \delta) \tag{3}$$

where ψ is a regression function within range $[0,1]$ and δ is the parameter of the function. The labeling $\mathbf{q}$ of all image blocks can be written as

$$\mathbf{q} = \psi(f_1, f_2, ... f_n, \mathbf{q}; \delta) \tag{4}$$

Given the depth values $\mathbf{q}$ and features f_1, f_2,...f_n of training data, the parameter δ is learned. In experiments, SVM is used as contextual prediction function. The function is trained for each image block on the basis of appearance features f_i and contextual features $\mathbf{q}N_i$. The contraction mapping, when learned is applied iteratively to the new structured inputs. For new structured input, the label q_i of a block $b_i \in \mathbf{I}$ is initialized with a value. It is taken as zero during implementation as it does not effect the accuracy of results.

4.1   Pre Processing and Appearance Features

The frames are extracted first from the 2D input video and converted to gray scale. These frames are divided into 8×8 blocks to reduce the computations. The appearance features of these input frames/images is extracted then. The examples of appearance features of image are maximum height of the block, width of the block, aspect ratio and mean wise averaging of the image blocks. The RGB values can also be taken as appearance features of images.

4.2   Neighborhood and Contextual Features

The neighborhood of each image block is identified then. The parameter M defines the span of neighborhood of image blocks. The neighboring blocks of ith block is defined by all the adjacent blocks to its left, right, top and bottom. In our experiments, we used M = 1 to 6 where M = 6 gives the best results. A normalized histogram of neighboring blocks is used as contextual feature $\mathbf{q}N_i$. For each block b_i, a $4 \times M \times C$ dimensional contextual feature vector is created. In this vector, 4 specifies the neighborhood in four directions (upper, lower, left and right), M is the span of the context and C is the number of class labels. In implementations, following values are used.

M = 6 = Number of neighbors a node can have in the neighborhood specification.
C = 255 = No of classes = No of histogram bins
Number of iterations in prediction = 3
Constant for labeling vector = 0.

4.3   Labeling of Blocks

All the image blocks are labeled in this step. Labeling means assigning depth values to each block of an image. The ground truth depth values are used as training labels resulting in fixed point iteration. At each iteration, the depth value for which convergence is achieved, is assigned to the corresponding block during testing process. The fixed-point model converges very quickly at the testing stage with only 2 to 3 iterations. We have used three number of iterations in our experiments as it gives good accuracy with less implementation time. We experimented with L1 regularized support vector machine (SVM-L1), provided in the Lib linear software package as the classifier. The SVM parameter is taken as '−s 5 −c 1' and the constant of labeling vector is taken as 0.

5 Stereo Generation

The stereo pairs are generated at the end from 2D original images and final obtained depth maps. Consider original 2D image as intermediate image and generate left and right views from this image. The right and the left images will be half shifted or warped toward the respective direction and all the artifacts introduced by the processing will be part of both images but halved visible. So, less artifacts are produced in this method. We applied this technique [9] for final stereo generation.

First, pre-processing of depth maps is done by shifting depth map by ZPS i.e. Zero Parallax Setting:

$$Z_c = \frac{Z_n - Z_f}{2} \tag{5}$$

where Z_n and Z_f are the nearest clipping plane and the farthest clipping plane of the depth map. In an 8-bit depth map, $Z_n = 255$ and $Z_f = 0$. After that, the depth map is further normalized with the factor of 255, so that the values of the depth map lie in the interval of $[-0.5, 0.5]$, values that are required by the image warping algorithm. Then, Gaussian filter is used for depth smoothing.

w = $3 \times \sigma$ where w is the filter window size
σ is standard deviation
Depth smoothing strength = Baseline/4

The stereo pair is obtained by transferring the original points at location (P_m, y) to left points (P_l, y) and right points (P_r, y):

$$P_l = P_m + \frac{Bf}{2Z} \tag{6}$$

$$P_r = P_m - \frac{Bf}{2Z} \tag{7}$$

where B is the baseline distance between two virtual cameras.
B = 5 percent of width of depth image
f is the focal length of camera
f is chosen as one in our experiments without any loss of generality.

6 Implementation Results

The input color video of Mickey mouse is taken as input here as shown in Fig. 2a. The video is in wmv format and the resolution of video frames is 960×1080.

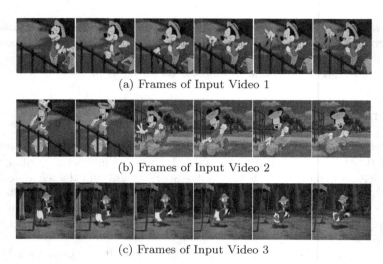

(a) Frames of Input Video 1

(b) Frames of Input Video 2

(c) Frames of Input Video 3

Fig. 2. Examples of 2D video frames to be converted to 3D video

The resolution of final 3D image is 1920×1080 where left and right frames are put side by side. The results of fixed point model for some video frames in form of accuracy (percentage) is shown in Table 1 as follows:

Table 1. Fixed point implementation results

Image	Accuracy with fixed point model	Accuracy with MRF fusion
1	83.57	73.6
2	83.45	73.72
3	82.71	74.90
4	82.39	74.00
5	83.00	74.72
6	82.38	73.73

Here, results are shown in Fig. 3 where (a) and (b) shows original input frames of input video in RGB format. Figure 3c shows depth from focus, (d) shows depth from motion (optical flow) and (e) shows the final 3D output which can be shown on 3DTV display using 3D glasses. The examples of other cartoon and human video frames are also shown in Fig. 4. The subjective test is conducted to show the quality of 3D images, that is, left and right image pair for every 3D image on 3DTV display by wearing 3D glasses. Five participants involved in the experiment to evaluate 3D quality. Their individual scores are mentioned in Table 2 using which MOS i.e. mean opinion score is calculated. The MOS is the arithmetic mean of all the individual scores and can range from 1 (worst) to 5 (best).

Table 2. Subjective quality measure

Video	Viewers score	MOS
Mickey	4,4,5,4,4	4.2
Donald	4,4,4,4,5	4.2
Human	4,3,5,4,4	4

(a) Input frame1 (b) Input frame2 (c) Focus depth (d) Motion depth (e) Final 3D

Fig. 3. Example of two 2D frames to final 3D output

(a) Input frames from different videos

(b) Corresponding depths from focus depth cue

(c) Corresponding depths from motion depth cue

Fig. 4. Examples of other video frames

7   Conclusion

The depth estimation is the key step for converting 2D to 3D images or videos. For high quality and accurate 3D output, the depth maps should be reliable. In this paper, depth extraction has been done using depth cue of single still image and motion cues also which use more than one image. Fixed point model uses ground truth depth maps and provides learning framework for accurate estimation of depths which gives higher accuracy than MRF and other fusion methods. Finally, stereo pair generation is done to get the final 3D output which can be viewed on any 3DTV display. The quality of 3D images is also evaluated using MOS which indicates good and reliable depth extraction. MRF has limitations in capturing neighborhood interactions which limits their modeling capabilities while fixed point learned function captures rich contextual/structural information and is easy to train and test. The model is much faster to train and balances the performance and learning time giving higher accuracy than MRF depth fusion.

References

1. Saxena, A., Sun, M., Ng, A.Y.: Make3d: learning 3d scene structure from a single still image. Trans. Pattern Anal. Mach. Intell. **31**, 824–840 (2009). IEEE
2. Konrad, J., Wang, M., Ishwar, P.: 2d-to-3d image conversion by learning depth from examples. In: Computer Society Conference on Computer Vision and Pattern Recognition Workshops (CVPRW), pp. 16–22. IEEE (2012)
3. Liu, C., Yuen, J., Torralba, A.: Nonparametric scene parsing: label transfer via dense scene alignment. In: Proceedings of the IEEE Conference on Computer Vision and Pattern Recognition, pp. 1972–1979. IEEE (2009)
4. Konrad, J., Wang, M., Ishwar, P., Wu, C., Mukherjee, D.: Learning-based, automatic 2d-to-3d image and video conversion. Trans. Image Process. **22**(9), 3485–3496 (2013). IEEE
5. Zhuo, S., Sim, T.: Defocus map estimation from a single defocused image. Pattern Recogn. **44**(9), 1852–1858 (2011)
6. Sun, D., Roth, S., Black, M.J.: Secrets of optical flow estimation and their principles. In: IEEE Conference on Computer Vision and Pattern Recognition (CVPR), pp. 2432–2439 (2010)
7. Xu, M., Chen, H., Varshney, P.K.: An image fusion approach based on markov random fields. Trans. Geosci. Remote Sens. **49**(12), 5116–5127 (2011)
8. Li, Q., Wang, J., Tu, Z.: Fixed-point model for structured labeling. In: Proceedings of the 30th International Conference on Machine Learning, Atlanta, Georgia, USA, vol. 28 (2013)
9. Kang, Y., Lai, Y., Chen, Y.: An effective hybrid depth-generation algorithm for 2D-to-3D conversion in 3D displays. J. Disp. Technol. **9**(3), 154–161 (2013)

Fast and Accurate Foreground Background Separation for Video Surveillance

Prashant Domadiya[1]([✉]), Pratik Shah[2], and Suman K. Mitra[1]

[1] Dhirubhai Ambani Institute of Information and Communication Technology,
Gandhinagar 382007, Gujarat, India
pmdomadiya@gmail.com, suman_mitra@daiict.ac.in
[2] Indian Institute of Information Technology, Vadodara, Gujarat, India
pratik@iiitvadodara.ac.in

Abstract. Fast and accurate algorithms for background-foreground separation are essential part of any video surveillance system. GMM (Gaussian Mixture Models) based object segmentation methods give accurate results for background-foreground separation problems, but are computationally expensive. In contrast, modeling with only a single Gaussian improves the time complexity with a reduction in the accuracy due to variations in illumination and dynamic nature of the background. It is observed that these variations affect only a few pixels in an image. Most of the background pixels are unimodal. We propose a method to account for dynamic nature of the background and low lighting conditions. It is an adaptive approach where each pixel is modeled as either unimodal Gaussian or multimodal Gaussians. The flexibility in terms of number of Gaussians used to model each pixel, along with *learning when it is required* approach reduces the time complexity of the algorithm significantly. To resolve problems related to false negative due to homogeneity of color and texture in foreground and background, a spatial smoothing is carried out by K-means, which improves the overall accuracy of proposed algorithm.

Keywords: Foreground background separation · Adaptive Gaussian mixture model · Video surveillance

1 Introduction

For any video surveillance system foreground background separation is a crucial step. Foreground (object) detection is a first stage of any computer vision based system including, but not limited to, activity detection, activity recognition, behavioral understanding, surveillance, medicare, parenting etc. Accuracy of such systems depend heavily on accuracy of object detection. Moreover, error in detection can affect system performance adversely. Object background segmentation from video is a challenging task, main challenges being variations in illumination, low lighting conditions, shadow effect, moving background and occluded foreground.

© Springer International Publishing Switzerland 2015
M. Kryszkiewicz et al. (Eds.): PReMI 2015, LNCS 9124, pp. 95–104, 2015.
DOI: 10.1007/978-3-319-19941-2_10

A naive frame differencing approach for object segmentation from video suffers from difficulty of setting up an appropriate threshold. Many times, even if a proper threshold is set, method fails due to the dynamic nature of background. In this method, a background frame is stored and subtracted from upcoming frame to detect movements. Sometimes, mean of intensity values over a series of N frames is also used to generate the difference image. This method is fast and handles noise, shadow and trailing effect but again it's accuracy depends on threshold value. It is observed that threshold depends on the movement of foreground, fast movement require high threshold and slow movement require low threshold.

Method proposed by Wren et al. in [1] models every pixel with a Gaussian density function. The method initializes mean by the intensity of first frame pixel and variance by some arbitrary value. Subsequently, the mean and variances are learnt from series of N frames of video for each pixel. This method can not cope with dynamic background. Method uses 2.5 times standard deviation as the threshold. This method can not cope with dynamic property of background. Intensity values of background pixels change with time due to the change in illumination. This results in miss-classification of background.

Another method, based on GMM, models each pixel independently as a mixture of at most K Gaussian components. Each component of GMM is either background or different environmental factors (may be movement of leaves of trees, moving objects, shadow etc.). This adaptive method, proposed by Stauffer and Grimson in [2], can handle dynamic nature of the background. During learning if there is no significant variation in intensity of some background pixels, then the resultant background components of those pixels shrink. As a result, when we try to classify such pixels based on identified Gaussians, a sudden change in illumination can cause miss-classification.

2   Related Works

A statistical classification approach was proposed by Lee [3] that improves the convergence rate of learning parameters without affecting the stability in computation. This method improves learning of parameters of each model via replacing retention parameter by learning rate. In [4], Shimanda et al., proposed a method which reduces number of components by merging similar components to reduces computational time. It handles intensity variation by increasing number of components (Gaussians). A hybrid method proposed by M. Haque et al. in [5], is a combination of probabilistic subtraction method and basic subtraction method. This method reduces miss-classifications and number of components to be maintained at each pixel by combining thresholds used in basic background subtraction (BBS) and Wren et al. [1]. This method can handle variation in illumination without increasing number of components.

To learn the parameters of Gaussian components, sampling-resampling based Bayesian learning approach was proposed, by Singh et al. in [6]. This method gives better result than simple learning methods as suggested by Stauffer and Grimson in [2], and the method does not depend upon the learning rate. But learning is slow.

All previous methods modelled every pixel by a fixed number (K) of Gaussian components. But we believe that it is not necessary to model every pixel with a fixed number of components. Moreover, we can not predict the number of components for each pixel. Some pixel can be modeled by less than K components while other may require more than K components. We propose a method which is based on a fact that intensity variation, shadow effect, movement of object, movement of leaf etc., do not affect uniformly each pixel. As a result the unaffected pixels may have single component and affected pixels may have different number of components (some may have 2 or 3 or more, complex background pixels can be modeled by 3 to 7 components). This method dynamically assigns the number of models to each pixel while learning. Further, the parameter learning is adaptive, in the sense that the parameters corresponding to Gaussian models are updated only when it is required. Finally, for classification, we use method proposed by Stauffer and Grimson [2] with a small change in threshold.

3   Proposed Method

We model each pixel as a Mixture of Gaussians (MoG).

$$X_j(t = N) = \{x_{j1}, x_{j2}, x_{j3}, ..., x_{jN}\} \tag{1}$$

denotes j^{th} pixel intensities for first N frames. Our algorithm works in stages: (a) learning, (b) background component identification, (c) classification and (d) smoothing. The block diagram of proposed method is shown in Fig. 1. The detailed description of each stage is presented next.

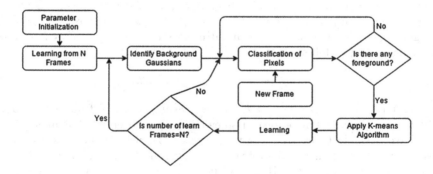

Fig. 1. Block diagram of the proposed method

3.1   Learning

In learning, we estimate parameters of Gaussians at every pixel from N frames. In Eq. (2), parameters $\mu_{ij}(t), \sigma_{ij}(t), W_{ij}(t)$ represent the mean, the standard deviation and weight of ith Gaussian at time t for j^{th} pixel respectively. S is the threshold used in BBS method, α and β are learning rates, B_j is the number of background components and K_j is the number of Gaussian components at j^{th} pixel. Moreover, $C_{ij}(t)$ denotes number of intensity matches accumulated at j^{th} pixel for i^{th} Gaussian up to time t.

Let's consider j^{th} pixel of first frame at time $t = 1$, $X_j(t = 1) = \{x_{j1}\}$ is first observation. We start with single Gaussian $K_j = 1$(initially $i = 1$) and initialize mean $\mu_{ij}(t) = x_{j1}$, standard deviation $\sigma_{ij}(t) = V_0$, learning rate α, number of observation $C_{ij}(t) = 1$. With a new observation x_{jn} of nth frame, we check whether it is matched to any of existing Gaussians as follows,

$$l = arg\ min_{\forall k:|x_{jn}-\mu_{kj}(t)|\leq max(3\sigma_{kj}(t),S)}\{W_{kj}(t)\eta(x_{jn};\mu_{kj}(t),\sigma_{kj}(t))\} \tag{2}$$

where, l is the component where a match is found and $\eta(x;\mu,\sigma)$ is the value of Gaussian function with mean μ and standard deviation σ at x.

After we find a match, we update parameter of lth Gaussian component of jth pixel according to given in method [3],

$$C_{lj}(t) \leftarrow C_{lj}(t) + 1 \tag{3}$$

$$\beta \leftarrow (1 - \alpha)/C_{lj}(t) + \alpha \tag{4}$$

$$\sigma_{lj}^2(t) \leftarrow (1 - \beta) * \sigma_{lj}^2(t) + \beta * (x_{jn} - \mu_{lj}(t))^2 \tag{5}$$

$$\mu_{lj}(t) \leftarrow (1 - \beta) * \mu_{lj}(t) + \beta * x_{jn} \tag{6}$$

$$W_{lj}(t) \leftarrow (1 - \alpha) * W_{lj}(t) + \alpha \tag{7}$$

where, β is learning rate. We do not put restriction on generation of new Gaussian components, number of Gaussians may increase if match is not found during learning. We control K_j after learning from N frames. We initialize new Gaussian components with $K_j = K_j + 1, C_{ij}(t) = 1, \sigma_{ij}(t) = V_0, \mu_{ij}(t) = x_{jn}, W_{ij}(t) = \alpha$. After generating new Gaussian(s) or updating existing Gaussian(s), we normalize weights according to following,

$$W_{kj}(t) \leftarrow W_{kj}(t)/\sum_{\forall k} W_{kj}(t), \ \forall k. \tag{8}$$

We do not put any restriction on K_j, after learning from N frames each pixel might have different number of Gaussians. But, it is observed that most of the pixels have a single Gaussian (due to background). Sensitivity S is used as a threshold in basic background subtraction (BBS) method. Accuracy and speed of this method depends on α. Higher value of α shrinks background component and it results in more number of components and it further reduces speed. If the value of α is selected on lower side, more time is required to learn parameters. To make learning independent of α, M. Haque et al. [5] replaces threshold $3\sigma_{ij}(t)$ by $max(3\sigma_{ij}(t), S)$. S is a minimum threshold that identifies pixel's membership with a component. Sensitivity parameter S prevents generation of unnecessary component due to shrinking of background components. Lower value of S gives better result as discussed in [7].

3.2 Identification of Background Gaussians

We model background pixels by mixture of Gaussains, 2 to 3 Gaussains are enough to model background but it is observed that pixels which are not affected by dynamic nature of background is modeled by only one Gaussian. Such single component is obviously due to background, there is no need to identify it with background components. To identify background components from multi-component models, we use the fact that, background components typically have higher weights and lower standard deviations, whereas foreground components have lower weights and higher standard deviations due to dynamic nature. Once learnt, we sort all components according to $W_{ij}(t)/\bar{\sigma}_{ij}(t)$ ratio (where, $\bar{\sigma}_{ij}(t) = 0.2989\sigma_{ij}^r(t) + 0.5870\sigma_{i2}^g(t) + 0.1140\sigma_{i3}^b(t)$ according to gray scale conversion from RGB). For background components this ratio is higher than foreground components. B_j is set of components which are identified as background. B_j is computed by,

$$B_j = \underset{b \in B_j}{argmin} \sum_{i=1}^{b} W_{ij}(t) > T \tag{9}$$

here T is threshold (we choose $T = 0.7$) and its value depends on complexity of background. Higher value causes misidentification of foreground components as background components and lower value causes misclassification of some background components.

Controling number of components (K_j): Since we have not put any restriction on number of components K_j during learning, the number of components keep on increasing. For classification, we need either background or foreground components. We identified B_j as set of background components so, we can set K_j as size of B_j. Discarding components other than B_j after learning from N frames, we are able to control the continuous increment in K_j.

3.3 Classification

To classify any new observation x_{jn}, we check whether it is part of any of B_j components. If it is part of any of background components then it will be considered as background otherwise foreground. To represent foreground and background in a frame we use an indicator function as suggested below,

$$x_{jn} = \begin{cases} 0, & \forall b \in B_j : |x_{jn} - \mu_{bj}(t)| \leq max(3\sigma_{bj}(t), S) \\ 1, & \text{otherwise.} \end{cases}$$

Learning when it is required: To handle dynamic nature of background it is required to learn parameters of components continuously. Continuous learning increases complexity of algorithm which in turn increases time for foreground detection. To reduce the complexity, we do not learn parameters at every time instance because learning is required only when there is some change in background (i.e. foreground present, illumination change, insertion/deletion of background). During learning, we may remove previously identified background components B_j, we also allow modification of parameters of that components or add extra components if required. After learning from N frames, during classification we allow generation of new components which might be due to change in background. Components having larger weights and smaller standard deviations are considered background components. This assumption reduces misclassification error.

Using K-means to reduce False Negatives: Pixel based MoG method can't detect foreground when foreground colors and textures are homogeneous and have low contrast with background. This situation results in false negative error. To improve on the results, we use the prior about the spatial smoothness of the foreground or the object in motion. We apply K-means algorithm to the portion of frame where foreground is detected. We choose $K_{sp} = 3$ (K_{sp} is number of classes in K-means) because there may be more than one object present in foreground or the object in foreground may have more than one color. Let's consider $Y = \{y_1, y_2, ..., y_n\}$ as the intensities (in R,G,B format) in the frame where the foreground is detected (shown in Fig. 2(c)). We follow a stepwise procedure to correct errors as follows,

Step-1: Apply K-mean algorithm on data set Y and get centers $D_m (m = 1, 2, 3)$ of all three classes.

Step-2: Sort all three centers based on number of observations in each clusters and remove the center which has least number of observations. We also remove empty cluster centers. Now m is number of centers which are not removed.

Step-3: Identify rectangular blocks that cover every detected object. To reduce the time complexity, we apply the spatial smoothing, suggested in next step, only on identified object blocks.

Step-4: Consider the set of identified background pixels inside the object block, denoted by $Z = \{z_1, z_2, ..., z_n\}$, we re-classify a pixel $z_k \in Z$ as foreground based on the following rule:

$$z_k = \begin{cases} 1, & \forall m, \; \|z_k - D_m\|^2 \leq th \\ 0, & \text{otherwise,} \end{cases}$$

where, th is threshold (for our experiment we have set $th = 10$). By using spatial information, we can also detect slow moving object whose some portion is wrongly classified by pixel based method. Results of proposed method on various datasets are presented in Fig. 2(e).

4   Experiments and Results

We have experimented with various datasets including *Wallflower* dataset [8] (frame size 160 × 120 24 Bits RGB) and compared results of our method qualitatively and quantitatively with state-of-the-art method proposed by Stauffer and Grimson [2] (with K = 3 and $\alpha = 0.01$). We have implemented proposed method and state-of-art methods in MATLAB-2009. For comparative timing analysis we run our codes on a system with Core-i3 1.80GHz processor and 4GB memory.

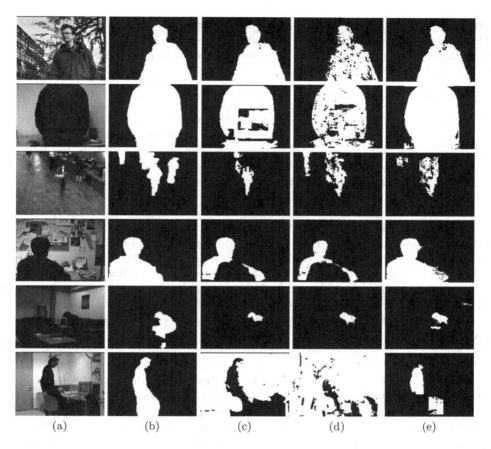

(a)   (b)   (c)   (d)   (e)

Fig. 2. (a) Original frames from video sequences, (b) Ground truth, (c) Results of method proposed by Stauffer and Grimson [2], (d) Results of hybrid method proposed by M. Haque et al. [5], and (e) Results of proposed method

4.1 Qualitative Analysis

To analyze effect of sensitivity S, we experimented with different value of S. It is observed that low S gives better results as suggested in [7]. We have selected $S = 15$, $N = 50$, $\alpha = 0.1$ and $V_0 = 10$ for simulations. Figure 2 shows a significant improvement in detection quality of our method as compared to methods suggested in [2,5]. Higher value of α provides better adaption in case of illumination change but results in generation of spurious and redundant components due to shrinking of components. There is a significant reduction in the dependency of number of Gaussian components on α after using S as threshold in both learning and classification as suggested by [5]. With S, we may choose high learning rate $\alpha = 0.1$ which adapts to change in background quickly. In *Light Switch* video, the improvement due to high learning rate is evident. Results also show improvement in detection for *Camouflage* and *Foreground Aperture* video data sets where the foreground intensities match background components. Spatial smoothing further improves the results by re-classifying misclassified background pixels.

4.2   Quantitative Analysis

Table 1 presents comparison of proposed method with method suggested in [2,5]. For comparison, we have used precision, recall and accuracy. Recall is defined as fraction of number of detected foreground pixels from actual foreground ($Racall = TP/(TP+FN)$). Precision is defined as fraction of number of detected foreground pixels which are actually foreground ($Precision = TP/(TP+FP)$). Accuracy indicates how accurate the detection is ($Accuracy = (TP+TN)/(TN+TP+FN+FP)$). TP is number of truly detected foreground pixels. TN is number of truly detected background pixels. FP is number of falsely detected foreground pixels. FN is number of falsely detected background pixels. Occlusion of foreground and background occurs in *Camouflage* and *Foreground Aperture* videos that is corrected by K-means in proposed method. So, recall and accuracy are high for proposed method in comparison with the methods proposed in [2,5]. Last row of Table 1 presents average precision, recall and accuracy. Compared to method in [2,5] proposed method gives high average precision, recall and accuracy.

For timing analysis, we present our experiment results for all videos of *Wallflower* data set. Table 2, shows results of timing analysis of proposed method, and the others proposed in [2] and in [5]. Our method significantly reduces learning time by not assigning same number of Gaussian components (K) to each pixel. Using threshold as S, we have restricted generation of redundant components. Our method requires on an average 0.229 (we learn parameters when it is require and varable K_j) sec per frame in comparison to 2.6155 s per frame by [2] and 2.9387 s per frame by method in [5]. It is observed that learning requires most of the time for all the methods. With the reduction in number of redundant

Table 1. Video sequence in table appear in the same order in which the results are presented in Fig. 2.

Video sequence	MoG [2]			Hybrid method [5]			Proposed method		
	Preci	Rec	Acc	Preci	Rec	Acc	Preci	Rec	Acc
Waving trees	98.61	99.58	93.04	97.30	87.72	81.32	98.21	99.13	93.15
Camouflage	98.34	88.86	77.54	96.25	88.82	69.09	98.50	99.95	88.11
Bootstrap	97.67	38.09	81.65	75.68	46.28	78.84	81.21	46.36	79.31
Foreground aperture	88.69	46.66	77.03	86.03	32.57	74.40	96.55	99.40	89.52
Time of day	100	17.18	89.23	100	23.86	89.29	87.18	32.23	88.60
Light switch	3.82	26.13	12.96	7.37	70.13	8.90	76.76	45.55	81.94
Average	81.18	52.75	71.90	77.11	58.23	66.97	89.74	70.43	86.77

Gaussians and *learning when it is required* concept, we are able to reduce the average learning time significantly.

Table 2. Time analysis of all video of *Wallflower* dataset (in s per frame). Video sequence in table appear in the same order in which the results are presented in Fig. 2.

Video sequence	MoG [2]		Hybrid method [5]		Proposed method		
	Learn	classi.	Learn	classi.	Learn	classi.	correc.
Waving trees	2.2940	0.0359	1.7823	1.9035	1.1478	0.0866	0.0679
Camouflage	2.2503	0.0315	1.3642	1.4865	0.7921	0.0610	0.0247
Bootstrap	2.2840	0.0352	1.7726	1.7678	0.9469	0.0714	0.0528
Foreground aperture	2.5748	0.0352	1.4305	1.5082	0.7535	0.0446	0.0135
Time of day	2.2903	0.0377	1.3601	1.4565	0.8697	0.0507	0.0172
Light Switch	2.2790	0.0325	1.4027	1.4681	0.9280	0.0393	0.1844

5   Conclusion and Future Work

In this work, we have proposed a method for background-foreground separation which is real-time and shows better accuracy. Other methods discussed are either time efficient but lack accuracy or are accurate but not time efficient. Proposed method is efficient in time and gives better results (both qualitatively and quantitatively) in many adverse conditions. Using spatial smoothness as a prior, we have shown that K-means successfully reduces false negative error when foreground and background intensities are homogeneous. It is interesting to note that this method can detect occluded portion of foreground with background. Time saving is essentially because of removal of redundant components from bookkeeping. Additionally we have used the fact that most of the pixels in a frame are background and allowed every pixel to have variable K_j number of Gaussians. The other significant factor for time saving is due to *learning when it is required* approach.

As an extension to this work, we would like to address a few other issues related to shadow in the image and illumination changes. Shadow in video introduces error in detection of object and it also adds up to generation of redundant components which in turn increases computational time. We would like to incorporate illumination invariance in our approach to make detection scheme more robust.

References

1. Wren, C., Azarbayejani A., Darrell, T., Pentland, A.: Pfinder: real-time tracking of the human body. In: IEEE Proceedings of the Second International Conference on Automatic Face and Gesture Recognition (1996)

2. Stauffer, C., Grimson, W.E.L.: Adaptive background mixture models forreal-time tracking. In: IEEE Computer Society Conference on Computer Vision and Pattern-Recognition,vol. 2 (1999)
3. Lee, D.-S.: Effective mixture learning for video background subtraction. IEEE Trans. Pattern Anal. Mach. Intell. **27**, 827–832 (2005)
4. Shimanda, A., Arita, D., Taniguchi, R.: Dynamic control of adaptive mixture of gaussians background models. In: IEEE International Conference on Video and Signal Based Surveillance (2006)
5. Haque, M., Murshed, M., Paul, M.: A hybrid object detection technique from dynamic background using Gaussian mixture models. In: IEEE 10th Workshop on Multimedia Signal Processing (2008)
6. Singh, A., Jaikumar, P., Mitra, S.K.: A sampling-resampling based Bayesian learning approach for object tracking. In: IEEE Sixth Indian Conference on Computer Vision, Graphics and Image Processing, ICVGIP (2008)
7. Nascimento, J.C., Marques, J.S.: Performance evaluation of object detection algorithms for video surveillance. In: Seventh IEEE Transactions on Multimedia, vol. 8, pp. 761–774 (2006)
8. Toyama, K., Krumm, J., Brumitt, B., Meyers, B.: Wallflower: principles and practice of background maintenance. In: The Proceedings of the Seventh IEEE International Conference on Computer Vision (1999)

Enumeration of Shortest Isothetic Paths
Inside a Digital Object

Mousumi Dutt[1]([✉]), Arindam Biswas[2], and Bhargab B. Bhattacharya[3]

[1] Indian Institute of Information Technology, Kalyani, India
duttmousumi@gmail.com
[2] Indian Institute of Engineering Science and Technology, Howrah, India
barindam@gmail.com
[3] Indian Statistical Institute, Kolkata, India
bhargab.bhatta@gmail.com

Abstract. The computation of a shortest isothetic path (SIP) between two points in an object is important in various applications such as robot navigation and VLSI design. However, a SIP between two grid points in a digital object laid on a uniform 2D isothetic square lattice may not be unique. We assume that each discrete path consists of a sequence of consecutive grid edges that starts from the source point and ends at the sink point. In this paper, we present a novel algorithm to calculate the *number* of such distinct shortest isothetic paths between two given grid points inside a digital object, with time complexity $O(S/g^2)$, where S is the total number of pixels in the digital object, and g is the grid size. The number of available SIPs also serves as a metric for shape registration.

1 Introduction

Shortest path problems have several variations depending on their applicability in diverse fields such as graph theory, computational geometry, operations research, image processing, and computer vision. These problems have wide range of applications in various fields like networking, robotics, geographical information systems (GIS), VLSI design, resource allocation and collection, and traveling salesman problem (TSP) with neighborhoods [1,6,11,19]. Many variations of shortest path problems related to the applications can be found [2,3,7, 16–18] in the literature.

In the isothetic domain, a *shortest isothetic path* (SIP) inside a digital object A (without any hole) laid on a grid $\mathbb{G}$ has already been studied [12,13]. The *family of shortest isothetic paths* (FSIP) is the set of regions where all possible shortest isothetic paths lie [14]. In this paper, we enumerate the total number of SIPs that pass through FSIP. A smaller number of paths is indicative of more obstacles and congestion in the object. In Fig. 1(a) the family of shortest isothetic paths between two grid points is shown. All possible SIPs between two points are shown in Fig. 1(b) (two extreme SIPs are in red and intermediate SIPs in yellow colors).

© Springer International Publishing Switzerland 2015
M. Kryszkiewicz et al. (Eds.): PReMI 2015, LNCS 9124, pp. 105–115, 2015.
DOI: 10.1007/978-3-319-19941-2_11

Fig. 1. (a) The family of shortest isothetic paths between two grid points for $g = 8$ inside A (**Dancer**); (b) All possible intermediate SIPs are shown in yellow lines; (c) Family of shortest isothetic paths between p and q (Color figure online).

A shortest path between two points in a simple polygon in the Euclidean plane is unique, whereas, a SIP is not. In [10], the recursive formulation for the number of cityblock, chessboard, and octagonal shortest paths between two points in 2D digital plane was proposed. The number of minimal paths in a digital image between every pair of points with respect to a particular neighborhood relation is presented in [9], where the image is considered as matrix and hence the algorithm contains matrix operations. Our algorithm determines the number of SIPs between the two grid points inside a digital image using a simple combinatorial method. If a few salient or control points are chosen at certain prominent positions of the object, the number of SIPs between these points serves as a good indicator of the shape of the object. In an object which is devoid of any hole, the number of SIPs is related to the Catalan number [8].

2   Definitions and Preliminaries

A *(digital) object* is a finite subset of $\mathbb{Z}^2$ consisting of one or more $k(= 4 \ or \ 8)$-*connected components* [15]. In our work, we consider the object as a single 8-connected component. The *background grid* is given by $\mathbb{G} = (\mathbb{H}, \mathbb{V})$, where $\mathbb{H}$ and $\mathbb{V}$ represent the respective sets of (equi-spaced) horizontal grid lines and vertical grid lines. The *grid size*, g, is defined as the distance between two consecutive horizontal/vertical grid lines. A *grid point* is the point between intersection of a horizontal and a vertical grid line. An *unit grid block* (UGB) is the smallest square having its four vertices as four grid points. The length between two consecutive grid points in an UGB is equal to the grid size. An *isothetic polygon* has its vertices as grid points and its edges lying on grid lines. An isothetic polygon or object without any hole is said to be *ortho-convex* or simply *convex*, if its intersection with any axis-parallel line produces either zero or one segment. The *inner isothetic cover* of A is the maximum-area isothetic polygon P that tightly inscribes A [4,5]. Since P is an isothetic polygon, it has 90^0 and 270^0 vertices, which are referred to as vertices of type **1** and type **3**, respectively. All other grid points lying on the border of P are called *edge points*. The sequence of vertices of P is such that A always lies left of each edge during its traversal. An edge of P defined by two consecutive vertices of type 1 is termed as a *convex edge* as it gives rise to a *convexity*. Similarly, a *concave edge* is defined by two consecutive type 3 vertices, which gives rise to a *concavity*.

An (simple) *isothetic path* $\pi_{(p,q)}$ from a grid point $p \in P$ to a grid point $q \in P$ is a sequence of 4-connected grid points such that all the constituent points of $\pi_{(p,q)}$ are distinct and lie on or inside P. For two given points in P, a/the *shortest isothetic path* (SIP) has the minimum length over all isothetic paths between them. A SIP is said to be *monotone* if it consists of moves only in one or two (orthogonal) directions. The union of all possible SIPs between two given points $p \in P$ and $q \in P$ defines the *family of SIPs* (FSIP), denoted by $\mathcal{F}_{(p,q)}$. The set $\mathcal{F}_{(p,q)}$ is equivalent to a (isothetic) region bordered by two SIPs between p and q, which are said to be its *extremum SIPs*. If we traverse the border of $\mathcal{F}_{(p,q)}$ from p to q such that each other SIP in $\mathcal{F}_{(p,q)}$ lies to the left during the traversal, then that path of traversal is one of the two *extremum SIPs*, which is termed as the *infimum SIP* and denoted by $\pi_{(p,q)}^{\text{inf}}$. Similarly, on traversing the border of $\mathcal{F}_{(p,q)}$ from q to p with each other SIP lying to the left during the traversal, we get the other *extremum SIP*, termed as the *supremum SIP* and denoted by $\pi_{(p,q)}^{\text{sup}}$. A sub-path common to $\pi_{(p,q)}^{\text{inf}}$ and $\pi_{(p,q)}^{\text{sup}}$ in $\mathcal{F}_{(p,q)}$ is called a *bridge*. FSIP between two points can be perceived as a set of convex isothetic polygons connected by bridges as shown in Fig. 1(c). A region can be assumed to be a set of *UGBs*. If we consider a bounding rectangle for the region then there will be some *UGBs* in the bounding rectangle, which do not belong to the region. Those are termed as *non-region UGBs*. The UGBs belonging to a region is termed as *region UGBs*.

The algorithm to compute a SIP between two grid points, p and q inside a inner isothetic cover [4,5] of a digital object has been proposed in [12,13]. The family of shortest isothetic paths can be obtained using the algorithm stated in [14]. The two extremum SIPs are obtained to determine the FSIP between two grid points (shown in Fig. 1(c)).

3   Characterization of the Total Number of SIPs

The number of shortest isothetic paths between two points, p and q, can be determined from the FSIP which consists of sets of regions and bridges in sequence. Let an FSIP consist of i convex isothetic polygons namely $R_1, R_2, \ldots, R_i$.

Lemma 1. *All shortest isothetic paths between two given grid points in a convex isothetic polygon without any hole are monotone.*

Corollary 1. *All possible shortest isothetic paths in a region of the family of shortest isothetic paths between two points are monotone.*

In Fig. 1(c), there are three regions and two bridges. Let the number of shortest isothetic paths in the regions $R_1, R_2,$ and R_3 (in other words, between p and p_1;

p_2 to p_3; and p_4 to q) be k_1, k_2, and k_3. From k_1, k_2, and k_3, the total number of shortest isothetic paths in a FSIP can be calculated. There are two bridges (i.e., only one possible path) from p_1 to p_2, and from p_3 to p_4.

Theorem 1. *If the number shortest isothetic paths in the regions $R_1, R_2, \ldots, R_i$ is $k_1, k_2, \ldots k_i$ respectively, then the total number of shortest isothetic paths between two points is $\lambda = k_1 \times k_2 \times \ldots \times k_i$.*

The weight of a region UGB is defined as the number of additional paths contributed by the UGB in order to reach the other end point of R_i. The total number of shortest isothetic paths between two end points of a region is obtained by summing the weights of all region UGBs in the bounding rectangle.

Lemma 2. *The number of possible shortest isothetic paths between two diagonal points in an UGB is '2'.*

Proof. An *UGB* is the bounding rectangle for the two points which are diagonally opposite in the corresponding *UGB*. Except the two monotone semi-borders of the UGB there are no more monotone paths between the two diagonal points. Thus, it can be concluded that the number of possible shortest isothetic paths between two diagonal points in an UGB is two (see Fig. 2(left)).   □

Here, to calculate the number of SIPs, we proceed from q. First, the number of possible paths for the UGB containing q as a grid point (called start UGB), is computed as two (the weight of the start UGB), according to Lemma 2. The total number of SIPs in a region is the sum of the weights of all UGBs in that region. Let the weight of all non-region UGBs be '0' (as it cannot be part of any SIP). The region UGBs sharing an edge with either of the two sides of the bounding rectangle intersecting at q (c_1q and qc_2) are termed as border UGBs in the corresponding region.

Lemma 3. *The weight of border UGBs in a region is '1'.*

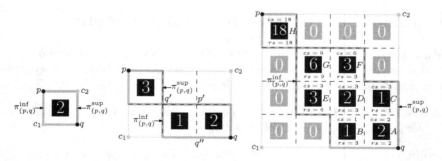

Fig. 2. Left: In any UGB, there are two possible SIPs. **Middle:** Weights of all region UGBs between p and q. **Right:** Calculation of weights using row-sum and column-sum.

Proof. Let there be one start UGB having weight '2' and a border UGB immediately left to it. The number of possible shortest isothetic paths between two diagonal grid points in an UGB is '2' (Lemma 2). The border UGB immediately left to it has also two possible paths between the two diagonal grid points. There is a common grid edge between these two UGBs ($p'q''$ in Fig. 2(middle)). Leaving the common edge we can say that there are two paths—$q'p'$ and $q'q''$. The two paths of the start UGB can be extended to q' by adding the grid edge $q'p'$ to its existing two shortest paths. Thus, there is one more shortest path from $q'q''$, which can be extended to q by adding the grid edge $q''q$. It is true for all border UGBs. Hence, border UGBs contributes to one additional path. □

Lemma 4. *The weight of an UGB, U, is the sum of the weights of all region UGBs lying bottom-right of it inside the corresponding bounding rectangle.*

Proof. Let the bottom-right and top-left corners of U be c' c'' respectively. Let the sum of the weights of all region UGBs bottom-right of U in the corresponding bounding rectangle be w, i.e., total number of shortest paths from q to cc' is w. According to Lemma 2, there are two possible paths in an UGB. Thus, total number of possible paths from q to c'' is $2 \times w$. So, we can say U will contribute w number of additional paths. Hence the weight of the corresponding UGB is w, equal to the sum of the weights of all UGBs bottom-right to it in the corresponding bounding rectangle. □

In order to find the weight of a UGB efficiently, a novel combinatorial technique is used—the concept of row-sum and column-sum. Row-sum (column-sum) of a region UGB is the sum of the weights of all region UGBs in a particular row (column) lying right (below) to the mentioned UGB. The weight of the mentioned UGB is also included in the sum (see Fig. 2(right)).

Theorem 2. *The weight, w, of a region UGB, U, except the start UGB and border UGBs is as follows.*

1. *when there is a region UGB, U_b, below U, then the weight of U is row-sum of U_b;*
2. *when there is a region UGB, U_r, right of U, then the weight of U is column-sum of U_r;*
3. *when there are non-region UGBs right and below U, then the weight of U is the sum of row-sum and column-sum of U_{br}, where U_{br} is the region UGB at immediate bottom–right position (diagonal) of U (but U_{br} is not the start UGB). When U_{br} is the start UGB, then the weight of U is either equal to its row-sum or column-sum.*

If U_b, U_r, or U_{br} is the border UGB, then one has to be deducted from the calculated weight.

Proof. Let there be a region UGB, U_b, below U. The weight of U is the sum of all region UGBs lying bottom-right to it in the corresponding bounding rectangle following Lemma 4. Let us consider the situation depicted in Fig. 3(left).

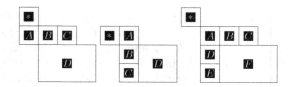

Fig. 3. The calculation of the weight of the UGB marked with '*' (proof of Theorem 2). First three conditions of Theorem 2 are shown in **left**, **middle**, and **right** respectively.

The weight, w, of U is $B + C + D$. According to Lemma 4, the weight of U_b is $A = D$. Thus, $w = B + C + A$, which is equal to the row-sum of U_b. Similarly, if Fig. 3(middle) is considered, then the weight, w, of U is equal to $B+C+D$ (following Lemma 4). We can say that the weight of U_r is $A = D$. Thus, $w = B+C+A$, which is equal to the column-sum of U_r. To prove the third case, let us consider the Fig. 3(right). The weight of U is $w = A + B + C + D + E + F$. The weight of U_{br} is $A = F$. Now, we can write, $w = (A+B+C) + (D+E+A)$, which is equal to the row-sum and column-sum of U_{br}.

If U_{br} is the start UGB, then its row-sum and column-sum both are equal to its weight. If corresponding weight of U is calculated as the sum of row-sum and column-sum, then the addition will include the weights twice and does not follow Lemma 4. Thus, the weight of U will be either equal to the row-sum or column-sum of U_{br}.

Now, consider the UGB D in Fig. 3(right), where U_b is a border UGB. According to condition 1, the weight of UGB D, will be 3—the row-sum of U_b. Following Lemma 4, we can say that the weight of UGB D should be equal to the weight of UGB A. Here, considering the UGB immediate bottom of D (which is border UGB) adds one more to the calculated weight. Thus, one should be deducted when U_b, U_r, or U_{br} is the border UGB.   □

Theorem 3. *The total number of possible shortest isothetic paths in a region is the sum of the weights of all UGBs belonging to the region.*

Proof. According to Lemma 2, the number of possible shortest isothetic paths between two diagonal points in an UGB is '2'. The weight of an UGB computed as the number of additional shortest isothetic paths which involve one of the two shortest paths leading to p. Thus, the sum of all weights of the UGBs in a region gives the total number of possible shortest isothetic paths in that region.   □

4   Algorithm to Determine the Total Number of SIPs

The algorithm ENUM-SIP computes the total number of SIPs, λ, between two given grid points. The given digital object, A, two grid points inside it, i.e., p and q, and the grid size g are taken as input for the algorithm. In Step 1, the set of regions and bridges are obtained in order using the algorithm of family of shortest isothetic paths using the procedure FSIP (detailed in [14]). Steps 3–9 describes the procedure of computing the number of shortest isothetic paths in each region. Here, i denotes the current region for which the number of possible shortest isothetic paths will be calculated. i is initialized to '1' in Step 4. The number of possible shortest isothetic paths in a region is computed using the procedure FIND-SIP-NUMBER and the value is stored k_i (Step 5). The value of k_i is multiplied with λ in Step 6 (see Theorem 1). The corresponding region is deducted from the list of regions in Step 7 of ENUM-SIP.

Algorithm
ENUM-SIP(A, p, q, g)

1. $R, B \leftarrow$ FSIP(A, p, q, g)
2. $\lambda \leftarrow 1$
3. do
4. $i \leftarrow 1$
5. $k_i \leftarrow$ FIND-SIP-NUMBER(R_i, g)
6. $\lambda \leftarrow \lambda \times k_i$
7. $R \leftarrow \{R\} - R_i$
8. $i \leftarrow i + 1$
9. while$(R \neq \emptyset)$
10. return λ

Procedure FIND-SIP-NUMBER(R_i, g)

1. $p', q' \leftarrow$ FIND-RECT(R_i)
2. $r, c \leftarrow$ CALCULATE-DIM(p', q', g)
3. $w \leftarrow$ INITIALIZE(r, c, R_i, g)
4. $k' \leftarrow 0, j \leftarrow r$
5. do
6. $l \leftarrow c$
7.   do
8.   if$(w[j][l] = \infty)$ then
9. $k' \leftarrow k' +$ CALCULATE-WEIGHT(w, r, c, j, l)
10.   if$(w[j][l + 1] \neq 0$ and $w[j][l] = 0)$ then
11.   break
12. $l \leftarrow l - 1$
13.   while$(l \neq 0)$
14. $j \leftarrow j - 1$
15. while$(j \neq 0)$
16. return k'

Procedure CALCULATE-WEIGHT(w, r, c, j, l)

1. if $j = r$ and $l = c$ then
2. $w[j][l] \leftarrow 2, rs_{j,l} \leftarrow 2, cs_{j,l} \leftarrow 2$
3. else if $j = r$ or $l = c$ then
4. $w[j][l] \leftarrow 1$
5.   if $j = r$ then
6. $rs_{j,l} \leftarrow rs_{j,l-1} + w[j][l], cs_{j,l} \leftarrow 1$
7.   else
8. $rs_{j,l} \leftarrow 1, cs_{j,l} \leftarrow cs_{j-1,l} + w[j][l]$
9. else
10.   if $j = r - 1$ and $l = c - 1$ then
11. $w[j][l] \leftarrow rs_{j-1,l-1}$
12.   else if $w[j-1][l] = 0$ and $w[j][l-1] = 0$ then
13. $w[j][l] \leftarrow rs_{j-1,l-1} + cs_{j-1,l-1}$
14.   else if $w[j-1][l] = 0$ or $w[j][l-1] \neq 0$ then
15. $w[j][l] \leftarrow rs_{j-1,l}$
16.   else if $w[j-1][l] \neq 0$ or $w[j][l-1] = 0$ then
17. $w[j][l] \leftarrow cs_{j,l-1}$
18.   if $j + 1 = r$ or $l + 1 = c$ then
19. $w[j][l] \leftarrow w[j][l] - 1$
20. $rs_{j,l} \leftarrow w[j][l] + rs_{j,l-1}, cs_{j,l} \leftarrow w[j][l] + rs_{j-1,l}$
21. return $w[j][l]$

Fig. 4. The algorithm and procedures to determine number of shortest isothetic paths.

The procedure FIND-SIP-NUMBER determines the total number of possible shortest isothetic paths in a region. The bounding rectangle (having p' and q' as two opposite corners) for the region is considered first (Step 1) using the procedure FIND-RECT. Let w be a matrix in which value of each cell corresponds to the weight of an UGB inside the bounding rectangle. The dimension of w is computed in Step 2 using the procedure CALCULATE-DIM. Now, r and c are the corresponding number of rows and columns. w is initialized in Step 3 with ∞ for region UGBs and with '0' for non-region UGBs. k' is initialized to '0' and j is initialized to the total number of rows, i.e., r (Step 4). The weight of each UGB is calculated and the sum of all weights of region UGBs is the total number of possible shortest isothetic paths in a region (computed in Steps 5–15) (follows Theorem 3). The weight of each row starting from bottom is calculated (from right to left). l is initialized to the number of columns for the evaluation of each row (Step 6). The weight of region UGBs are computed by calling the procedure CALCULATE-WEIGHT (Steps 8–9). The calculation stops in a row when there is a non-region UGB (Steps 10–11). There will be no more region UGBs in a given row as the regions are convex. It is to be noted that this calculation can be optimized more if the border of supremum SIP is followed. In that case only region UGBs will be accessed. l and j are decremented by '1' in Step 12 and Step 14 respectively.

The procedure CALCULATE-WEIGHT calculates the weight along with its row-sum and column-sum of each of the UGBs. Steps 1–2 computes the weight, row-sum, and column-sum of the start UGB. The weights, row-sums, and column-sums of the border UGBs are computed in Steps 3–8. For other region UGBs, the weight, row-sum, and column-sum are calculated following the Theorem 2 (Steps 9–20). The weight of the corresponding UGB is returned (Step 21).

4.1   Time Complexity

The time complexity can be explained in terms of the total number of UGBs in the FSIP between p and q. Let there be i regions; the procedure FIND-SIP-NUMBER will then be called i times. The time taken to evaluate the number of SIPs in a region is $O(r \times c)$ (r and c be the number of rows and columns corresponding to the region), i.e., the number of UGBs in the region. In other words, $(r \times c)$ is equal to the number of region UGBs. The time taken by the procedure CALCULATE-WEIGHT is $O(1)$. Hence the total time complexity is $O(i \times r \times c)$, where $(i \times r \times c)$ is the total number of UGBs in FSIP. The time taken to access the grid-blocks of UGBs in FSIP is proportional with $O(S/g^2)$ in the worst case, where S is the total number of object pixels, and g is the grid size. Let S' be the number of pixels contained in FSIP, i,e, in general $S' < S$. Thus, we can say $O(i \times r \times c) \simeq O(S/g^2) \simeq O(S'/g^2)$, where $O(S'/g^2)$ is the UGBs in FSIP (depend on the position of p and q). The time taken to calculate FSIP is $O(n/g \log n/g)$ that includes preprocessing time, where n is the number of pixels on the contour of A, and g is the grid size. Hence, the total time complexity is $(O(n/g \log n/g) + O(S'/g^2))$. Note that there exist an exponential number of paths between two points. Thus, to report all possible SIPs, we need exponential time. In this work, we report the number of such SIPs between two points in $O(S/g^2)$ time (worst case), but do not explicitly list them.

5   Experimental Results

The proposed algorithm is implemented in C in Ubuntu 10.4, Kernel version 2.6.32-21-generic. It is tested on several datasets containing various digital images of different shapes and forms. Results for three digital objects (**Flower**, **Wrench**, and **Star**) are shown in Fig. 5. Control points are placed on the digital objects as per the rules stated in [14]. Number of SIPs are determined from one point (green color) to other points (blue color) in the digital objects (considered in anticlockwise manner). The corresponding data, i.e., area of the object, isothetic distance between two points, $|pq|$, length of SIP, $|\pi_{p,q}|$, area of FSIP, and number of SIPs, δ, are shown in Table 1. It may be noted here that for similar values of $|\pi_{p,q}|$, the number of SIPs, δ, varies significantly depending upon the complexity of the FSIP. The number of SIPs has significant role in shape analysis of digital object, which can be explored further in future.

Fig. 5. Experimental results on digital objects—**Flower**, **Wrench**, and **Star** (Color figure online).

Table 1. Data for the results shown in Fig. 5.

Object, A	Area of A	$|pq|$	$|\pi_{p,q}|$	Area of FSIP	# δ
Flower	154039	250	610	21911	9611031
		680	680	29881	1530444304
		670	670	26571	2490054094
		460	600	17603	2459367
Wrench	65374	200	200	5401	110052
		230	270	4971	2913
		850	850	31831	2401189960
		730	730	21831	3152889414
Star	91758	420	420	6121	32890
		600	600	31501	1869737929
		600	600	31401	3444584329
		420	420	8421	511244

6   Conclusion

A shortest isothetic path (SIP) between two points may not be unique. We have presented an algorithm to enumerate the number of SIPs between two points in a digital object. The enumeration of SIPs is facilitated by an incremental analysis of paths in each unit grid block (UGB) of a region. Once the number of SIPs in each region is obtained, the total number of SIPs can be computed by taking up their product. The time complexity of the algorithm is determined by the number of UGBs within the FSIP. Estimation of SIP's may find important applications in robot navigation and shape analysis of the objects.

References

1. Arkin, E., Mitchell, J., Piatko, C.: Minimum-link watchman tours. Inf. Process. Lett. **86**, 203–207 (2003)
2. de Berg, M.: On rectilinear link distance. Comput. Geom. Theor. Appl. **1**(1), 13–34 (1991)
3. de Berg, M., van Kreveld, M., Nilsson, B.J., Overmars, M.H.: Finding shortest paths in the presence of orthogonal obstacles using a combined L_1 and link metric. In: Gilbert, J.R., Karlsson, R. (eds.) SWAT 1990. LNCS, vol. 447, pp. 213–224. Springer, Heidelberg (1990)
4. Biswas, A., Bhowmick, P., Bhattacharya, B.B.: TIPS: on finding a tight isothetic polygonal shape covering a 2D object. In: Kalviainen, H., Parkkinen, J., Kaarna, A. (eds.) SCIA 2005. LNCS, vol. 3540, pp. 930–939. Springer, Heidelberg (2005)
5. Biswas, A., Bhowmick, P., Bhattacharya, B.B.: Construction of isothetic covers of a digital object: a combinatorial approach. J. Vis. Commun. Image Represent. **21**(4), 295–310 (2010)
6. Chin, W.P., Ntafos, S.: The zookeeper route problem. Inf. Sci. **63**(3), 245–259 (1992)
7. Clarkson, K.L., Kapoor, S., Vaidya, P.: Rectilinear shortest paths through polygonal obstacles in $O(n(\log n)^2)$ time. In: SoCG 1987, pp. 251–257. ACM, NY (1987)
8. Cohen, D.I.A.: Basic Techniques of Combinatorial Theory. Wiley, NY (2007)
9. Das, P.P.: An algorithm for computing the number of the minimal paths in digital images. Pattern Recognit. Lett. **9**(2), 107–116 (1989)
10. Das, P.P.: Counting minimal paths in digital geometry. Pattern Recognit. Lett. **12**(10), 595–603 (1991)
11. Dumitrescu, A., Mitchell, J.S.B.: Approximation algorithms for TSP with neighborhoods in the plane. J. Algorithms **48**(1), 135–159 (2003)
12. Dutt, M., Biswas, A., Bhowmick, P., Bhattacharya, B.B.: On finding shortest isothetic path inside a digital object. In: Barneva, R.P., Brimkov, V.E., Aggarwal, J.K. (eds.) IWCIA 2012. LNCS, vol. 7655, pp. 1–15. Springer, Heidelberg (2012)
13. Dutt, M., Biswas, A., Bhowmick, P., Bhattacharya, B.B.: On finding a shortest isothetic path and its monotonicity inside a digital object. Ann. Math. Artif. Intell. (2014, in press)
14. Dutt, M., Biswas, A., Bhowmick, P., Bhattacharya, B.B.: On the family of shortest isothetic paths in a digital object—an algorithm with applications. Comput. Vis. Image Underst. **129**, 75–88 (2014)
15. Klette, R., Rosenfeld, A.: Digital Geometry: Geometric Methods for Picture Analysis. Morgan Kaufmann, San Francisco (2004)

16. Larson, R.C., Li, V.O.: Finding minimum rectilinear distance paths in the presence of barriers. Networks **11**, 285–304 (1981)
17. Li, F., Klette, R.: Finding the shortest path between two points in a simple polygon by applying a rubberband algorithm. In: Chang, L.-W., Lie, W.-N. (eds.) PSIVT 2006. LNCS, vol. 4319, pp. 280–291. Springer, Heidelberg (2006)
18. Lozano-Perez, T., Wesley, M.A.: An algorithm for planning collision-free paths among polyhedral obstacles. CACM **22**(10), 560–570 (1979)
19. Tan, X., Hirata, T.: Finding shortest safari routes in simple polygons. Inf. Process. Lett. **87**(4), 179–186 (2003)

Modified Exemplar-Based Image Inpainting via Primal-Dual Optimization

Veepin Kumar[✉], Jayanta Mukhopadhyay,
and Shyamal Kumar Das Mandal

Indian Institute of Technology Kharagpur, Kharagpur 721302, West Bengal, India
{veepinkmr,sdasmandal}@cet.iitkgp.ernet.in, jay@cse.iitkgp.ernet.in
http://www.iitkgp.ac.in/

Abstract. In this paper we present a modified exemplar based image inpainting technique to remove objects from digital images. Traditional exemplar based image inpainting techniques do not take into account similarity among patches to be filled with neighbors inside the hole. This gives visually incoherent results. To correct this problem we formulate image inpainting as a global energy optimization problem. We use primal-dual schema of linear programming for optimization. We also modify the criteria for determining priority among candidate patches to be inpainted by introducing one *'edge length'* term which propagates linear structures better than the existing techniques. Results show the effectiveness of our method compared to other recent methods.

Keywords: Image restoration · Inpainting · Exemplar · Linear programming · Metric labeling

1 Introduction

Inpainting is the process of filling damaged portions of an image, or removing any portion of image and filling it such that it looks like an original image. Various applications of image inpaintng include photograph restoration, occlusion removal, image enhancement, etc. Inpainting methods developed so far can be broadly classified into structure based and texture based methods.

Structure based techniques are based on variational methods and solving a set of partial differential equations [1,2]. They are good for inpainting non textured and smaller regions. They interpolate the geometric structure of an image (e.g. level lines, edges, etc.) in the region to be inpainted. They are local in nature as they use information available only at the boundary between a known and an unknown region. However, they introduce some blur in the inpainted region.

Exemplar based methods give relatively good results for large target regions. But, these methods fill the hole by finding most similar patches from the rest of the image iteratively [3,4]. They give good results for texture or repetitive patterns. They are non-local as they search whole image to find the best exemplar. These methods may fail to synthesize a geometry, if there are no examples of it in the image.

© Springer International Publishing Switzerland 2015
M. Kryszkiewicz et al. (Eds.): PReMI 2015, LNCS 9124, pp. 116–125, 2015.
DOI: 10.1007/978-3-319-19941-2_12

Most of the exemplar based methods are greedy in the sense that each patch is filled only once, and after filling it is not checked again for better reconstruction. This may sometimes produce visually inconsistent results as they do not take into account consistency between neighboring patches in the inpainted region. To overcome this difficulty, some inpainting techniques [5–7] formulate the task as solving a discrete global energy optimization problem.

In this paper, we present exemplar based image inpainting as a global energy optimization problem. We use primal-dual optimization schema of the linear programming problem to achieve global optimization. Our cost function consists of one self cost and one neighbor cost term. Self cost term ensures consistency between the boundary pixels, and neighbor cost term ensures visual consistency among neighbors in the inpainting domain. To our knowledge primal-dual optimization schema has not been used previously for image inpainting. We also introduce a new parameter named 'edge length' in determining priority for selecting candidate patches in exemplar based inpaintng [3]. This helps in propagating linear structures more effectively compared to existing techniques. Results demonstrate the effectiveness of our method.

In next section we describe our modification to the priority term of exemplar based inpainting technique [3]. In Sect. 3, we describe image inpaintng by primal-dual optimization. Typical results of the proposed inpainting algorithm are presented and discussed in Sect. 4. Conclusions are drawn in Sect. 5, highlighting future research directions that may come out of this work.

2   Modified Exemplar Based Inpainting

In exemplar based inpainting [3] we determine priority among candidate patches in the target region. The patch having maximum priority is inpainted first. Priority '$P(p)$' for each patch is given by:

$$P(p) = C(p)D(p). \tag{1}$$

where $C(p)$ is called the confidence term, and $D(p)$ is called the data term. They are given by:

$$C(p) = \frac{\sum\limits_{q \in \psi_p \cap (I - \Omega)} C(q)}{\mid \psi_p \mid}, \quad D(p) = \frac{\mid \nabla I_p^{\perp}.n_p \mid}{\alpha}. \tag{2}$$

where $\mid \psi_p \mid$ is the area of the patch to be inpainted denoted by ψ_p as shown in Fig. 1a. α is the normalization factor, n_p is an orthogonal unit vector to the fill front and $\perp$ denotes the orthogonal operator.

We find that Eq. (1) gives equal priority to the two points 'A' and 'B' as shown in Fig. 2. Clearly, point 'B' should be given more priority in order to propagate linear structure inside the hole in a better way. To take care of this situation, we introduce 'edge length' term defined as:

$$E(p) = \frac{\sum\limits_{q \in \psi_p \cap (I - \Omega)} I(q)}{\mid \psi_p \mid}. \tag{3}$$

Fig. 1. (From left to right) (a) Diagram showing an image, with target region ω, its contour $\delta\omega$, source region ϕ, target patch and candidate patch. (b) Diagram showing two candidate patches for a target patch. (c) Diagram showing neighboring patches for a target patch.

where:

$$I(q) = \mid I_x(q) \mid + \mid I_y(q) \mid . \tag{4}$$

In the above equations, I_x and I_y are respectively intensity gradients in x and y directions. This *'edge length'* term gives a measure of number of pixels, which are part of an edge, and belong to known part of the candidate patch to be filled. Thus, it gives more priority to that edge whose length is more. The modified priority term is given by:

$$\tilde{P}(p) = C(p)D(p)E(p). \tag{5}$$

Fig. 2. (From left to right) (a) White line is to be inpainted. Both points A and B are given equal priority according to [3] but point B is given more weightage due to the $E(p)$ term in our technique. (b) Result of inpainting due to [3]. (c) Result of inpainting by our modified technique.

Figures 2a–c demonstrate the effectiveness of $E(p)$ term to generate linear structures.

3   Inpainting by Primal-Dual Optimization

We apply the modified exemplar based inpainting algorithm to the image which is to be inpainted. While filling each patch Ψ_p during modified exemplar based inpainting we find first two exemplars $\Psi_{\hat{p}_1}, \Psi_{\hat{p}_2}$ which are most similar to Ψ_p according to minimum distance criterion, as shown in Fig. 1b.

We pose the inpainting problem as a metric labeling problem [8]. Here patches in target region correspond to objects, and the best two exemplars $\Psi_{\hat{p}_1}$ and $\Psi_{\hat{p}_2}$

correspond to its two candidate labels. We further consider the integer programming formulation of the metric labeling problem introduced in [8]. For our inpainting problem it becomes:

$$min(\sum_{\psi_p \in V, \psi_1 \in L} c_{\psi_p, \psi_1} \mu_{\psi_p, \psi_1} + \sum_{(\psi_p, \psi_q) \in E} \sum_{\psi_1, \psi_2 \in L} d_{\psi_1 \psi_2} \mu_{\psi_p \psi_q, \psi_1 \psi_2}). \tag{6}$$

$$s.t. \sum_{\psi_1} \mu_{\psi_p, \psi_1} = 1 \qquad \forall \psi_p \in V. \tag{7}$$

$$\sum_{\psi_1} \mu_{\psi_p, \psi_1} = 1 \qquad \forall \psi_2 \in L, (\psi_p, \psi_q) \in E. \tag{8}$$

$$\sum_{\psi_2} \mu_{\psi_p \psi_q, \psi_1 \psi_2} = \mu_{\psi_p, \psi_1} \qquad \forall \psi_1 \in L, (\psi_p, \psi_q) \in E. \tag{9}$$

$$\mu_{\psi_p, \psi_1}, \mu_{\psi_p \psi_q, \psi_1 \psi_2} \in \{0, 1\} \quad \forall \psi_p \in V, (\psi_p, \psi_q) \in E, \{\psi_1, \psi_2\} \in L. \tag{10}$$

where, L is a set of labels containing ψ_1 and ψ_2. ψ_1 corresponds to first best matching exemplar patch and ψ_2 corresponds to second best matching exemplar patch. V is a set of vertices, and E is a set of edges of a graph (V, E). The patches to be filled inside the target region (ψ_p, ψ_q, etc.) correspond to the set of vertices or nodes. Set of edges consists of pairs of neighboring vertices. We have considered four connected neighborhood. Distance between the centers of two neighboring patches is w, where $w \times w$ is the patch size as shown in Fig. 1c. Let "$\psi_p \sim \psi_q''$" denotes either $(\psi_p, \psi_q) \in E$ or $(\psi_q, \psi_p) \in E$. μ_{ψ_p, ψ_1} is 1 when vertex ψ_p is labeled ψ_1, otherwise it is set to 0. Similarly, $\mu_{\psi_p \psi_q, \psi_1 \psi_2}$ is 1 when ψ_p is labeled ψ_1 and ψ_q is labeled ψ_2, otherwise it is set to 0. c_{ψ_p, ψ_1} denotes the cost of assigning label ψ_1 to node ψ_p. It is given by the sum of squared differences of the already filled pixels of the two patches ψ_p and ψ_1. Let $d_{\psi_1 \psi_2}$ denotes the neighborhood cost of assigning label ψ_1 to node ψ_p and label ψ_2 to node ψ_q. It is given by the sum of squared differences of the already filled pixels of the two patches ψ_1 and ψ_2.

Constraint expressed in Eq. (10) is relaxed to $\mu_{\psi_p, \psi_1} \geq 0$ and $\mu_{\psi_p \psi_q, \psi_1 \psi_2} \geq 0$, so that the above integer program becomes a linear program. Dual problem [9] of this linear program is given below:

$$max \sum_{\psi_p} \xi_{\psi_p}$$

$$s.t. \quad \xi_{\psi_p} \leq c_{\psi_p, \psi_1} + \sum_{\psi_q : \psi_q \sim \psi_p} \xi_{\psi_p \psi_q, \psi_1} \quad \forall \psi_p \in V, \psi_1 \in L. \tag{11}$$

$$and \quad \xi_{\psi_p \psi_q, \psi_1} + \xi_{\psi_q \psi_p, \psi_2} \leq d_{\psi_1 \psi_2} \quad \forall (\psi_1, \psi_2) \in L, (\psi_p, \psi_q) \in E. \tag{12}$$

Here, ξ_{ψ_p} is the dual variable for each vertex ψ_p. $\xi_{\psi_p \psi_q, \psi_1}$ and $\xi_{\psi_q \psi_p, \psi_1}$ are two dual variables for each pair of neighboring vertices (ψ_p, ψ_q) and any label ψ_1.

We define an auxiliary variable $ht^{\xi}_{\psi_p, \psi_1}$ called "height variable" for any label ψ_1 as:

$$ht^{\xi}_{\psi_p, \psi_1} \equiv C_{\psi_p, \psi_1} + \sum_{\psi_q : \psi_q \sim \psi_p} \xi_{\psi_p \psi_q, \psi_1}. \tag{13}$$

This variable gives a measure of the cost of assigning a label ψ_1 to a node ψ_p.

We use the following primal dual schema [9] to design our algorithm:

Primal-Dual Schema: *Generate a sequence of pairs of integral-primal, dual solutions* $\{\mu^k, \xi^k\}_{k=1}^t$ *until the elements* $\mu = \mu^t$ *and* $\xi = \xi^t$ *of the last pair of the sequence are both feasible, and satisfy the relaxed primal complementary slackness conditions.*

We have used "PD1 algorithm" [10] of the primal-dual schema. The relaxed primal complementary slackness conditions for our pair of primal-dual linear program are given below.

$$\xi_{\psi_p \psi_q, \psi_1} + \xi_{\psi_q \psi_p, \psi_1} = 0. \tag{14}$$

$$\xi_{\psi_p} = min_{\psi_1} ht^{\xi}_{\psi_p, \psi_1}. \tag{15}$$

$$ht^{\xi}_{\psi_p, \mu_{\psi_p}} = min_{\psi_1} ht^{\xi}_{\psi_p, \psi_1}. \tag{16}$$

The feasibility condition for our algorithm, which can be derived from dual constraint given in Eq. (12), is given below.

$$\xi_{\psi_p \psi_q, \psi_1} \leq d_{\psi_p \psi_q, \psi_1}/2 \quad \forall \psi_1 \in L, \psi_p \sim \psi_q. \tag{17}$$

where, $d_{\psi_p \psi_q, \psi_1}$ denotes the neighborhood cost of assigning first best matching exemplar patches to nodes ψ_p and ψ_q. Now, our solution has to satisfy Eqs. (14) – (17).

In the PD1 algorithm, we generate a series of primal-dual pairs of solutions, one primal-dual pair per iteration. At each iteration we make sure that Conditions expressed by Eqs. (14), (15) and (17) are automatically satisfied by the current primal-dual pair. Condition expressed by Eq. (16) requires that the height of label ψ_1 assigned to any node ψ_p must be lower than that of all other labels at that node. Let ψ_p be a node for which this condition fails i.e. suppose height of some label ψ_2 is less than that of the currently applied label ψ_1. To satisfy Eq. (16), we need to raise height of label ψ_2 up to that of label ψ_1 by increasing one of the balance variables $\{\xi_{\psi_p \psi_q, \psi_2}\}_{\psi_q: \psi_q \sim \psi_p}$ according to Eq. (13). But as we increase $\xi_{\psi_p \psi_q, \psi_2}$, neighbor variable $\xi_{\psi_q \psi_p, \psi_2}$ decreases according to Eq. (14). Thus, the height of label ψ_2 at the neighboring vertex ψ_q decreases. This may result in making height of label ψ_2 at ψ_q lower than the height of currently applied label at this node, thus violating Eq. (16). We observe that any update of the balance variables can be simulated by pushing flow through an appropriately constructed capacitated graph. The optimal update can be achieved by pushing maximum flow through that graph. The computational steps are briefly discussed below:

1. We start the iterative algorithm by assigning the first best exemplar patch $\Psi_{\hat{p}_1}$ to each patch Ψ_p in the target region.
2. We compare the height of both the labels (exemplar patches) $\Psi_{\hat{p}_1}$ and label $\Psi_{\hat{p}_2}$ for each patch Ψ_p. If height of label $\Psi_{\hat{p}_2}$ is less than that of label $\Psi_{\hat{p}_1}$ for any patch, then we need to rearrange label heights such that the label we

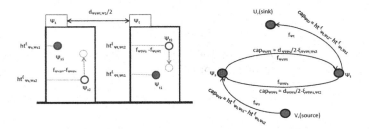

Fig. 3. (From left to right) (a) An arrangement of labels (represented by circles) for a graph G with 2 vertices Ψ_s, Ψ_t and an edge $\Psi_s \Psi_t$. The labels are $\Psi_{\hat{p}1}$ and $\Psi_{\hat{p}2}$. The dashed arrows show how the labels will move after adding flows calculated by max-flow algorithm and dashed circle indicate final position of those labels. (b) Shows the corresponding graph that will be used for the update of the dual variables.

assign to a node should have minimum height at that node. This is to obtain better visual consistency with neighboring patches.

3. We construct a directed graph to update label's heights. The graph G is augmented by two external nodes - the source 'V' and the sink 'U' as shown in Fig. 3. Let this graph be called $G^{\mu,\xi}$. All other nodes of graph $G^{\mu,\xi}$ which are also the nodes of graph G are known as internal nodes.

Interior Edges: Corresponding to each edge $(\psi_p, \psi_q) \in G$, there are two interior edges $\psi_p \psi_q$ and $\psi_q \psi_p$ in graph $G^{\mu,\xi}$. $f_{\psi_p \psi_q}$ is the amount of flow leaving ψ_p through $\psi_p \psi_q$, and it gives increase in balance variable $\xi_{\psi_p \psi_q, \psi_2}$. Also, $f_{\psi_q \psi_p}$ is the amount of flow entering ψ_p through $\psi_q \psi_p$, and it gives the decrease in $\xi_{\psi_p \psi_q, \psi_2}$. Thus, total change in $\xi_{\psi_p \psi_q, \psi_2}$ is given by:

$$\xi'_{\psi_p \psi_q, \psi_2} = \xi_{\psi_p \psi_q, \psi_2} + f_{\psi_p \psi_q} - f_{\psi_q \psi_p}. \tag{18}$$

Capacity $cap_{\psi_p \psi_q}$ of an interior edge $\psi_p \psi_q$ represents the maximum allowed increase of the $\xi_{\psi_p \psi_q, \psi_2}$ variable. These capacities are given by:

$$cap_{\psi_p \psi_q} = cap_{\psi_q \psi_p} = 0 \ \ if \ \Psi_p = \psi_2 \ or \ \Psi_q = \psi_2. \tag{19}$$

otherwise if $\Psi_p \neq \psi_2$ and $\Psi_q \neq \psi_2$,

$$\xi_{\psi_p \psi_q, \psi_2} + cap_{\psi_p \psi_q} = d_{\psi_p \psi_q, \psi_1}/2 = \xi_{\psi_q \psi_p, \psi_2} + cap_{\psi_q \psi_p}. \tag{20}$$

Exterior Edges: Each external node is connected to either the sink (If $ht^{\xi}_{\psi_p, \Psi_{\hat{p}2}} > ht^{\xi}_{\psi_p, \Psi_{\hat{p}1}}$) or source node (If $ht^{\xi}_{\psi_p, \Psi_{\hat{p}2}} \leq ht^{\xi}_{\psi_p, \Psi_{\hat{p}1}}$) through an external edge. Capacities of external edges depend on following three cases.

Case 1: $ht^{\xi}_{\psi_p, \Psi_{\hat{p}2}} < ht^{\xi}_{\psi_p, \Psi_{\hat{p}1}}$: The flow f_{ψ_p} passing through this edge represents the total increase in height of label $\Psi_{\hat{p}2}$:

$$ht^{\xi'}_{\psi_p, \Psi_{\hat{p}2}} = ht^{\xi}_{\psi_p, \Psi_{\hat{p}2}} + f_{\psi_p}. \tag{21}$$

where:

$$f_{\psi_p} = f_{\psi_p \psi_q} - f_{\psi_q \psi_p}. \tag{22}$$

Capacity of $\psi_s \psi_p$ is determined by the fact that we want to increase the height of $\Psi_{\hat{p}_2}$ only upto the height of $\Psi_{\hat{p}_1}$, so:

$$cap_{\psi_s \psi_p} = ht^\xi_{\psi_p, \Psi_{\hat{p}_1}} - ht^\xi_{\psi_p, \Psi_{\hat{p}_2}}. \tag{23}$$

Case 2: $ht^\xi_{\psi_p, \Psi_{\hat{p}_2}} > ht^\xi_{\psi_p, \Psi_{\hat{p}_1}}$. The flow f_{ψ_p} passing through this edge represents the total decrease in height of label $\Psi_{\hat{p}_2}$:

$$ht^{\xi'}_{\psi_p, \Psi_{\hat{p}_2}} = ht^\xi_{\psi_p, \Psi_{\hat{p}_2}} - f_{\psi_p}. \tag{24}$$

Capacity of $\psi_p \psi_t$ is determined by the fact that the maximum decrease in height of $\Psi_{\hat{p}_2}$ can be upto the height of $\Psi_{\hat{p}_1}$, so:

$$cap_{\psi_s \psi_p} = ht^\xi_{\psi_p, \Psi_{\hat{p}_2}} - ht^\xi_{\psi_p, \Psi_{\hat{p}_1}}. \tag{25}$$

Case 3: $ht^\xi_{\psi_p, \Psi_{\hat{p}_2}} = ht^\xi_{\psi_p, \Psi_{\hat{p}_1}}$. Here, we want to keep the height of $\Psi_{\hat{p}_2}$ fixed. So, $f_{\psi_p} = 0$. By convention we set capacity of the edge $\psi_s \psi_p$ $cap_{\psi_s \psi_p}$ equal to 1.

After construction of the graph, a maximum flow algorithm [11] is applied to it to get flows through the edges. These flows are used to update the height of label $\Psi_{\hat{p}_2}$ as:

$$ht^{\xi'}_{\psi_p, \Psi_{\hat{p}_2}} = ht^\xi_{\psi_p, \Psi_{\hat{p}_2}} + f_{\psi_p \psi_q} - f_{\psi_q \psi_p}. \tag{26}$$

4. Based on the resulting heights we update the primal variables by assigning new labels to the vertices of G. If the new height of label $\Psi_{\hat{p}_2}$ is greater than that of label $\Psi_{\hat{p}_1}$, we assign label $\Psi_{\hat{p}_2}$ as the new label of node ψ_p (because the active label at ψ_p should be the lowest at ψ_p, (refer to Eq. (16)). This means that we have filled the current patch of the target region with the second best exemplar patch. In order to maintain visual consistency among filled patches, patches with filling priority (refer to Eq. (5)) greater than the patch ψ_p are assigned label ψ_2 i.e. are filled with second best exemplar patch.
5. Now the priorities of patches corresponding to boundary pixels of the target region which includes the patch ψ_p changes as it has been assigned label $\Psi_{\hat{p}_2}$. So, we again run the base inpainting algorithm to the remaining target region and calculate 'first two best matching exemplars for each patch to be filled'.
6. We repeat steps 1 to 5 for the remaining target region.

We keep on repeating steps 1 to 6 till the algorithm converges.

4   Results and Discussion

We applied our algorithm to remove objects from images. Number of iterations required depends upon image size, size of object to be removed, and the patch size. We have experimented with several images. A few typical results of our experimentation are presented here. We also compare the results with those obtained by the techniques reported in [3,7]. The technique reported in [3] was implemented by us and the code for technique reported in [7] was provided by the author[1]. We took patches of size 9×9. While presenting these results we have shown outputs of modified inpainting algorithm using edge length as a factor in the computation of priority. Subsequently, the final result through primal-dual optimization algorithm is shown. It is observed that there is an improvement in the reconstruction quality in successive stages of above processing.

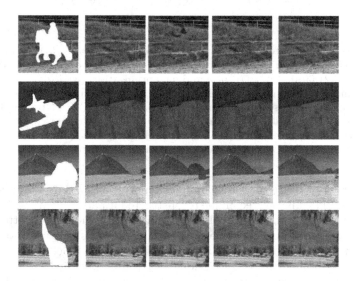

Fig. 4. (From left to right) (a) Original image with mask. (b) Result of [3]. (c) Result of [7]. (d) Result of our modified exemplar based inpainting using edge-length measure in the priority computation. (e) Result of PD1 primal dual optimization.

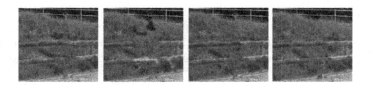

Fig. 5. (From left to right) Zoomed versions of inpainted regions of (b), (c), (d), (e) of first row for clarity.

---

[1] We are thankful to Mr. Yunqiang Liu for providing code for his paper [7].

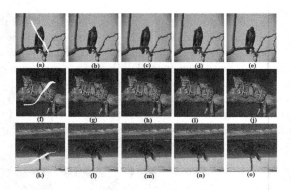

Fig. 6. Results for scratch inpainting. (a) Image with mask. (b) Criminisi's [3] (41.53 dB, 19.06 s). (c) Liu's [7] (43.25 dB, 2.09 s). (d) Modified exemplar based inpainting (41.66 dB, 21.09 s). (e) PD1 primal-dual optimization (41.75 dB, 156.87 s). (f) Image with mask. (g) Criminisi's [3] (35.43 dB, 35.43 s). (h) Liu's [7] (37.89 dB, 3.46 s). (i) Modified exemplar based inpainting (36.16 dB, 36.16 s). (j) PD1 primal-dual optimization (36.68 dB, 145.47 s). (k) Image with mask. (l) Criminisi's [3] (40.42 dB, 18.09 s). (m) Liu's [7] (43.89 dB, 2.03 s). (n) Modified exemplar based inpainting (40.42 dB, 18.61 s). (o) PD1 primal-dual optimization (41.70 dB, 74.21 s).

In the first row of Fig. 4, the techniques reported in [3, 7] do not reconstruct the wooden stick, while our technique reconstructed it partially. In Fig. 5, zoomed versions of inpainted regions of this set of images are shown for better visualization. In the second row of images of Fig. 4, a spike of shadow is produced by the technique reported in [3], and a rectangular shadow block is also observed in the output obtained by the technique in [7]. Our modified exemplar based inpainting using edge-length in the priority computation reduces this effect marginally. But, the overall optimization process reduces it considerably. In the third row, there is a faulty reconstruction of a portion of river and mountain by the techniques reported in [3, 7], while our technique produces a better quality of reconstruction. In the fourth row, there appears an abrupt change in boundaries between tree bushes and dirt terrain in the inpainted region by the techniques in [3, 7], while the proposed technique provides a smoother transition. We applied our technique to remove scratches from images as shown in Fig. 6. Here, scratch (white region in Fig. 6) becomes the target region and rest of image becomes the source region. The same algorithm is applied as explained in Sects. 2 and 3. We computed peak signal to noise ratio (PSNR) of the reconstructed images as:

$$PSNR = 10 \times log_{10}(\frac{255 \times 255}{MSE}). \tag{27}$$

where, MSE is mean squared error. For reference image (original image I) and reconstructed image (I_{re}) of size $m \times n$ it is given by:

$$MSE = \frac{1}{3 \times m \times n} \sum_{i=0}^{2} \sum_{j=0}^{m-1} \sum_{k=0}^{n-1} [I(i,j,k) - I_{re}(i,j,k)]^2. \tag{28}$$

Our technique gave better results than technique in [3], while results of technique in [7] are better than our technique. But, our technique outperforms the technique in [7] for object removal as discussed in the beginning of this section.

5   Conclusion and Future Work

In this paper we first proposed a modification of the exemplar based image inpainting technique [3]. We have introduced a new "edge length" term in the priority equation of [3]. This propagates the linear structures in a better way. Next, we have applied the "PD1 primal dual linear programming approximation" with two labels to the image inpainting problem. In future, we would like to extend our technique for handling more labels.

References

1. Masnou, S.: Disocclusion: a variational approach using level lines. IEEE Trans. Image Process. **11**, 68–76 (1981)
2. Bertalmio, M., Sapiro, G., Caselles, V., Ballester, C.: Image inpainting. In: Proceedings of the 27th Annual Conference on Computer Graphics and Interactive Techniques, pp. 417–424. ACM Press/Addison-Wesley Publishing Co. (2000)
3. Criminisi, A., Pérez, P., Toyama, K.: Region filling and object removal by exemplar-based image inpainting. IEEE Trans. Image Process. **13**, 1200–1212 (2004)
4. Efros, A.A., Leung, T.K.: Texture synthesis by non-parametric sampling. In: The Proceedings of the Seventh IEEE International Conference on Computer Vision, 1999, pp. 1033–1038. IEEE (1999)
5. Komodakis, N., Tziritas, G.: Image completion using efficient belief propagation via priority scheduling and dynamic pruning. IEEE Trans. Image Process. **16**, 2649–2661 (2007)
6. Wexler, Y., Shechtman, E., Irani, M.: Space-time completion of video. IEEE Trans. Pattern Anal. Mach. Intell. **29**, 463–476 (2007)
7. Liu, Y., Caselles, V.: Exemplar-based image inpainting using multiscale graph cuts. IEEE Trans. Image Process. **22**, 1699–1711 (2013). IEEE
8. Chekuri, C., Khanna, S., Naor, J.S., Zosin, L.: Approximation algorithms for the metric labeling problem via a new linear programming formulation. In: Proceedings of the Twelfth Annual ACM-SIAM Symposium on Discrete Algorithms, pp. 109–118. Society for Industrial and Applied Mathematics (2001)
9. Chandra, S., Jayadeva, Mehra, A.: Numerical Optimization with Applications. Alpha Science International, United Kingdom (2009)
10. Komodakis, N., Tziritas, G.: A new framework for approximate labeling via graph cuts. In: IEEE International Conference on Computer Vision, vol. 2, pp. 109–118. IEEE Computer Society (2005)
11. Gibbons, A.: Algorithmic Graph Theory. Cambridge University Press, New York (1985)

A Novel Approach for Image Super Resolution Using Kernel Methods

Adhish Prasoon[1]([✉]), Himanshu Chaubey[1], Abhinav Gupta[1],
Rohit Garg[1], and Santanu Chaudhury[2]

[1] Samsung Research and Development Institute Delhi, Noida, India
adhishprasoon@gmail.com
[2] Department of Electrical Engineering, Indian Institute of Technology Delhi,
Delhi, India

Abstract. We present a learning based method for image super resolution problem. Our approach uses kernel methods to build an efficient
representation and also to learn the regression model. For constructing an
efficient set of features, we apply Kernel Principal Component Analysis
(Kernel-PCA) with a Gaussian kernel on a patch based data-base constructed from 69 training images up-scaled using bi-cubic interpolation.
These features were given as input to a non-linear Support Vector Regression (SVR) model, with Gaussian kernel, to predict the pixels of the high
resolution image. The model selection for SVR was performed using grid
search. We tested our algorithm on an unseen data-set of 13 images.
Our method out-performed a state-of the-art method and achieved an
average of 0.92 dB higher Peak signal-to-noise ratio (PSNR). The average improvement in PSNR over bi-cubic interpolation was found to be
3.38 dB.

1 Introduction

The aim of the super resolution methods is to increase the resolution of low
resolution (LR) images. The motivations behind increasing interest in super
resolution methods are their potential to improve the resolution of the images
taken by the low cost imaging devices and also to exploit the full capability of
the high resolution (HR) displays. Recently machine learning techniques have
been increasingly used to solve the problem of image super-resolution. Kernel
methods [1] have been successfully employed for various categorization tasks in
image processing domain. In this paper we present a novel approach to solve
the image super resolution problem using two such methods; kernel principal
component analysis (kernel-PCA) and support vector regression (SVR). Below,
we discuss some other important contributions which use machine learning techniques for solving the image super resolution problem. Freeman et al. [2] use
Markov Random Fields and Belief Propagation to predict the high resolution
image corresponding to a low resolution image. Chang et al. [3] propose a method
inspired by manifold learning using Locally Linear Embedding (LLE). Their
method assumes that the similar manifolds are formed in the low resolution and

© Springer International Publishing Switzerland 2015
M. Kryszkiewicz et al. (Eds.): PReMI 2015, LNCS 9124, pp. 126–135, 2015.
DOI: 10.1007/978-3-319-19941-2_13

the high-resolution patch space. Yang et al. [4] proposes a sparse representation based method. Their method is inspired by the fact that a sparse linear combination of elements from over-complete dictionary can represent image patches. They find a sparse representation for the patches of the low resolution image and use this representation to obtain corresponding high resolution image. Due to the excellent results and a robust technique, Yang et al.'s method [4] is certainly one of the state-of-the-art methods. We compare our results to their method. The reason we chose their method to be the benchmark for our study is that apart from being one of the state-of-the-art methods, they have provided their code [5], the training data and their pre-learned dictionaries which can be directly used to have fair comparison between the two methods. Apart from the above methods, SVR has also been used for the purpose of super-resolution. Ni et al. [6] apply SVR in the DCT domain. An et al. [7] have also used SVR for image super-resolution. However, our method is different from their method [7] as our feature set is entirely different. We have applied kernel-PCA for feature extraction while [7] utilizes simply the raw pixels along with center pixel's gradient with weights as their features. Ni et al. [6] also apply SVR for image super resolution but the framework and motivation of their usage of SVR is entirely different from our method. Yuan et al. [8] use sparse solution of Kernel-PCA applied on HR image patches to determine the coefficients which can represent higher-order statistics of HR image structures. Further Yuan et al. [8] constructs dictionaries and maps an LR image patch to the coefficients of the respective HR image patch. Kernel-PCA is also a part of our method. However, we have applied it on LR image patches in order to extract a powerful set of features which are eventually fed to yet another kernel based method (Support Vector Regression).

The next section explains our method in detail. We start the next section with a brief introduction to kernel-PCA and support vector regression followed by the detailed explanation of our super resolution method. In Sect. 3, we discuss our experiments and present the results. We conclude the paper in Sect. 4.

2   Method

Kernel trick [1] is one of the most powerful tools in machine learning and has been the backbone of many algorithms. In this work we combine two such algorithms, Kernel Principal Component Analysis [9] (kernel PCA) and Support Vector Regression (SVR) to solve the problem of single-image super-resolution. In the next two subsections we give a brief introduction to kernel PCA and Support Vector Regression (SVR) [10,11] respectively. Subsequently, we will explain our method which uses kernel-PCA and SVR for single image super-resolution.

2.1   Kernel PCA

Before moving on to kernel PCA let us have a brief overview of PCA. PCA is a linear method which is often used for feature extraction. The idea behind it is to transform a dataset having correlated features in such a way that the new features become linearly uncorrelated. Let P be an $d \times n$ sized matrix, where each column represents an example with d features and n be the number of training examples. Let $p^{(i)}$ be the ith example. Here we assume that data-set has zero mean across all the examples, for each feature. Let Q be the covariance matrix given as

$$Q = \frac{1}{n} \sum_{i=1}^{n} p^{(i)} (p^{(i)})^{'} \tag{1}$$

where $(p^{(i)})^{'}$ is the transpose of vector $p^{(i)}$. Let E be the $d \times d$ sized matrix, where jth column represents jth eigen-vector $e^{(j)}$ of Q. The new transformed data matrix with uncorrelated features is given as

$$P_N = E^{'} * P \tag{2}$$

Although Principal Component Analysis (PCA) is an appropriate method for dimensionality reduction and feature extraction if the data is linearly-separable, in many cases, where the data in not linearly separable, we need a non-linear feature extraction method. So, if we apply a non-linear mapping to the input features, the data have higher chance to be linearly separable in the new feature space. Let $\mathcal{F}$ be the non-linear mapping, which transforms the input $p^{(i)}$ to a very high-dimensional space. The new covariance matrix can be given as

$$Q_N = \frac{1}{n} \sum_{i=1}^{n} \mathcal{F}(p^{(i)})(\mathcal{F}(p^{(i)}))^{'} \tag{3}$$

For transforming the non-linearly mapped data the same way as PCA, we need to calculate the eigen-vectors of the covariance matrix Q_N. Unfortunately, calculating the eigen-vectors and eigen-values of such a huge matrix is computationally not feasible. However, it has been shown [9] that the projections of the data points onto the eigen-vectors in the new high dimensional space can be obtained without even calculating the high dimensional features $\mathcal{F}(p^{(i)})$; using the *kernel-trick* [1]. Let $e_N^{(j)}$ be the jth eigen-vector of the covariance matrix Q_N. Let $p^{(t)}$ be a new data-point and $\mathcal{F}(p^{(t)})$ be the corresponding data-point in the high-dimensional space. As proved in [9], projection of $\mathcal{F}(p^{(t)})$ onto $e_N^{(j)}$, can be written as

$$\langle e_N^{(j)}, \mathcal{F}(p^{(t)}) \rangle = \sum_{i=1}^{n} v(i)^{(j)} \langle \mathcal{F}(p^{(t)}), \mathcal{F}(p^{(i)}) \rangle \tag{4}$$

where $v(i)^{(j)}$ is the ith element of the column vector $v^{(j)}$, and $v^{(1)}, .., v^{(j)}, ..v^{(n)}$ are the solutions of the equation

$$\beta v = K_\kappa v \tag{5}$$

K_κ being an $n \times n$ sized matrix whose (i,j)th element is given by $\langle \mathcal{F}(\boldsymbol{p}^{(i)})\mathcal{F}(\boldsymbol{p}^{(j)})\rangle$. The kernel trick says [1] that the dot product in the high dimensional space (obtained using a non-linear mapping) can be calculated using a kernel function. A kernel function is a dual input function which follows the Mercer's condition. Further, the methods which use kernel function and kernel-trick are classified as kernel methods in machine learning. Let k_κ be the kernel function, given as $k_\kappa(x,y) = \langle \mathcal{F}(\boldsymbol{x})\mathcal{F}(\boldsymbol{y})\rangle$. It can be seen that the two Eqs. (4) and (5) are written in terms of the dot products in the non-linear feature space, which eventually are nothing but the kernel function evaluations for the input values. Replacing, the dot product with the kernel function in Eq. (4), we have

$$\langle e_N^{(j)}, \mathcal{F}(\boldsymbol{p}^{(t)})\rangle = \sum_{i=1}^{n} v(i)^{(j)} k_\kappa(\mathcal{F}(\boldsymbol{p}^{(t)}), \mathcal{F}(\boldsymbol{p}^{(i)})) \tag{6}$$

Further, the matrix K_κ has be to modified in such a way that the data in the high dimensional space becomes approximately centered and the eigen-vectors of the covariance matrix Q_N are normalized, [9,12]. In the next section we give a brief overview of another popular kernel based method, i.e. support vector regression.

2.2   Support Vector Regression

Support Vector Regression [10,13] was developed as an extension of Support Vector Machines [11], which are one of the most powerful algorithms in machine learning and have excellent generalization ability.

Let $\{(\boldsymbol{x}_1, y_1), (\boldsymbol{x}_2, y_2), ...(\boldsymbol{x}_n, y_n)\}$ be the training dataset of n examples with $\boldsymbol{x}_1, \boldsymbol{x}_2, ...\boldsymbol{x}_n$ being n input variables and $y_1, y_2, ...y_n$ being corresponding target variables. The dual optimization function for linear SVR training can be written as

$$\max_{\boldsymbol{\alpha}, \boldsymbol{\alpha}^*} \left[\sum_{i=1}^{n} (\alpha_i - \alpha_i^*)(y_i) - \frac{1}{2} \sum_{i,j=1}^{n} (\alpha_i - \alpha_i^*)(\alpha_j - \alpha_j^*)\langle \boldsymbol{x}_i, \boldsymbol{x}_j\rangle - \sum_{i=1}^{n} (\alpha_i + \alpha_i^*)(\epsilon_i) \right]$$

$$\text{subject to } \sum_{i=1}^{n} (\alpha_i - \alpha_i^*) = 0 \ , \ C \geq \alpha_i, \alpha_i^* \geq 0 \ , \ i = 1, \ldots, n \tag{7}$$

where $\boldsymbol{\alpha}, \boldsymbol{\alpha}^*$ are the dual variables whose ith elements are given as α_i, α_i^*; C is the regularization parameter and ϵ represents the maximum permissible training error. Now to convert linear SVR to a non-linear SVR, let ϕ be the nonlinear mapping applied to the inputs. The new optimization problem can be written as

$$\max_{\boldsymbol{\alpha},\boldsymbol{\alpha}^*} \left[\sum_{i=1}^{n}(\alpha_i - \alpha_i{}^*)(y_i) - \frac{1}{2}\sum_{i,j=1}^{n}(\alpha_i - \alpha_i{}^*)(\alpha_j - \alpha_j{}^*)\langle\phi(\boldsymbol{x}_i),\phi(\boldsymbol{x}_j)\rangle - \right.$$
$$\left. \sum_{i=1}^{n}(\alpha_i + \alpha_i{}^*)(\epsilon_i) \right] \quad (8)$$
$$\text{subject to } \sum_{i=1}^{n}(\alpha_i - \alpha_i{}^*) = 0 \ , \ C \ge \alpha_i, \alpha_i{}^* \ge 0 \ , \ i = 1,\dots,n$$

Now, as the above equation is written in terms of dot products in the high dimensional feature space, the kernel trick is exploited to solve the dual optimization problem without even calculating the new features in the high dimensional space. Let k_s be the kernel function defined as $k_s(\boldsymbol{x},\boldsymbol{y}) = \langle\phi(\boldsymbol{x}),\phi(\boldsymbol{y})\rangle$. Thus finally inserting the kernel function in place of $\langle\phi(\boldsymbol{x}),\phi(\boldsymbol{y})\rangle$, the above equation can be written as

$$\max_{\boldsymbol{\alpha},\boldsymbol{\alpha}^*} \left[\sum_{i=1}^{n}(\alpha_i - \alpha_i{}^*)(y_i) - \frac{1}{2}\sum_{i,j=1}^{n}(\alpha_i - \alpha_i{}^*)(\alpha_j - \alpha_j{}^*)k_s(\phi(\boldsymbol{x}_i),\phi(\boldsymbol{x}_j)) - \right.$$
$$\left. \sum_{i=1}^{n}(\alpha_i + \alpha_i{}^*)(\epsilon_i) \right] \quad (9)$$
$$\text{subject to } \sum_{i=1}^{n}(\alpha_i - \alpha_i{}^*) = 0 \ , \ C \ge \alpha_i, \alpha_i{}^* \ge 0 \ , \ i = 1,\dots,n$$

The solution of the dual optimization function is finally used for predicting the output value of a new test pattern. In the following, we discuss our approach which applies the two kernel methods discussed above (kernel PCA and SVR), for image super-resolution.

2.3 Application of Kernel PCA and SVR to Single Image Super-Resolution

For any regression task to perform well we need features with high discrimination ability and a strong regression method. In our method for image super resolution we use kernel methods, i.e. kernel PCA for extracting a strong feature set and SVR for predicting the intensity values of the high resolution (HR) image. Figure 1 depicts our method in detail.

Feature Extraction. Due to high sensitivity of human vision system to luminance changes in comparison to the color changes, we converted the images from RGB color space to $YCbCr$ color space. We applied our method on the Y component only, while Cb and Cr components are directly taken from up-scaled images using bi-cubic interpolation. For extracting an efficient set of features, we determine a mapping obtained by applying kernel PCA on a patch based

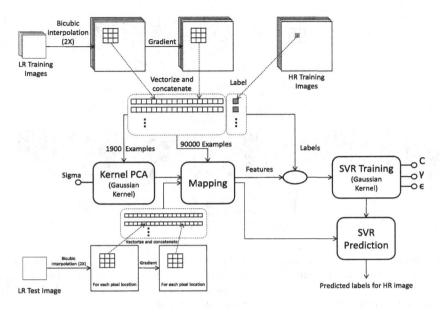

Fig. 1. Block diagram for the proposed method

training data-set constructed using 69 low resolution images. Each of the low resolution images was first up-scaled by a scaling factor of 2 using bi-cubic interpolation. We extracted patches of size 3×3, centered around the locations randomly selected from the up-scaled image. The Y values at the same locations in the high resolution images were also saved to be used as the target values for training our regression model. Moreover, patches were also extracted from the same locations in gradient image of up-scaled image (using Sobel's operator). Thus for each randomly selected location, we extracted a patch pair. Both the patches of each patch-pair were vectorized and concatenated to get our 18 dimensional input examples. Overall, from 69 images, we extracted 90000 examples which were used to learn our regression models. Earlier, we have seen that kernel PCA needs to perform the eigen-value analysis on the gram matrix $\boldsymbol{K}_{\kappa}$, which is of size $n \times n$, n being the total number of examples. Thus, in order to reduce the memory and computational complexity, we performed kernel PCA on a much smaller data-set of 1900 examples to obtain the solutions of the Eq. (5). Finally, these solutions were used to extract features from 90000 examples (as in Eq. (6)). The kernel we used for feature extraction was Gaussian kernel, $k_{\kappa}(\boldsymbol{x}, \boldsymbol{y}) = \exp(-\frac{\|\boldsymbol{x}-\boldsymbol{y}\|^{2}}{2\sigma^{2}})$, $\boldsymbol{x}, \boldsymbol{y} \in \mathbb{R}^{18}$. The standard deviation σ was chosen to be the mean of the pair-wise distance of all the possible example pairs. Total 18 features were extracted to construct our training data-set, which was eventually given as input to learn the SVR model.

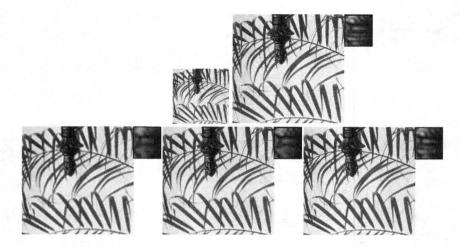

Fig. 2. Comparison of results obtained for the "Leaves" image. Left image in the upper row shows the LR and the right image shows the ground truth HR image. The left most image in the lower row shows the result obtained using bicubic interpolation. The middle image in the lower row shows results obtained using sparse representation [4]. The right most image in the lower row shows results obtained by our method. The black rectangular box indicates one of the regions of interest for the viewers to focus. It has been zoomed-in and showed adjacent to the top of the right edge of the images.

SVR Model Selection. We use non-linear SVR to learn our regression model. We used LIBSVM for non linear SVR training [14]. The kernel we used was again a Gaussian kernel $k_s(\boldsymbol{x}, \boldsymbol{y}) = \exp(-\gamma \|\boldsymbol{x} - \boldsymbol{y}\|^2)$, $\boldsymbol{x}, \boldsymbol{y} \in \mathbb{R}^{18}$. For choosing the optimum set of regularization parameter C, kernel width parameter γ and ϵ parameter for the ϵ-SVR, we performed grid search using 3 validation images. C was varied as $\{2^0, 2^2, 2^4....2^{16}\}$ while γ and ϵ parameter were varied as $\{2^{-6}, 2^{-5}, 2^{-4}, ...2^3\}$ and $\{0.1, 0.2, 0.4, 0.8\}$.

Prediction. When a new LR image is presented for prediction, we up-scale (scaling factor 2) it using bi-cubic interpolation and also take the gradient of the up-scaled image. For each location of the image, we extract 3×3 patch around the location, from the up-scaled image and from its gradient image too. These two patches are vectorized and concatenated to obtain an 18 dimensional vector. Kernel-PCA mapping obtained using 1900 examples (please refer Sect. 2.3) is then used to calculate the new 18 dimensional feature vector which was eventually given as input to the SVR model (learnt using 90000 transformed training examples, refer Sect. 2.3) for prediction (please see Fig. 1).

3   Experiments and Results

We tested our method on 13 unseen test images. These images were neither part of the training images nor part of the validation images. We compared our

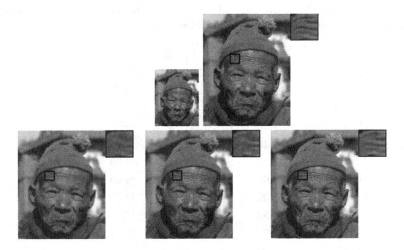

Fig. 3. Comparison of results obtained for the "Old Man" image. Left image in the upper row shows the LR and the right image shows the ground truth HR image. The left most image in the lower row shows the result obtained using bicubic interpolation. The middle image in the lower row shows results obtained using sparse representation [4]. The right most image in the lower row shows results obtained by our method. The black rectangular box indicates one of the regions of interest for the viewers to focus. It has been zoomed-in and showed adjacent to the top of the right edge of the images.

Fig. 4. Comparison of results obtained for the "Bike" image. Left image in the upper row shows the LR and the right image shows the ground truth HR image. The left most image in the lower row shows the result obtained using bicubic interpolation. The middle image in the lower row shows results obtained using sparse representation [4]. The right most image in the lower row shows results obtained by our method. The black rectangular box indicates one of the regions of interest for the viewers to focus. It has been zoomed-in and showed adjacent to the top of the right edge of the images.

Table 1. Comparison of results achieved by bi-cubic interpolation, sparse representation method [4] and our proposed method. Results are in dB.

Sr. No.	Image name	PSNR- bicubic	PSNR-[4]	PSNR- Proposed method
1	Hat	31.73	34.02	34.86
2	Parthenon	28.11	29.35	30.33
3	Parrot	31.38	33.92	34.92
4	Butterfly	27.46	31.23	32.60
5	Flower	30.45	32.83	33.76
6	Leaves	27.44	31.38	32.70
7	Bike	25.65	28.18	29.43
8	Baby	30.37	32.45	33.40
9	Chip	32.82	36.96	37.36
10	Pimpled Girl	32.96	35.26	35.66
11	Lena	32.79	35.04	35.86
12	Girl	34.74	35.58	36.18
13	Old Man	32.30	33.94	35.11
	Average	30.63	33.09	34.01

method with one of the state-of-the-art methods [4] (sparse representation based method). For fair comparison we used the same 69 training images as used by [4]. The authors have provided their code [5], as well as their pre-trained dictionary (please refer [4,5] for details) for the scaling factor of 2. The evaluation was based on Peak Signal to Noise Ratio (PSNR) for the Y channel. The PSNR is calculated as follows:

$$PSNR = 10 \log_{10}(\frac{255^2}{E_{ms}})$$
(10)

where E_{ms} is the mean squared error. On all the 13 images, our method outperformed the sparse representation based methods [4] as well as the bi-cubic interpolation. The average improvement over sparse representation method [4] was found to be 0.92 dB. Moreover the average improvement over bi-cubic interpolation was found to be 3.38 dB. However, when compared to the bi-cubic interpolation the better PSNR was achieved at the cost of higher execution time. The execution time for our method, for a 128×128 LR image, on a system with 3.40 GHz processor and 8 GB RAM was 225 s. Table 1 shows the results achieved by bi-cubic interpolation, sparse representation method [4] and our proposed method. For visualization purpose we present some of our test LR images with their ground truth HR images, results obtained by bicubic interpolation, sparse representation method [4] and our proposed method in Figs. 2, 3 and 4.

4   Conclusion

We presented a novel method for image super resolution problem. For any categorization task to be successful, we need features having high discrimination power and also a strong regression/classification method having excellent generalization ability. We deployed two machine learning techniques which exploit the power of kernel-trick. We tested our algorithm on a hold-out data-set of 13 images. Our method outperformed a state-of-the-art method [4]. That we achieved higher PSNR than a state-of-the-art method on each of the 13 test images, shows the generalization power of our method.

References

1. Schölkopf, B., Smola, A.J.: Learning with Kernels: SupportVector Machines, Regularization, Optimization, and Beyond. MIT press, Cambridge (2002)
2. Freeman, W.T., Pasztor, E.C., Carmichael, O.T.: Learning low-level vision. Int. J. Comput. Vision **40**, 25–47 (2000)
3. Chang, H., Yeung, D.Y., Xiong, Y.: Super-resolution through neighbor embedding. In: Proceedings of the 2004 IEEE Computer Society Conference on Computer Vision and Pattern Recognition, 2004, vol. 1, p. 1. IEEE (2004)
4. Yang, J., Wright, J., Huang, T.S., Ma, Y.: Image super-resolution via sparse representation. IEEE Trans. Image Process. **19**, 2861–2873 (2010)
5. Yang, J., Wright, J., Huang, T.S., Ma, Y.: (Image Super-resolution via Patch-wise Sparse Recovery http://www.ifp.illinois.edu/jyang29/ScSR.htm)
6. Ni, K.S., Nguyen, T.Q.: Image superresolution using support vector regression. IEEE Trans. Image Process. **16**, 1596–1610 (2007)
7. An, L., Bhanu, B.: Improved image super-resolution by support vector regression. In: The 2011 International Joint Conference on Neural Networks (IJCNN), pp. 696–700. IEEE (2011)
8. Yuan, T., Yang, W., Zhou, F., Liao, Q.: Single image super-resolution via sparse kpca and regression. In: 2014 IEEE International Conference on Image Processing (ICIP), pp. 2130–2134. IEEE (2014)
9. Schölkopf, B., Smola, A., Müller, K.R.: Kernel principal component analysis. In: Artificial Neural Networks ICANN 1997, pp. 583–588. Springer (1997)
10. Vapnik, V.: The Nature of Statistical Learning Theory. Springer Science and Business Media, New York (2000)
11. Cortes, C., Vapnik, V.: Support-Vector Networks. Mach. Learn. **20**, 273–297 (1995)
12. Weinberger, K.Q., Sha, F., Saul, L.K.: Learning a kernel matrix for nonlinear dimensionality reduction. In: Proceedings of the Twenty-first International Conference on Machine Learning, p. 106. ACM (2004)
13. Smola, A.J., Schölkopf, B.: A tutorial on support vector regression. Stat. Comput. **14**, 199–222 (2004)
14. Chang, C.C., Lin, C.J.: Libsvm: a library for support vector machines. ACM Trans. Intell. Syst. Technol. (TIST) **2**, 27 (2011)

Generation of Random Triangular Digital Curves Using Combinatorial Techniques

Apurba Sarkar[1]([✉]), Arindam Biswas[2], Mousumi Dutt[3],
and Arnab Bhattacharya[4]

[1] Department of Computer Science and Technology,
Indian Institute of Engineering Science and Technology, Howrah, India
as.besu@gmail.com
[2] Department of Information Technology,
Indian Institute of Engineering Science and Technology, Howrah, India
barindam@gmail.com
[3] Department of Information Technology,
Indian Institute of Information Technology, Kalyani, India
duttmousumi@gmail.com
[4] Department of Computer Science and Engineering,
Indian Institute of Technology, Kanpur, India
arnabb@iitk.ac.in

Abstract. This work presents an algorithm to generate simple closed random triangular digital curves of finite length imposed on a background triangular grid. A novel timestamp-based combinatorial technique is incorporated to allow the curve to grow freely without intersecting itself. The algorithm runs in linear time as a fixed set of vertices are consulted to find the next direction and since it does not require backtracking. The proposed algorithm is implemented and tested exhaustively.

Keywords: Random digital curves · Testing of algorithms · Combinatorial technique · Triangular curve

1 Introduction

The problem of generating simple random polygon has attracted lot of researchers not only because of its theoretical interest but also for its application in many areas of computer science. Random polygon has two main areas of application, firstly in testing correctness of geometric and graph algorithms [2,6,7] and secondly evaluating efficiency, i.e., estimating consumption of CPU-Time of algorithms that operate on polygons. Sometimes, it is difficult to obtain a large set of practically relevant input data for an algorithm to run on, hence next best choice would be to run the algorithm on random input. To that end an algorithm that generates random polygon can supply random polygon as input to algorithms that expect polygons as input. Random curves may also be useful in graphical applications [4] which intends to generate textures of the nature: like clouds and landforms. In [1,4,5,8,10,11] several works on generating random

© Springer International Publishing Switzerland 2015
M. Kryszkiewicz et al. (Eds.): PReMI 2015, LNCS 9124, pp. 136–145, 2015.
DOI: 10.1007/978-3-319-19941-2_14

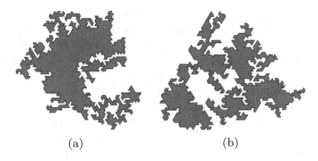

(a)   (b)

Fig. 1. Two instances of triangular digital random curves with grid length 4

polygon and random curves have been reported. A heuristic approach for generation of simple polygons is presented in [8], which also presents several tests to verify the uniformity of random simple polygon generator. An approach for generating a random x-monotone polygon on a given set of vertices is reported by Zhu et al. [11]. Auer et al. [1] analyzed heuristics to generate simple and star-shaped polygons on a given set of points. Two methods for generating random orthogonal polygon with given set of vertices are presented in [10], one of them uses an inflate-cut technique, whereas the other employs a constraint programming approach and modelled the problem as constraint satisfaction problem. In most of these works, a polygon is generated from a random set of vertices. Two recent work on generating random digital closed curve (4- and 8- connected) is presented in [3] and in [9]. However, to the best of our knowledge, there are no proposed algorithms to generate random closed curves in triangular grid.

A random curve is called triangular when it consists of edges which lie in either $\mathbb{L}_0$ or $\mathbb{L}_{60}$ or $\mathbb{L}_{120}$ as defined in Definition 1. In this paper, a linear-time algorithm is proposed based on combinatorial technique devoid of any backtracking. The algorithm generates random simple triangular closed curve imposed on the background triangular grid where the resolution of the generated curve are controlled by varying the grid length. A move from a given grid point is chosen randomly from a set of 'safe' directions calculated on the basis of the occupancy and orientation of some designated neighbors of the current vertex. Two instances of random triangular curves generated by the proposed algorithm are shown in Fig. 1.

The paper is organized as follows. Required definitions are presented in Sect. 2. The basic principle behind formulation of the combinatorial rules for traversal is discussed in Sect. 3. The method for generation of random digital curves are presented in Sect. 4. Section 5 presents the experimental results with analysis and the conclusion is presented in Sect. 6.

2   Definitions

Definition 1 *(triangular grid). A triangular grid (henceforth simply referred as grid)* $\mathbb{T} := (\mathbb{L}_0, \mathbb{L}_{60}, \mathbb{L}_{120})$ *consists of three sets of parallel grid lines, which* *are inclined at $0°$, $60°$, and $120°$ w.r.t. x-axis. The grid lines in $\mathbb{L}_0, \mathbb{L}_{60}, \mathbb{L}_{120}$ correspond to three distinct coordinates, namely α, β, γ. Three grid lines, one each from $\mathbb{L}_0, \mathbb{L}_{60}, \mathbb{L}_{120}$, intersect at a (real) grid point. The distance between two consecutive grid points along a grid line is termed as grid size, g. A line segment of length g connecting two consecutive grid points on a grid line is called grid edge.* A portion of the triangular grid is shown in Fig. 2. It has six distinct regions called sextants, each of which is well-defined by two rays starting from $(0,0,0)$. For example, Sextant 1 is defined by the region lying between $\{\beta = \gamma = 0, \alpha \geq 0\}$ and $\{\alpha = \gamma = 0, \beta \geq 0\}$. One of α, β, γ is always 0 in a sextant. For example, $\gamma = 0$ in Sextant 1 and Sextant 4. For a given grid point, p, there are six neighboring UGTs, given by $\{T_i : i = 0, 1, \ldots, 5\}$. The three coordinates of p are given by the corresponding moves along a/the shortest path from $(0,0,0)$ to p, measured in grid unit. For example, $(1,2,0)$ means a unit move along $0°$ followed by two unit moves along $60°$, starting from $(0,0,0)$. The grid point p can have six neighbor grid points, whose direction codes are given by $\{d : i = 0, 1, \ldots, 5\}$.

Definition 2 *(triangular curve). A (finite) closed curve $\mathcal{C}$ imposed on the triangular grid $\mathbb{T}$ is termed as a triangular curve if its sides coincide with lines in $\mathbb{L}_0, \mathbb{L}_{60},$ and $\mathbb{L}_{120}$. It is represented by the (ordered) sequence of its vertices, which are grid points on the triangular grid. It can also be represented by a sequence of directions and a start vertex. Its interior is defined as the set of points with integer coordinates lying inside it. A triangular curve is said to be simple if it does not intersect itself.*

The objective is to generate random simple closed triangular digital curves imposed on the background triangular grid $\mathbb{T}$ where the vertices are grid points.

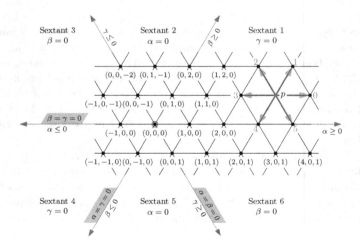

Fig. 2. Portion of a triangular canvas, the UGTs $\{T_0, T_1, \ldots, T_5\}$ incident at a grid point p, and the direction codes $\{0, 1, \ldots, 5\}$ of neighboring grid points of p

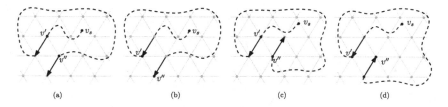

(a)   (b)   (c)   (d)

Fig. 3. (a) and (b) curves are in same direction, start of $\mathcal{C}$ is unreachable, (c) and (d) in opposite direction, $\mathcal{C}$ can traverse to its start point

3   Basic Principle

The proposed triangular random digital curve, $\mathcal{C}$, selects a random direction from a set of "safe" directions, S, calculated from its current position at the grid point $p(i, j, k)$. The set S is computed on the basis of the presence (and orientation) of $\mathcal{C}$ in designated set of vertex $V_d \in N_6(p)$[1]. The curve grows itself by choosing a random direction from the set S and eventually concludes at the start point, producing a closed non-intersecting curve with its side lying on $\mathbb{L}_0, \mathbb{L}_{60}$, and $\mathbb{L}_{120}$. To compute S, a combinatorial technique is used which guarantee that $\mathcal{C}$ never moves into a situation where there is no way for it to meet the start vertex without intersecting itself. The proposed algorithm is designed based on the observation that under no condition, either of incoming or outgoing direction of any two grid points on $\mathcal{C}$, separated by unit grid length will be same as the other. When two points on $\mathcal{C}$ are unit grid length apart, they describe a closed region. The start point may lie inside the closed region or out side of it but in any case if the directions are same, then the $\mathcal{C}$ propagates into a region where the start point does not lie, thus moves into a dead end. As shown in Fig. 3(a) the out going direction at v' and v'' are same and the start point is in the closed region defined by v' and v''. In Fig. 3(b) the incoming direction at v' and v'' are same and the start point is in the closed region defined by v' and v''. In either of the cases, the curve can never meet the start point without intersecting itself. On the other hand, if the directions at those two points are different there is always a way for the curve to eventually meet v_s (Fig. 3(c) and (d)). Since there are six possible directions available in triangular grid (Fig. 2), the notion of direction being same or opposite is not so obvious. This is made simpler by considering only two directions namely clockwise and anticlockwise and it is determined by the incoming and outgoing direction at the point p under consideration. Another fact is that when the curve $\mathcal{C}$ is at a particular point p, there could be more than one visited neighbor of p at $N_6(p)$. Out of all those visited neighbors, the one through which $\mathcal{C}$ has entered in $N_6(p)$ is considered to determine S. To find the vertex through which $\mathcal{C}$ has entered in $N_6(p)$, a timestamp t_i is assigned in

---

[1] Two points p and q are said to be 6-connected in a set S if and only if there exists a sequence $\langle p = p_0, p_1, \ldots, p_n = q \rangle \subseteq S$ such that $p_i \in N_6(p_{i-1})$ for $1 \leqslant i \leqslant n$. The 6-neighborhood of a point (x, y, z) is given by $N_6(x, y, z) = \{(x', y') : \max(|x - x'|, |y - y'|, |z - z'|) = 1\}$.

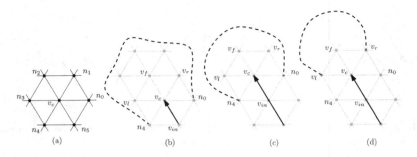

Fig. 4. (a) 6 neighbors (b) trap involving n_4 and n_0, (c) trap involving n_3 and n_1

increasing order to each vertex v_i as C grows. If v_i is generated earlier than v_j then t_i is less than t_j.

To start with, a random start grid point v_s of the curve is chosen randomly and a timestamp of $t_s = 0$ is assigned to it. The curve grows itself from a given (current) point, v_c, by computing set of safe direction S from v_c based on the occupancy and orientation of C at the earliest visited amongst the designated neighbours of v_c. Then C chooses its next direction of propagation randomly from S, denoted by d_c. As the curve proceeds along d_c to the new grid point from v_c, the timestamp of the new point along d_c is increased by one. The curve repeats this procedure until it reaches to the start point v_s and concludes.

Since random digital triangular curve of finite length is of our interest, the algorithm is designed to execute on a finite canvas which is parallelogram in shape. So, a boundary condition is imposed such that if C takes a clockwise (anticlockwise) turn when it meets the boundary for the first time, then it will take a clockwise (anticlockwise) turn in all subsequent cases whenever it meets the boundary again. The methods for generating triangular digital curves are described in the following section.

4 Generation of Simple Random Triangular Digital Curves

To generate simple random triangular digital curves of finite length following the principle described in Sect. 3, some rules are formulated. This section explains those rules in detail. Time complexity is discussed at the end of this section.

Let the set $N = \{n_0, n_1, \ldots, n_5\}$ denotes the six neighbors of a point in triangular grid as shown in Fig. 4(a) and let the neighbor along direction d is n_d. Also, let v_c be the current point, and v_l, v_f, and v_r be the left, front, and right points (Fig. 4(b)). An array **b** of size 6 is considered to signify the 6 directions initialized to 1, which means all directions are initially permissible. Also, if a neighbour n_d is already visited then **b**[d] is invalidated. However the start vertex is always considered to be not visited, allowing the curve to meet the start vertex whenever it wishes to. The array **b** is used throughout to compute the set of

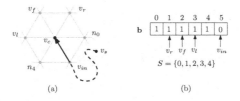

Fig. 5. Case I: (a) None of $\{v_l, v_f, v_r\}$ are visited, (b) Direction array **b**

permissible direction from the current point v_c until the curve concludes at start point v_s.

To generate random triangular digital curves, only the neighbors v_l, v_f, and v_r of v_c are consulted to find S, thus designated set of vertices V_d consist of $\{v_l, v_f, v_r\}$. The reason is to avoid potential traps involving neighbors of the current point, v_c. Without loss of generality, let the curve $\mathcal{C}$ reaches v_c from v_{in} as shown in Fig. 4(b). This means that it has already checked the existence of a possible trap involving either n_4 or v_c or n_0, whichever is visited earlier, and v_{in}. Now, for a safe move from v_c, possible loop between v_{in} and the earliest visited of $\{v_l, v_f, v_r\}$ need to be checked. Suppose after checking the possible loop described in Fig. 4(b) the curve now reaches to v_c as shown in Fig. 4(c) with the new labels of vertices. Again to make safe move from v_c the curve needs to check possible loop between earliest visited of $\{v_l, v_f, v_r\}$ and v_{in} as shown in Fig. 4(d). Depending on the occupancy and orientation of v_l, v_r, and v_f, there exist three different cases as stated below.

Case I: *None of* v_l, v_f *and* v_r *are visited.* The curve can move in any direction except in the direction of the visited neighbors.
For example, in Fig. 5(a), none of $\{v_l, v_f, v_r\}$ are visited. **b**[5] is assigned 0 as the curve has come to v_c via n_5 (marked as v_{in} in figure) making it invalid. Thus, the curve can randomly choose one of the directions from $S = \{0, 1, 2, 3, 4\}$ to make further progress, shown as cells with 1 in it (Fig. 5(b)).
Case II: *At least one in* $\{v_l, v_f, v_r\}$ *is visited.* Let v_m from $\{v_l, v_f, v_r\}$ has the lowest timestamp, and v_{m+1}, the grid point next to v_m on $\mathcal{C}$, lies in $N_6(v_c)$. For each visited neighbor n_d of v_c, **b**[d] is marked as invalid. Then, a traversal is made in **b** from v_m in the direction of v_m to v_{m+1} (the immediate next cell in **b**) upto the farthest visited vertex in $\{v_l, v_f, v_r, v_{in}\} \setminus v_m$ setting all the location encountered in **b** to 0 during the traversal. The indices in **b** with value 1 constitute the set of permissible directions, S.

Figure 6 depicts one such situation in which v_r and v_f is visited, v_r has lower timestamp and its next grid point v_{m+1} (v_f in this case) lies in $N_6(v_c)$. Figure 6(b) shows the initial marking; location **b**[1], **b**[2] and **b**[5] are marked 0 as $v_{in}(= n_5)$, $v_f(= n_2)$ and $v_r(= n_1)$ are already visited. The traversal is made in the direction from v_r to v_{m+1}, in **b** till v_{in} thereby setting **b**[3]=0 and **b**[4]=0, others being already set to 0, thus $S = \{0\}$. The array **b** is treated as circular during the traversal. One interesting thing to observe

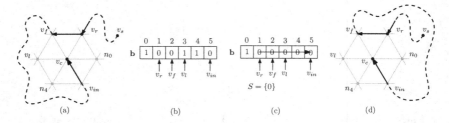

Fig. 6. Illustration of Case II (a) v_r is visited and has lowest timestamp, (b) initial marking of **b** (c) final marking of **b** (d) v_r is visited: another scenario of (a)

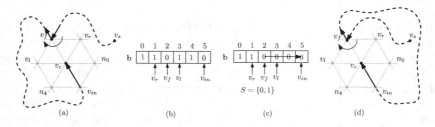

Fig. 7. Illustration of Case III (a) v_f is visited and has clockwise movement (b) Initial marking of **b** (c) Final marking of **b** (d) another alternative of (a)

is that the curve can traverse to v_c following two paths as shown in Fig. 6 (i) keeps the start point v_s inside (Fig. 6(d)) the closed region defined by v_f and v_c and (ii) keeps the start point v_s outside (Fig. 6(d)) the closed region defined by v_f and v_c. However, in both the cases the only permissible direction is 0.

Case III: *At least one in $\{v_l, v_f, v_r\}$ is visited. Let v_m from $\{v_l, v_f, v_r\}$ has the lowest timestamp, and v_{m+1}, the grid point next to v_m on $\mathcal{C}$, does not lie in $N_6(v_c)$.* First, the visited neighbours of v_c are marked 0 in **b**. Then, a traversal is made from v_m in forward (backward) direction if there is a clockwise (anticlockwise) turn at v_m till the farthest visited vertex in $\{v_l, v_f, v_r, v_{in}\} \setminus v_m$ thereby setting all the location encountered in **b** to 0 during the traversal. The indices in **b** with value 1 constitute the set of permissible directions, S, from v_c.

In Fig. 7(a) $v_m = v_f$ and $v_{m+1} \notin N_6(v_c)$ and there is an clockwise turn at v_m. Hence, a forward traversal is made from v_f to v_{in} which is the farthest visited vertex in $\{v_l, v_r, v_{in}\}$, as v_l and v_r are not visited at all. Thus, $S = \{0, 1\}$.

Figure 8(a) shows another situation in which all of $\{v_l, v_f, v_r\}$ are visited, v_l has the lowest timestamp and there is an anti-clockwise turn at v_l. Hence, a backward traversal is made from v_l to v_{in} since it is the farthest visited vertex in $\{v_l, v_r, v_{in}\}$ along backward direction. Thus, $S = \{4\}$.

The steps for generating the random curve are captured in the form of pseudocode as follows

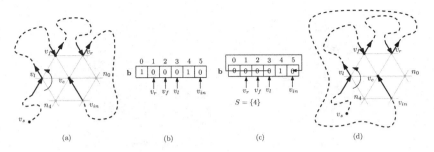

Fig. 8. Another illustration of Case III (a) v_l, v_f, v_r are all visited, v_l has lowest timestamp and has clockwise movement (b) Initial marking of **b** (c) Final marking of **b** (d) another alternative of (a)

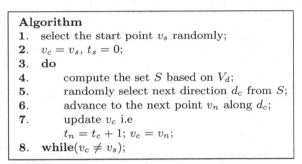

where $V_d = \{v_l, v_f, v_r\}$, S is the set of permissible directions, v_c is the current vertex, d_c is incoming direction to v_c, and t_i is the timestamp of the ith vertex.

Time Complexity: Every time the curve progresses it computes the set of safe directions, S, from the current point, v_c. To compute S, a fixed number of vertices (v_l, v_f, v_r) are checked. The number of grid points on the generate curve is of $O(|\mathcal{C}|/g)$ and the computation of S at each grid point requires constant amount of time. Thus, the running time of algorithm is linear in the number of grid points on the length of the generated curve which is given by $O(|\mathcal{C}|/g)$.

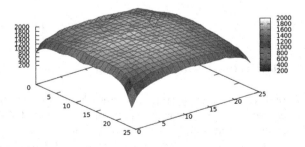

Fig. 9. Test of randomness of the generated curves

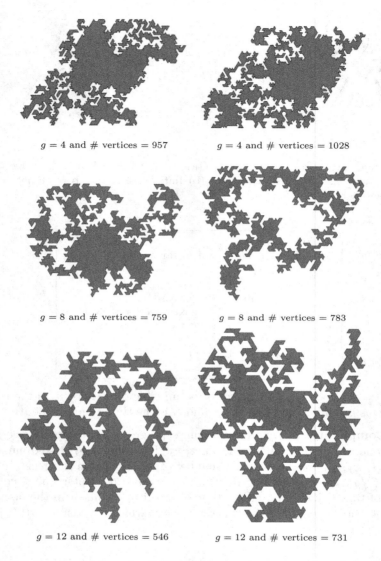

$g = 4$ and # vertices = 957 $g = 4$ and # vertices = 1028

$g = 8$ and # vertices = 759 $g = 8$ and # vertices = 783

$g = 12$ and # vertices = 546 $g = 12$ and # vertices = 731

Fig. 10. Instances of triangular digital random curves on a canvas of size 50×50

5   Experimental Results and Analysis

The proposed algorithm is implemented in C in Ubuntu 12.04, 64-bit, kernel version 3.5.0-43-generic, the processor being Intel i5-3570, 3.4 GHz FSB. Six instances of triangular random curves are shown in Fig. 10, first row shows two random curves for grid length, $g = 4$ whereas the second and third rows show two random curves each for $g = 8$ and $g = 12$ respectively. The figures depict randomness of the curves and also the number of vertices for each curve. The resolution of the curves can be controlled by varying the grid length g.

In order to test the randomness of the algorithm, we have generated $10,000$ curves on a fixed canvas size and counted number of times a grid point in the canvas lies on $\mathcal{C}$. Figure 9 shows the plot with the coordinates of the grid points along the x and y axes whereas the z-axis represents the frequency. It shows that each of the grid points in the canvas has almost equal frequency (flat at the top) indicating the equal probability of occurrences.

6 Conclusions

A combinatorial technique to generate random triangular digital curve is presented in this paper. The algorithm is linear on the length (in number of grid points) of the generated curve and does not require backtracking. The randomness of the generated curve is obvious from the experimental results. In this work, we have used a regular canvas to generate the curve, however, with some modification the algorithm can be used to generate random curve on any arbitrary canvas. The algorithm can further be tuned to generate random paths inside a digital object which can be useful for practical applications.

References

1. Auer, T., Held, M.: Heuristics for the generation of random polygons. In: Proceedings of the Canadian Conference on Computational Geometry, pp. 38–44 (1996)
2. Bhowmick, P., Bhattacharya, B.B.: Fast polygonal approximation of digital curves using relaxed straightness properties. IEEE Trans. PAMI **29**(9), 1590–1602 (2007)
3. Bhowmick, P., Pal, O., Klette, R.: Linear-time algorithm for the generation of random digital curves. In: Proceedings of the 2010 Fourth Pacific-Rim Symposium on Image and Video Technology, pp. 168–173 (2010)
4. Dailey, D., Whitfield, D.: Constructing random polygons. In: Proceedings of the 9th ACM SIGITE Conference on Information Technology Education, pp. 119–124 (2008)
5. Epstein, P.: Generating geometric objects at random. Master's thesis, CS Dept., Carleton University, Canada (1992)
6. Klette, R., Rosenfeld, A.: Digital Geometry: Geometric Methods for Digital Picture Analysis. Morgan Kaufmann, San Francisco (2004)
7. Rennesson, I., Luc, R., Degli, J.: Segmentation of discrete curves into fuzzy segments. Electron. Notes Discrete Math. **12**, 372–383 (2003)
8. Rourke, J., Virmani, M.: Generating random polygons, TR:011. CS Dept. Smith College, Northampton (1991)
9. Sarkar, A., Biswas, A., Dutt, M., Bhattacharya, A.: Generation of random digital curves using combinatorial techniques. In: Ganguly, S., Krishnamurti, R. (eds.) CALDAM 2015. LNCS, vol. 8959, pp. 286–297. Springer, Heidelberg (2015)
10. Tomas, A.P., Bajuelos, A.L.: Generating random orthogonal polygons. In: Conejo, R., Urretavizcaya, M., Pérez-de-la-Cruz, J.-L. (eds.) CAEPIA/TTIA 2003. LNCS (LNAI), vol. 3040, pp. 364–373. Springer, Heidelberg (2004)
11. Zhu, C., Sundaram, G., Snoeyink, J., Mitchell, J.S.B.: Generating random polygons with given vertices. CGTA **6**, 277–290 (1996)

Image Retrieval

Tackling Curse of Dimensionality for Efficient Content Based Image Retrieval

Minakshi Banerjee[(✉)] and Seikh Mazharul Islam

RCC Institute of Information Technology, Canal South Road,
Beliaghata, Kolkata 700015, West Bengal, India
mbanerjee23@gmail.com, mislam2911@gmail.com

Abstract. This paper proposes a content based image retrieval (CBIR) technique for tackling curse of dimensionality arising from high dimensional feature representation of database images and search space reduction by clustering. Kernel principal component analysis (KPCA) is taken on MPEG-7 Color Structure Descriptor (CSD) (64-bins) to get low-dimensional nonlinear-subspace. The reduced feature space is clustered using Partitioning Around Medoids (PAM) algorithm with number of clusters chosen from optimum average silhouette width. The clusters are refined to remove possible outliers to enhance retrieval accuracy. The training samples for a query are marked manually and fed to One-Class Support Vector Machine (OCSVM) to search the refined cluster containing the query image. Images are ranked and retrieved from the positively labeled outcome of the belonging cluster. The effectiveness of the proposed method is supported with comparative results obtained from (i) MPEG-7 CSD features directly (ii) other dimensionality reduction techniques.

Keywords: Kernel principal component analysis · Partitioning around medoids · Outliers detection · Support vector clustering · One-class support vector machine

1 Introduction

Developing efficient content-based image retrieval (CBIR) [3,11] techniques have proven to be a challenging research area for facilitating tremendous demand of accessing digital image libraries in real life applications. CBIR is used in place of traditional image retrieval technique which employs manual keyword annotations. It combines different image processing, pattern recognition, computer vision techniques, human-computer interactions etc. to improve retrieval accuracy. Visual contents of an image such as color, shape, texture and spatial layout etc. are extracted automatically and represented in terms of multi-dimensional feature vectors [1,11]. Similar images are returned based on measuring similarities between the feature vectors of the submitted query image and those stored in the database. Owing to the impact of multimedia retrieval, ISO/IEC has launched MPEG-7, which provides a collection of specific standard descriptors

© Springer International Publishing Switzerland 2015
M. Kryszkiewicz et al. (Eds.): PReMI 2015, LNCS 9124, pp. 149–159, 2015.
DOI: 10.1007/978-3-319-19941-2_15

[9] for evaluation of new image retrieval schemes. While there have been ample amount of publications in the area of CBIR, the performance of image retrieval systems is still not satisfactory due to semantic gap, which is created from the discrepancies between the computed low-level features (color, texture, shape, etc.) and users conception of an image. Iterative refinement of results through relevance feedback [12] has been popular for minimizing semantic gap in CBIR. An efficient image search method using small codes for large image databases is proposed in [13] which converts Gist descriptor to a compact binary code for representing images and their neighborhood structures.

Recently, automatic image annotation (AIA) is used as a distinctive method for bridging semantic gap. Bag-of-visual-words model which can represent images at object level and provides spatial information to build a visual dictionary between low level features and high-levels semantics for AIA is proposed in [8]. Scalable object recognition with vocabulary tree is also proposed in [6]. One important issue in AIA is high dimensional features which should be selected in right number and right features for efficient annotation. Performance of the classifiers may degrade when feature dimension is too high. Also large number of features may better represent the discriminative properties of visual contents. But this leads to dimensionality curse problem [5]. Proper dimensionality reduction methods must be employed to preserve the distinctive properties so that performance is not degraded. The popular dimensionality reduction algorithms include Principal Component Analysis (PCA), Kernel Principal Component Analysis (KPCA), Linear discriminant analysis (LDA), Factor analysis (FA), and Laplacian eigenmaps (LEM) etc. The kernel PCA [10] which is a nonlinear form of PCA, can efficiently compute principal components in high dimensional feature spaces. It captures more information than PCA as it is related to the input space by some nonlinear mapping. Image retrieval using Laplacian eigenmap [5] produces efficient results by preserving local neighborhood information of data points but suffers from the problem of new data point, that is if query image is absent in database. Although efficient feature reduction for a particular query is an important criteria to amend curse of dimensionality, to speed up the retrieval process data clustering is also an effective way. However selection of number of clusters for an unknown database is an important issue. A Fuzzy Clustering approach to CBIR is proposed in [7] but it makes no attempt in removing groups of images that are irrelevant to the search query, nor re-ranking the search results.

The motivation behind the proposed method lies in addressing the task of dimensionality reduction, search space reduction by efficient clustering and proper searching and retrieving from reduced search space by labeling them efficiently as positive (relevant) and negative (non-relevant) images so that the overall performance is improved. To pursue these motivations the proposed method uses kernel principal component analysis for dimensionality reduction, Partitioning Around Medoids (PAM) [4] algorithm for clustering. The clusters are further processed using Support Vector Clustering (SVC) [15] to remove possible outliers from the cluster containing the query. One-class support vector machine [2] is proposed for classification as this classifier is biased to the learned concept of a particular category. Training samples are generated from displayed results obtained using KPCA-reduced CSD feature and L_1 similarity distance. The proposed method is compared with others dimensionality reduction techniques, namely, Principal Component Analysis (PCA), Factor analysis (FA), and Laplacian eigenmaps (LEM). The remaining sections are organized as follows. Section 2 gives the mathematical background of Kernel PCA and one-class SVM. Proposed method is highlighted in Sect. 3. Experimental discussion and results are reported in the Sect. 4. Section 5 concludes.

2   Mathematical Preliminaries

2.1   Kernel Principal Component Analysis (KPCA)

Suppose $x_i \in R^l, i = 1, 2, ...n$ are n observations. The basic idea of KPCA [10], is as follows. First, the samples are mapped into some potentially high-dimensional feature space F by following.

$$\emptyset : R^l \to F, x_i \to \emptyset(x_i), (i = 1, 2, ...n) \tag{1}$$

where $\emptyset$ is a nonlinear function. Now these mapped samples in F using a kernel function k occur in terms of dot product are given by following equation.

$$k(x_i, x_j) = (\emptyset(x_i), \emptyset(x_j)) \tag{2}$$

For simple notation assume $\sum_{i=1}^{n} \emptyset(x_i) = 0$. To find eigenvalues $\lambda > 0$ and associated eigenvectors $\upsilon \in F \setminus \{0\}$ which are computed by:

$$\upsilon = \sum_{i=1}^{n} \alpha_i \emptyset(x_i) \text{ where } \alpha_i(i = 1, ..., n) \text{ are coefficients.} \tag{3}$$

Then the eigenvalue equation is $n\lambda\alpha = K\alpha$, where α denotes a column vector with entries $\alpha_1, ..., \alpha_n$ and K is called called kernel matrix. Kernel principal components of a test point t are computed by projecting $\emptyset(t)$ onto the kth eigenvectors υ^k is:

$$(\upsilon^k.\emptyset(t)) = \sum_{i=1}^{n} \alpha_i^k(\emptyset(x_i), \emptyset(t)) = \sum_{i=1}^{n} \alpha_i^k k(x_i, t), \text{where } \lambda_k(\alpha^k.\alpha^k) = 1 \tag{4}$$

Kernel function for our experiment is:

$$k(x_i, x_j) = \exp(-\|x_i - x_j\|^2/2\sigma^2), \text{ where } \sigma > 0 \tag{5}$$

2.2   One Class Support Vector Machine (OCSVM)

The OCSVM [2] maps input data into a high dimensional feature space using a kernel and iteratively finds the maximal margin hyperplane which best separates the training data from the origin. The OCSVM may be viewed as a regular two-class SVM where all the training data lies in the first class, and origin is taken as only member of the second class. Thus, the hyperplane (or linear decision boundary) corresponds to the classification rule is:

$$f(x) = \langle w, x \rangle + b \tag{6}$$

where w is the normal vector and b is a bias term. The OCSVM solves an optimization problem to find the rule with maximal geometric margin. To assign a label to a test sample x if $f(x) < 0$ then the sample is non-relevant otherwise relevant.

Kernels:
The optimization problem of OCSVM is obtained by solving the dual quadratic programming problem, is given by:

$$min_\alpha \frac{1}{2} \sum_{ij} \alpha_i \alpha_j K(x_i, x_j), \text{ s.t. } 0 \le \alpha_i \le \frac{1}{vl} \text{ and } \sum_i \alpha_i = 1 \tag{7}$$

where α_i is a lagrange multiplier (or weight on example i such that vectors associated with non-zero weights are called support vectors and solely determine the optimal hyperplane), v is a parameter that controls the trade-off between maximizing the distance of the hyperplane from the origin and the number of data points contained by the hyper-plane, l is the number of points in the training dataset, and $K(x_i, x_j)$ is the kernel function. By using the kernel function to project input vectors into a feature space, we allow for nonlinear decision boundaries. Given a feature map:

$$\emptyset : X \to R^N \tag{8}$$

where $\emptyset$ maps training vectors from input space X to a high-dimensional feature space, we can define the kernel function as:

$$K(x, y) = \langle \emptyset(x), \emptyset(y) \rangle \tag{9}$$

The commonly used kernels are linear, radial basis function, and sigmoid.

3   Proposed Method

The proposed method is explained in Fig. 1 and the algorithmic steps are also shown below.

3.1   Feature Extraction

The MPEG-7 Color Structure Descriptor (CSD) (64-bins) is extracted for each image. Number of times a particular color is contained within the structuring element is counted using $c_0, c_1, c_2, ..., c_{M-1}$ quantized colors. A color structure histogram can then be denoted by $h(m), m = 0, 1, 2, ..., M-1$ where the value in each bin represents the number of structuring elements in the image containing one or more pixels with color c_m. The hue-min-max-difference (HMMD) color space is used in this descriptor [9].

Algorithm 1. Steps in the proposed method

Input: A query Image
Output: Similar Images
Step1: Extract the MPEG-7 CSD (64-bins) features of all the images in database.
Step2: Apply KPCA to map feature vectors in lower-dimensional space.
Step3: Cluster the reduced dataset obtained in **Step2** using PAM algorithm. Select the cluster containing the query by feature matching.
Step4: Training samples collection - Display 36 nearest images with respect to query image using L_1-norm in KPCA - reduced space and mark relevant images to get training set.
Step5: Test samples collection - Apply SVC to the cluster containing the query obtained in **Step3** to remove possible outliers.
Step6: Train OCSVM using samples obtained in **Step4**.
Step7: Input test samples obtained in **Step5** in OCSVM for prediction (prediction label 1 is for positive sample and 0 for negative sample).
Step8: Consider all positive samples obtained in **Step7** automatically using the labels of prediction.
Step9: Select original CSD (64-bins) feature vectors corresponding to all positive samples and calculate the similarity value by L_1 norm. Sort these similarity values in ascending order.
Step10: Display the ranked images obtained in **Step9**.

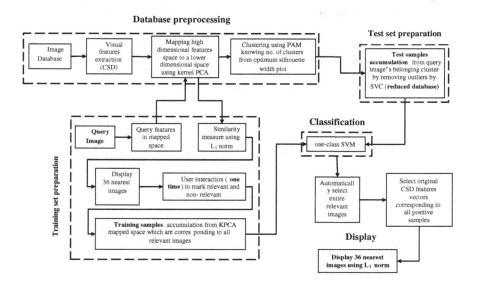

Fig. 1. Proposed method

3.2   Number of Clusters Selection Using PAM

Searching similar images in few nearest clusters are more effective than sequential search in large database. The PAM [4] clustering algorithm is usually less sensitive to outliers than k-means. It groups similar objects by minimizing sum of the dissimilarities of all the objects to their nearest medoids. The aim is to find the clusters $C_1, C_2, ..., C_k$ that minimize following target function:

$$\sum_{i=1}^{k} \sum_{r \in C_i} d(r, m_i) \text{ where for each } i \text{ the medoid } m_i \text{ minimizes } \sum_{r \in C_i} d(r, m_i) \quad (10)$$

where $d(r, m_i)$ indicates dissimilarity between r and m_i. The numbers of clusters are determined from optimum average silhouette width plot. For every point i, the silhouette width $silw(i)$ is calculated as follows: Let $p(i)$ be the average dissimilarity between i and other all points of the partition to which i lies. If i is only object in its belonging partition then $silw(i) = 0$ barring further computation. For all remaining partitions C we get $d(i, C)$ = average dissimilarity of i to all objects of C. The smallest of these $d(i, C)$ is $q(i) = min_{\forall C}(d(i, C))$ indicates the dissimilarity between i and its nearest partition obtained from minimum leads to Eq. 11 and $max(avg(silw_C), \forall C)$ determines the number of clusters. For example, if number of clusters is k (where $k = 2, 3, 4, ..., 25$) then silhouette width for every point is computed and average is found. Finally, the cluster number which gives the maximum average silhouette width plot is selected. In Fig. 2 it is obvious that the number of clusters is 3 by taking upto 25 clusters.

$$silw(i) = (q(i) - p(i))/max(p(i), q(i)) \quad (11)$$

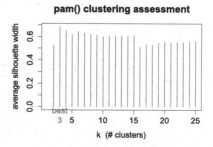

Fig. 2. Computation of number of clusters from optimum average silhouette width

3.3   Outliers Detection Criteria of Support Vector Clustering (SVC)

Let $\{x_i\}$ be a dataset with dimensionality d and x be a data point. SVC [15] computes a sphere of radius R and center a containing all these data. The computation of such a smallest sphere is obtained from solving minimization problem

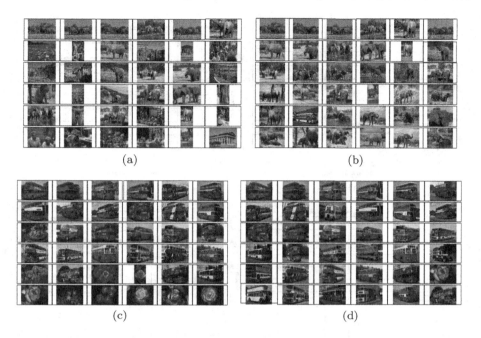

(a)   (b)

(c)   (d)

Fig. 3. (a) and (c) represent MPEG-7 CSD(64-bins) with L_1 norm, relevant/scope ratios are 16/36 and 20/36 respectively, (b) and (d) represent proposed method, relevant/scope ratios are 35/36 and 35/36 respectively, with KPCA projected dimension=1, OCSVM based prediction from refined cluster containing the query. The top-left most image is the query.

considering Lagrangian formulation which produces following expression.
$\|x - a\|^2 = (x.x) - 2\sum_{i=1}^{N} \alpha_i(x.x_i) + \sum_{i=1}^{N}\sum_{j=1}^{N} \alpha_i\alpha_j(x_i.x_j) \leq R^2$, where α_i is the Lagrangian multipliers. To test a data x for outlier, the necessary condition is $\alpha_i \geq 0$.

Computational Issues of the Proposed Method. The proposed method, at first, applies KPCA to reduce original data set in $O(n^3)$ times, where n is the number of data points, then the clustering algorithm PAM operates on this reduced dataset at least in quadratic times. The complexity concerned with SVC is $O(n^2d)$ if the number of support vectors is $O(1)$ (where d is the number of dimensions of a data point). OCSVM algorithm uses sequential minimal optimization to solve the quadratic programming problem, and therefore takes time $O(dL^3)$, where L is the number of entities in the training dataset. The total complexity is involved mainly in SVC and OCSVM which are applied on reduced features and search space.

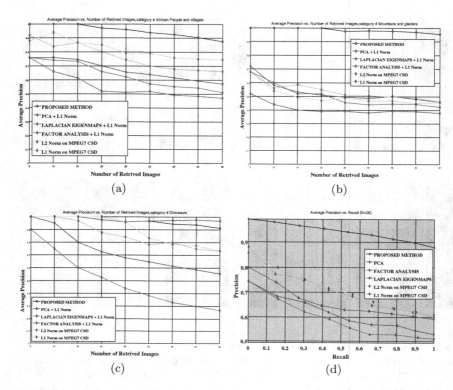

Fig. 4. (a), (b) and (c) represent Average precision vs. Number of retrieved images. Category in (a) African people and villages (b) Mountains and glaciers (c) Dinosaurs and (d) represents Average precision vs. Average recall on overall database.

4   Experiments and Results

The experimental results have been demonstrated from Figs. 3 to 4 using 1000 SIMPLIcity database [14] (http://wang.ist.psu.edu) which is divided into 10 Semantic groups including African people and villages, Buses, Dinosaurs, Elephants etc. Each group comprises of 100 sample images. Experiments are carried out extensively on each group. The class label of each image in the database are not predefined and label of training samples for classification by OCSVM is based on visual judgement of similarity which is obtained by marking relevant and non-relevant images. Therefore the performance of retrieval is evaluated on the finally displayed images based on the visual judgement of similarity instead of classification scores by predefined labels. The performances metrics used are Recall rate and Precision rate. Let n_1 be the number of images retrieved in top n positions that are close to a query image. Let n_2 be the number of images in the database similar to the query. Evaluation standards, Recall rate (R) is given by $n_1/n_2 \times 100\%$ and Precision rate (P) are given by $n_1/n \times 100\%$.

KPCA is used to reduce the CSD dataset taking Gaussian function with $sigma = 0.2$ as a parameter of kernel function using kernlab R language Package. In the classification stage by OCSVM the training of classifier is done with radial basis function using LIBSVM package with $c = 0.01$, $gama = 0.00001$. Finally the predicted relevant images of OCSVM have been ranked using $L_1 - norm$ considering original CSD features of predicted set. The result of Fig. 3(a) and (b) indicate a query from category elephant having diverse background. Figure 3(a) shows result with MPEG-7 CSD (64-bins) features using L_1-norm as a similarity measure. Figure 3(b) represents proposed method with KPCA where the projected dimension is equal to one. The query image is in top-left corner. The comparative results suggest that the precision obtained in case of Fig. 3(a) within the scope of 36 images is 44.44 %. For the same query the proposed method gives 97.22 % precision where similar images have attained higher rank compared to Fig. 3(a), in the reduced search space. The effectiveness of the proposed method is also observed in Fig. 3(c) for the category of Bus. Here color structure descriptors have proven to be good features for retrieving buses not only for similar shapes but also for similar colors. Precision and ranking are improved in Fig. 3(d) compared to Fig. 3(c). Intuitively, nonlinear KPCA based feature mapping captures discriminating information of the CSD (64-bins) into a lower dimensional subspace. The results are compared with other unsupervised dimensionality reduction methods namely, PCA with L_1-norm, Laplacian eigenmaps with L_1-norm, Factor analysis with L_1-norms, and with MPEG-7 CSD (64-bins) dataset with L_2 and L_1 norms. Average precision is chosen as a comparative measure because it gives a good measure within the displayed scope and it is not always possible to know the number of relevant image in advance for a particular query in an unknown database. The proposed method gives 95 % average precision within the scope of 30 images for the category of African people. This is shown in Fig. 4(a) which performs far well than other techniques. In case of the category of mountain as shown in Fig. 4(b) the proposed method gives 98 % average precision within the scope of 30 images and performs far well than other methods. For the category of dinosaur as shown in Fig. 4(c), Factor analysis gives marginally better result. Particularly for this category the images are covered by white background and lacking prominent color structure information which may be a possible reason for this performance. Considering images from all categories the average precision vs. recall graph on the overall database is plotted in Fig. 4(d) which is giving more than 95 % precision before recall reaches at 0.5. This proves the effectiveness of the proposed method.

5   Conclusion

In this paper we have proposed an efficient content-based image retrieval system to address the issues of dimensionality reduction and search space reduction followed by possible outlier removal, without degrading the retrieval performance. With the proposed method the MPEG-7 Color Structure Descriptor (CSD) in reduced dimensions is performing very well for many categories compared to other retrieval systems which combine several descriptors together. As a further scope of research we intend to make the system more generic considering several newer datasets for different retrieval applications. We would also like to combine others MPEG-7 features and extend the proposed method for object retrieval by unifying textual cues.

References

1. Agarwal, S., Verma, A., Dixit, N.: Content based image retrieval using color edge detection and discrete wavelet transform. In: 2014 International Conference on Issues and Challenges in Intelligent Computing Techniques (ICICT), pp. 368–372 (2014)
2. Chen, Y., Zhou, X.S., Huang, T.: One-class SVM for learning in image retrieval. Int. Conf. Image Process. **2001**, 34–37 (2001)
3. Jaworska, T.: Application of fuzzy rule-based classifier to CBIR in comparison with other classifiers. In: 11th International Conference on Fuzzy Systems and Knowledge Discovery (FSKD), pp. 119–124 (2014)
4. Kaufman, L., Rousseeuw, P.J.: Partitioning Around Medoids (Program PAM). Wiley, New York (2008)
5. Lu, K., He, X., Zeng, J.: Image retrieval using dimensionality reduction. In: Zhang, J., He, J.-H., Fu, Y. (eds.) CIS 2004. LNCS, vol. 3314, pp. 775–781. Springer, Heidelberg (2004)
6. Nistér, D., Stewénius, H.: Scalable recognition with a vocabulary tree. In: IEEE Conference on Computer Vision and Pattern Recognition, pp. 2161–2168 (2006)
7. Ooi, W., Lim, C.: A fuzzy clustering approach to content-based image retrieval. In: Workshop on Advances in Intelligent Computing, pp. 11–16 (2009)
8. Philbin, J., Chum, O., Isard, M., Sivic, J., Zisserman, A.: Object retrieval with large vocabularies and fast spatial matching. In: IEEE Conference on Computer Vision and Pattern Recognition (CVPR 2007) (2007)
9. Salembier, P., Sikora, T.: Introduction to MPEG-7: Multimedia Content Description Interface. Wiley, New York (2002)
10. Schölkopf, B., Smola, A., Müller, K.R.: Nonlinear component analysis as a kernel eigenvalue problem. Neural Comput. **10**(5), 1299–1319 (1998)
11. Smeulders, A.W.M., Worring, M., Santini, S., Gupta, A., Jain, R.: Content-based image retrieval at the end of the early years. IEEE Trans. Pattern Anal. Mach. Intell. **22**(12), 1349–1380 (2000)
12. Su, J.H., Huang, W.J., Yu, P., Tseng, V.: Efficient relevance feedback for content-based image retrieval by mining user navigation patterns. IEEE Trans. Knowl. Data Eng. **23**(3), 360–372 (2011)
13. Torralba, A., Fergus, R., Weiss, Y.: Small codes and large image databases for recognition. In: IEEE Conference on Computer Vision and Pattern Recognition, pp. 1–8. IEEE (2008)
14. Wang, J., Li, J., Wiederhold, G.: Simplicity: semantics-sensitive integrated matching for picture libraries. IEEE Trans. Pattern Anal. Mach. Intell. **23**(9), 947–963 (2001)
15. Yang, J., Estivill-Castro, V., Chalup, S.: Support vector clustering through proximity graph modelling. In: Proceedings of the 9th International Conference on Neural Information Processing, ICONIP 2002, pp. 898–903 (2002)

Face Profile View Retrieval Using Time of Flight Camera Image Analysis

Piotr Bratoszewski[✉] and Andrzej Czyżewski

Multimedia Systems Department, Faculty of Electronics, Telecommunications and Informatics,
Gdansk University of Technology, Gdansk, Poland
{bratoszewski,andcz}@sound.eti.pg.gda.pl

Abstract. Method for profile view retrieving of the human face is presented. The depth data from the 3D camera is taken as an input. The preprocessing is, besides of standard filtration, extended by the process of filling of the holes which are present in depth data. The keypoints, defined as the nose tip and the chin are detected in user's face and tracked. The Kalman filtering is applied to smooth the coordinates of those points which can vary with each frame because of the subject's movement in front of the camera. Knowing the locations of keypoints and having the depth data the contour of the user's face a profile retrieval is attempted. Further filtering and modifications are introduced to the profile view in order to enhance its representation. Data processing enhancements allow emphasizing minima and maxima in the contour signals leading to discrimination of the face profiles and enable robust facial landmarks tracking.

Keywords: Depth image · Signal processing · Profile view · Keypoints tracking

1 Introduction

The profile view of human face has multiple applications and robust methods of its retrieving are regarded as highly necessary. Historically, one of the first serious applications of the profile view were the mug shot photographs of the persons after being arrested and they are dated back to 1844. Mug shot photography consists of frontal and profile photography of a person suspected of a crime and the purpose of those is to create a record and identify criminals. Hence, the profile view is considered to contain enough biometric data to be successfully used in face recognition domain.

Furthermore, the profile view is known to be used in lip-reading applications. The Audio-Visual Speech Recognition Systems (AVSR) use the visual data to improve the accuracy of AVSR system in the noisy environments. To name a few applications: systems of emotion recognition or facial actions, methods for face modeling and texturizing and more.

However, authors of this work are focused on the way of retrieving of the profile view rather than on the variety of applications. Retrieval of the profile view can be achieved by using either one RGB camera placed on the side of the user or two cameras in front of the subject – in the stereo configuration. The first solution is considered to be both inconvenient (e.g. user of the mobile computer requires the eye contact with the

© Springer International Publishing Switzerland 2015
M. Kryszkiewicz et al. (Eds.): PReMI 2015, LNCS 9124, pp. 159–168, 2015.
DOI: 10.1007/978-3-319-19941-2_16

platform in which the camera is built into) and resource-intensive as it needs to employ additional background subtraction techniques which are susceptible to light conditions and deal poorly with non-static backgrounds. The stereo configuration of RGB cameras increases the overall cost of the system as for good depth estimation high quality sensors must be used and additional processing must be applied in order to extract the depth information from the image. Therefore, authors use the newer technology, namely the Time of Flight (TOF) 3D camera which provides the depth data on its output using integrated dedicated sensors and microprocessors for the depth calculation. TOF cameras are less sensitive to lighting conditions, hence the background subtraction is based on a proper thresholding rather than on the modeling of the background, making it possible to extract the foreground robustly, also when the background is non-static. One of main disadvantages of TOF sensors is low spatial resolution, however it is improving over time. Utilizing depth sensors for the task of profile view retrieval is more convenient than RGB sensor, as the TOF camera acquires the depth information of user's head directly and can be mounted for example in a laptop or any mobile platform in order to gather data while the user is simply using the device. The process of gathering and analyzing of such data will be presented in a more detailed way in next sections.

The paper is organized as follows: in the next section authors present their related work to the subject of profile view retrieving and usage. Section 3 describes the methods applied in this work. Section 4 presents the final results of conducted experiments and Sect. 5 concludes the paper.

2   Related Work

The significant interest in automatic face profile view retrieval has been reported in literature. There are many studies concerning supporting of the Automatic Speech Recognition using visual signal in noisy environments. Kumar et al. proposed usage of the profile view for lip reading [1]. Their study involves usage of RGB camera standing on the side of the subject and the blue color background for easy foreground extraction using color thresholding. Authors of the mentioned paper use the dictionary consisting of 150 words and their system was speaker-dependent. The WER gain from using the profile view geometrical features was 39.6 % in environment contaminated by noise of SNR equal to -10 dB. More researches concerning profile view in field of AVSR can be found in the literature. Worth noticing are the works of Lucey and Potamianos [2], Pass et al. [3] and Navarathna et al. [4].

Dalka et al., present another approach to AVSR system where the visual features are based on Active Appereance Models (AAM) [5]. Geometrical features of lips image acquired by the RGB camera pointing the subject are proposed. The database used by authors in their work is described in detail in the paper by Kunka et al. [6]. The language corpus adopted in this database is based on the thorough studies of the natural English characteristics by Czyzewski et al. who prepared the language sample that reflects the vowel and consonant frequencies in natural speech for the purposes of training the AVSR systems [7]. The result of the study by Dalka et al. proves that the lip contour detection algorithm is reliable and accurate for visual speech recognition tasks.

Zhou and Bhanu proposed human recognition system based on face profile views in video [8]. Their approach uses high resolution face profile images constructed from low resolution videos from a side view camera. Authors used their own method for feature extraction known as curvature estimation and the classifier applied to match profile views is based on the dynamic time warping method. On the average, more than 70 % of persons were correctly recognized by usage of face profile.

A large study on profile view keypoints detection and tracking is presented by Pantic and Patras [9]. Authors studied the role of facial expressions that provide a number of social signals while people communicate to each other. They proposed a system for automatic facial actions recognition based on 15 keypoints found in the profile view image from RGB camera. The average recognition rate of the 27 facial actions tested in the work was equal to 86.6 %.

Major of studies to date are concerned on profile view retrieved by the RGB cameras. Therefore, authors of this work found it necessary to use the new type of data provided by depth imaging cameras in order to retrieve the facial profile view of the subjects.

3   Methods

In this paper the workflow depicted in Fig. 1 is used for retrieving the profile view of the user's face. After the frame acquisition, the preprocessing is needed to smooth the depth map of the user's face and to filter out the artifacts present in the image. Having the preprocessed frame, the nose and chin detection are performed in order to fit the face bisection line in the next stage. Knowing the line which passes through the center of the face, the profile view based on the depth data can be retrieved and the facial landmarks can be tracked. A more detailed description of all modules is presented in further subparagraphs.

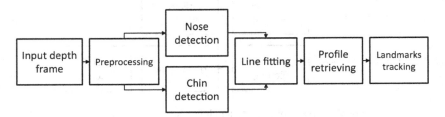

Fig. 1. Workflow for face profile retrieving method

3.1   Image Acquisition

In this work the Softkinetic DepthSense 325 Time-of-Flight camera was used. This camera enables imaging of the depth of the scene. Depth frames are acquired at 60 frames per second with resolution of 320 per 240 pixels, whereas data are transferred to PC using the USB interface. For the application of accurate face profile retrieving the acquired depth map is too noisy, hence, the preprocessing had to be applied, as is discussed in the following paragraph.

3.2 Preprocessing

The preprocessing is focused on denoising of the input data without degrading the useful signal. The preprocessing, which stages are depicted in Fig. 2, is split into two main subprocesses – the first one is a classical filtering using median and bilateral filters [10, 11], the second one is a so-called hole filling process [12, 13]. As it can be seen in Fig. 6. On the right, the depth signal is accompanied with large amount of noise. Therefore it must be subjected to an appropriate filtration. Two filtration methods were chosen – median and bilateral ones. The kernel of median filter used is equal to 5, so the examined neighborhood is of size 5×5 pixels with the purpose to remove the pepper noise from the signal. A bilateral filter is chosen to perform a more thorough filtration as bilateral filters are known to be able to denoise the signal preserving the information concerning the shape of photographed image edges. The weights of the bilateral filter are calculated as in Eq. 1 which is derived from the work of Barash [14]:

$$w\left(i,j,k,l\right) = e^{\left(-\frac{(i-k)^2+(j-l)^2}{2\sigma_d^2} - \frac{\|I(i,j) - I(k,l)\|^2}{2\sigma_r^2}\right)}, \tag{1}$$

where i, j are pixel locations being filtered, k, l are the neighboring pixels I denotes the pixel intensity and σ_d and σ_r represent smoothing parameters. In experiments described later on, the neighborhood size was set equal to 9×9 pixels and the sigma color σ_r was set to 20, whereas the sigma space σ_d was set to 8. The values of constants and neighborhood sizes in median and bilateral filtration methods were chosen empirically for optimal balance between noise removal and information preservation. Values had to be chosen with respect to moderately low depth image resolution of 320 per 240 pixels.

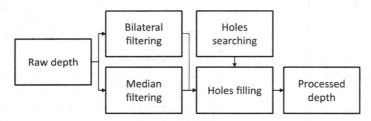

Fig. 2. Process of filtering and holes-filling of depth data

The hole filling (morphological processing) is a necessary stage as image data holes is a common issue of depth imaging cameras based on near-infrared light. It is mainly caused by the highly reflective surfaces like polished metals or the glass, hence, in our application it mainly occurs when the user wears glasses. Another source of these artifacts is fast movement, as the TOF cameras do not handle it very well. The depth holes phenomena is depicted in Fig. 3 as well as the result of holes filling process.

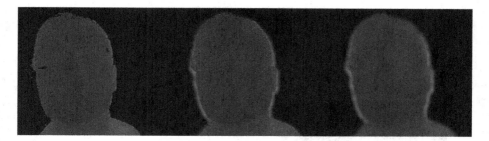

Fig. 3. Holes-filling process; left: raw depth frame, center: face with filled holes, right: filtered data used for hole-filling

The authors approach to the process of holes filling consists of following steps:

- Creating a binary mask of the user's silhouette
- Skimming through every pixel of depth map inside of the binary mask to find irrelevant pixels with values equal to 0 or 1 (holes can be seen in Fig. 3 on the left)
- Values of these pixels are replaced with the values of strongly filtered depth map which does not contain any hole (depicted in Fig. 3 on the right)

This process results in a depth image of the user without image holes (Fig. 3. in the center) with accurate depth data. Only the holes locations are filled with approximation values from the filtered depth data.

3.3   Keypoints Detection and Tracking

Keypoints are defined in this paper as points representing the tip of the nose and the chin. Those points are found in template matching process, using the artificially prepared models of the nose and the chin which are depicted in Fig. 4. In order to create those models, the patches of nose and chin regions were extracted from the depth images of the subjects. Furthermore, they were artificially modified in order to enhance the contrast between the close and far pixels using level based transform which lowers the intensity of dark pixels and increases the value of bright pixels. The process of localization is illustrated in Fig. 5. The template matching method utilizes the normalized cross correlation coefficient in order to find the similar part to template in the image. This method is described in more detail in work of Briechle and Hanebeck [15] and it adopts the following equation to calculate the values of the coefficient (Eq. 2):

$$R(x,y) = \frac{\sum_{x',y'}(T'\left(x',y'\right) * I'(x+x',y+y'))}{\sqrt{\sum_{x',y'}(T'(x',y')^2 * \sum_{x',y'} I'(x+x',y+y')^2}} \tag{2}$$

where I denotes the image, T – template, $x' = 0...w - 1$, $y' = 0..h - 1$, w is width and h is the height of the template patch used.

The template matching was performed twice on the same image frame I, firstly for nose model as template T secondly with chin model as a template T. The left part of

Fig. 4. Nose tip and chin models

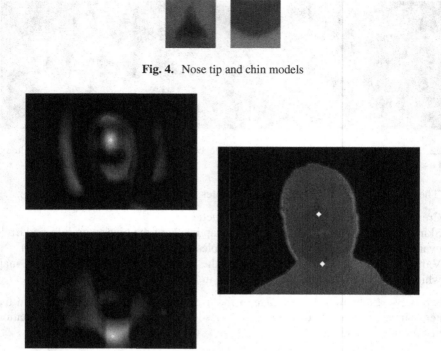

Fig. 5. Nose tip and chin detection results

Fig. 5 depicts the greyscale map showing where the most probable areas of the searched templates are present in the frame. The higher the intensity of the probability map, the larger the value R from Eq. 2 is. The right part of the Fig. 5 presents the result of template matching process with keypoints localized on nose tip and chin. Keypoints are detected in every frame independently, thus the smoothing of the rapid changes of their locations is needed and the Kalman filter [16, 17] was adopted for that purpose.

3.4 Line Fitting and Face Profile View Retrieving

Given the nose tip and chin coordinates, the equation of a line passing through those points can be found. The calculation as in Eq. 3 produces the bisection line of the user's face (which is depicted in Fig. 6):

$$y = \frac{y_2 - y_1}{x_2 - x_1}x + \left(y_2 - \frac{y_2 - y_1}{x_2 - x_1}x_2\right),$$

(3)

where x_1, y_1 and x_2, y_2 denote the coordinates of detected chin point and nose tip, respectively. Equation 2 and the locations of the keypoints enable the calculation of the line leading from the bottom to the top of image frame, through the keypoints. Thus, the whole face of the subject is taken into the profile retrieving process.

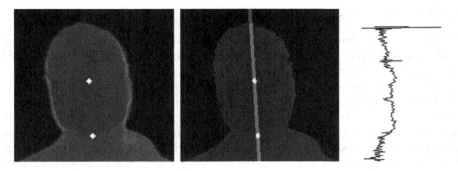

Fig. 6. Result of nose-chin line fitting and profile retrieving from depth data

Having the trajectory of this line and the depth data the profile view of the person can be reconstructed. The width of this line was set to 4 pixels so the final contour will be the mean value of 4 pixel columns from the center of user's face.

The profile view shown in Fig. 6 is a result of retrieving the profile view from the raw input of the TOF camera. This signal is corrupted by the noise, therefore it must undergo a further filtration and modification process in order to extract the characteristic features of considered face. This process is depicted in Fig. 7.

Fig. 7. Filtration of the profile view

The contour signal presented in Fig. 7 left is retrieved from raw depth data the TOF camera provides. The center contour in Fig. 7 is achieved from the depth after median filtration. The right contour is retrieved from median and bilateral filtered. Both median and bilateral filtration are described in a more detailed way in Sect. 3.2 of this paper. The contour shown on the right of Fig. 7, besides of being median and bilateral filtered in 2D domain, provides a subject of 1D median filtering with window size of 5 elements, chosen experimentally. Additionally, the following transformation [1] (Eq. 4) was used in order to emphasize the minima and maxima in the shape of face contour:

$$T = \sqrt{x' + y'} \tag{4}$$

where $x' = x/\max(x)$ and $y' = y/\max(y)$.

3.5   Landmarks Tracking in Profile View

Having the profile view of the subject's face extracted, it is possible to track the facial landmarks by analyzing its minima and maxima. Positions of 9 points in total are tracked i.e. eyebrow, eye, nosetip, nostril, upper and lower lip, corner of mouth, chin depression and chin. For landmarks searching the 1 dimension non maximum suppression method (1D-NMS) is employed. Details concerning the algorithm can be found in the work of Nuebeck and Gool [18]. This method is chosen in order to limit the multiple local minima and maxima which may occur in the close neighborhood while subject utters or changes his or her facial expression. The window size in which two minima or maxima cannot occur is set to 3 pixels. It is the smallest window possible in the 1D-NMS method, nonetheless, in our approach it allows to get stable and accurate landmark positions. The results of landmark searching and tracking are presented in Sect. 4.

4   Results

The image processing methods presented in this paper result in retrieving the filtered signal of subjects' faces contour. In Fig. 8 results of profile view retrieving method application for 4 persons is presented. It is clearly visible that every individual has its own characteristic contour of the face.

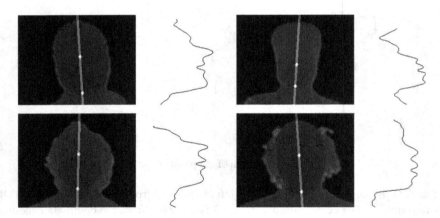

Fig. 8. Four different subjects with corresponding profile views

In Fig. 9 the result of facial landmarks tracking is presented. Thanks to properly filtered shape of the profile view it is possible to search for its minima and maxima positions which correspond with facial landmarks. It is visible that in the three image frames extracted from subject's utterance all 9 landmarks are tracked accurately during the speech. Knowing the landmark positions further parameters may extracted such as geometrical relationships (i.e. static parameters), individual and relative points accelerations (i.e. dynamic parameters), etc.

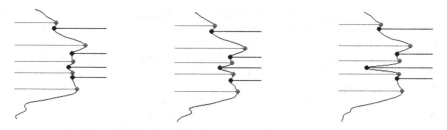

Fig. 9. Profile view facial landmarks tracking during speech

5 Conclusions

The method for profile view of the user's face using the depth data from Time of Flight camera was presented. The method results in filtered contour of the face seen from the profile. The contour signal is algorithmically processed in order to emphasize its distinctive features. The emphasis produces a signal that contains the discriminative features of individual's face. Such a signal may serve as an input to many useful applications, as: Lip-reading in AVSR systems, biometric systems for face recognition, automatic speech therapy systems or facial action recognition systems, to name a few examples.

Additionally, the method of the automatic tracking of facial landmarks locations was presented. Landmarks include 9 characteristic points in total, such as: eyebrows, nose tip, lips, chin depression, chin, etc. This functionality could result in programming API for the future researchers who could use it in the application of their choice which would employ landmarks' coordinates as an input.

Owing to the keypoints tracking of the nose tip and chin and Kalman filtration the proposed method can adapt to subject's movement in front of the device in use. In comparison with other methods that use 3D cameras, our method, thanks to chin and nose tip tracking does not require to have only the subject's face present in the scene as well as the face does not need to be the closest object to the camera. Proposed method is not resource intense and is able to work on-line, with the framerate of 25 fps (on Intel i7 2nd gen processor).

Acknowledgments. This work was supported by the grant No. PBS3/B3/0/2014 Project ID 246459 entitled "Multimodal biometric system for bank client identity verification" co-financed by the Polish National Centre for Research and Development.

References

1. Kumar, K., Chen, T., Stern, R.M.: Profile view lip reading. In: ICASSP (2007)
2. Lucey, P., Potamianos, G.: Lipreading using profile versus frontal views. In: 2006 IEEE 8th Workshop on Multimedia Signal Processing, pp. 24–28 (2006)
3. Pass, A., Zhang, J., Stewart, D.: An investigation into features for multi-view lipreading. In: 17th IEEE International Conference on Image Processing (ICIP 2010), pp. 2417–2420 (2010)

4. Navarathna, R., Dean, D., Sridharan, S., Fookes, C., Lucey, P.: Visual voice activity detection using frontal versus profile views. In: 2011 International Conference on Digital Image Computing Techniques and Applications (DICTA), pp. 134–139 (2011)
5. Dalka, P., Bratoszewski, P., Czyzewski, A.: Visual lip contour detection for the purpose of speech recognition. In: ICSES, Poland (2014)
6. Kunka, B., Kupryjanow, A., Dalka, P., Szczodrak, M., Szykulski, M., Czyzewski, A.: Multimodal english corpus for automatic speech recognition. In: Signal Processing Algorithms, Architectures, Arrangements and Application, Poland (2013)
7. Czyżewski, A., Kostek, B., Ciszewski, T., Majewicz, D.: Language material for english audiovisual speech recognition system development. J. Acoust. Soc. Am. 134(5), 4069 (2013). (abstr.) plus Proceedings of Meetings on Acoustics, No. 1, vol. 20, pp. 1 – 7, San Francisco, USA, 2.12.2013 – 6.12.2013
8. Zhou, X., Bhanu, B.: Human recognition based on face profiles in video. In: IEEE Computer Society Conference on Computer Vision and Pattern Recognition (2005)
9. Pantic, M., Patras, I.: Dynamics of facial expression: recognition of facial actions and their temporal segments from face profile image sequences. IEEE Trans. Syst. Man Cyber. 2(2) (2006)
10. Tomasi, C., Manduchi, R.: Bilateral filtering for gray and color images. In: IEEE International Conference on Computer Vision, Bombay, India (1998)
11. Nagao, M., Matsuyama, T.: Edge preserving smoothing. Comput. Graphics Image Proc. 9, 394–407 (1979)
12. Yang, N., Kim, Y., Park, R.: Depth hole filling using the depth distribution of neighboring regions of depth holes in the kinect sensor. In: ICSPCC, Honk Kong, pp. 658–661 (2012)
13. Kim, J., Piao, N., Kim, H., Park, R.: Depth hole filling for 3-d reconstruction using color and depth images. In: ISCE, South Korea (2014)
14. Barash, D.: Bilateral filtering and anisotropic diffusion: towards a unified viewpoint. In: Kerckhove, M. (ed.) Scale-Space 2001. LNCS, vol. 2106, pp. 273–280. Springer, Heidelberg (2001)
15. Briechle, K., Hanebeck, U.D.: Template matching using fast normalized cross correlation. In: Proceeding of the SPIE, Optical Pattern Recognition XII, vol. 4387(95) (2001)
16. Kalman, R.E.: A new approach to linear filtering and prediction problems. ASME J Basic Eng. 82, 35–45 (1960)
17. Szwoch, G., Dalka, P., Czyżewski, A.: Resolving conflicts in object tracking for automatic detection of events in video. Elektronika: konstrukcje, Technologie, zastosowania 52(1), 52–54 (2011). ISSN 0033-2089
18. Neubeck, A., Van Gool, L.: Efficient non-maximum suppression. In: 18th International Conference on Pattern Recognition (ICPR 2006), vol. 3, pp. 850–855 (2006)

Context-Based Semantic Tagging
of Multimedia Data

Nisha Pahal$^{(\boxtimes)}$, Santanu Chaudhury, and Brejesh Lall

Department of Electrical Engineering, Indian Institute of Technology Delhi,
New Delhi, India
nisha23june@gmail.com, {santanuc,brejesh}@ee.iitd.ac.in

Abstract. With the rapid growth of broadcast systems and ease of
accessing internet services, lots of information is available and accessible
on the web. The information available in multimedia documents may
have different *context* and *content*. Since, interpretation of multimedia
content cannot be free of context so tagging on the basis on context is
indispensable for dealing with this problem. Tagging plays an important
role in retrieving multimedia data as now-a-days most of the videos are
retrieved based on text describing them and not by the actual context
embodied in them. So, in this paper we have proposed a scheme for
tagging multimedia data based on the contents and context as identified from web-based resources. The hierarchical LDA (hLDA) is used to
model the context information while Correspondence-LDA (Corr-LDA)
is used to model the content information of multimedia data. Finally,
multimedia data is tagged with the relevant contents and context information on the basis of Context-Matching Algorithm. These tags can then
be used by search engines for increasing precision and recall of multimedia search results.

Keywords: Context · Tagging · hLda · SIFT · Corr-LDA

1 Introduction

Due to the availability of high speed digital cameras and smart phones with
high-capacity storage, the multimedia data specifically, images and videos have
gained importance in everybody's lives. Lots of new videos and images are being
uploaded over the internet every second. A major challenge is handling this
rapidly growing large volume of multimedia data on the web. This multimedia
data encodes information in context to any event and this information needs to
be tagged correspondingly. Another issue is that a lot of images and videos on the
web are incorrectly tagged. Any description or title tagged with that multimedia
data provides information about that data. This information is nothing but
meta-data i.e. data about data. The available multimedia search engines rely on
this meta-data to search for videos/ images. This meta-data information may be
too subjective and may lose its formal semantic meaning. So, it is required to

© Springer International Publishing Switzerland 2015
M. Kryszkiewicz et al. (Eds.): PReMI 2015, LNCS 9124, pp. 169–179, 2015.
DOI: 10.1007/978-3-319-19941-2_17

associate meaningful semantic tags with available multimedia data. This makes the multimedia retrieval process more efficient and accurate.

In this paper we have proposed a tagging framework that tags multimedia data with semantically relevant context and content information. Unlike other approaches, in this approach we have utilized our proposed tagging framework to associate a query video or an image with semantically relevant context and content information, exploiting multi-modal data - visual content of the frames and textual content available on the web. We have established the validity of our approach using experimental results. The remainder of the paper is structured as follows: Sect. 2 discusses the related work. In Sect. 3 the context-based tagging model is introduced. Section 4 contains the Problem Formulation. In this section, initially the problem is defined and then an approach for modeling the context information is discussed. Also, image content classification and merging of context and content information with multimedia data based on contextual meaning is presented. Section 5 contains experimental results and finally the conclusion and references are provided.

2   Related Work

In this section, we discuss some of the approaches that exist today for tagging the multimedia data. Siersdorfer et al. [7,8] presented the different tag propagation methods for automatically obtaining richer video annotations. Their approach provides the user with additional information about videos which leads to enhanced feature representations for applications like automatic data organization and search. Another approach provided by Nguyen [6] exploits available unstructured data and hidden topic models to infer surrounding contexts for better text and image retrieval. This helps in bridging the semantic gap in Web Mining and Information Retrieval. A description model of video metadata for semantic analysis taking into account various contextual factors was presented by Steinmetz et al. [1]. They considered the contextual information of video metadata for the purpose of Named Entity Recognition (NER) and for calculating a confidence value so as to bring metadata items in a specific order and to use them as context items for the disambiguation process. In another work by Giannakidou et al. [2] the authors provided an unsupervised model for efficient and scalable mining of multimedia social-related data. Their model jointly relies on social, semantic and content knowledge which can be used for tag clusters. And, the tag clusters produced can be used for semantics extraction and knowledge mining required for automated multimedia content analysis. Aytar [3], utilized the semantic word similarity measures for video retrieval. Their approach provided an effective way of using high level semantic relations in video retrieval problem by establishing a bridge between high level visual and semantic relations. Their method for modeling context was trivial as they didn't consider the spatial information due to the complexity of segmentation and localization tasks in TRECVID data sets. So, to bridge the semantic gap between low-level content-based and high-level context-based techniques, hybrid

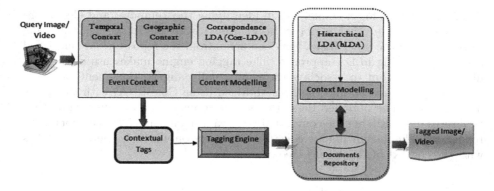

Fig. 1. Tagging framework

approaches have been proposed. A music retrieval system that combines content-based acoustic similarity and context-aware semantic descriptions was presented by Miotto et al. [4]. The system represents a music collection using a hidden Markov model with the purpose to build a music retrieval system that combines content-based acoustic similarity and context-aware semantic descriptions. Various approaches of content-based multimedia retrieval were discussed by Lew et al. [5]. The authors in this paper described various aspects of content-based multimedia information retrieval along with some major challenges for the future.

Although the existing work follows different approaches but none of them aspire to tag multimedia data with the relevant content and context information. Our multimedia tagging approach is different and novel from the work discussed above. It aggregates data from diverse web-based resources and associates a description of the multimedia data with the query video or an image on the basis of contextual meaning.

3   Context-Based Tagging Model

Context refers to some situations or conditions that make the meaning of the content of any web document more clear and specific. Context-based search for multimedia retrieval is an insightful technique which revolutionizes the efficiency of search engines. The multimedia documents with almost the same content may have totally different contexts and thus, may refer to different events. An event is always associated with several inherent contexts like temporal, geographical or entity based. Since context gets defined with respect to an event instance, therefore, identification of context information facilitates in identification of an event that is taking place in multimedia document. The tagging model is depicted in Fig. 1.

The *Event Context* refers to the event related information that is associated with the multimedia data. This includes the time and place related information of an event which for example help in identifying the Temporal and Geographical context. Some other tags can be identified from the image content using Corr-LDA. On the other hand, the documents from the repository are classified contextually using hLDA approach. The tagging engine makes use of context-matching algorithm to associate the contextually aligned documents and the tags (temporal, spatial and the ones obtained using Corr-LDA) with the query image/ video. The obtained tags are given as input to the tagging engine that facilitates in identifying the contextually relevant documents from the document repository. The documents are considered relevant only if their context matches with the context of the input video or an image.

4   Problem Formulation

Given an image or video, our objective is to annotate the video or an image with the relevant context and content information on the basis of semantic meaning. The contextual information is modeled using hierarchical LDA (hLDA) as explained in Sect. 4.1. For content classification, the Correspondence LDA (hereafter, Corr-LDA) is used to classify videos into different categories corresponding to different events as explained in Sect. 4.2. To associate query image or video with the relevant context and content information on the basis of contextual meaning, Context Matching Algorithm has been devised as described in Sect. 4.3.

4.1   Context Modeling

We have constructed a hierarchical event based context model for an image/ video. In this model, a set of web documents are organised in the form of a hierarchy. Different paths along the hierarchy refer to different sets of contexts. Thus, each path refers to some specific set of web documents. The contexts of the topics of higher levels are more generalised than the lower level ones. The proposed hierarchy consists of *Event-Type* at the lowest level. *Event-Topic* is the super class of the event-type. At the top level we have *Event-Context* class as shown in Fig. 2. For example, *cutting of birthday cake* is an event at the lowest level which belongs to Event-Topic *birthday* event. It further belongs to the Event-Context class *Family*. This hierarchical model is pre-constructed. However, we need to build such hierarchy automatically based on resources available. As an application we have focused on events in the public domain. Accordingly, we chose news websites for building the context models to associate with an image or video of news events. We have built a corpus of news documents classified and clustered at *Google News Website*. On the clustered set of documents we have applied hierarchical LDA (hLDA) [11] to discover a hierarchy of topic models. For example, we have created a repository of all news article pertaining to the domain *entertainment* and *automobile*. Using these documents we get a set of topics.

For obtaining this hierarchical context from text news, hierarchical topic modeling is used which is based on nested Chinese Restaurant Process (CRP) [9].

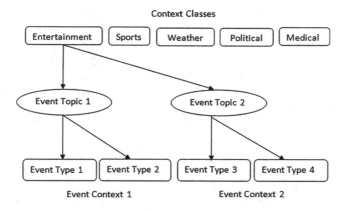

Fig. 2. Hierarchical context classification model

One of the foremost advantages of using hLDA is that it does not rely on a fixed tree structure for the representation of documents. As the documents are added to the model, the structure of the hierarchy keeps on changing to add new levels and topics. Once the hierarchy is obtained, each document in a repository corresponds to a distribution over the L distinct levels and is drawn by choosing an L-level path. The words are drawn from the L topics corresponding to the documents along that path. All the lower-level documents share the topic associated with the highest-level root document.

Assuming a data set consisting of a corpus of documents wherein each document comprises of a collection of words, and a word is considered to be an item in a vocabulary. The fundamental assumption is that the words in a document are generated according to a mixture model where the mixing proportions are random and document specific [9]. A multinomial variable z, and an associated set of distributions over words is represented as p(w | z, β), where β is a parameter. The topics, one distribution for each possible value of z, are the basic mixture components in this model. The document specific mixture distribution that temporally assumes K possible topics in the corpus where, z ranges over K possible values and θ is a K-dimensional vector is represented by:

$$p(w \mid \theta) = \sum_{i=1}^{K} \theta_i p(w \mid z = i, \beta_i) \tag{1}$$

where, $\text{Dir}(\alpha)$ is the Dirichlet distribution for parameter α and multinomial(θ) is the multinomial distribution for vector θ. The vector θ represents the document specific mixing proportions corresponding to the different components.

The test database consists of a set of 85 documents corresponding to automobile and entertainment category. From these we chose three events, namely *car launch event, celebrity event* and *music event*, and pre-processed them to remove stopwords. When we ran hLDA model on this data we obtained a hierarchy of topics (context) as shown in Fig. 3.

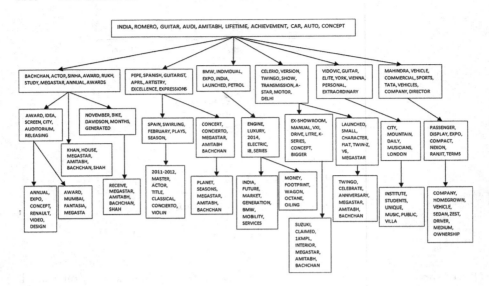

Fig. 3. Hierarchy of event context

4.2 Image Content Classification

An image is characterized by temporal and geographic context. Such temporal and geographic context links an image to image content. Hence, for tagging an image we need to formulate a strategy for identification of event context of an image. Image classification plays a key role in many applications such as video surveillance, video tagging, image retrieval and web content analysis. To classify the content information we have used Corr-LDA. First the SIFT features are extracted and then, a dictionary of visual words is obtained. This data is used for image content classification as explained in the next sub-section.

Space-Based SIFT Feature. First the SIFT feature points are obtained for each image. For each SIFT feature point, SIFT descriptors are constructed. The size of each SIFT feature vector is 128. These descriptors are quantized into visterms. In our experiments, we have trained Corr-LDA model using a dataset of 460 images and considered 35 as the total number of clusters. The K-means clustering is then applied and visual vocabulary is constructed which is represented by an NxM matrix, where each row corresponds to an image and each column represents a visterm. Unlike LDA model that allows documents to be represented as bag-of-words model in text analysis, the bag-of-visterms (BOV) model allows representation of an image as an orderless sequence of visual terms called visterms. The feature descriptors are matched with the vocabulary and a histogram is obtained for each image. Finally, the basic annotations of images are predicted based on image's content using Corr-LDA. We now present a brief introduction of Corr-LDA and how we have used it in our framework.

Table 1. Content classification

Event	Actual No. of Images/ Frames	Relevant Frames Tagged	Accuracy
Music Artist	588	476	.81
Celebrity	601	530	.88
Car Launching	594	495	.83

Correspondence LDA (Corr-LDA). The Corr-LDA model extracts conditional relationships between the set of image regions and set of words. In this, the features of an image are first obtained and then the corresponding words are generated. Let the size of dictionary be M. Assume that an image comprises of N finite regions. While annotating the images, first descriptors for various image regions are generated and then for each of the textual annotations, a region is selected. That is, for each of the M words, one of the regions is chosen from the image and subsequently a word is drawn conditioned on the topic that generates the selected region [10]. The Corr-LDA model was trained using the dataset of the type which one expects to find in entertainment/ automobile video. A total of 16 annotations were used to train the Corr-LDA model. Example of the annotations are guitar, car, person, stage, screen, sky, tree, trophy etc. The test images are then provided as input to predict the basic annotation based on image's content.

Tags: Car, Screen, Person, Stage   Tags: Person, Mike, Trophy, Screen

Tags: Person, Guitar

Fig. 4. Results of corr-lda

If $r=\{r_1, r_2, ...r_N\}$ denotes the set of image features, w denotes the set of associated words, $z=\{z_1, z_2, ...z_N\}$ is the set of latent variables, $y=\{y_1, y_2, ...y_M\}$ is the set of equiprobable indexing variables and θ is the Dirichlet random variable, then the joint probability distribution is given as follows:

$$p(r, w, \theta, z, y) = p(\theta \mid \alpha)(\prod_{n=1}^{N} p(z_n \mid \theta)p(r_n \mid z_n, \mu, \sigma)).(\prod_{m=1}^{M} (p(y_m) \mid N)p(w_m \mid y_m, z, \beta)) \quad (2)$$

Conditioned on N and M, Corr-LDA specifies the joint distribution on image regions, caption words and latent variables.

We have sampled videos into frames at the rate of 1 frame/ second to generate the images for our database. The annotations for the various video frames whose $p(annotation \mid image)$ is more than a threshold value (0.1 in our case) are selected. Some of the annotations as obtained using Corr-LDA and the classification accuracy are shown in Fig. 4 and Table 1.

4.3   Merging of Context and Content

In this section, the query image image or video is tagged with the relevant context and content information on the basis of their contextual meaning. To achieve this objective we assume V to be a set of I video frames or images $V=\{v_1, v_2, v_3, ..., v_I\}$, C to be a set of J contents $C = \{c_1, c_2, c_3, ..., c_J\}$ and S be a set of K contexts $S = \{c_1, c_2, c_3, ..., c_K\}$. The content information $c_J \in C$ along with the tags obtained using Corr-LDA and the temporal and geographical context are associated to an image. This information gets tagged with a query video or an image $v_i \in V$ if their context matches i.e. S_J is equal to S_I. The proposed *Context Matching Algorithm* for associating a query video or an image with context and content information is explained below:

Algorithm 1. Algorithmic Steps

---

1 Input: Query Video or Image (ζ_i)
2 Output: Tagged Video or Image (ζ_t), where (ζ_t) comprises of annotations as resulted from Corr-LDA (ϕ), temporal and geographical context i.e. the information about when and where the multimedia data was uploaded (ρ) and the relevant text news from the corpus of documents (σ).
3 **Description:**
   1: The database of text files Γ is created and hLDA is applied over it.
   2: The output of hLDA is parsed.
   3: Hierarchy of hLDA gives different context along different paths. Let the total no of paths be Π such that $\eta \in \Pi$.
   4: Apply the Depth First Traversal to traverse every path of the hierarchy.
   5: Map the annotations ϕ (derived from Corr-LDA corresponding to the image/video frames) and tags ρ (temporal and geographical context associated to an image/ video frames) i.e. ($\phi + \rho$) onto η.
   6: The path η is selected if the context of ζ_i is matched with the context of the path η.
   7: Find the set of documents $\sigma \in \Gamma$ which correspond to the above path η.
   8: Tag ζ_i with ζ_t.

Table 2. Tagging accuracy

Event	Relevant Videos Tagged	Actual No. of Videos	Accuracy
Music Artist	32	40	.80
Celebrity	34	40	.85
Car Launching	33	40	.82

The basic annotations of images (frames) as obtained after applying Corr-LDA along with the tags (date, location etc.) provided with the video are matched with the topics present in event-context hierarchy as obtained using hLDA. Based on the contextual meaning the content and context information gets tagged to query video or an image. For each context class there is a separate lexicon which contains the frequently used keywords. For example, in weather news, keywords like rain, storm, temperature are common. These contexts are nothing but the topics that we obtained in hLDA, so topic-document probability can be obtained and document based on score of their probability can be linked to a query video or an image. The information like the video broadcast time, source and news event date is tagged with the video, to refine tagging and search processes. The system makes a news video repository, which allows users to search news videos.

Thus, following tags can be associated with a video:

- The date and location of the video
- A list of annotations as obtained using Corr-LDA
- A text summary of the video

Further, key frames can be stored and used to return search results.

5 Experimental Results

Initially, news videos were downloaded from YouTube website and frames were extracted from them. We have trained the Corr-LDA using 460 images. The training images were manually annotated using a list of 16 annotations (like person, car, sky, guitar, grass, tree, screen, trophy etc.). The calculated histogram vectors corresponding to each image feature along with the manual annotations are given as an input for training of the Corr-LDA. Approximately, 60 frames per video were extracted (typical video duration was 1-2 minutes). The histogram vectors for the test images have been calculated similarly. These histogram vectors are given as test input to the Corr-LDA. Using the image-annotation probabilities for the test images, only those annotations are considered which have a significant probability i.e., have image-annotation probability greater than a threshold value. In our case, we have considered the threshold value as 0.1. The key frames and hence the videos are classified into three given categories *viz* music artist event, celebrity event and car launch event. The category into

Tags: February, 2014, New Delhi, Car, BMW, i8, Car Launch Event
BMW India presented the new generation of individual mobility by
unveiling the BMW i8, the all-new BMW X5 and the all-new BMW
M6 Gran Coupe at the 12th Auto Expo 2014...

Tags: January, 2014, Mumbai, Person, Trophy, Celebrity Event
Lifetime Achievement Award to Amitabh Bachchan, was presented
by Vidhu Vinod Chopra and Shatrughan Sinha...

Fig. 5. Car launch event february 2014   **Fig. 6.** Celebrity event january 2014

which most of the frames extracted from the video are classified is chosen as the video category. For extracting context for news videos, our system crawled GoogleNews website to retrieve text news related to the three categories mentioned above. Then we obtained a hierarchy of news context using hLDA on text news. The query video or image is tagged with the context and content information on the basis of contextual meaning using Context Matching Algorithm. The context and content information thus gets tagged to a query video or an image.

We have considered 40 videos per category for experimentation out of which in the paper 2 illustrations have been presented as depicted in Figs. 5 and 6. The tagging accuracy is shown in Table 2.

6   Conclusion

In this paper, we have proposed an approach for tagging multimedia data based on its context and content information. The query video or an image is tagged automatically without any manual intervention on the basis of contextual meaning. We have discussed an application for tagging multimedia data that utilizes context matching algorithm for associating a resultant video or an image with the appropriate content and context information.

References

1. Steinmetz, N., Sack, H.: Semantic multimedia information retrieval based on contextual descriptions. In: Philipp, P., Oscar, C., Valentina, P., Laura, H., Sebastian, R. (eds.) ESWC 2013. LNCS, pp. 382–396. Springer, Heidelberg (2013)
2. Giannakidou, E., Kompatsiaris, I., Vakali, A.: Semsoc: Semantic, social and content-based clustering in multimedia collaborative tagging systems. In: 2008 IEEE International Conference on Semantic Computing, pp. 128–135, IEEE (2008)

3. Aytar, Y.: Semantic video retrieval using high level context. In: Semantic Video Retrieval Using High Level Context (Doctoral dissertation, University of Central Florida Orlando, Florida) (2008)
4. Miotto, R., Orio, N.: A Probabilistic approach to merge context and content information for music retrieval. In: Proceedings of ISMIR, pp. 15–20 (2010)
5. Lew, M.S., Sebe, N., Djeraba, C., Jain, R.: Content-based multimedia information retrieval: State of the art and challenges. ACM Trans. Multimed. Comput. Commun. Appl. 2(1), 1–19 (2006)
6. Nguyen, C.T.: Bridging semantic gaps in information retrieval: context-based approaches. In: ACM VLDB, p. 10 (2010)
7. Siersdorfer, S., San Pedro, J., Sanderson, M.: Automatic video tagging using content redundancy. In: Proceedings of the 32nd International ACM SIGIR Conference on Research and Development in Information Retrieval, pp. 395–402, ACM (2009)
8. Pedro, J.S., Siersdorfer, S., Sanderson, M.: Content redundancy in YouTube and its application to video tagging. ACM Trans. Inf. Syst. (TOIS) 29(3), 47–58 (2011). Article No. 13
9. Blei, D.M., Griffiths, T L., Jordan, M.I., Tenenbaum, J.B.: Hierarchical topic models and the nested chinese restaurant process. In: NIPS, Vol. 16 (2004)
10. Blei, D. M., Jordan, M. I.: Modeling annotated data. In: Proceedings of the 26th Annual International ACM SIGIR Conference on Research and Development in Informaion Retrieval, pp. 127–134, ACM (2003)
11. Blei, D.M., Griffiths, T.L., Jordan, M.I.: The nested chinese restaurant process and bayesian nonparametric inference of topic hierarchies. J. ACM (JACM) 57(2), 7 (2010)

Image Tracking

Real-Time Distributed Multi-object Tracking in a PTZ Camera Network

Ayesha Choudhary[1]([✉]), Shubham Sharma[2], Indu Sreedevi[2],
and Santanu Chaudhury[3]

[1] School of Computer and System Sciences, Jawaharlal Nehru University,
New Delhi, India
ayeshac@mail.jnu.ac.in
[2] Department of Electronics and Communication Engineering,
Delhi Technological University, New Delhi, India
shubh2494@gmail.com, s.indu@rediffmail.com
[3] Department of Electrical Engineering, Indian Institute of Technology Delhi,
New Delhi, India
santanuc@ee.iitd.ac.in

Abstract. A visual surveillance system should have the ability to view an object of interest at a certain size so that important information related to that object can be collected and analyzed as the object moves in the area observed by multiple cameras. In this paper, we propose a novel framework for real-time, distributed, multi-object tracking in a PTZ camera network with this capability. In our framework, the user is provided a tool to mark an object of interest such that the object is tracked at a certain size as it moves in the view of various cameras across space and time. The pan, tilt and zoom capabilities of the PTZ cameras are leveraged upon to ensure that the object of interest remains within the predefined size range as it is seamlessly tracked in the PTZ camera network. In our distributed system, each camera tracks the objects in its view using particle filter tracking and multi-layered belief propagation is used for seamlessly tracking objects across cameras.

Keywords: Distributed multi-camera tracking · Real-time tracking · PTZ camera network · Collaborative multi-object tracking · Belief propagation

1 Introduction

A real-time video surveillance system consisting of a PTZ (pan, tilt, zoom) camera network requires seamless tracking of multiple objects in the scene. Moreover, particular objects of interest, such as suspects, may be required to be tracked at a certain dimension in each frame so that important information related to that object is continuously retained. In general, it is possible that the object of interest can become so small that a lot of information about the object is lost. On the other hand, the object of interest can come so close to a camera that the

© Springer International Publishing Switzerland 2015
M. Kryszkiewicz et al. (Eds.): PReMI 2015, LNCS 9124, pp. 183–192, 2015.
DOI: 10.1007/978-3-319-19941-2_18

object becomes too large and blocks the view of the camera, hiding important information. In this paper, we propose a novel framework for real-time, distributed multi-object tracking in PTZ camera network that also addresses the situation mentioned above. The pan/tilt capability of the cameras along with camera handoff ensures seamless tracking of the objects in the scene. We leverage on the zoom capability of the cameras to ensure that the objects of interest are tracked at a certain size as the objects move across space and time in the camera network. The main contributions of our framework are: (a) multiple objects are seamlessly tracked across space and time in the camera network; (b) the user is provided with a tool to mark an object of interest, such that, the object of interest can be seamlessly tracked at a certain predefined size throughout the area under observation; (c) the user is notified when the object of interest leaves the area under observation.

Distributed PTZ camera networks are well-suited for wide area surveillance [2]. However, such a system is complex because network topology changes as cameras pan, tilt or zoom to seamlessly track the objects. We assume a distributed system with an underlying communication network such that each camera can communicate with every other camera either directly or indirectly. We assume that the camera network is calibrated and each camera has the list of its network neighbors. We define the network neighbors of a camera as those cameras that have overlapping or contiguous views in some pan/tilt/zoom position of the camera. When a camera receives a message from any of its network neighbors, it takes the decision to pan/tilt/zoom so that the object can be seamlessly tracked at the required size. Data fusion between cameras viewing a common region and across cameras needs to be addressed to enable seamless tracking. We apply belief propagation at multiple levels for data fusion.

In our framework, we assume that there are priority areas, that are pre-specified. These priority areas also include the entry and exit locations in the area under observation. Placement of cameras in this case plays an important role. We apply the optimal placement algorithm [6] to place the cameras in such a manner that the priority areas are observed at all times. Since the cameras that view the entry/exit areas are static for a certain time period, these cameras apply background subtraction [12] to detect objects that enter the area under observation. Based on the detected object, we initialize the particle filter tracker [16] in these cameras. The camera then communicates the particle filter estimates of the detected object to all its network neighbors. The system ensures that as the object moves in the area under observation, it is continuously tracked at all times by at least one camera. In the next section we discuss the related work.

2 Related Work

In recent times, research on multi-camera tracking in camera networks consisting of static cameras as well as PTZ camera networks has been gaining importance [1,3]. More recently, research on active camera systems using distributed processing is gaining importance since they are better suited for wide area surveillance [4]. Various distributed computer vision algorithms are discussed in [10,13–15]. A system consisting of static and PTZ cameras was proposed in [8] for surveillance of a parking lot. It is a hierarchical framework and uses the active camera for tracking a suspicious object at higher resolution. Authors in [4] apply distributed optimization in the game theoretic framework for controlling PTZ cameras in a wide area distributed camera network. The aim is to optimize solutions for various dynamic scene analysis problems. Moreover, the cameras collaborate among themselves to ensure that all objects are seamlessly tracked. The concept of multi-player learning in games have also been used in [11], for distributed collaboration among neighboring cameras viewing a common target for multi-object tracking in a PTZ camera network. In comparison to these systems, our framework consists of only PTZ cameras and each camera zooms in or out as required to track the pre-specified target at a certain resolution. Our framework provides a user interaction layer, to enable the user to mark objects of interest as they enter into the scene. The user can also specify the size at which the objects of interest should be tracked. Moreover, we use particle filter based tracking in each camera independently and use its parameters in multi-layered belief propagation for collaborative tracking of multiple objects in the area under observation. Authors in [9], proposed a method for controlling PTZ cameras to obtain high resolution face images of targets at opportune points in time for each camera in a distributed PTZ camera network. Our work is essentially different from this as it tracks the whole body of the targets of interest at a pre-specified size requiring the camera to zoom in or zoom out while the object is in its view. Moreover, if the camera is tracking more than one target of interest, it collaborates with the neighboring cameras to ensure that at least one camera is tracking the object at the required size.

3   Particle Filter Based Tracking Framework

Particle filter is a Monte Carlo method that is simple, yet capable of approximating complex models. Let the total number of particles be N and the total number of components be M. Then, $X_t = x_t^{(i)}{}_{i=1}^N$ be the particles and the particle weights are $W_t = w_t^{(i)}{}_{i=1}^N$. Then, the mixture filtering distribution is of the form given in [16],

$$p(x_t|z^t) = \sum_{k=1}^{M} \pi_{k,t} \sum_{i \in I_k} w_t^{(i)} \delta_{X_t^{(i)}}(x_t) \tag{1}$$

where, $\delta_b(.)$ is the Dirac delta function with mass at b and I_k is the set of indices of the particles belonging to k^{th} mixture component. These particles are updated sequentially, and the new weights are recalculated at each step. The new particle set P^t has to be computed in such a manner that it is a sample set from $p(x_t|z^t)$ given that the particle set P^{t-1} is a sample set from $p(x_t|z^{t-1})$. Each component evolves independently in the tracking module and therefore, the particle representation of each mixture component also evolves independently.

3.1 Measurement Module

In our framework, all entry locations are priority areas and therefore, continuously observed by at least one camera, that is static for a certain time period. As an object enters the area under observation, it is detected using background subtraction [12] and represented by its bounding box. The reference color model of the object is created at the time it is first detected in the manner discussed below.

Let $B_i = \{x_i, y_i, w_i, h_i\}$ denote the bounding box of the object of interest, where, (x_i, y_i) is the center of the bounding box and (w_i, h_i) denote the width and height of the bounding box respectively. Similar to [5], we consider the Hue-Saturation-Value (HSV) color histogram of the bounding box to represent the measurement model that is robust with respect to illumination changes. The HSV histogram consists of N bins where $b_t(p) \in \{1, 2, \ldots, N\}$ is the bin index at the color value $y_t(p)$ at pixel location p in frame t. The HSV histogram is formulated for the pixels inside the bounding box and the kernel density estimate is $H(x_t) \triangleq \{h(n; x_t)\}$, $n = 1, 2, \ldots, N$ of the color distribution at time t is given by

$$h(n; x_t) = \alpha \sum_{d \in B_i} \delta[b_t(p) - n] \tag{2}$$

For tracking, in each frame the color model of the previous frame is treated as the reference color model, H^*, to overcome the variations in the background as the objects moves in the scene. Similar to [5], the distance between the reference color model and the color model of the current frame is calculated using the Bhattacharya distance given by Eq. 3:

$$d(H^*, H(x_t)) = \left[1 - \sum_{n=1}^{N} \sqrt{h * (n; x(t-1)) h(n; x_t)} \right]^{\frac{1}{2}} \tag{3}$$

The likelihood distribution that is required for particle filter tracking is obtained using the distance between the current and previous HSV color histograms given by Eq. 4:

$$p(z_t | x_t) = \gamma e^{-\beta d^2 (H^*, H(x_t))} \tag{4}$$

where, γ and β are normalizing constants.

3.2 Single Camera Tracking

In our distributed framework, each camera tracks the objects in its view based on its own image measurements. These measurements are measured as described in Sect. 3.1 and used to initialize the particle filter tracker. Let $x_{o:t} = \{x_0, x_1, \ldots, x_t\}$ be the state vector and $z_{0:t}$ be the observation vectors up to time t. Then, for a particular object O_i, the posterior probability distribution is given by Eq. 5

$$p_i(x_t|z_{0:t}) = \frac{p(z_t|x_t)p(x_t|z_{0:t-1}))}{p(z_t|z_{0:t-1})}$$

$$= p(z_t|x_t)\int p(x_t|x_{t-1})p(x_{t-1}|y_{0:t-1})dx_{t-1} \qquad (5)$$

Given that there are M objects in a camera's view, then the posterior distribution $p(x_t|z_{0:t})$ is modeled as an M-component non-parametric mixture model given by Eq. 6

$$p(x_t|z_{0:t}) = \sum_{i=1}^{M} \pi_{i,t}p_i(x_t|z_{0:t}) \qquad (6)$$

where, the weights $\pi_{i,t}$ are such that $\sum_{j=1}^{M} \pi_{j,t} = 1\forall t$. As can be easily seen from Eq. 5,

$$\pi_{i,t} = \frac{\pi_{i,t-1}\int p_i(z_t|x_t)p_i(x_t|z_{0:t-1})dx_t}{\sum_{j=1}^{M} \pi_{j,t-1}\int p_j(z_t|x_t)p_j(x_t|z_{0:t-1})dx_t} \qquad (7)$$

There are M different likelihood distributions $p_k(x_t|z_{0:t}), k = 1, 2, \ldots, M$, one for each object in the cameras view. Usually, in multi-camera tracking, it is assumed that all the M objects are being viewed by all the cameras, however, in our framework, this does not hold true. Since the camera network is spread in the area under observation, it is not necessary that even two cameras will be viewing the same area. Therefore, each camera computes these likelihood distributions for the objects that are in that camera's view. The 3D position of the object gives the identity of the object since that is a unique feature for each object. Only one object can be in a 3D position at a time. Moreover, each camera will have its own uncertainty in measurement, so the 3D position is taken to be same if it is within a predefined threshold.

4   Collaboration for Multi-camera Tracking

In this section, we assume that each camera is capable of tracking multiple objects in its view. The same object may or may not be tracked by more than one camera simultaneously. However, since the camera network is calibrated, the 3D position of the object can always be calculated. Since each object is represented by its bounding box and its center (x, y), each camera computes whether an object will get out of its view or not. In general, a camera pans and/or tilts to keep the object in its view, however, there are limitations on the maximum pan/tilt that a camera can perform. Therefore, when an object is about to get out of the view of the camera, that camera sends a message to all its neighboring cameras about this object. The message contains all the information, such as, current 3D position, size of the bounding box, the track till the current point in time, the probability estimates till that time as well as the predicted 3D position of that object.

A camera is a neighbor of another camera if an object can get out of one camera's view and get into the other camera's view, or if both the cameras have overlapping views in some pan/tilt position of both the cameras. When a camera receives a message from its neighbor, it checks whether it is tracking the same object or not. If it is tracking the same object, it continues to do so. If the message is about a new object, the camera checks whether the object has entered its view or not. To check the identity of the object, we use the 3D location of the center of the bounding box. The camera first checks whether it is already tracking that object. It computes the distance between the 3D position received in the message with the 3D position of the objects that it is currently tracking. If this distance is within a threshold with one of the objects in the camera's view, it identifies that the message received is for an object that is currently being tracked and continues to track that object. In case, the distance is not within the threshold, it checks whether a new object has entered its view. To do so, the camera that receives the message, computes the image coordinates of the 3D position of the object, and considers a bounding box around that image position. It then forms an HSV color histogram of the pixels in that bounding box and compares the distance between the histogram received in the message with the computed histogram. If the distance is within a predefined threshold, it assumes that the object has been identified. It continues to track the object using the information present in the message.

Belief propagation is used to compute the probabilities in the new view, based on the probabilities received from the camera that was previously tracking it. Let C_k be the camera that has received messages about an object from multiple cameras, $j = 1, 2, \ldots, r$, where r could be 1 or more than 1. Then, each C_j in its message also sends the predicted value of the object, that is, $x_{t,j}$, $j = 1, 2, \ldots, r$. Let the target state in C_k be $x_{t,k}$ and its state in each C_j be $x_{t,j}$, $j = 1, 2, \ldots, r$.

Let $z_{t,j}, j = 1, 2, \ldots, r$ denote the observation in C_j at time t. Then, $Z_t = \{z_{t,1}, \ldots, z_{t,r}\}$ be the multi-camera observation at time t. This implies that $Z^t = \{Z^1, Z^2, \ldots, Z^r\}$ are the multi-camera observations till time t. Then, the message from camera C_j to C_k is

$$m_{kj}(x_{t,k}) \leftarrow p_j(z_{t,j}|x_{t,j})\psi_{k,j}^t(x_{t,k}, x_{t,j})$$
$$\times \int p(x_{t,j}|x_{t-1,j})p(x_{t-1,j}|Z^{t-1})dx_{t-1,j}dx_{t,j} \tag{8}$$

Then, the belief is computed by,

$$p(x_{t,k}|Z^t) \propto \Pi_{j=1,\ldots,r}m_{kj}(x_{t,k})$$
$$\times \int p(x_{t,k}|x_{t-1,k})p(x_{t-1,k}|Z^{t-1})dx_{t-1,k} \tag{9}$$

where, $p(x_{t,k}|x_{t-1,k})$ is computed as discussed above and $p(x_{t-1,k}|Z^{t-1})$ is set to 1, since the object was not in this camera during that time period.

Therefore, even if the camera has not seen the object of interest before it will get tracked using the history from its neighboring cameras.

5   Zooming into an Object

In our framework, we give the user the capability to observe any person such that the size of the object in any camera's view is within a predefined range, as he/she moves across the camera network. The user can mark the object of interest when he/she enters the area under observation. Then, as the object moves across the camera network, along with the message that each camera sends to its neighbors, a tag is also sent and the range of the size of the tracked object.

This ensures that each camera can change its pan/tilt and zoom parameters to continuously track the object such that the size of that object in that camera's view is within the predefined range. It is not necessary that the object remains in the center of the image, therefore, the camera does not need to pan/tilt continuously. Instead, the camera needs to change its pan/tilt and zoom to be able to track the object at the required size. By size of the object, we imply the size of the bounding box. In many cases, if the camera zooms without bringing the object to its center, then it may lose the object from its zoomed view. Therefore, before zooming, the camera pans and/or tilts to bring the object to the center of its view and then zooms into be able to view the person at the predefined size. The camera only pans and/or tilts next when the object is about to get out of its view or if the size of the object goes out of the desired range.

Suppose that object O_i is currently in view of camera C_j and about to get in the view of its neighbor, C_k. Then, C_k also receives the predicted $3D$ position of O_i. Once C_k checks that O_i is within its view, it checks on the size of the object. If the size is outside the required range, then first the camera C_k pans/tilts to bring the object into its center and then zooms. This is to ensure that the object is not lost from the camera's view after zooming. Panning by angle α is rotation about the Y-axis by α and tilting by angle β is rotation around the X-axis by β. Let (X_i, Y_i, Z_i) be the position of O_i in C_k. Then, as discussed in [7],

$$\alpha = -\arctan \frac{X_i}{Z_i} \tag{10}$$

and,

$$\beta = \arctan \frac{Y_i}{Z_i \cos \alpha - X_i \sin \alpha} \tag{11}$$

After the target object is centered, the camera zooms in by $\delta = f'_k - f_k$ where, f'_k is the focal length of C_k after zoom in.

Then, the zoom f'_k is computed as

$$f'_k = \frac{aH_i}{H} Z_i \tag{12}$$

where, a is the ratio of the current height h_i to the desired height H_i, $\frac{h_i}{H_i} \leq a \leq 1$ and H is taken to be the average human height. And, tracking is resumed after adjusting the size of the target. Since this does not take too much time, the tracking is smooth despite the transition.

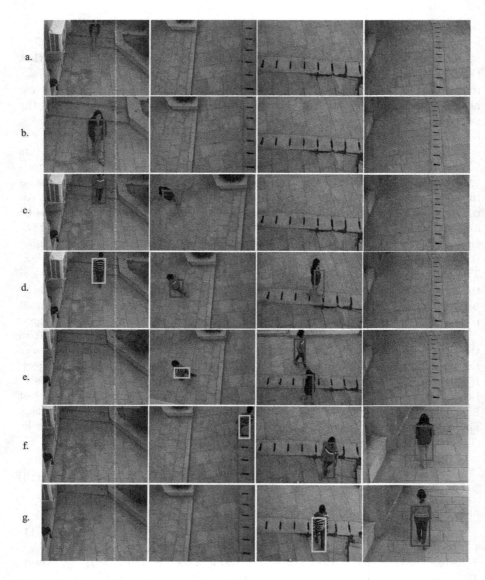

Fig. 1. Each row shows the state of the four cameras C1, C2, C3, C4, respectively at different time stamps. Two objects of interest O1 (red) and O2 (blue) are in the scene. In(a), O1 enters at C1 and the user marks it as an object of interest. (b) C1 zooms to bring O1 to the predefined size. (c) O2 enters, since C1 is about to zoom, it will lose O1. C1 communicates to its neighbor C2 and C2 pans to bring O1 in its view. (d) O3 enters and O1 and O2 are tracked by C3 and C2 respectively. (e) O1 and O2 are in same cameras but moving in different directions. Therefore, in (f) C4 pans and zooms to track O1, while C3 continues to track O2. (g) O1 has exited the scene, that is informed to the user, O2 is also an object of interest and therefore, tracked at the zoom level (Color figure online).

6 Experimental Results

We perform various experiments using four PTZ SONY EVI D70 cameras, $C1$, $C2$, $C3$ and $C4$. In the scene, camera $C1$ views the entrance and camera $C2$ views the exit. These are the two priority areas. We show one of the scenarios of our experimentation that covers the all aspects of our framework. In Fig. 1, the user marks two objects of interest $O1$ (red) and $O2$ (blue) when they enter the view of camera $C1$. In both cases, the camera zooms to track the objects at the predefined size. In Fig. 1(e), both the objects of interest are in the view of the same camera but moving in different directions. Since both $O1$ and $O2$ need to be tracked at all times, $C3$ sends a message about $O1$ to camera $C4$ and then, $C4$ pans, tilts and zooms to continue tracking the $O1$ at the required size.

7 Conclusion

In this paper, we have proposed a novel framework for real-time, distributed, multi-object tracking in a PTZ camera network. In our framework, the user is given the ability to mark an object of interest to track it across cameras such that the size of the object remains within a pre-specified range. If the size of the object reduces or increases beyond this range, the camera zooms in or out, as required, to bring the object's size within the range. We have used particle filter based tracking for tracking objects in each camera and multi-layered belief propagation for seamlessly tracking objects across cameras. The pan, tilt and zoom capabilities of each camera are used whenever required for seamlessly tracking all the objects in the scene.

References

1. Black, J., Ellis, T.: Multi-camera image tracking. Image Vis. Comput. **24**(11), 1256–1267 (2006)
2. Choudhary, A., Sharma, G., Chaudhury, S., Banerjee, S.: Distributed calirbration of a pan-tilt camera network using multi-layered belief proapagation. In: Proceedings of IEEE Workshop on Camera Networks in conjunction with CVPR (2010)
3. Collins, R., Lipton, A., Fujiyoshi, H., Kanade, T.: Algorithms for cooperative multisensor surveillance. Proc. IEEE **89**(10), 1456–1477 (2001)
4. Ding, C., Song, B., Morye, A., Farrell, J., Roy-Chowdhury, A.: Collaborative sensing in a distributed PTZ camera network. IEEE Trans. Image Process. **21**(7), 3282–3295 (2012)
5. Hue, P.P.C., Gangnet, J.V.M.: Color-based probabilistic tracking. In: Heyden, A., Sparr, G., Nielsen, M., Johansen, P. (eds.) ECCV 2002. LNCS, vol. 2350, pp. 661–675. Springer, Heidelberg (2002)
6. Indu, S., Chaudhary, S., Mittal, N., Bhattacharya, A.: Optimal sensor placement for surveillance of large spaces. In: Indian Conference on Vision, Graphics and Image Processing (ICVGIP) (2008)
7. Lu, Y., Payandeh, S.: Cooperative hybrid multi-camera tracking for people surveillance. Can. J. Electr. Comput. Eng. **33**(3/4), 145–152 (2008)

8. Micheloni, C., Foresti, G.L., Snidaro, L.: A network of cooperative cameras for visual surveillance. Vis. Image Signal Process. **15**(2), 205–212 (2005)
9. Morye, A., Ding, C., Roy-Chowdhury, A., Farrell, J.: Distributed constrained optimization for bayesian opportunistic visual sensing. IEEE Trans. Control Syst. Technol. **22**(6), 2302–2318 (2014)
10. Song, B., Ding, C., Kamal, A.T., Farrell, J.A., Roy-Chowdhury, A.K.: Distributed camera networks. Signal Process. Mag. **28**(3), 20–31 (2011)
11. Soto, C., Song, B., Roy-Chowdhury, A.K.: Distributed multi-target tracking in a self-configuring camera network. In: Proceedings of IEEE Conference on Computer Vision and Pattern Recognition (CVPR) (2009)
12. Stauffer, C., Grimson, W.: Adaptive background mixture models for real-time tracking. In: IEEE International Conference on Computer Vision and Image Proccessing (1999)
13. Taj, M., Cavallaro, A.: Distributed and decentralized multicamera tracking. Signal Process. Mag. **28**(3), 46–58 (2011)
14. Tron, R., Vidal, R.: Distributed computer vision algorithms. Signal Process. Mag. **28**(3), 32–45 (2011)
15. Tron, R., Vidal, R.: Distributed computer vision algorithms through distributed averaging. In: Proceedings of IEEE Conference on CVPR, pp. 57–63 (2011)
16. Vermaak, J., Perez, A.D.P.: Maintaining multi-modality through mixture tracking. In: International Conference on Computer Vision (2003)

Improved Simulation of Holography Based on Stereoscopy and Face Tracking

Łukasz Dąbała[(✉)] and Przemysław Rokita

Warsaw University of Technology, Nowowiejska 15/19, 00-655 Warsaw, Poland
L.Dabala@mion.elka.pw.edu.pl, pro@ii.pw.edu.pl

Abstract. To meet the requirements of the market, people are improving communication with the virtual reality systems. We propose a method for simulating holography, where person can see the object from different points of view. To achieve such effect we used stereoscopy in combination with face tracking, what enables us to manipulate content on the screen. Despite heavy computation load we were able to maintain interactivity of the whole system.

1 Introduction

People always wanted to recreate real world in virtual environment. Such projection gives the possibility to manipulate elements from the surrounding without modifying reality. There are many ways to move the content in the virtual environment, but the simple manipulation of everything on screen with keyboard and mouse is not enough.

Solution for this are virtual reality systems. Each of them must consist of at least three elements. The first one is computer, heart of the system, which is responsible for every calculation. Secondly, the hardware for communicating with the user. This can be handled in many ways, for example with mouse and keyboard, camera, microphone. The third element is equipment, that presents user the result of his/her manipulation. There are many possibilities for showing results, just as it is for taking the input for the system, the simplest one is screen of the monitor. In this work we would like to create such system in combination with stereoscopy, as the enrichment of the experience for the user.

Stereoscopy will give the possibility to recreate depth, so user will be more involved into the virtual world. Stereo vision by itself, cannot be enough to immerse person in computer generated content, so we would like to simulate the holography. For this reason, the part of our system is a simple camera device, that will be used for tracking position of the user. That will help with involving user more into the manipulation and giving the impression, that everything can be seen from different points of view.

2 Background and Previous Work

In this section, there will be presented background for stereo perception as well as an overview of the previous solutions for view-dependent rendering.

© Springer International Publishing Switzerland 2015
M. Kryszkiewicz et al. (Eds.): PReMI 2015, LNCS 9124, pp. 193–201, 2015.
DOI: 10.1007/978-3-319-19941-2_19

2.1 Stereopsis

One of the most complicated systems in human organism is visual system. To achieve good perception it combines information from many different cues (for example occlusion or perspective). The most influential one is stereopsis [6,9]. It has a really strong influence on viewing experience. Human get information about the depth and 3d structure of objects from binocular vision. Because of the position of eyes, binocular vision results in two slightly different images, which are projected to the retinas of the eyes. Usually the difference between images is only in horizontal direction and it is called horizontal disparity. It can happen, that vertical disparities are also present (for example during watching content with reflections or refractions [11]). Excessive disparities can lead to visual discomfort, what results in poor user experience.

2.2 View-Dependent Systems

Previously there were some attempts of creating view-dependent systems. In [10] they tracked user position by using two cameras. The result of the tracking was used as an input for rendering. The system worked in real time, but it did not consider steroscopy. Interesting work has been showed in [5]. In developed system they used stereoscopic rendering and simple camera for face tracking. They focused on reducing distortions and saving as correct 3d perspective as possible. It was done by changing frustum from symmetric to assymetric. In the second work, they considered perception element. But in the end, they considered only movement in horizontal and vertical direction.

There were also some systems that used depth cameras. In one of them [3] authors used depth sensor from Microsoft KINECT to perform the head tracking. It was quite novel approach to this problem, because device they used solved most of the problems with detection of the head's orientation. The output of their algorithm can be projected on a flat surface, which then can be explored by a user. In another approach presented in [13] they improved depth perception by tracking position of the user and creating motion parallax for the rendered image. Their solution was confirmed by creating some interesting applications, that used pseudo 3d effects.

If it comes to the virtual reality systems, the most advanced is probably the one presented in [7]. They proposed a solution for changing a whole room into the augmented reality environment. By using the set of cameras and projectors, they cover whole room with the output pixels, and track user movement. Thanks to the tracking they are able to dynamically adapt the content of the room. They showed new way of creating and presenting augmented reality and in the same way, they created new area that can be explored.

3   Our Method

We are developing further the method presented in [2], so in this section we present short review of the previous method and what has been added to make the solution better.

3.1   Camera Arrangement

In our setup a camera was placed above the monitor, so like in every laptop. The device we used, was really simple, without any depth sensors. During using of our system, user can move his head freely, without any move restrictions.

3.2   Rendering

To simplify rendering part we used deferred shading. Because some objects (reflecting or refracting) can cause more problems [6, Ch. 12] than benefits, we considered only diffuse objects. In this case only thing, that we should be aware of accommodation and vergence conflict. What is connected also to diffuse rendering is limiting too large disparities, because otherwise it will be impossible to see the image in 3d. In our rendering we used image based lighting [4] in addition to the pre-computed ambient occlusion [14].

3.3   Face Tracking

Every view-dependent system uses some kind of tracking. Information, that comes from tracking movements, gives the possibility to manipulate the environment according to the gestures or position of the user. This is what we are aiming at, tracking technique that we used gives us the potentiality to simulate holography. Imitating the effect consists of showing user content on the screen from different perspectives.

The most common technique for showing 3d content is using glasses of different types. Such method has its drawbacks. Because of the glasses and the fact, that they are 'coloured', it is really hard to find position of the eyes and use them for tracking. Sometimes it is nearly impossible to find regions of interest, but to not handle such problem, we remained with face tracking. That gives us more robust solution, because it allows user for wearing glasses also during detection step.

For face detection we used Viola-Johns framework [12]. Because of the way it is working, lighting conditions should be really good. There should not be any over- and unsaturated regions. In the framework all features for detection are represented by combination of sequences of rectangles. Every such region consist of two kind of pixels: white and shaded ones. The value for feature is the difference of the sums of pixels that are inside white regions and shaded ones. By using a method called integral image, it is possible to calculate such feature in constant time, by only four table reads. Integral image can be precalculated by summing values in the grid, so the value in point (x, y) is a sum of all values that are above and to the left of (x, y). For selecting features and training classifiers AdaBoost was used. We used cascade classifiers, because every single part works only on the date from the previous one, so it rejects wrong solution very quickly.

After detection step, we need to perform tracking step. For this we used a method based on motion, called Lucas-Kanade optical flow [8]. The method assumes constant neighbourhood of the pixel and solves equation for it for the flow using least squares criterion.

3.4   View-Dependent Part

The origin of the camera in the scene is changed, to show objects from different perspecitve, so the user will have the impression of the holography. The scene is represented by a virtual sphere, where all objects are in inside the sphere and camera is moving on the surface of it (Fig. 1).

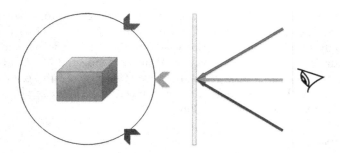

Fig. 1. Scene representation in our algorithm. Different colors of arrows are representing different positions of the viewer, that is watching the screen. Chevrons are representing corresponding camera positions in the scene.

Our alogrithm consists of three paths, like it is shown in Fig. 2. Green one is responsible for calculating movement in XY direction. The yellow path regulates the size of the radius of the virtual sphere. Thanks to this, we are able to make objects bigger or smaller, depending how far from the camera the user is. The last red part takes into account rotations of the user's head. That gives us the possibility to manipulate stereo matrices to add vertical disparity, in amount that depends on the angle.

Movement of the user in vertical or horizontal direction is simple to handle. All information, that is needed to change position of the camera comes directly from the optical flow. We calculate sum of differences in X and Y direction of the tracked points from current frame and previous one. Next, we directly use that values to change the position of the camera, and make it look at the center of the virtual sphere. To remove flickering from the presented content, small movements of the user are just ignored.

What can be done further is changing distance to the observed object. This can be done, by expanding or reducing the length of the radius of virtual sphere. Here, the optical flow, does not give enough information, because it is limited to vertical and horizontal direction. To do this, there is a need to calculate differences between consecutive frames. The problem that arises here is quality of input images, which depends on the video frames coming from the camera. To limit the error, that can happen during this step, consecutive frames are subtracted from each other. For the further calculation we have textures with differences between current frame and the previous one, and between previous one the penultimate. Too small values in such textures are filtered out. Next, the

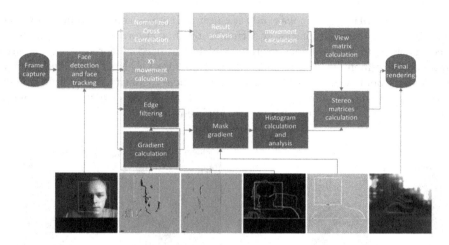

Fig. 2. Flow representing our algorithm.

cross correlation is done between these. The result of this step is an image with disparities in X direction, because vertical coefficient is not important. What needs to be done with the result is checking the sign of the sum of such texture. Moving towards the camera will result in negative values, in opposite direction in positive values. Negative values will shrink radius of the virtual sphere, positive ones will make it bigger.

Stereoscopic component of the application gives us the reason for handling new type of the user movement - head rotations. The result of the operation will be addition of small vertical disparity to the final image. Too much of such disparity can result in bad experiences during 3d watching, but a small can enhance it. What should be done in this step is calculating, how much user has turned his head in XY plane. Input for this step are images that are free of the camera capture errors. First, we calculate gradient in both directions. Such gradient will be calculated for a whole image, which will result in giving values that are not important. To remove them, we calculate the edge image for interesting region and improve it's quality by using erosion and dilation. Such mask is used to remove unnecessary regions. Next, we calculate histogram for such image. That gives us the possibility to predict rotation of head. The highest value in the histogram is the most probable angle of rotation. To prevent numeric errors, we remove values below some threshold, and use this value to add vertical disparity to the image.

4 Results and Discussion

We have created a virtual-reality system for simulating holography, by using stereoscopy and face tracking. We were aiming for creating view-dependent system with limited cost, so with the hardware that everybody can possibly have.

In our case, additional part was only a camera. It gives the possiblity to watch an object from different places and thanks to this, improve user's experience. In comparison to the previous system, which was used as a basic [2], we have improved movement in depth, made algorithm more stable and added stereoscopic component, which resulted in adding vertical disparity to the final image.

In Fig. 3 is shown movement in XY direction. As can be seen, by moving head, we are able to see the object from different angles.

Fig. 3. XY movement - in the first row normal position, in the second row has changed.

Z direction movement can be seen in Fig. 4. By moving further from the camera it is possible to see the object as it is further from the user. The same applies to moving closer. The closer to the camera person will be, the closer the object is.

The most complicated transformation is shown in Fig. 5. By rotating head, small vertical disparity has been added, which can enhance experience in watching 3d content on the screen.

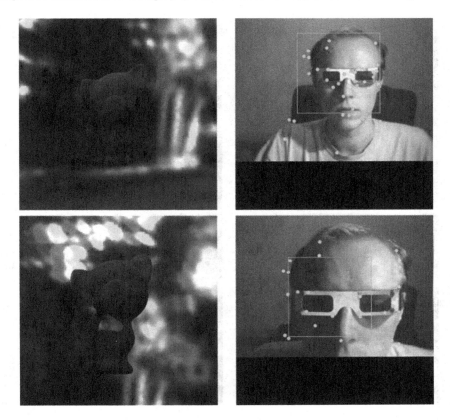

Fig. 4. Z movement - in the first row normal position, in the second row has changed.

We have not limited movement in any direction, so too dynamic movements can cause flipping the sphere, so the object will be seen from the other side. What is more, moving too close or too far away will change disparity too much, so it will be very difficult to verge on the object. This can be misleading for the user. The simplest solution is limiting the movement or correcting the disparity, but it also can cause some difficulties in understanding how user can influence on the rendered scene.

Even with that much processing our system, is quite responsive. We have not tried raytracing as a method of rendering, because it is very time consuming method and it will add more calculations between consecutive frames.

Face tracking technique is now a problem, because it seems that this is the part which is really time consuming. This is the part, that is done on the CPU, so probably moving it to the GPU will make system more responsive. Another problem is dependancy between the detection, tracking and working of whole system. If one of the parts goes wrong, the whole system will not work correctly, but of course, it is inevitable, because face tracking is the heart of the system.

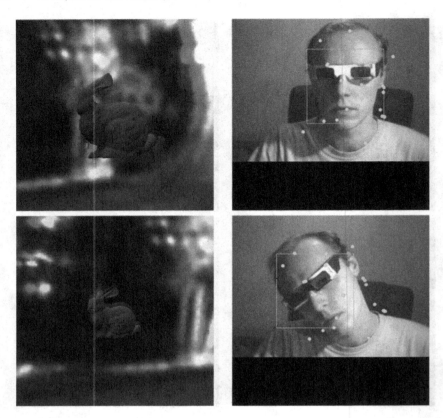

Fig. 5. Head rotations - in the first row normal position, in the second row has changed.

In the future we plan to move face detection and tracking to the GPU to improve speed of the algorithm, because now the delay is visible. We would like to add raytracing to the current pipeline to handle reflections and refractions. Such addition can cause more problems that are connected to the visual discomfort in stereo 3D (for example rivalry). There are now some algorithms [1], that can cope up with this inconveniences, so we plan to add them to the pipeline. After adding raytracing we will be able to give user the chance to see really complicated scenes with many reflections (for example many mirrors), or refractions (materials like glass). There is a possibility to improve the current solution.

References

1. Dąbała, Ł., Kellnhofer, P., Ritschel, T., Didyk, P., Templin, K., Myszkowski, K., Rokita, P., Seidel, H.-P.: Manipulating refractive and reflective binocular disparity. Comput. Graph. Forum (Proc. Eurographics 2012) **33**(2) (2014)
2. Dąbała, Ł., Rokita, P.: Simulated holography based on stereoscopy and face tracking. In: Chmielewski, L.J., Kozera, R., Shin, B.-S., Wojciechowski, K. (eds.) ICCVG 2014. LNCS, vol. 8671, pp. 163–170. Springer, Heidelberg (2014)

3. Garstka, J., Peters, G.: View-dependent 3D projection using depth-image-based head tracking. In: 8th IEEE International Workshop on Projector Camera Systems PROCAMS, pp. 52–57 (2011)
4. Greene, N.: Environment mapping and other applications of world projections. IEEE Comput. Graph. Appl. **6**(11), 21–29 (1986)
5. Nguyen Hoang, A., Tran Hoang, V., Kim, D.: A real-time rendering technique for view-dependent stereoscopy based on face tracking. In: Murgante, B., Misra, S., Carlini, M., Torre, C.M., Nguyen, H.-Q., Taniar, D., Apduhan, B.O., Gervasi, O. (eds.) ICCSA 2013, Part I. LNCS, vol. 7971, pp. 697–707. Springer, Heidelberg (2013)
6. Howard, I.P., Rogers, B.J.: Perceiving in Depth, Volume 2: Stereoscopic Vision. OUP, USA (2012)
7. Jones, B., Sodhi, R., Murdock, M., Mehra, R., Benko, H., Wilson, A., Ofek, E., MacIntyre, B., Raghuvanshi, N., Shapira, L.: Roomalive: magical experiences enabled by scalable, adaptive projector-camera units. In: Proceedings of the 27th Annual ACM Symposium on User Interface Software and Technology, UIST 2014, pp. 637–644, ACM, New York (2014)
8. Lucas, B.D., Kanade, T.: An iterative image registration technique with an application to stereo vision. In: Proceedings of the 7th International Joint Conference on Artificial Intelligence - Volume 2, IJCAI 1981, pp. 674–679. Morgan Kaufmann Publishers Inc., San Francisco (1981)
9. Palmer, S.E.: Vision Science: Photons to Phenomenology. The MIT Press, Cambridge (1999)
10. Slotsbo, P.: 3D interactive and view dependent stereo rendering (2004)
11. Tyler, C.W., Likova, L.T., Atanassov, K., Ramachandra, V., Goma, S.: 3D discomfort from vertical and torsional disparities in natural images (2012)
12. Viola, P., Jones, M.: Rapid object detection using a boosted cascade of simple features. pp. 511–518 (2001)
13. Zhang, C., Yin, Z., FlorÃlncio, D.A.F.: Improving depth perception with motion parallax and its application in teleconferencing. In: MMSP, pp. 1–6. IEEE (2009)
14. Zhukov, S., Lones, A., Kronin, G.: An ambient light illumination model. In: Proceedings of EGSR, pp. 45–55 (1998)

Head Pose Tracking from RGBD Sensor
Based on Direct Motion Estimation

Adam Strupczewski, Błażej Czupryński[✉], Władysław Skarbek,
Marek Kowalski, and Jacek Naruniec

Warsaw University of Technology, Warsaw, Poland
blazej.czuprynski@gmail.com

Abstract. We propose to use a state-of-the-art visual odometry technique for the purpose of head pose estimation. We demonstrate that with small adaptation this algorithm allows to achieve more accurate head pose estimation from an RGBD sensor than all the methods published to date. We also propose a novel methodology to automatically assess the accuracy of a tracking algorithm without the need to manually label or otherwise annotate each image in a test sequence.

Keywords: Head pose estimation · Head pose tracking · Camera pose estimation · RGBD camera · Kinect · Visual odometry

1   Introduction and Related Work

Head pose estimation is a very popular topic as it directly links computer vision to human-computer interaction. In recent years a lot of research has been conducted in this field. The focus of most contemporary approaches seem to be Active Appearance Model-like approaches such as [1,15] or hybrid approaches based on model tracking [9,13]. Despite impressive performance, we argue that the accuracy of these methods in terms of head pose estimation accuracy (rotation, translation) still leaves room for improvement. Most importantly, all the popular approaches assume that the RGB camera input is available, but do not consider depth input data. Recent developments in the field of depth sensors suggest that they will soon be deployed on a large scale, such as RGB cameras are today. Therefore, using this additional information for basic computer vision challenges such as head pose tracking, seems to be important.

We show how the method developed primarily be Kerl et al. [10,11] for camera pose estimation can be used for head pose tracking. We believe that this approach presents new opportunities and benefits compared to previous approaches. We demonstrate that this method has superior accuracy to the current state-of-the-art. Furthermore, we propose a generic scheme of adopting a camera pose estimation method to the head pose estimation scenario.

© Springer International Publishing Switzerland 2015
M. Kryszkiewicz et al. (Eds.): PReMI 2015, LNCS 9124, pp. 202–212, 2015.
DOI: 10.1007/978-3-319-19941-2_20

The contributions of this work are as follows:

1. A derivation on how to use the visual odometry algorithm proposed in [10] for head pose tracking.
2. A new method to automatically evaluate the accuracy of a head pose tracking algorithm.

1.1 Related Work in Head Pose Estimation

Head pose estimation is a very important topic in the field of human-computer interaction. It has many applications as a component of HCI systems such as eye gaze tracking or facial expression analysis, but also directly for controlling interface elements such as the cursor. The first vision-based approaches utilized a single RGB camera to estimate the head pose. Recently however, the focus of researchers has shifted to depth sensors. Starting with Microsoft Kinect, depth cameras have become increasingly available, which further encourages research in this area.

A good overview of head pose estimation using visible light cameras can be found in [16]. A follow-up of this study focused on high accuracy techniques was presented in [4]. The most accurate RGB based methods are related to tracking. The current state-of-the-art in this field are hybrid methods using a combination of static pose estimation and tracking, with the usage of reference templates [13,15].

The depth-based head pose tracking methods are all relatively new. One of the pioneering studies [14] infers pose from facial geometry. In order to make this work, facial landmarks such as the nose tip are first detected, and smoothed with a Kalman filter. The relative position of the facial landmarks allows to calculate the head pose. Despite high accuracy reported in the paper, it is questionable whether such can be achieved in real world scenarios under unconstrained head movements and varying lighting. Another statistical framework was later presented without the requirement of such high resolution input data [2]. The accuracy of the system seems to be good with a reported accuracy around 3 pixels related to manually tagged facial landmarks in images (in the presented set-up this is below 2 mm). Other statistical head pose estimation approaches using depth cameras have also appeared [5,17], but they are only capable of performing coarse pose estimation, focused on robustness not accuracy.

A somewhat different type of approach has been proposed in [6]. The authors propose to perform depth based tracking and intensity based tracking separately, and later fuse them with an extended Kalman filter. A set of features is detected and tracked in the intensity domain, as well as in the pure depth image. The reported accuracy of around 3 degrees is quite impressive. In other work, [12] propose to track the head using an elaborated variant of ICP with head movement prediction. While being fast, the system seems less accurate than other related work. Last but not least, a new approach to using Active Appearance Models has been recently proposed [20]. It is an improved AAM tracking with linear 3D morphable model constraints and claims to have a point tracking accuracy of 3–6mm. It is a very good choice for tracking initialization, but does not reach the full potential of frame-to-frame tracking accuracy that we will present.

1.2   Related Work in RGBD Camera Pose Estimation

The traditional algorithm for RGBD pose estimation is the Iterative Closest Point (ICP) algorithm, which aligns two point clouds. One of the first popular Simultaneous localization and mapping (SLAM) methods, Kinect Fusion [8], is based on this approach. ICP is however not very accurate, even when adopted in a coarse-to-fine scheme. A much more accurate method for small baselines has been recently proposed by Steinbrucker et al. [21]. Assuming that the scene geometry is known, the camera pose estimation can be denoted as the minimization of the following total intensity error

$$min_\xi \int_\Omega |I_1(\mathbf{x}) - I_2(\pi g_\xi(h\mathbf{x}))|^2 \, dx \qquad (1)$$

where ξ represents the six degrees of freedom for rigid body motion, g_ξ is the rigid body motion (of the camera), $\mathbf{x}$ is a point in homogenous coordinates, $h\mathbf{x}$ is the 3D structure and π is projection. This can be solved by linearization and a Gauss-Newton approach. It turns out that on contemporary hardware such minimization, and in turn camera motion estimation, can be performed in real time even for high camera resolutions. Extensions to the basic idea include parallel minimization of the depth discrepancy as well as a probabilistic error model [11].

We use the described work in our derivation in Sect. 2. We propose to solve a dual problem to the classical one: instead of estimating the camera pose we estimate the pose of the tracked object, in our case the human head.

2   Direct Head Pose Tracking Approach

We consider the problem of object pose tracking, i.e. the affine motion estimation using a single RGB-D camera. The camera can be used to find the instant motion with respect to its Cartesian coordinate system for selected objects which are detected in the image - in our case a human head. The below derivation is based on the work of [10,11,21] extending their presentation by the complete EM scheme including weight computation for Iteratively Reweighted Least-Squares (IRLS), and by the formulas (3), (5), (8), and (9).

2.1   Spatial and Pixel Trajectories of Tracked Objects

Object affine motion can be approximated by discrete trajectories of its points represented in homogeneous Cartesian coordinates and defined recursively

$$\boldsymbol{X}_{t+1} \approx (I_4 + \hat{s}_t) \qquad (2)$$

where the twist vector s defines the instant motion matrix $\hat{s}$. The cummulated rotation R_t and translation T_t can be recovered from the matrix $G_{t+1} = (I_4 + \hat{s}_t)G_t$, $G_0 = I_4$, requiring identification of the twist s_t for each discrete time.

The spatial trajectories are viewed by the RGB-D camera as pixel trajectories. Having the depth function $D(x, y)$ and the intrinsic camera parameter matrix $K \in \mathbb{R}^{3 \times 3}$, we can re-project the pixel (x_t, y_t) from the image frame I_t to the spatial point (X_t, Y_t, Z_t), find the point $(X_{t+1}, Y_{t+1}, Z_{t+1})$ after instant motion determined by the twist $s_t = (a_t, v_t)$), and project onto the pixel (x_{t+1}, y_{t+1}) into the image frame I_{t+1} (where $K' \in \mathbb{R}^{2 \times 3}$ is the upper part of K):

$$
\begin{bmatrix} X_t \\ Y_t \\ Z_t \\ 1 \end{bmatrix} = D_t(x_t, y_t)K^{-1} \begin{bmatrix} x_t \\ y_t \\ 1 \end{bmatrix} \rightarrow \begin{bmatrix} x_{t+1} \\ y_{t+1} \end{bmatrix} = K' \begin{bmatrix} X_t - (a_t)_z Y_t + (a_t)_y Z_t + (v_t)_x \\ -(a_t)_y X_t + (a_t)_x Y_t + Z_t + (v_t)_z \\ (a_t)_z X_t + Y_t - (a_t)_x Z_t + (v_t)_y \\ -(a_t)_y X_t + (a_t)_x Y_t + Z_t + (v_t)_z \end{bmatrix}
$$

(3)

2.2   Probabilistic Framework for Object Pose Tracking

The twist vector s can be found by Iterated Re-Weighted Least Square method with weights derived from probabilistic optimization given residuals $\mathbf{r}_i$ (differences between warped and observed intensities/depths)

$$
s_{MAP} = \arg \min_s \left[-\sum_{i=1}^n \log p(\mathbf{r}_i | s) - \log p(s) \right]
$$

(4)

When d-variate Student t-distribution is assumed for d-dimensional residuals, the weights can be found by EM iterations. Let ν be fixed and initially $\mu^{(1)} = \mathbf{0}_d$, $\Sigma^{(1)} = I_d$. Then at the iteration j, the parameters w, μ, Σ are updated:

1. the i-th residual $\mathbf{r}_i$ gets the weight: $w_i^{(j)} \leftarrow \dfrac{\nu + d}{\nu + (\mathbf{r}_i - \mu^{(j)})^T (\Sigma^{(j)})^{-1}(\mathbf{r}_i - \mu^{(j)})}$

2. the mean value: $\mu^{(j+1)} \leftarrow \dfrac{\sum_{i=1}^n w_i^{(j)} \mathbf{r}_i}{\sum_{i=1}^n w_i^{(j)}}$

3. the covariance matrix: $\Sigma^{(j+1)} \leftarrow \dfrac{\sum_{i=1}^n w_i^{(j)} (\mathbf{r_i} - \mu^{(j+1)})(\mathbf{r_i} - \mu^{(j+1)})^T}{\sum_{i=1}^n w_i^{(j)}}$

The stop condition occurs if both the mean vector and the weighted covariance matrix stabilize all their components. However, it should be noted that if the weight $w_i^{(j)}$ is small, the algorithm can be stopped earlier and thus speeded up. The optimization step can keep a linear form if the weights w_i satisfy the requirements $w_i A \mathbf{r}_i \doteq \frac{\partial \log p(\mathbf{r_i}|s)}{\partial \mathbf{r_i}}$, for a certain matrix $A \in \mathbb{R}^{d \times d}$.

Assuming the normal prior probability distribution $p(s) = \text{Norm}_s$ $(s_{pred}, \Sigma_{pred})$ and approximating the residual function $\mathbf{r}_i$ by its linear part we get the linear equation for the optimized step Δs:

$$
\left[\Sigma_{pred}^{-1} + \sum_{i=1}^n w_i \frac{\partial \mathbf{r}_i^T}{\partial s} \Sigma^{-1} \frac{\partial \mathbf{r}_i}{\partial s^T} \right] \Delta s = -\Sigma_{pred}^{-1}(s_0 - s_{pred}) - \sum_{i=1}^n w_i \frac{\partial \mathbf{r}_i^T}{\partial s} \Sigma^{-1} \mathbf{r}_i(s_0)
$$

(5)

2.3 Residual Derivatives Based on Intensity and Depth Constancy

Since an RGBD camera produces 2D photometric data and 3D spatial data we can track the object relying on the depth D constancy along the 3D point trajectory and the luminance L constancy along the pixel trajectory. In this case the vectorial residual has the form:

$$\mathbf{r}_i(\mathbf{s}) = \begin{bmatrix} L'(x_i', y_i') - L(x_i, y_i) \\ D'(x_i', y_i') - Z' \end{bmatrix} \tag{6}$$

The twist derivative is the combination of two twist derivatives: luminance and depth

$$\frac{\partial \mathbf{r}_i^T}{\partial \mathbf{s}} = \left[\frac{\partial L'(x_i', y_i')}{\partial \mathbf{s}}, \left(\frac{\partial D'(x_i', y_i')}{\partial \mathbf{s}} - \frac{\partial Z'}{\partial \mathbf{s}} \right) \right] \tag{7}$$

The luminance derivative is a special case of the constancy formula for photometric attributes:

$$\frac{\partial L_i'^T}{\partial \mathbf{s}} = \left[\frac{\partial \mathbf{x}'}{\partial \mathbf{s}}, \frac{\partial \mathbf{y}'}{\partial \mathbf{s}} \right] \begin{bmatrix} \frac{\partial L'^T(x', y')}{\partial x'} \\ \frac{\partial L'^T(x', y')}{\partial y'} \end{bmatrix} = \begin{bmatrix} \hat{\mathbf{X}}_i \\ I_3 \end{bmatrix} \begin{bmatrix} Z' & 0 \\ 0 & Z' \\ -X' & -Y' \end{bmatrix} \frac{K^{*T}}{Z'^2} \begin{bmatrix} \frac{\partial L'^T(x', y')}{\partial x'} \\ \frac{\partial L'^T(x', y')}{\partial y'} \end{bmatrix} \tag{8}$$

where we define $K^* \doteq \begin{bmatrix} f_x & g_x \\ 0 & f_y \end{bmatrix}$. The depth is differentiated along the pixel trajectory and along the 3D point trajectory:

$$\frac{\partial \left[D_i'^T - Z_i'^T \right]}{\partial \mathbf{s}} = \begin{bmatrix} \hat{\mathbf{X}}_i \\ I_3 \end{bmatrix} \begin{bmatrix} Z' & 0 \\ 0 & Z' \\ -X' & -Y' \end{bmatrix} \frac{K^{*T}}{Z'^2} \begin{bmatrix} \frac{\partial D'^T(x', y')}{\partial x'} \\ \frac{\partial D'^T(x', y')}{\partial y'} \end{bmatrix} - [Y_i, -X_i, 0, 0, 0, 1]^T \tag{9}$$

Using the Gauss Newton algorithm and formulas 5 and 7 we can align two RGBD images by minimizing the photometric and geometric error. Because of using the probabilistic model and IRLS, the method is robust to small deviations and outliers. In order to perform head pose estimation relative to a fixed-position camera sensor, we propose to crop out the face image from the whole scene. Once only facial pixels are provided, the problem comes down to estimating rigid body motion as described above.

2.4 Head Region Extraction

In order to crop out the face, we use a face detection algorithm based on the boosting scheme. For feature extraction we use an extended Haar filters set along with HOG-like features. For weak classification we use logistic regression and simple probability estimations. The strong classifier is based on the Gentle Boosting scheme.

Our facial landmark detection method is based on a recently popular cascaded regression scheme [18], where starting from an initial face shape estimate S^0 the shape is refined in a fixed number of iterations. At each iteration t an increment that will refine the previous pose estimate S^{t-1} is found by regressing a set of features:

$$S^t = S^{t-1} + R^t \Phi^t(I, S^{t-1}), \tag{10}$$

where R^t is the regression matrix at iteration t and $\Phi^t(I, S^{t-1})^t$ is a vector of features extracted at landmarks S^{t-1} from image I. Figure 1 summarizes the whole process of facial feature alignment. Our method was trained using parts of the 300-W [19] and Multi-PIE [7] datasets.

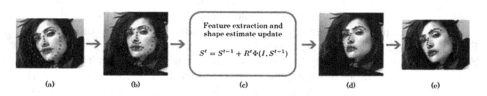

Fig. 1. The process of facial feature alignment. (a) The mean shape is aligned with the landmarks from the previous video frame. (b) The shape and image are transformed to a canonical scale and orientation. (c) Cascaded regression is performed for a predefined number of iterations. (d) The resulting shape. (e) An inverse transformation is applied to the shape to obtain its location in the original image.

3   Experiments

3.1   System Implementation

We have developed our head pose tracking system on top of the dense visual odometry tracking algorithm as implemented by the authors of [10]. The original DVO tracker is designed to estimate the camera motion. As this is dual to the estimation of an object's pose relative to a motionless camera, we crop out face pixels from the whole image to get a consistent face pose estimation.

In order to provide the initial region of interest we run our face detection algorithm described in Sect. 2.4 in the first frame. Once we have a coarse rectangular face area, depth-based segmentation can be performed. We begin with finding the mean head depth by averaging the depth values in a small window in the center of the face ROI. Next we perform depth thresholding to remove pixels far away from the mean head depth. The suitable depth range has been determined experimentally. During tracking the coarse region of interest is moved using estimated head motion. The depth thresholding is repeated in the same way for each subsequent frame. An example of coarse rectangular ROI and final segmented head is presented in Fig. 2. Once the head is segmented from the images the motion is computed using DVO tracker. The motion is estimated in a frame-to-frame manner and accumulated to obtain the transformation to the first frame.

3.2 Test Methodology

Unfortunately, popular databases for head pose estimation, such as the Boston University dataset [3], do not suit our needs. We require a dataset where the RGB and depth stream are registered simultaneously and the user performs various head movements. Because of the lack of such datasets, we have decided to record our own. We have recorded ten sequences of five different people, each about one minute long. Each sequence begins with a frontal head pose, after which unconstrained, significant head movements are performed. Typically the sequences contain large head rotations along three axes (yaw, pitch, roll).

Recording the ground truth of a person's head pose is a very difficult task. Firstly, it requires specialized, expensive equipment such as magnetic trackers. Secondly, the accuracy of such a dataset is inherently limited by the accuracy of the equipment used. We propose a novel method to measure how well the head pose is estimated, which overcomes both of the mentioned limitations. We propose to estimate the motion of projected facial landmarks found in the input data. The algorithm can be outlined as follows.

For each image:

1. Estimate head pose in the image.
2. Use the estimation to transform the head pose to the initial view.
3. Calculate the projection of the transformed facial image.
4. Calculate the distance between the projected coordinates of landmarks. in the first and current image - this is proportional to the tracking accuracy.

As facial features we propose to use the eye inner and outer corners, and lip corners. The facial features can be tagged manually, but what is even more important, they can be determined automatically by a face landmark localisation algorithm. This is the biggest advantage of the proposed methodology - long sequences can be reliably evaluated without the necessity to perform tedious and error-prone manual tagging of each frame. Figure 2 shows the type of performed head movements in one of the test sequences (top), along with warped segmented faces using the described algorithm (bottom).

3.3 Results

The results on sequences that we have recorded are shown in Table 1. The set of six chosen facial features (inner and outer eye corners, lip corners) was used to calculate the errors in the top row. For comparison, the bottom row shows the average errors for all points (68) detected by the facial landmark alignment algorithm. The average errors in pixels have been calculated for all the frames in each test sequence. We have used an RGB camera resolution of 1920x1080 pixels

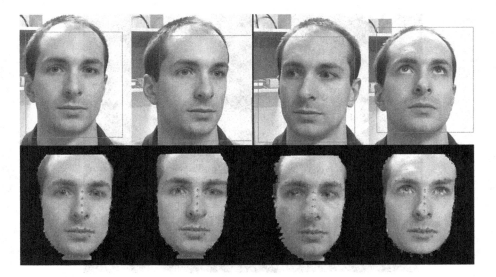

Fig. 2. Head pose dataset. Landmark locations from the reference frame are marked by green circles, those from the current warped frame are marked by red crosses.

and recorded people who were about 1 m away, which resulted in the face images having around 100 pixels between the eyes. For depth recording we have used the Kinect 2 sensor having 512x424 resolution in the image plane and 1 mm depth resolution. The proposed algorithm works in real-time on a CPU.

Table 1. Head pose estimation accuracy on ten sequences [pixels]

Sequence	Seq 1	Seq 2	Seq 3	Seq 4	Seq 5	Seq 6	Seq 7	Seq 8	Seq 9	Seq 10
Subset error	3.78	3.48	2.77	4.4	3.37	4.15	2.76	3.69	4.38	4.62
Full set error	4.92	4.85	4.59	5.48	4.07	4.74	4.21	4.61	5.61	5.82

The measured errors of around 3–4 pixels demonstrate very high accuracy of the proposed head pose estimation method. Because the proposed tracking algorithm uses a probabilistic model and IRLS, it can handle outliers and non-rigid face motion by downvoting the pixels inconsistent with global motion.

We wish to point out, that the facial landmark localisation algorithm used in our experiments is sensitive to illumination and the inaccuracy of the facial landmark localization algorithm is a part of the measured errors in Table 1. In our opinion, judging from a set of manually tagged images, the proposed head pose estimation algorithm has better accuracy than indicated by the measurements in Table 1 by at least 1 pixel. This is better than the accuracy measured in [2].

Figure 3 shows the trajectory of head motion that was performed similarly in each test sequence. The trajectory is displayed as a virtual camera trajectory

Fig. 3. Head pose trajectory presented as virtual camera trajectory

relative to a motionless head for better visualization. We would like to note that
the trajectory finishes in roughly the same place as it started even though the
motion is estimated in a frame-to-frame fashion. This proves that there is very
small tracking drift, which is a strong advantage of the proposed algorithm.

4   Conclusions

We have presented a novel, highly accurate approach to head pose tracking.
Based on the obtained accuracy measurements, as well as observations of live
performance when stabilizing the head image, we conclude that the presented
method surpasses the current state-of-the-art in the field of RGBD head pose
tracking. The real-time algorithms for camera motion estimation proposed in
[10] can be easily adapted to the scenario of head pose tracking and used for
various use cases such as eye gaze tracking or face frontalization. We have also
presented a new head pose tracking accuracy evaluation methodology which can
be easily used to assess algorithm performance.

A good direction for future work is exploring the possibility of using
keyframes and thus removing accumulated tracking error from the proposed
head pose estimation method.

References

1. Baltrušaitis, T., Robinson, P., Morency, L.: 3D constrained local model for rigid
   and non-rigid facial tracking. In: IEEE Conference on Computer Vision and Pattern
   Recognition (2012)

2. Cai, Q., Gallup, D., Zhang, C., Zhang, Z.: 3D deformable face tracking with a commodity depth camera. In: Daniilidis, K., Maragos, P., Paragios, N. (eds.) ECCV 2010, Part III. LNCS, vol. 6313, pp. 229–242. Springer, Heidelberg (2010)
3. La Cascia, M., Sclaroff, S., Athitsos, V.: Fast, reliable head tracking under varying illumination: an approach based on registration of texture-mapped 3D models. IEEE Trans. Pattern Anal. Mach. Intell. **22**(4), 322–336 (2000)
4. Czupryński, B., Strupczewski, A.: High accuracy head pose tracking survey. In: Ślęzak, D., Schaefer, G., Vuong, S.T., Kim, Y.-S. (eds.) AMT 2014. LNCS, pp. 407–420. Springer, Heidelberg (2014)
5. Fanelli, G., Weise, T., Gall, J., Van Gool, L.: Real time head pose estimation from consumer depth cameras. In: Mester, R., Felsberg, M. (eds.) DAGM 2011. LNCS, vol. 6835, pp. 101–110. Springer, Heidelberg (2011)
6. Gedik, O.S., Alatan, A.A.: Fusing 2D and 3D clues for 3D tracking using visual and range data. In: 2013 16th International Conference on Information Fusion (FUSION), pp. 1966–1973. July 2013
7. Gross, R., Matthews, I., Cohn, J., Kanade, T., Baker, S.: Multi-pie. In: 8th IEEE International Conference on Automatic Face Gesture Recognition, FG 2008, pp. 1–8. September 2008
8. S. Izadi, Kim, D., Hilliges, O., Molyneaux, D., Newcombe, R., Kohli, P., Shotton, J., Hodges, S., Freeman, D., Davison, A., Fitzgibbon, A.: Kinectfusion: real-time 3D reconstruction and interaction using a moving depth camera. In: Proceedings of the 24th Annual ACM Symposium on User Interface Software and Technology, UIST 2011, pp. 559–568. ACM, New York (2011)
9. Jang, J., Kanade, T.: Robust 3D head tracking by online feature registration. In: The IEEE International Conference on Automatic Face and Gesture Recognition (2008)
10. Kerl, C., Sturm, J., Cremers, D.: Dense visual slam for RGB-D cameras. In: Proceedings of the International Conference on Intelligent Robot Systems (IROS) (2013)
11. Kerl, C., Sturm, J., Cremers, D.: Robust odometry estimation for RGB-D cameras. In: Proceedings of the IEEE International Conference on Robotics and Automation (ICRA). May 2013
12. Li, S., Ngan, K.N., Sheng, L.: A head pose tracking system using RGB-D camera. In: Chen, M., Leibe, B., Neumann, B. (eds.) ICVS 2013. LNCS, vol. 7963, pp. 153–162. Springer, Heidelberg (2013)
13. Liao, W., Fidaleo, D., Medioni, G.: Robust, real-time 3D face tracking from a monocular view. EURASIP J. Image Video Process. **2010** (2010)
14. Malassiotis, S., Strintzis, M.G.: Robust real-time 3D head pose estimation from range data. Pattern Recognit. **38**(8), 1153–1165 (2005)
15. Morency, L., Whitehill, J., Movellan, J.: Generalized adaptive view-based appearance model: integrated framework for monocular head pose estimation. In: 8th IEEE International Conference on Automatic Face Gesture Recognition (2008)
16. Murphy-Chutorian, E., Trivedi, M.M.: Head pose estimation in computer vision: a survey. IEEE Trans. Pattern Anal. Mach. Intell. **31**, 607–626 (2009)
17. Padeleris, P., Zabulis, X., Argyros, A.A.: Head pose estimation on depth data based on particle swarm optimization. In: 2012 IEEE Conference on Computer Vision and Pattern Recognition Workshops (CVPRW), pp. 42–49. June 2012
18. Ren, S., Cao, X., Wei, Y., Sun, J.: Face alignment at 3000 FPS via regressing local binary features. In: 2014 IEEE Conference on Computer Vision and Pattern Recognition (CVPR), pp. 1685–1692. June 2014

19. Sagonas, C., Tzimiropoulos, G., Zafeiriou, S., Pantic, M.: A semi-automatic methodology for facial landmark annotation. In: 2013 IEEE Conference on Computer Vision and Pattern Recognition Workshops (CVPRW), pp. 896–903. June 2013
20. Smolyanskiy, N., Huitema, C., Liang, L., Anderson, S.: Real-time 3D face tracking based on active appearance model constrained by depth data. Image Vis. Comput. **32**(11), 860–869 (2014)
21. Steinbruecker, F., Sturm, J., Cremers, D.: Real-time visual odometry from dense RGB-D images. In: Workshop on Live Dense Reconstruction with Moving Cameras at the International Conference on Computer Vision (ICCV) (2011)

Pattern Recognition

A Novel Hybrid CNN-AIS Visual Pattern Recognition Engine

Vandna Bhalla[(✉)], Santanu Chaudhury, and Arihant Jain

Indian Institute of Technology Delhi, New Delhi, India
{vbhalla.du,arihant.jms}@gmail.com, santanuc@ee.iitd.ernet.in
http://www.iitd.ac.in

Abstract. Machine learning methods are used today mostly for recognition problems. Convolutional Neural Networks (CNN) have time and again proved successful for many image processing tasks primarily for their architecture. In this paper we propose to apply CNN to small data sets like for example, personal photo albums or other similar environs where the size of training dataset is a limitation, within the framework of a proposed hybrid CNN-AIS model. We use Artificial Immune System Principles to enhance the small size of training data set. A layer of Clonal Selection is added to the local filtering and max pooling of CNN Architecture. The proposed Architecture is evaluated using the standard MNIST dataset by limiting the data size and also with a small personal data sample belonging to two different classes. Experimental results show that the proposed hybrid CNN-AIS based recognition engine works well when the size of training data is limited in size.

Keywords: CNN · Clonal Selection (CS) · Artificial Immune Systems (AIS) · Small data size · Diversity

1 Introduction

Today all object recognition approaches use machine-learning methods. Larger the dataset better is the performance. Labeled datasets like NORB, Caltech-101/256 and CIFAR-10/100 with tens of thousands of images are in todays scenario considered small and LabelMe and Image Net with millions of images are preferred. A simple recognition task also requires datasets of size of the order of tens of thousands of images. It is always assumed that objects in realistic settings show a lot of variability. Hence it is essential to have larger training sets to learn to recognize them with almost all current technologies. To learn from thousands of objects from millions of images, a model with a large learning capacity with powerful processing is required. CNN have shown very good image classification performance which can be attributed to their ability to learn rich image representation compared to hand crafted low level features used in other image classification techniques. Recent publications indicate that deep hierarchical NN improve pattern classification. In fact Deep NN are fully exploited to

© Springer International Publishing Switzerland 2015
M. Kryszkiewicz et al. (Eds.): PReMI 2015, LNCS 9124, pp. 215–224, 2015.
DOI: 10.1007/978-3-319-19941-2_21

their best potential when they are wide with many maps per layer and deep with many layers. We too saw through our experiments that properly trained wide and deep CNNs can outperform all previous methods. Learning CNNs requires a huge number of annotated training images. This property prevents application of CNNs to scenarios with limited training data. We present an innovative, adaptive, self-learning, and self-evolving hybrid recognition engine, which works well with small sized training data. The model uses the intelligent information processing mechanism of Artificial Immune System (AIS) and helps Convolutional Neural Network (CNN) generate a robust feature set taking the small set of input training images as seeds. Our model performs visual pattern learning using a heterogeneous combination of supervised CNN and Clonal Selection (CS) principles of AIS. It can be extended to perform classification tasks with limited training data particularly in the context of personal photo collections where for each training sample different points of view are gathered in parallel using clonal selection. This is very different from populating datasets with artificially generated training examples [9] by randomly distorting the original training images with randomly picked distortion parameters. The many shortcomings of small size data sets have been widely recognized by Pinto [14]. Small training data has not been given much prominence in recent research.

Specific contribution of this paper is as follows: A hybrid Convolutional Neural Network-Artificial Immune System (CNN-AIS) Recognition Engine Architecture designed to work with modest sized training data. This is detailed in Sect. 3. The model was tested on well-known MNIST digit database and showed remarkable success. The current best error rate of 0.21 % on the MNIST digit recognition task approaches human performance. We have got very good results with considerably smaller number of training samples. In addition we get comparable accuracy with the state of the art methods when the data size is increased to large numbers. The results are presented in Sect. 4. For completeness we assessed our CNN-AIS model by applying it on a personal photo collection and successfully accomplished classification for two categories of classes. Section 5 gives an application of this system in real world with results.

2   Related Work

Our model is inspired by many related works on image classification and deep learning which we briefly discuss here. The general structure of the deep convolutional neural network (CNN) was introduced in early nineties [6]. This deep convolutional neural network architecture, called LeNet, is still being used today with a lot of consistent improvements to the individual components within the architecture. An important idea of the CNN is that the feature extraction and classifier were unified in a single structure. The model was proposed for handwritten digit recognition and achieved a very high success rate on MNIST dataset [7]. But it demands substantial amount of labeled data for training (60,000 for MNIST). Though the results are promising and exciting but the bothersome part is that millions of annotated images are needed for each new visual recognition task. Also the size of input is very small (28×28) with no background clutter, illumination change etc. which is an integral part of normal pictures/images. In fact for most realistic vision applications this is not the case. The multiple processing layers of machine learning systems extract more abstract, invariant features of data and have higher classification accuracy than the traditional shallower classifiers. These deep architectures have shown promising performances in image [7] language [19] and speech [10]. Ranzato et al. [15] trained a large CNN for object detection (Caltech 101 dataset) but obtained poor results though it achieved perfect classification on the training set. The weak generalization power of CNN when the number of training data is small and the number of free parameter is large is a case of over fitting or over parametrization. The success of object recognition algorithm to a large extent depends on features detected. The features should have the most distinct characteristics among different classes while retaining invariant characteristics within a class. Other biologically inspired models like HMAX [16] use hardwired filter and use hard Max functions to compute the responses in the pooling layer. The problem was that it was unable to adapt to different problem settings.

Transfer learning is one technique to conquer the shortfall of training samples for some categories by adapting classifiers trained for other categories. One such method [12] proposes to transfer image representations learned with CNNs on large datasets to other tasks with limited training datasets. This method fails to recognize spatially co occurring objects. The false positives in their results correspond to samples closely resembling target object classes. Recognition of very small or very large objects could also fail.

Successful algorithms have been built on top of handcrafted gradient response features such as SIFT and histograms of oriented gradients (HOG). These are fixed features and are unable to adjust to model the intricacies of a problem. Traditional hand designed feature extraction is laborious and moreover cannot process raw images while the automatic extraction mechanism can fetch features directly. In [1,2] supervised classifiers such as CNNs, MLPs, SVMs and K-nearest Neighbors are combined in a Mixture of Experts approach where the output of parallel classifiers is used to produce the final result. One such recognition system, CNSVM [11], is a classifier built as a single model with SVM and CNN. The CNN is trained using the back-propagation algorithm and the SVM is trained using a non-linear regression approach. The work, again, requires large training data. A comparison of Support Vector Machine, Neural Network, and CART algorithms using limited training data points was done though the data was for the land-cover classification [17]. SVM generated overall accuracies ranging from 77% to 80% for training sample sizes from 20 to 800 pixels per class, compared to 67–76% and 6273% for NN and CART, respectively. CNNs though efficient at learning invariant features from images, do not always produce optimal classification and SVMs with their fixed kernel function are unable to learn complicated invariance. Our approach is different and we propose a single coupled architecture for training and testing using deep CNN and AIS principles.

3   Convolutional Neural Network-Artificial Immune System (CNN-AIS) Model

We use deep convolutional neural networks as wide and deep trained networks are better than most other methods. Our proposed Architecture integrates Clonal Selection (CS) principles from Artificial Immune System (AIS) with deep Convolutional Neural Networks (CNN) in a novel way. We will briefly introduce the AIS theory and the basic CNN structure that we have used in our model. Subsequently the architecture of the hybrid CNN-AIS trainable recognition engine is presented.

Artificial Immune Systems (AIS). Artificial Immune System use Clonal Selection and Negative Selection principles imitating the Human Biological Immune System. The main task of the immune system is to defend the organism against pathogens. In the human body the B-cells with different receptor shapes try to bind to antigens. The best fit cells proliferate and produce clones which mutate at very high rates. The process is repeated and it is likely that a better B-cell (better solution) might emerge. This is called Clonal Selection. These clones have mutated from the original cell at a rate inversely proportional to the match strength. Two main concepts are particularly relevant for our framework. (i) Generation of Diversity: The B cells produce antibodies for specific antigens. Each B cell makes a specific antibody, which is expressed from the genes in its gene library. The gene library does not contain genes that define antibodies for every possible antigen. Gene fragments in the gene library randomly combine and recombine and produce a huge diverse range of antibodies. This helps the immune system to make the precise antibody for an antigen it may never have encountered previously. (ii) Avidity: Refers to the accrued strength of various diverse affinities of individual binding interaction. Avidity (functional affinity) is the collective strength of multiple affinities of an antigen with various antibodies. Based on this biological process, quite a few Artificial Immune System, (AIS), [4,5], have been developed in the past. Castro developed the Clonal Selection Algorithm (CLONALG) [3] on the basis of Clonal Selection theory of the immune system. It was proved that it can perform pattern recognition. The CLONALG algorithm can be described as follows: 1. Randomly initialize a population of individual (M); 2. For each pattern of P, present it to the population M and determine its affinity with each element of the population M; 3. Select n of the best highest affinity elements of M and generate copies/clones of these individuals proportionally to their affinity with the antigen which is the pattern P. The higher the affinity, the higher the number of clones, and vice-versa; 4. Mutate all these copies with a rate proportional to their affinity with the input pattern: the higher the affinity, the smaller the mutation rate; 5. Add these mutated individuals to the population M and reselect m of these matured individuals to be kept as memories of the systems; 6. Repeat steps 2 to 5 until a predefined optimal criterion is met.

Convolutional Neural Network (CNN). A Convolutional Neural Network [9] is a multilayer feed forward artificial neural network with a deep supervised learning architecture. The ordered architectures of MLPs progressively learn the higher level features with the last layer giving classification. Two operations of convolutional filtering and down sampling alternate to learn the features from the raw images and constitute the feature map layers. The weights are trained by a back propagation algorithm using gradient descent approaches for minimizing the training error. We have used Stochastic Gradient Approach as it prevents getting stuck in poor local minima. A simplified CNN was presented in [13] which we have used for our work instead of using the rather complicated LeNet-5 [8]. The model has five layers.

CNN-AIS Model. The architecture of our hybrid CNN-AIS model was designed by adding an additional layer of Artificial Immune System (AIS) based Clonal Selection (CS) in the traditional Convolutional Neural Network (CNN) structure, Fig. 1. The model is explained layer wise. **Convolutional Layer:** A 2D filtering between input images n, and a matrix of kernels/weights K produces the output I where $I_k = \Sigma_{i,j,k}$ M $(n_i * K_j)$ where M is a table of input output relationships. The kernel responses from the inputs connected to the same output are linearly combined. As with MLPs a scaled hyperbolic tangent function is applied to every I. **Sub sampling Layer:** Small invariance to translation and distortion is accomplished with the Max-Pooling operation. This is for faster convergence and improves generalization as well. **Fully Connected Layer - I:** The input to this layer is a set of feature maps from the lower layer which are

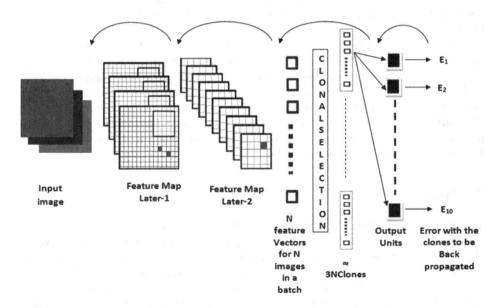

Fig. 1. The architecture of the hybrid model

combined into a 1-dimensional feature vector and subsequently passed through an activation function. **Clonal selection Layer:** This is the new additional layer that we propose in our architecture and it is the second last layer. The purpose of this layer is to generate additional input data for the final MLP layer. Additional data helps the MLP to train better which in turn leads to better trained kernels at the convolutional layer. This layer receives its input from the fully connected layer-I in the form of 1-D feature vector for all the images (n) in the current running batch. Each feature vector in the Feature set undergoes Cloning, Mutation and Crossover according to the rules of Clonal Selection to generate additional features that satisfy the minimum threshold criteria and resemble the particular class. The number of clones is calculated by

$$CNum = \eta \text{ x affinity (Feature Vector1, Feature Vector2) ... (i)},$$

where η is the cloning constant. Higher the affinity of match the greater the clone stimulus gets, the more the cloning number is. On the contrary, the number is less, which is consistent with biological immune response mechanism. Mutation frequency is defined as Rate, which is calculated by

$$Rate = \alpha \ 1/ \text{ affinity (Feature Vector1, Feature Vector2) ... (ii)}$$

where α is mutation constant. In accordance with (ii), the higher the affinity of match, the smaller the clone stimulus gets, the lower the mutation frequency is. On the contrary, the mutation frequency is higher. Hence from n initial feature sets we now have (n × CNum) features. These newly generated feature vectors are grouped into batches and individually fed to the output layer and the subsequent error is back propagated to train the kernels of the CNN. Hence from the seeds of a few representative images of each class a bigger set is evolved using Clonal Selection principles of Artificial Immune System. End of training phase yields a set of representative features, which we call antibodies, from each class of size much larger than the original dataset and a trained CNN. Though we start with random values of feature sets (antibodies) for each class but eventually they converge to their optimal values. **Output Layer (Fully Connected Layer-II):** This layer has one output neuron per class label and acts as linear classifier operating on the 1-dimensional feature vector set computed from the CS layer

4   Results

We performed tests on MNIST dataset. The results are tabulated in Table 1 which show the remarkable improvement that our hybrid model achieves when the training dataset is limited. We observe around 10–15 % improvement over traditional CNN when applied for small data size. When the data set is large, then too, CNN+ AIS achieves 0.66 % error rate giving an improvement of 0.04 %. Table 2 compares our performance with some other distortion techniques used to increase the data size found in literature. Very recently error rate upto

0.21 % has been achieved using DropConnect [18]. So for large datasets too our model approaches the achieved accuracy. Our model enhances accuracy by extending data set at feature level using CS principles rather than at the input level like for example in affine distortions. Our model improves accuracy for applications where the training data set is scarce. This inspired us to apply it for personal photo album where the training data is small. This application is discussed in the next section. We reiterate that our model gives very good results for small as well as large training data sizes unlike other models in literature.

Table 1. Test results

Training data size	300	500	1000	2000	5000	15000	30000	60000
CNN	70 (%)	89 (%)	91.6 (%)	94 (%)	96.4 (%)	98.3 (%)	98.9 (%)	99.3 (%)
CNN+AIS	85 (%)	91.9 (%)	94 (%)	96.02 (%)	97.9 (%)	98.9 (%)	99 (%)	99.34 (%)
Improvement	**15 (%)**	**3 (%)**	**2.4 (%)**	**2 (%)**	**1.5 (%)**	**0.6 (%)**	**0.1 (%)**	**0.04 (%)**

Table 2. Comparison of results with some other techniques

S.No	Algorithms	Technique	Error
1	2-layer MLP(MSE)	Affine Distortion	1.6 (%)
2	SVM	Affine + Thickness	1.4 (%)
3	Tangent Dist.	Affine Distortion	1.1 (%)
4	LeNet5(MSE)	Affine Distortion	0.8 (%)
5	Boost.LeNet4(MSE)	Affine Distortion	0.7 (%)
6	**CNN+AIS**	**Clonal Selection Principle**	**0.66(%)**

5 Application: Personal Photo Album

The CNN-AIS generates a robust and diverse pool of feature vectors and a trained CNN for any class. We tested this model for a personal collection of photos for two classes Picnic (A) and Conference (N). For every testing image (the antigen), the trained CNN-AIS model computes the feature vector and compares this with the feature set pool of that class, Fig. 2. If the number of matches of the test image with the various feature sets of that class and the combined affinities exceed the threshold then the testing image is placed in that class. These emulate the antibodies in a human body recognizing an antigen.

A two phase testing mechanism is used for classification. The first phase matches the test image feature with the 3N feature sets (antibodies) of each of the classes. The total number of antibodies lying above the threshold for matching is counted (C) for each class. All classes providing a minimum number of C are

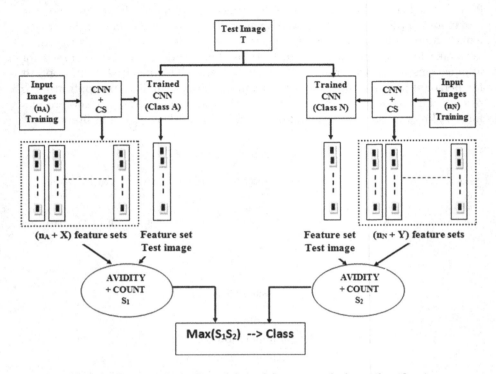

Fig. 2. Application: hybrid model used for personal photo classification

Fig. 3. Sample images from application dataset

qualified for phase 2 testing. The second phase calculates Avidity for each class which is the mean strength of multiple affinities of all qualified antibodies in C with the testing image antigen. It is calculated by taking the mean of individual scores (calculated using inner product measure) of matching of test image with each antibody above the threshold for each qualified class. This score is labeled avidity. The class is eventually decided on the basis of the combined scores S = (Count + Avidity). The experimental results are summarized in Table 3. Here TA is the True Acceptance, TR is True Rejection and Num is the number of images used. The sample dataset used for our experiments is shown in Fig. 3. Despite the diversity in the dataset and the small size of training data set, our model gives good results.

Table 3. Result analysis of personal photo album

Category	Training Num	Clones	TA(Num)	TR (Num)	Aggregate
Picnic (A)	40	160	86 (%)(50)	90 (%)(150)	88 (%)
Conference(N)	40	160	83 (%)(50)	88 (%)(150)	85 (%)

6   Conclusion

The AIS layer shows a marked improvement in recognition when training data is limited. Unlike other methods where additional data was generated at the input level, here the artificial data is generated at the feature level which is computationally fast and more accurate. The proposed model is showing promising results on personal photo albums and this can be extended to other applications where availability of data is scarce. A new class can be added to the existing set of classes dynamically replicating the behavioral aspects of self-learning and self evolving of human immune system. Even when one has an apparently enormous data set, the number of data for some particular cases of interest can be small. In fact, data sometimes exhibits a property known as the long tail, which means that a few things (e.g., words) are very common, but most things are quite rare. For example, 20 % of Google searches each day have never been seen before. So the problem of addressing small sample sizes is very relevant even in the big data era.

References

1. Abdelazeem, S.: A greedy approach for building classification cascades. In: Seventh International Conference on Machine Learning and Applications, ICMLA 2008, pp. 115–120. IEEE (2008)
2. Borji, A.: Combining heterogeneous classifiers for network intrusion detection. In: Cervesato, I. (ed.) ASIAN 2007. LNCS, vol. 4846, pp. 254–260. Springer, Heidelberg (2007)

3. De Castro, L.N., Von Zuben, F.J.: Learning and optimization using the clonal selection principle. IEEE Trans. Evol. Comput. **6**(3), 239–251 (2002)

4. Dudek, G.: An artificial immune system for classification with local feature selection. IEEE Trans. Evol. Comput. **16**(6), 847–860 (2012)

5. Hart, E., Timmis, J.: Application areas of ais: the past, the present and the future. Appl. Soft Comput. **8**(1), 191–201 (2008)

6. Knerr, S., Personnaz, L., Dreyfus, G.: Handwritten digit recognition by neural networks with single-layer training. IEEE Trans. Neural Netw. **3**(6), 962–968 (1992)

7. Krizhevsky, A., Sutskever, I., Hinton, G.E.: Imagenet classification with deep convolutional neural networks. In: Pereira, F., Burges, C., Bottou, L., Weinberger, K. (eds.) Advances in Neural Information Processing Systems. Curran Associates Inc., Red Hook (2012)

8. Lauer, F., Suen, C.Y., Bloch, G.: A trainable feature extractor for handwritten digit recognition. Pattern Recognit. **40**(6), 1816–1824 (2007)

9. LeCun, Y., Bottou, L., Bengio, Y., Haffner, P.: Gradient-based learning applied to document recognition. Proc. IEEE **86**(11), 2278–2324 (1998)

10. Mohamed, A.-R., Sainath, T.N., Dahl, G., Ramabhadran, B., Hinton, G.E., Picheny, M.A.: Deep belief networks using discriminative features for phone recognition. In: 2011 IEEE International Conference on Acoustics, Speech and Signal Processing (ICASSP), pp. 5060–5063. IEEE (2011)

11. Nagi, J., Di Caro, G.A., Giusti, A., Nagi, F., Gambardella, L.M.: Convolutional neural support vector machines: hybrid visual pattern classifiers for multi-robotsystems. In: 2012 11th InternationalConference on Machine Learning and Applications (ICMLA), vol. 1, pp. 27–32. IEEE (2012)

12. Oquab, M., Bottou, L., Laptev, I., Sivic, J.: Learning and transferring mid-level image representations using convolutional neural networks. In: 2014 IEEE Conference on Computer Vision and Pattern Recognition (CVPR), pp. 1717–1724. IEEE (2014)

13. Pan, W., Bui, T.D., Suen, C.Y.: Isolated handwritten farsi numerals recognition using sparse and over-complete representations. In: 10th International Conference on Document Analysis and Recognition, ICDAR 2009, pp. 586–590. IEEE (2009)

14. Pinto, N., Cox, D.D., DiCarlo, J.J.: Why is real-world visual object recognition hard? PLoS Comput. Biol. **4**(1), e27 (2008)

15. Ranzato, M., Huang, F.J., Boureau, Y.L., LeCun, Y.: Unsupervised learning of invariant feature hierarchies with applications to object recognition. In: IEEE Conference on Computer Vision and Pattern Recognition, CVPR 2007, pp. 1–8. IEEE (2007)

16. Serre, T., Wolf, L., Bileschi, S., Riesenhuber, M., Poggio, T.: Robust object recognition with cortex-like mechanisms. IEEE Trans. Pattern Anal. Mach. Intell. **29**(3), 411–426 (2007)

17. Shao, Y., Lunetta, R.S.: Comparison of support vector machine, neural network, and cart algorithms for the land-cover classification using limited training data points. ISPRS J. Photogram. Remote Sens. **70**, 78–87 (2012)

18. Wan, L., Zeiler, M., Zhang, S., Cun, Y.L., Fergus, R.: Regularization of neural networks using dropconnect. In: Proceedings of the 30th International Conference on Machine Learning (ICML-2013), pp. 1058–1066 (2013)

19. Yu, D., Wang, S., Karam, Z., Deng, L.: Language recognition using deep-structured conditional random fields. In: 2010 IEEE International Conference on Acoustics Speech and Signal Processing (ICASSP), pp. 5030–5033. IEEE (2010)

Modified Orthogonal Neighborhood Preserving Projection for Face Recognition

Purvi Koringa[1]([✉]), Gitam Shikkenawis[1], Suman K. Mitra[1],
and S.K. Parulkar[2]

[1] DA-IICT, Gandhinagar 382007, Gujarat, India
{201321010,201221004,suman_mitra}@daiict.ac.in
[2] BARC, Mumbai, India
skparul@barc.gov.in

Abstract. In recent times most of the face recognition algorithms are based on subspace analysis. High dimensional image data are being transformed into lower dimensional subspace thus leading towards recognition by embedding a new image into the lower dimensional space. Starting from Principle Component Analysis(PCA) many such dimensionality reduction procedures have been utilized for face recognition. Recent edition is Neighborhood Preserving Projection (NPP). All such methods lead towards creating an orthogonal transformation based on some criteria. Orthogonal NPP builds a linear relation within a small neighborhood of the data and then assumes its validity in the lower dimension space. However, the assumption of linearity could be invalid in some applications. With this aim in mind, current paper introduces an approximate non-linearity. In particular piecewise linearity, within the small neighborhood which gives rise to a more compact data representation that could be utilized for recognition. The proposed scheme is implemented on synthetic as well as real data. Suitability of the proposal is tested on a set of face images and a significant improvement in recognition is observed.

Keywords: Dimensionality reduction · Data visualization · Face recognition

1 Introduction

Subspace based methods are the recent trend for face recognition/identification problem. Face recognition appears as one of the most challenging problems in machine learning, and computer vision systems. Early recognition methods are based on the geometric features. Mostly local facial features, such as shape, size,

---

P. Koringa acknowledges Board of Research in Nuclear Science, BARC, India for supporting this research.
G. Shikkenawis is acknowledging Tata Consultancy Services for providing the financial support for carrying out this work.

© Springer International Publishing Switzerland 2015
M. Kryszkiewicz et al. (Eds.): PReMI 2015, LNCS 9124, pp. 225–235, 2015.
DOI: 10.1007/978-3-319-19941-2_22

structure and location of eyes, nose, mouth, chin, ears in each person's face happened to be popular identification features in face recognition. On the contrary, subspace based methods use the intrinsic data manifold present in the face images.

Though, face image seems to be high dimensional data, it is observed that it lies in comparatively very low linear or non-linear manifold [1,2]. This leads to develop face recognition systems based on subspace arising from data dimensionality reduction. The basic idea is to find a linear or non-linear transformation to map the image to a lower dimensional subspace which makes the same class of data more compact for the convenience of classification. Such underlying manifold learning based face recognition methods have attracted considerable interests in recent years. Some of the examples are Principal Component Analysis (PCA) [1], Linear Discriminant Analysis (LDA) [3], Locality Preserving Projection (LPP) [4,5] and Neighborhood Preserving Embedding (NPE) [6]. Techniques such as PCA and LDA tend to preserve mostly global geometry of data (image in the present context). On the other hand, techniques such as LPP and NPE preserve local geometry by a graph structure, based on nearest neighborhood information.

The linear dimensionality reduction method Orthogonal Neighborhood Preserving Projection (ONPP) proposed in [2] preserves global geometry of data as well as captures intrinsic dependency of local neighborhood. ONPP is linear extension of Locally Linear Embedding (LLE) [7] which uses a weighted nearest neighborhood graph to preserve local geometry by representing each data point as linear combination of its neighbors. It simply embeds sample points into lower dimensional space without having any mechanism of reconstructing the data. ONPP uses the same philosophy as that of LLE and projects the sample data onto linear subspace but at the same time suggests a procedure to reconstruct data points. A variant of ONPP, Discriminative ONPP (DONPP) proposed in [8] takes into acount both, intraclass as well as interclass geometry. In this paper, a modified ONPP algorithm is proposed and its performance is compared with existing ONPP and DONPP algorithms. In particular, a z-shaped function based criterion is used to compute the coefficient of linear combination of neighbors of each data point. Note that, when the algorithm is applied to face recognition, face images are considered as data points. The modified algorithm is tested on synthetic as well as real data. To show its efficiency, the algorithm is tested on well-known face databases like AR [9], ORL [10], and UMIST [11]. Results of the proposed algorithm are comparable to that of the existing one in some cases and are significantly better in other cases.

The paper is organized in five sections. In the next section, ONPP and DONPP algorithms are explained in detail, followed by the modification on ONPP suggested in Sect. 3. Section 4 consists of experimental results and Sect. 5 concludes the performance of suggested algorithm on various types of databases.

2   Orthogonal Neighborhood Preserving Projection (ONPP)

ONPP [2] is a linear extension of Locally Linear Embedding. LLE is a nonlinear dimensionality reduction technique that embeds high dimension data samples on lower dimensional subspace. This mapping is not explicit in the sense that embedding is data dependent. In LLE, intrinsic data manifold changes with the inclusion or exclusion of data points. Hence, on inclusion of a new data point, embedding of all existing data points changes. This prevented subspace based recognition of unknown sample point, as this unknown sample point can not be embedded into the existing lower dimensional subspace. Lack of explicit mapping thus makes LLE not suitable for recognition. ONPP resolves this problem and finds the explicit mapping of the data in lower dimensional subspace through a linear orthogonal projection matrix. In presence of this orthogonal projection matrix, new data point can be embedded into lower dimensional subspace.

Let $\mathbf{x_1}, \mathbf{x_2}, ..., \mathbf{x_n}$ be given data points form m-dimensional space ($\mathbf{x_i} \in \mathbf{R}^m$). So the data matrix is $\mathbf{X} = [\mathbf{x_1}, \mathbf{x_2}, ..., \mathbf{x_n}] \in \mathbf{R}^{m \times n}$. The basic task of subspace based methods is to find an orthogonal/non-orthogonal projection matrix $\mathbf{V}^{m \times d}$ such that $\mathbf{Y} = \mathbf{V}^T \mathbf{X}$, where $\mathbf{Y} \in \mathbf{R}^{d \times n}$ is the embedding of $\mathbf{X}$ in lower dimension as d is assumed to be less than m.

ONPP is a two step algorithm where in the first step each data point is expressed as a linear combination of its neighbors. In the second step the data compactness is achieved through a minimization problem.

For, each data point $\mathbf{x_i}$, nearest neighbors are selected in two ways. In one way, k neighbors are selected by Nearest Neighbor (NN) technique where k is a suitably chosen parameter. In another way, neighbors could be selected which are within ε distance apart from the data point. Let $\mathcal{N}_{x_i}$ be the set of k neighbors. First, data point $\mathbf{x_i}$ is expressed as linear combination of its neighbors as $\sum_{j=1}^{k} w_{ij} \mathbf{x_j}$ where, $\mathbf{x_j} \in \mathcal{N}_{x_i}$. The weight w_{ij} are calculated by minimizing the reconstruction errors i.e. error between $\mathbf{x_i}$ and linear combination of $\mathbf{x_j} \in \mathcal{N}_{x_i}$.

$$\arg\min \mathcal{E}(\mathbf{W}) = \frac{1}{2} \sum_{i=1}^{n} \| \mathbf{x_i} - \sum_{j=1}^{k} w_{ij} \mathbf{x_j} \|^2 \tag{1}$$

subject to $\sum_{j=1}^{k} w_{ij} = 1$.

Corresponding to point $\mathbf{x_i}$, let $\mathbf{X_{N_i}}$ be a matrix having $\mathbf{x_j}$ as its columns, where $\mathbf{x_j} \in N_{x_i}$. Note that $\mathbf{X_{N_i}}$ includes $\mathbf{x_i}$ as its own neighbor. Hence, $\mathbf{X_{N_i}}$ is a $m \times k{+}1$ matrix. Now by solving the least square problem $(\mathbf{X_{N_i}} - \mathbf{x_i}\mathbf{e}^T)\mathbf{w_i} = \mathbf{0}$ with a constraint $\mathbf{e}^T \mathbf{w_i} = \mathbf{1}$, a closed form solution, as shown in Eq. (2) is evolved for $\mathbf{w_i}$. Here, $\mathbf{e}$ is a vector of ones having dimension $k \times 1$ same as $\mathbf{w_i}$.

$$\mathbf{w_i} = \frac{\mathbf{G}^{-1}\mathbf{e}}{\mathbf{e}^T \mathbf{G}^{-1}\mathbf{e}} \tag{2}$$

where, $\mathbf{G}$ is Gramiam matrix of dimension $k \times k$. Each entry of $\mathbf{G}$ is calculated as $g_{pl} = (\mathbf{x_i} - \mathbf{x_p})^T (\mathbf{x_i} - \mathbf{x_l}), \; for \; \forall \mathbf{x_p}, \mathbf{x_l} \in \mathcal{N}_{x_i}$

Next step is dimensionality reduction or finding the projection matrix V as stated earlier. The method basically seeks the lower dimensional projection $\mathbf{y_i} \in \mathbf{R}^d$ of data point $\mathbf{x_i} \in \mathbf{R}^m$ ($d << m$) with some criteria. The criteria imposed here assumes that the linear combination of neighbors $\mathbf{x_j}$s which reconstruct the data point $\mathbf{x_i}$ in higher dimension would also reconstruct $\mathbf{y_i}$ in lower dimension with corresponding neighbors $\mathbf{y_j}$s along with same weight as in higher dimensional space.

Such embedding is obtained by minimizing the sum of squares of reconstruction errors in the lower dimensional space. Hence, the objective function is given by

$$\arg\min \mathcal{F}(\mathbf{Y}) = \sum_{i=1}^{n} \| \mathbf{y_i} - \sum_{j=1}^{n} w_{ij}\mathbf{y_j} \|^2 \qquad (3)$$

subject to, $\mathbf{V}^T\mathbf{V} = \mathbf{I}$ (orthogonality constraint).

This optimization problem results in computing eigen vectors corresponding to the smallest d eigen values of matrix $\tilde{\mathbf{M}} = \mathbf{X}(\mathbf{I}-\mathbf{W})(\mathbf{I}-\mathbf{W}^T)\mathbf{X}^T$. ONPP explicitly maps $\mathbf{X}$ to $\mathbf{Y}$, which is of the form $\mathbf{Y} = \mathbf{V}^T\mathbf{X}$, i.e. each new sample $\mathbf{x_l}$ can now be projected to lower dimension by just a matrix-vector product $\mathbf{y_l} = \mathbf{V}^T\mathbf{x_l}$.

ONPP can also be implemented in supervised mode where the class labels are known. Face recognition, character recognition etc. are problems where supervised mode is better suited. In supervised mode, data points $\mathbf{x_i}$ and $\mathbf{x_j}$ belonging to the same class are considered neighbors to each other thus $w_{ij} \neq 0$ and $w_{ij} = 0$ otherwise. In supervised technique, the parameter k (number of nearest neighbors) need not to be specified manually, it is automatically set to number of data samples in particular class.

Considering the undersampled size problem where the number of samples n is less than dimension m, $m > n$. In such scenario, the matrix $\tilde{\mathbf{M}} \in \mathbf{R}^{m \times m}$ will have maximum rank $n - c$, where c is number of classes. In order to ensure that the resulting matrix $\tilde{\mathbf{M}}$ will be non-singular, one may employ an initial PCA projection that reduces the dimensionality of the data vectors to $n - c$. If $\mathbf{V_{PCA}}$ is the dimensionality reduction matrix of PCA, then on performing the ONPP the resulting dimensionality reduction matrix is given by $\mathbf{V} = \mathbf{V_{PCA}}\mathbf{V_{ONPP}}$.

ONPP considers only intraclass geometric information, a variant of ONPP proposed in [8], takes into account interclass information as well to improve classification performance, is known as Discriminative ONPP (DONPP). For a given sample $\mathbf{x_i}$, its $n_i - 1$ neighbors having same class labels are denoted by $\mathbf{x_i}, \mathbf{x_{i^1}}, \mathbf{x_{i^2}}, ..., \mathbf{x_{i^{n_i-1}}}$ and its k nearest neighbors having different class labels are denoted by $\mathbf{x_{i_1}}, \mathbf{x_{i_2}}, ..., \mathbf{x_{i_k}}$. Thus, neighbors of sample $\mathbf{x_i}$ can be described as $\mathbf{X_i} = [\mathbf{x_i}, \mathbf{x_{i^1}}, \mathbf{x_{i^2}}, ..., \mathbf{x_{i^{n_i-1}}}, \mathbf{x_{i_1}}, \mathbf{x_{i_2}}, ..., \mathbf{x_{i_k}}]$ and its low-dimensional projection can be denoted as $\mathbf{Y_i} = [\mathbf{y_i}, \mathbf{y_{i^1}}, \mathbf{y_{i^2}}, ..., \mathbf{y_{i^{n_i-1}}}, \mathbf{y_{i_1}}, ..., \mathbf{y_{i_k}}]$.

In projected space, it is expected that the sample and its neighbors having same label preserve local geometry, while neighbors having different labels are projected as far as possible from the sample to increase interclass distance. This can be achieved by optimizing Eq. (3) in addition to Eq. (4).

$$\arg\max \mathcal{F}(\mathbf{Y_i}) = \sum_{p=1}^{k} \| \mathbf{y_i} - \mathbf{y_{i_p}} \|^2 \qquad (4)$$

Considering Eqs. (3) and (4) simultaneously, the optimization problem for sample x_i can be written as

$$\arg\min \mathcal{F}(\mathbf{Y_i}) = (\| \mathbf{y_i} - \sum_{j=1}^{n_i-1} w_{ij}\mathbf{y_j} \|^2 - \beta\sum_{p=1}^{k} \| \mathbf{y_i} - \mathbf{y_{i_p}} \|^2) \qquad (5)$$

where, β is scaling factor between $[0, 1]$. This minimization problem simplifies into an eigenvalue problem, and projection matrix $\mathbf{V_{DONPP}}$ can be achieved by eigenvectors corresponding to smallest d eigenvalues.

3   Modified Orthogonal Neighbourhood Preserving Projection (MONPP)

ONPP is based on two basic assumptions, first it assumes that a linear relation exists in a local neighborhood and hence any data point can be represented as a linear combination of its neighbors. Secondly it assumes that this linear relationship also exists in the projection space. The later assumption gives rise to a compact representation of the data that can enhance the classification performance. The data compactness would be more visible in the case when the first assumption is strongly valid. While experimenting with synthetic data, as shown in Fig. 2, it has been observed that data compactness is not clearly revealed. The main drawback could be the strict local linearity assumption. Focusing on this, we are trying to incorporate some kind of non-linear relationship of a data point with its neighbors. The proposed algorithm is termed as Modified ONPP.

In this proposed modification, a z-shaped function is used to assign weights to nearest neighbors in the first stage of ONPP. Note that in ONPP, the weight matrix $\mathbf{W}$ is calculated by minimizing the cost function in Eq. (1), which is a least square solution (2). In the least square solution, weights of neighbors are inversely proportional to the distance of the neighbors from the point of interest. We are looking for a situation where the neighbors closest to the point of interest would get maximum weight and there after the weights will be adjusted non-linearly (through Z-shaped function) as the distance increases. After a certain distance the weights will be very low. In particular, instead of assuming a linear relationship, a piecewise linear relationship is incorporated through the z-shaped function. This piecewise linear relationship is leading towards some kind of non-linear relationship.

$$
\mathcal{Z}(x; a, b) = \begin{cases} 1 & \text{if } x \leq a \\ 1 - 2\left(\frac{x-a}{a-b}\right)^2 & \text{if } a \leq x \leq \frac{a+b}{2} \\ 2\left(\frac{x-b}{a-b}\right)^2 & \text{if } \frac{a+b}{2} \leq x \leq b \\ 0 & \text{Otherwise} \end{cases} \tag{6}
$$

Parameters a and b locate the extremes of the sloped portion of the curve and can be set to 0 and maximum within class distance (i.e. maximum pair wise distance between data samples belonging to the same class) respectively, as shown in Fig. 1. In case of unsupervised mode, a k-NN algorithm could be implemented before assigning the weights and hence the parameters a and b of Eq. (6) can be adjusted.

Finally, Eq. (7) is used to assign weight to each neighbor $\mathbf{x_j}$ corresponding to $\mathbf{x_i}$. Note that this equation is same as Eq. (2), where $\mathbf{G}^{-1}$ is replaced by $\mathbf{Z}$. The new weights are

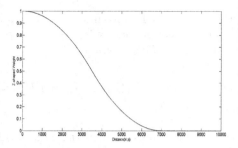

Fig. 1. Z-shaped weight function for Range [0, Maximum within class distance], illustrated for max distance of 7000 unit

$$\mathbf{w_i} = \frac{\mathbf{Ze}}{\mathbf{e}^T \mathbf{Ze}} \tag{7}$$

where, elements of this $\mathbf{Z}$ matrix are defined as

$$Z_{pl} = \mathcal{Z}(d_p; a, b) + \mathcal{Z}(d_l; a, b) \ for, \ \forall \mathbf{x_p}, \mathbf{x_l} \in \mathcal{N}_{\mathbf{x_i}} \tag{8}$$

here, $\mathcal{Z}(d_k; a, b)$ is calculated using Eq. (6) where, d_k is the Euclidean distance between $\mathbf{x_i}$ and it's neighbor $\mathbf{x_k}$. Parameters a and b are obtained as discussed earlier.

Next step computes projection matrix $\mathbf{V} \in \mathbf{R}^{m \times d}$ whose column vectors are smallest d eigen vectors of matrix $\tilde{\mathbf{M}} = \mathbf{X}(\mathbf{I} - \mathbf{W})(\mathbf{I} - \mathbf{W}^T)\mathbf{X}^T$. Embedding of $\mathbf{X}$ on lower dimension $\mathbf{Y}$ is achieved by $\mathbf{Y} = \mathbf{V}^T \mathbf{X}$.

4   Experiments and Results

The suggested Modified ONPP is used for two well-known synthetic datasets along with a digit data [12], low dimensional projection of these data sets is compared with ONPP. MONPP has also been implemented extensively for various well-known face databases and the results are compared with that of the existing ONPP algorithm and a variant of ONPP, DONPP.

4.1   Synthetic Data

The modified algorithm is implemented on two synthetic datasets *viz Swissroll* (Fig. 2(a)) and *S-curve* (Fig. 2(e)) to visualize their two dimensional plot. These two datasets reveal a linear relationship within the class as well as between the classes when unfolded. Dimensionality reduction techniques such as PCA when applied to the data fail to capture this intrinsic linearity. However, dimensionality reduction techniques such as LPP [4], NPE [6] try to capture the local geometry and retain it into the projection space are expected to perform better. These algorithms give rise to a compact representation of the data without much distorting the shape of the data. To implement existing ONPP and proposed Modified ONPP, 1000 data points are randomly sampled from three dimensional manifold

(Fig. 2(b) and (f)) to build the orthogonal transformation matrix($\mathbf{V}$). Note that similar experiment has been performed in [2] to show the suitability of ONPP over LPP. As suitability of ONPP over LPP has already been shown, we are not showing any results of LPP. From Fig. 2(c),(d) and (g),(h), it is clear that the 2D representations of both *Swissroll* and *S-curve* seem to be much better for MONPP. To explore how ONPP and MONPP work with varied values of k, experiments have been conducted and results are shown in Fig. 3. Note that repeated experiments with a fixed k may not guarantee to generate same results. It is observed that projection using ONPP algorithm depends on k, variation in k results in huge variation in its lower dimensional representation. However, projection using MONPP is more stable with varying values of k. Larger values of k imply larger area of local neighborhood. It is possible that larger local area does not posses linearity. The linearity assumption of ONPP thus is invalid here. So the non-linearity present in moderately large local area is well-captured in MONPP and is reflected in the results.

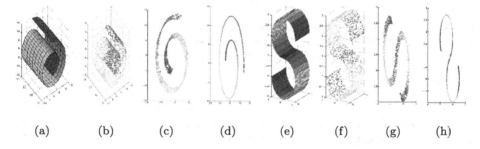

(a)   (b)   (c)   (d)   (e)   (f)   (g)   (h)

Fig. 2. Swissroll: original 3D data(a), sampled data(b), 2D projection obtained by ONPP(c) and MONPP(d). S-curve: original 3D data(e), sampled data(f), 2D projection obtained by ONPP(g) and MONPP(h). k(number of NN) is set to 6.

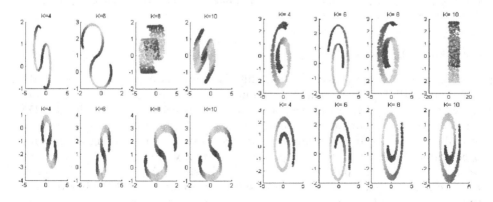

Fig. 3. 2D projection of S-curve(left) and Swissroll(right) with various k (number of NN) values with ONPP(top) and MONPP(bottom)

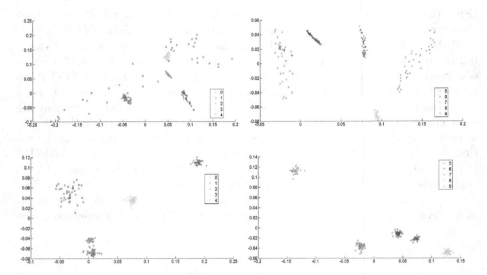

Fig. 4. 2D projection of digit data using ONPP and MONPP, top row shows performance of ONPP algorithm, Bottom row shows performance of MONPP algorithm, where, '+' denotes 5, 'o' denotes 6, '*' denotes 7, 'Δ' denotes 8, '$\diamond$' denotes 9.

4.2 Digit Data

The MNIST database [12] of handwritten digits is used to compare data visualization of both the algorithms. Randomly 40 data samples from each class (digit) are taken and projected on 2-D plane using ONPP and proposed Modified ONPP. The results are shown in Fig. 4, it can be clearly observed that the data is compact and well separated when MONPP is applied. It seems that there is a wide range of variations in digit '1' and that is reflected in Fig. 4(top-left). But the same digit '1' is more compact in the 2-D representation of MONPP (Fig. 4(bottom-left)). Similar argument is true for digits '7' and '9'. Overall, better compactness is evident for all digits in case of MONPP.

4.3 Face Data

The algorithm is also tested on three different face databases *viz* AR [9], ORL [10], and UMIST [11]. To maintain the uniformity, face images of all databases are resized to 38×31 pixels, thus each image is considered as a point in 1178 dimensional space. For each database, randomly 50 % of face images are selected as training samples and remaining are used for testing. The training samples are used to find the lower dimensional projection matrix $\mathbf{V}$. The test samples are then projected on this subspace and are recognized using a *NN-classifier*. The main intention of these experiments is to check the suitability of MONPP based image representation for face recognition and hence a simple classifier such as NN is used. To ensure that the results achieved are not biased to the randomized selection of training-testing data, the experiments are repeated twenty times with different

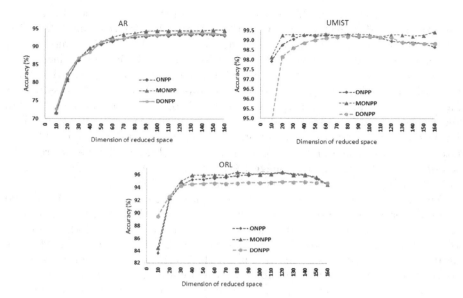

Fig. 5. Comparison of average performance of ONPP, DONPP and MONPP on AR database UMIST database ORL database

Table 1. Comparison of performance in the light of recognition score (in %) of ONPP, DONPP and MONPP

	ONPP		DONPP		MONPP	
Database	Average	Best (at dim d)	Average	Best (at dim d)	Average	Best (at dim d)
AR	93.50	96.25 (100)	93.80	96.83(140)	94.13	97.50 (110)
UMIST	98.95	99.05 (30)	99.50	99.23(60)	98.00	99.50 (20)
ORL	93.06	98.00 (120)	94.95	98.42(120)	95.90	99.00 (60)

randomization. Experiments are also conducted for different values of d (dimension of reduced space) ranging from 10 to 160 (at an interval of 10). The best as well as average recognition rates are reported here for all databases.

Average recognition results for AR, UMIST and ORL databases using ONPP, DONPP and MONPP are shown in Fig. 5. It can be observed that MONPP performs better than ONPP and DONPP across almost all values of d. Average recognition accuracies and best recognition accuracies along with the corresponding dimensions using ONPP, DONPP and MONPP for all three databases are reported in Table 1.

5   Conclusion

Subspace based methods for face recognition have been a major area of research and already proven to be more efficient. In this regard, Orthogonal Neighborhood

Preserving Projection (ONPP) is assumed to handle the intrinsic non-linearity of the data manifold. The first step of ONPP deals with a linear model building within local neighborhoods. This linearity assumption may not be valid for a moderately large neighborhood. In the present work, this linear model is thus replaced by a notion of non-linearity where a piecewise linear model (z-shaped) is used instead. The suitability of the proposal is tested on non-linear synthetic data as well as a few benchmark face databases. Significant and consistent improvement in data compactness is observed for synthetic data where manifold is surely nonlinear. This signifies the suitability of the present proposal to handle non-linear manifold of the data. On the other hand, noticeable improvement is obtained for the face recognition problem. The modification suggested over existing ONPP though very simple but overall improvement in face recognition results is very encouraging.

Another way to handle non-linear manifold is to use kernel method of manifold learning. Kernel versions of subspace methods such as PCA [13], OLPP [14], ONPP [2] have already been proposed. A kernel version of the current Modified ONPP could be the possible future direction of work. Either the current form of the proposal associating with discriminating method [8] or its kernel version is expected to exploit the non-linear data manifold that is present in the face database in the sense of variations in facial expression. Recognizing facial expression along with faces would be a challenging task for future.

References

1. Turk, M., Pentland, A.: Eigenfaces for recognition. J. Cogn. Neurosci. **3**(1), 71–86 (1991)
2. Kokiopoulou, E., Saad, Y.: Orthogonal neighborhood preserving projections: a projection-based dimensionality reduction technique. IEEE Trans. Pattern Anal. Mach. Intell. **29**(12), 2143–2156 (2007)
3. Lu, J., Plataniotis, K.N., Venetsanopoulos, A.N.: Face recognition using lda-based algorithms. IEEE Trans. Neural Netw. **14**(1), 195–200 (2003)
4. He, X., Niyogi, P.: Locality preserving projections. In: Thrun, S., Saul, L., Schölkopf, B. (eds.) Advances in Neural Information Processing Systems, pp. 153–160. MIT Press, Cambridge (2004)
5. Shikkenawis, G., Mitra, S.K.: Improving the locality preserving projection for dimensionality reduction. In: Third International Conference on Emerging Applications of Information Technology (EAIT), pp. 161–164. IEEE (2012)
6. He, X., Cai, D ., Yan, S., Zhang, H.J.: Neighborhood preserving embedding. In: Tenth IEEE International Conference on Computer Vision, ICCV 2005, vol. 2, pp. 1208–1213. IEEE (2005)
7. Roweis, S.T., Saul, L.K.: Nonlinear dimensionality reduction by locally linear embedding. Science **290**(5500), 2323–2326 (2000)
8. Zhang, T., Huang, K., Li, X., Yang, J., Tao, D.: Discriminative orthogonal neighborhood-preserving projections for classification. IEEE Trans. Syst. Man Cybern. B Cybern. **40**(1), 253–263 (2010)
9. Martinez, A.M.: The ar face database. CVC Technical report, **24**, (1998)
10. Ferdinando, S., Harter, A.: Parameterisation of a stochastic model for human face identification. In: Proceedings of 2nd IEEE Workshop on Applications of Computer Vision. AT&T Laboratories Cambridge, December 1994

11. Graham, D.B., Allinson, N.M.: Characterising virtual eigensignatures for general purpose face recognition. In: Face Recognition, pp. 446–456. Springer (1998)
12. LeCun, Y., Bottou, L., Bengio, Y., Haffner, P.: Gradient-based learning applied to document recognition. Proc. IEEE **86**(11), 2278–2324 (1998)
13. Schölkopf, B., Smola, A., Müller, K.R.: Kernel principal component analysis. In: Artificial Neural Networks ICANN 1997, pp. 583–588. Springer (1997)
14. Feng, G., Hu, D., Zhang, D., Zhou, Z.: An alternative formulation of kernel lpp with application to image recognition. Neurocomputing **69**(13), 1733–1738 (2006)

An Optimal Greedy Approximate Nearest Neighbor Method in Statistical Pattern Recognition

Andrey V. Savchenko$^{(\boxtimes)}$

Laboratory of Algorithms and Technologies for Network Analysis,
National Research University Higher School of Economics,
Nizhny Novgorod, Russia
avsavchenko@hse.ru

Abstract. The insufficient performance of statistical recognition of composite objects (images, speech signals) is explored in case of medium-sized database (thousands of classes). In contrast to heuristic approximate nearest-neighbor methods we propose a statistically optimal greedy algorithm. The decision is made based on the Kullback-Leibler minimum information discrimination principle. The model object to be checked at the next step is selected from the class with the maximal likelihood (joint density) of distances to previously checked models. Experimental study results in face recognition task with FERET dataset are presented. It is shown that the proposed method is much more effective than the brute force and fast approximate nearest neighbor algorithms, such as randomized kd-tree, perm-sort, directed enumeration method.

Keywords: Statistical pattern recognition · Approximate nearest neighbor · Kullback-Leibler divergence · Directed enumeration method · Face recognition

1 Introduction

The problem of small sample size is crucial in pattern recognition of complex objects (e.g., images) [1]. In fact, most of known algorithms in this case are equivalent to the nearest neighbor (NN) method with appropriate similarity measure [2]. If the number of classes is large (hundreds or even thousands of classes), the performance of NN's exhaustive search is not enough for real-time processing. It seems, conventional fast approximate NN image *retrieval* methods [3] can be applied, e.g. AESA (Approximating and Eliminating Search Algorithm) [4], composite kd-tree [5], randomized kd-tree [6], recent variations of Locality-Sensitive Hashing [7], etc. Unfortunately, these techniques usually cannot be efficiently used in *recognition* tasks as the latter are significantly different from retrieval in terms of

(1) quality indicators (accuracy in recognition and recall in retrieval): 3–5 % losses in accuracy/recall of retrieval techniques are inappropriate for many recognition tasks;
(2) similarity measures in recognition tasks are much more complex [8] in comparison with conventional Minkowski or cosine distance in retrieval. Image retrieval

M. Kryszkiewicz et al. (Eds.): PReMI 2015, LNCS 9124, pp. 236–245, 2015.
DOI: 10.1007/978-3-319-19941-2_23

methods are with similarity measures which satisfy metric properties (sometimes, triangle inequality and, usually, symmetry) [4, 9]. They are known to show good performance only if the first NN is quite different from other models;

(3) classification methods (1-NN in recognition and k-NN in retrieval);

(4) database size (medium in recognition and very-large in retrieval). Performance of approximate NN algorithms is comparable with brute-force for medium-sized training sets (thousands of classes). To decrease the recognition speed for such training sets, other methods, e.g., ordering permutations (perm-sort) [10] and directed enumeration method (DEM) [11] has recently been proposed.

Final issue is the heuristic nature of most approximate NN methods. It is usually impossible to prove that particular algorithm is optimal and nothing can be done to improve it. In this paper we propose an alternative solution on the basis of the statistical approach - while looking for the NN for particular query object, conditional probability of belonging of previously checked models to each class is estimated. The next model from the database is selected from the class with the maximal probability.

The rest of the paper is organized as follows. In Sect. 2 we recall the Kullback-Leibler minimum discrimination principle [12] in statistical pattern recognition. In Subsect. 2.2 we briefly review the baseline method (DEM). In Sect. 3 the novel Maximum-Likelihood DEM (ML-DEM) is proposed. In Sect. 4 experimental study results are presented in face recognition with FERET dataset. Finally, concluding comments are given in Sect. 5.

2 Materials and Methods

2.1 Statistical Pattern Recognition

In the pattern recognition task it is required to assign the query object X to one of $R > 1$ classes [2]. Each class is specified by the given model object X_r, $r \in \{1, ..., R\}$. First stage is feature extraction. In this paper we use the statistical approach and assume that each class is characterized with its own probabilistic distribution of appropriate features. Let's focus on the most popular discrete case, in which the features can take $N > 1$ different values. Hence, the distribution of rth class is defined as a histogram $H_r = [h_{r;1}, h_{r;2}, ..., h_{r;N}]$ estimated based on the X_r. The same procedure of histogram evaluation $H = [h_1, h_2, ..., h_N]$ is repeated for the query object X.

If the prior probabilities of each class are equal, the maximal likelihood criterion [2] can be used to test statistical hypothesis W_r, $r \in \{1, ..., R\}$ about distribution H:

$$\max_{r \in \{1,...,R\}} f_r(X), \tag{1}$$

where the likelihood of rth class $f_r(X)$ is estimated as follows

$$f_r(X) = \prod_{i=1}^{N} (h_{r;i})^{n \cdot h_i}. \tag{2}$$

Here it is assumed that the query object X contains n simple features to estimate the histogram H. Thus, the decision (1) is equivalent to the Kullback-Leibler minimum information discrimination principle [12]

$$\min_{r\in\{1,\dots,R\}} \rho(X, X_r),\tag{3}$$

where

$$\rho(X, X_r) = \rho_{KL}(H, H_r) = \sum_{i=1}^{N} h_i \cdot \ln\frac{h_i}{h_{r;i}}.\tag{4}$$

is the Kullback-Leibler divergence between densities H and H_r.

2.2   Baseline: Directed Enumeration Method

It is known that the performance of brute force implementation of criterion (3) can be rather low. To speed-up recognition process, fast approximate NN algorithms can be used. As a baseline approximate NN method we use the DEM [11] which was based on the metric properties of the Kullback-Leibler divergence and regards the models' similarity $\rho_{i,j} = \rho(X_i, X_j)$ as an average information from an observation to distinct class i from an alternative class j. At the preliminarily step, the model distance matrix $\mathbf{P} = [\rho_{i,j}]$ is calculated as it is done in the AESA [3]. This time-consuming procedure should be repeated only once for a particular task and training set.

Original DEM used the following heuristic: if there exists a model X_v for which $\rho(X, X_v) < \rho_0 \ll 1$, then condition holds $|\rho(X, X_r) - \rho_{v,r}| \ll 1$ with high probability for an arbitrary r-th model. Hence, criterion (3) can be simplified

$$\rho(X, X_v) < \rho_0 = const.\tag{5}$$

This equation defines the termination condition of the approximate NN method. If false-accept rate (FAR) β is fixed, then ρ_0 is evaluated as a β-quantile of the distances between images from distinct classes $\{\rho_{i,j} | i \in \{1, \dots, R\}, j \in \{1, \dots, i-1, i+1, \dots, R\}\}$ [11]. According to the DEM [11], at first, the distance $\rho(X, X_{r_1})$ to randomly chosen model X_{r_1} is calculated. Next, it is put into the priority queue of models sorted by the distance to X. The highest priority item X_i is pulled from the queue and the set of models $X_i^{(M)}$ is determined from

$$\left(\forall X_j \notin X_i^{(M)}\right) \left(\forall X_k \in X_i^{(M)}\right) \quad \Delta\rho(X_j) \geq \Delta\rho(X_k)\tag{6}$$

where $\Delta\rho(X_j) = |\rho_{i,j} - \rho(X, X_j)|$ is the deviation of $\rho_{i,j}$ relative to the distance between X and X_j. For all models from the set $X_i^{(M)}$ the distance to the query object is calculated and the condition (5) is verified. After that, every previously unchecked model from this set is put into the priority queue. The method is terminated if for one model object condition (5) holds or after checking for $E_{\max} = const$ models.

As we stated earlier, this method is heuristic as most popular approximate NN algorithms. However, the probability that the model is the NN of X can be directly calculated for the Kullback-Leibler discrimination by using its asymptotic properties. Let's describe this idea in detail in the next section.

3 Maximum-Likelihood Directed Enumeration Method

In this section we primarily focus on *greedy* algorithms: it explores an each step the model which is the NN of the query object X with the highest probability. It is known [12] that if an object X has distribution H_v, then the distance $2n \cdot \rho(X, X_v)$ is asymptotically distributed as a χ^2 with $(N - 1)$ degrees of freedom. Similarly, $2n \cdot \rho(X, X_r)$, $r \neq v$ has asymptotic non-central χ^2 distribution with $(N - 1)$ degrees of freedom and noncentrality parameter $2nK \cdot \rho_{v,r}$. If N is high, then, by using the central limit theorem, we obtain the normal distribution of the distance $\rho(X, X_r)$:

$$N\left(\rho_{v,r} + (N-1)/(2n); \left(\sqrt{8n \cdot \rho_{v,r} + 2(N-1)}/(2n)\right)^2\right). \tag{7}$$

At first, based on the asymptotic distribution (7) we replace the step (6) of the original DEM to the procedure of choosing the maximum likelihood model. Let's assume that the models $X_{r_1}, \ldots, X_{r_l}$ have been examined before the l-th step. We choose the next most probable model $X_{r_{l+1}}$ with the maximum likelihood method:

$$r_{l+1} = \operatorname*{arg\,max}_{v \in \{1,\ldots,R\} - \{r_1,\ldots,r_l\}} \prod_{i=1}^{l} f(\rho(X, X_{r_i}) | W_v). \tag{8}$$

where $f(\rho(X, X_{r_i}) | W_v)$ is the conditional density (likelihood) of the distance $\rho(X, X_{r_i})$ if the hypothesis W_v is true. By using asymptotic distribution (7), the likelihood in (8) can be written in the following form

$$
\begin{aligned}
f(\rho(X, X_{r_i}) | W_v) &= \frac{2n}{\sqrt{2\pi \cdot (8n \cdot \rho_{v,r_i} + 2(N-1))}} \\
&\quad \times \exp\left[-\frac{(2n \cdot (\rho(X, X_{r_i}) - \rho_{v,r_i}) - (N-1))^2}{8n \cdot \rho_{v,r_i} + 2(N-1)}\right] \\
&= \frac{2n}{\sqrt{2\pi}} \exp\left[-\frac{1}{2}\ln(8n \cdot \rho_{v,r_i} + 2(N-1))\right] \\
&\quad \times \exp\left[-\frac{(2n \cdot (\rho(X, X_{r_i}) - \rho_{v,r_i}) - (N-1))^2}{8n \cdot \rho_{v,r_i} + 2(N-1)}\right]
\end{aligned}
\tag{9}
$$

By several transformations of (9) and assuming that the number of simple features is much higher the number of parameters $n \gg N$, expression (8) is written as

$$r_{l+1} = \underset{\mu \in \{1,\ldots,R\}-\{r_1,\ldots,r_l\}}{\arg\min} \sum_{i=1}^{l} \varphi_\mu(r_i). \tag{10}$$

where

$$\varphi_\mu(r_i) \approx \left(\rho(X, X_{r_i}) - \rho_{\mu,r_i}\right)^2 / (4\rho_{\mu,r_i}). \tag{11}$$

This equation is in good agreement with the heuristic from the original DEM [11] - the closer are the distances $\rho(X, X_{r_i})$ and ρ_{μ,r_i} and the higher is the distance between models X_μ and X_{r_i}, the lower is $\varphi_\mu(r_i)$.

Next, the termination condition (5) is tested for the model $X_{r_{l+1}}$. If the distance $\rho(X, X_{r_{l+1}})$ is lower than a threshold ρ_0 or the number of calculated distances exceed $E_{\max}$, then the search procedure is stopped on the $L_{checks} = l + 1$ step. Otherwise the model $X_{r_{l+1}}$ is put into the set of previously checked models and the procedure (10) and (11) is repeated.

Let us return to the initialization of our method. We would like to choose the first model X_{r_1} to obtain the decision (5) in a shortest (in terms of number of calculations L_{checks}) way. An average probability to obtain the decision is maximized at the second step:

$$r_1 = \underset{\mu \in \{1,\ldots,R\}}{\arg\max} \frac{1}{R} \sum_{v=1}^{R} P\left(\varphi_v(\mu) \le \min_{r \in \{1,\ldots R\}} \varphi_r(\mu)\Big| W_v\right). \tag{12}$$

To estimate the conditional probability in (12) we use again the asymptotic distribution (7). The first model to check X_{r_1} is obtained from the following expression

$$r_1 = \underset{\mu \in \{1,\ldots,R\}}{\arg\max} \sum_{v=1}^{R} \prod_{r=1}^{R} \left(\frac{1}{2} + \Phi\left(\frac{\sqrt{n}}{2}\Big|\sqrt{\rho_{r,\mu}} - \sqrt{\rho_{v,\mu}}\Big|\right)\right). \tag{13}$$

where $\Phi(\ \cdot\ )$ is the cumulative density function of the normal distribution.

Thus, the proposed ML-DEM (10), (11) and (13) is an optimal (maximal likelihood) greedy algorithm for an approximate NN search. The ML-DEM is different from the baseline DEM in initialization (14) and in the rule of choosing the next model (10) and (11). In the DEM $M > 1$ models are chosen (6) and in the proposed ML-DEM only one model is selected (10). Only the termination condition (5) is the same for both DEM and ML-DEM. In fact, the proposed method can be applied not only with the Kullback-Leibler discrimination (4), but with an arbitrary similarity measure. However, the property of statistical optimality is preserved only for similarity measure (4).

4 Experimental Results

Our experimental study deals with face recognition problem [13] with color FERET dataset. All 2720 frontal photos were converted to grayscale intensity images. Random cross-validation repeated 100 times was applied. Each time $R = 1420$ randomly chosen images of 994 persons populate the database (i.e. a training set), other 1300 photos of the same persons formed a test set. Faces were detected with the OpenCV library. After that the median filter with window size (3×3) is applied to remove noise in detected faces. The facial image is divided into a regular grid of $K \times K$ blocks, where $K = 10$. Next the HOGs (histograms of oriented gradients) $H_r(k_1, k_2)$ with $N = 8$ bins are calculated for each block (k_1, k_2) [14]. We assume, that each HOG is normalized, so that it may be treated as a probability distribution [14] in (4). The distance in the nearest neighbor rule (3) is calculated as follows [9]

$$\rho(X, X_r) = \sum_{k_1=1}^{K} \sum_{k_2=1}^{K} \min_{|\Delta_1| \le \Delta, |\Delta_2| \le \Delta} \rho(H(k_1, k_2), H_r(k_1 + \Delta_1, k_2 + \Delta_2)) \quad (14)$$

with the mutual alignment of the histograms in the Δ-neighborhood in order to take into account the small spatial deviations due to misalignment after face detection. In (14) we use the Kullback-Leibler divergence (4) between the HOGs and the homogeneity-testing probabilistic neural network (HT-PNN) which showed high face recognition rate and is equivalent to the statistical approach if the pattern recognition problem is referred as a task of testing for homogeneity of segments [15].

The error rate obtained with the NN rule and similarity measure (1) with Kullback-Leibler and the HT-PNN distances is shown in Table 1 in the format average error rate $\pm$ its standard deviation. Here, first, alignment of HOGs (22) with $\Delta = 1$ improves the recognition accuracy. And, second, we experimentally support the claim [15] that the error rate for the Kullback-Leibler distance (4) is higher when compared with the HT-PNN.

Table 1. Face recognition error rate, criterion (3) and (14)

	$\Delta = 0$	$\Delta = 1$
Kullback-Leibler divergence	8.9 ± 1.3	7.0 ± 1.3
HT-PNN	7.8 ± 1.2	6.6 ± 1.3

In the next experiment we compare the performance of the proposed ML-DEM with brute force (3), original DEM [11], and several approximate NN methods from FLANN [5] and NonMetricSpaceLib [16] libraries showed the best speed, namely

1. Randomized kd-tree from FLANN with 4 trees [6]
2. Composite index from FLANN which combines the randomized kd-trees (with 4 trees) and the hierarchical k-means tree [5].
3. Ordering permutations (perm-sort) from NonMetricSpaceLib which is known to decrease the recognition speed for medium-sized databases [10].

We evaluate the error rate (in %) and the average time (in ms) to recognize one test image with a modern laptop (4 core i7, 6 Gb RAM) and Visual C++ 2013 Express compiler (×64 environment) and optimization by speed. We explore an application of parallel computing [17] by dividing the whole training set into T = const non-over-lapped parts. Each part is processed in its own thread implemented by using the Windows ThreadPool API. We analyze both nonparallel ($T = 1$) and parallel ($T = 8$) cases. After several experiments the best (in terms of recognition speed) value of parameter M of original DEM (6) was chosen $M = 64$ for nonparallel case and $M = 16$ for parallel one. To obtain threshold ρ_0, the FAR is fixed to be $\beta = 1$ %. Parameter E_{max} was chosen to achieve the recognition accuracy which is not 0.5 % less than the accuracy of brute force (Table 1). If such accuracy can not be achieved, E_{max} was set to be equal to the count of models R. The average recognition time per one test image (in ms) is shown in Table 2.

Table 2. Average recognition time (ms.)

Distance/ features	Kullback-Leibler divergence				HT-PNN			
	$\Delta = 0$		$\Delta = 1$		$\Delta = 0$		$\Delta = 1$	
	$T = 1$	$T = 8$	$T = 1$	$T = 8$	$T = 1$	$T = 8$	$T = 1$	$T = 8$
Brute force	12.9	2.8	99.1	26.6	19.4	5.5	146.1	38.7
Randomized KD tree	11.9	2.6	91.2	21.4	16.4	4.3	129.4	30.4
Composite	12.0	2.6	91.5	22.5	16.7	4.3	129.9	35.2
Perm-sort	4.0	2.1	31.0	12.9	7.8	2.4	43.7	14.3
DEM	5.34	1.3	52.7	12.7	7.3	2.3	52	16.1
ML-DEM	2.8	0.8	24.2	10.0	5.3	1.4	24.9	5.8

Here randomized and composite kd-trees do not show superior performance even over brute force as the number of models in the database is not very high. However, as it was expected, perm-sort method is characterized with 2–3.5 times lower recognition speed in comparison with an exhaustive search. Moreover, perm-sort seems to be better than the original DEM for nonparallel case ($T = 1$), though the DEM's parallel implementation is a bit better. The most important conclusion here is that the proposed ML-DEM shows the highest speed in all experiments. The results of the HT-PNN's usage are very similar, though the error rate here is 0.5–1 % lower (see Table 1). In this case the original DEM is slightly faster than the perm-sort for conventional distance ($\Delta = 0$) but is not so effective for alignment ($\Delta = 1$). FLANN's kd-trees are 10–15 % faster than the brute force. And again, the proposed ML-DEM is the best choice here especially for most complex case ($T = 8$, $\Delta = 1$) for which only 6 ms (in average) is necessary to achieve 93 % accuracy.

To clarify the difference in performance of the original DEM and the proposed ML-DEM, we show the dependence of the error rate and the number of checked models $L_{checks}/R \cdot 100$ % on E_{max} in Fig. 1a, b respectively. Here the speed of convergence to an optimal solution for the ML-DEM is much higher than the same indicator of the

DEM (Fig. 1a). Even when $E_{max} = 0.1 \cdot R$ we can get an appropriate solution. Figure 1b proves that, as expected, the proposed ML-DEM is better than the DEM in terms of the number of calculated distances L_{checks}. However, additional computations of the ML-DEM which include the calculations for every non previously checked model, are quite complex. Hence, the difference in performance with the DEM and other approximate NN methods is high only for very complex similarity measures (e.g., in case of $\Delta = 1$).

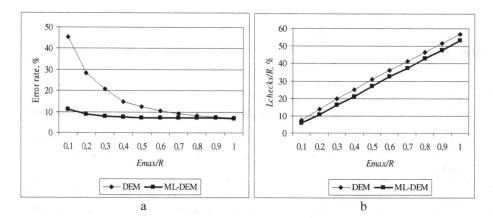

Fig. 1. Dependence of: (a) error rate; and (b) count of models checks per database size $L_{checks}/R \cdot 100$ % on E_{max}/R, Kullback-Leibler discrimination, $\Delta = 1$

5 Conclusion

In this paper we demonstrated that using the asymptotic properties (7) of the Kullback-Leibler divergence in the proposed ML-DEM gives very good results in image recognition with medium-sized database, reducing the recognition speed by more than 2.5–6.5 times in comparison with brute force and by 1.2–2.5 times in comparison with other approximate NN methods from FLANN and NonMetricSpaceLib libraries. In contrast to the most popular fast algorithms, our method is not heuristic (except the termination condition (5)). Moreover, it does not build data structure based on an algorithmic properties of applied similarity measure (e.g., triangle inequality of Minkowski metric in the AESA [4], Bregman ball for Bregman divergences [9]). The proposed ML-DEM is an optimal (maximum likelihood) greedy method in terms of the number of distance calculations for NN rule (3) with the sum (14) of Kullback-Leibler discriminations (4). Moreover, the ML-DEM can be successfully applied with other distances, e.g. the HT-PNN [15].

The main direction for further research of the proposed method is its modification in case of simple similarity measures. For now, the complexity of extra computation at each step of the ML-DEM (10) and (11) is rather high. Hence, the difference in performance with original DEM and popular approximate NN methods is significant only for very complex similarity measure. One possible solution is a pivot-based indexing [10] and ordering all models with respect to their log-likelihoods (10) and (11).

Acknowledgements. The article was prepared within the framework of the Academic Fund Program at the National Research University Higher School of Economics (HSE) in 2015–2016 (grant № 15-01-0019) and supported within the framework of a subsidy granted to the HSE by the Government of the Russian Federation for the implementation of the Global Competitiveness Program.

References

1. Tan, X., Chen, S., Zhou, Z.H., Zhang, F.: Face recognition from a single image per person: a survey. Pattern Recogn. **39**(9), 1725–1745 (2006)
2. Theodoridis, S., Koutroumbas, K.: Pattern Recognition, 4th edn. Elsevier Inc., Burlington (2009)
3. Arya, S., Mount, D.M., Netanyahu, N.S., Silverman, R., Wu, A.Y.: An optimal algorithm for approximate nearest neighbor searching fixed dimensions. J. ACM **45**(6), 891–923 (1998)
4. Vidal, E.: An algorithm for finding nearest neighbours in (approximately) constant average time. Pattern Recogn. Lett. **4**(3), 145–157 (1986)
5. Muja, M., Lowe, D.G.: Fast approximate nearest neighbors with automatic algorithm configuration. In: 4th International Conference on Computer Vision Theory & Applications (VISAPP), pp. 331–340 (2009)
6. Silpa-Anan, C., Hartley, R.: Optimised KD-trees for fast image descriptor matching. In: International Conference on Computer Vision & Pattern Recognition, pp. 1–8 (2008)
7. He, J., Kumar, S., Chang, S.: On the difficulty of nearest neighbor search. In: 29th International Conference on Machine Learning (ICML-2012), pp. 1127–1134 (2012)
8. Savchenko, A.V.: Nonlinear transformation of the distance function in the nearest neighbor image recognition. In: Zhang, Y.J., Tavares, J.M.R.S. (eds.) CompIMAGE 2014. LNCS, vol. 8641, pp. 261–266. Springer, Heidelberg (2014)
9. Cayton, L.: Efficient Bregman range search. In: Bengio, Y., Schuurmans, D., Lafferty, J.D., Williams, C.K.I., Culotta, A. (eds.) Advances in Neural Information Processing Systems, vol. 22, pp. 243–251. (2009)
10. Gonzalez, E.C., Figueroa, K., Navarro, G.: Effective proximity retrieval by ordering permutations. IEEE Trans. Pattern Anal. Mach. Intell. **30**(9), 1647–1658 (2008)
11. Savchenko, A.V.: Directed enumeration method in image recognition. Pattern Recogn. **45** (8), 2952–2961 (2012)
12. Kullback, S.: Information Theory and Statistics. Dover Publications, Mineola (1997)
13. Chellappa, R., Du, M., Turaga, P., Zhou, S.K.: Face tracking and recognition in video. In: Li, S.Z., Jain, A.K. (eds.) Handbook of Face Recognition, pp. 323–351. Springer, London (2011)
14. Dalal, N., Triggs, B.: Histograms of oriented gradients for human detection. In: International Conference on Computer Vision & Pattern Recognition, pp. 886–893 (2005)
15. Savchenko, A.V.: Probabilistic neural network with homogeneity testing in recognition of discrete patterns set. Neural Netw. **46**, 227–241 (2013)

16. Boytsov, L., Naidan, B.: Engineering efficient and effective non-metric space library. In: Brisaboa, N., Pedreira, O., Zezula, P. (eds.) SISAP 2013. LNCS, vol. 8199, pp. 280–293. Springer, Heidelberg (2013)
17. Savchenko, A.V.: Real-time image recognition with the parallel directed enumeration method. In: Chen, M., Leibe, B., Neumann, Bernd (eds.) ICVS 2013. LNCS, vol. 7963, pp. 123–132. Springer, Heidelberg (2013)

Ear Recognition Using Block-Based Principal Component Analysis and Decision Fusion

Alaa Tharwat[1]([⊠]), Abdelhameed Ibrahim[2], Aboul Ella Hassanien[3], and Gerald Schaefer[4]

[1] Electrical Engineering Department, Suez Canal University, Ismailia, Egypt
engalaatharwat@hotmail.com
[2] Computer Engineering and Systems Department,
Mansoura University, Mansoura, Egypt
[3] Faculty of Computers and Information, Cairo University, Giza, Egypt
[4] Department of Computer Science, Loughborough University, Loughborough, UK
gerald.schaefer@ieee.org

Abstract. In this paper, we propose a fast and accurate ear recognition system based on principal component analysis (PCA) and fusion at classification and feature levels. Conventional PCA suffers from time and space complexity when dealing with high-dimensional data sets. Our proposed algorithm divides a large image into smaller blocks, and then applies PCA on each block separately, followed by classification using a minimum distance classifier. While the recognition rates on small blocks are lower than that on the whole ear image, combining the outputs of the classifiers is shown to increase the recognition rate. Experimental results confirm that our proposed algorithm is fast and achieves recognition performance superior to that yielded when using whole ear images.

Keywords: Ear recognition · Principal component analysis (PCA) · Feature fusion · Classifier fusion

1 Introduction

Ear recognition systems are a relatively recent biometric technique, and are challenging to implement in practice due to difficulties controlling occlusions, pose, illumination etc. Ears have played a significant role in forensic science, especially in the United States, where an ear classification system based on manual measurements has been used for more than 40 years [5]. Using a collection of over 10,000 ears, they were found to be distinguishable based on only 12 measurements.

Chang *et al.* [3] compared ear recognition with face recognition using a standard PCA technique on face and ear images, and reported accuracies of 71.6 % and 70.5 % for ear and face recognition, respectively. They also presented results with varying ligthing which resulted in lower recognition accuracies of 64.9 % and 68.5 % for face and ear images, respectively. Combining ear and face images lead

© Springer International Publishing Switzerland 2015
M. Kryszkiewicz et al. (Eds.): PReMI 2015, LNCS 9124, pp. 246–254, 2015.
DOI: 10.1007/978-3-319-19941-2_24

to a significant improvement and an accuracy of 90.9 %. Kumar and Zhang [7] employed different feature extraction methods and different classification algorithms, namely feed-forward artificial neural networks and three classifiers based on a nearest neighbour rule. The experiments they performed yielded recognition rates ranging from 76.5 % to 94.1 %. Alaa *et al.* [13] used feature combination to improve the performance of an ear recognition system, and achieved recognition rates between 85.9 % and 96.1 %.

Principal component analysis (PCA) is widely used for dimensionality reduction, feature extraction, compression, visualiation, and other tasks. PCA finds the c principal orthonormal vectors which describe an eigenspace. In many applications, c is much smaller than the original dimensionality of the data, while the computation of PCA can be implemented using eigenvalue decomposition (EVD) of the covariance matrix of the data matrix [16]. However, PCA requires relatively high computational complexity and memory requirements, especially for large datasets [1,11].

To address this, Golub and van Loan [4] used Jacobi's method which diagonalises a symmetric matrix and requires about $O(d^3 + d^2n)$ computations, where n represents the number of feature vectors or samples and d^d represents the dimensionality of the vectors. Roweis [9] proposed an expectation maximisation (EM) algorithm for PCA, which is shown to be computationally more effective compared to the EVD method for PCA. However, calculating PCA based on EM is still expensive, while the EM algorithm may not converge to the global maximum but only a local one, and is dependent on the initialisation. The power method can also be used to find leading eigenvectors, and is less expensive, but can compute only one most leading eigenvector [10]. Also, Skarbek [12] and Liu *et al.* [8] proposed eigenspace merging where it is not necessary to store the covariance matrix of previous training samples.

In this paper, we propose a fast and accurate ear recognition system through fusion at classification level and feature level and PCA on subimages. Our algorithm aims to decrease the dimensionality and hence decrease the complexity of a PCA-based algorithm by dividing the ear image into small blocks. PCA is then applied on the image blocks separately and classification performed using a minimum distance classifier. The outputs of these classifiers at abstract, rank, and score level are combined, while we also investigate combining the block features at the feature level. Experimental results confirm that our proposed algorithm is fast and achieves recognition performance superior to that yielded when using whole ear images.

The rest of the paper is organised as follows. In Sect. 2, we summarise some of the background on principal component analysis and current fusion methods. Section 3 then details our proposed algorithm. Section 4 gives experimental results, while Sect. 5 concludes the paper.

2   Background

2.1   Principal Component Analysis

Principal component analysis (PCA) is a popular linear subspace method that finds a linear transformation which reduces the d-dimensional feature vectors to h-dimensional feature vectors with $h < d$. It is possible to reconstruct the h-dimensional feature vectors from the d-dimensional reduced representation with some finite error known as reconstruction error. Of the resulting h basis vectors, the first one is in the direction of the maximum variance of the given data, while the remaining basis vectors are mutually orthogonal and maximise the remaining variance. Each basis vector represents a principal axis. These principal axes can be obtained by the dominant/leading eigenvectors (i.e. those with the largest associated eigenvalues) of the measured covariance matrix of the original data matrix. In PCA, the original feature space is characterised by these basis vectors and the number of basis vectors is usually much smaller than the dimensionality d of the feature space [12].

For an ear image $\Gamma(M \times N)$, where M and N are the width and height of the image, it is first transformed into a vector of length $M \times N$. The feature matrix of K training ear images is then given by $\Gamma = [\Gamma_1, \Gamma_2, \ldots, \Gamma_K]$ and the average of the training set is calculated as $\psi = \frac{1}{K} \sum \Gamma_i$. The average is subtracted, i.e. $\phi_i = \Gamma_i - \psi$, and the data matrix created as $A = [\phi_1, \phi_2, \ldots, \phi_K]((M \times N) \times K)$.

The covariance matrix of A is then calculated as

$$C = AA^T. \tag{1}$$

Next, the eigenvalues (λ_k) and eigenvectors (V_k) of C are computed and the eigenvectors sorted according to the corresponding eigenvalues. Dimensionality reduction is then achieved by retaining only the top h eigenvectors to yield a projection matrix P. For an ear image T (of the same size as the training images), it is first mean-normalised by $\phi_T = T - \psi$, and then transformed into the "eigen-ear" components, i.e. projected into ear space, by

$$\omega = P^T \phi_T. \tag{2}$$

Considering the computational complexity of PCA [15], for an ear image set of K images of dimensions $M \times N$, calculating the mean image is of $O(KMN)$ and subtracting it from the data matrix also of $O(KMN)$, while the complexity of calculating the covariance matrix is $O(K(MN)^2)$. Identifying the eigenvalues and eigenvectors of the covariance matrix then requires $O(K(MN)^3)$, whereas sorting the eigenvectors according to their eigenvalues can be done in $O(K(MN)log_2(MN))$ (using a merge sort algorithm). Since only the first h eigenvectors are considered, computation of the reduced eigenspace is carried out in $O(hMN)$. Finally, projecting the images into this eigenspace requires $O(KLMN)$. Consequently, the overall computational complexity is

$$\begin{aligned} O_{\text{PCA}} &= O(KMN) + O(KMN) + O(K(MN)^2) + O(K(MN)^3) + \\ &\quad O(K(MN)log_2(MN)) + O(hMN) + O(KhMN) \\ &= O(K(MN)^3). \end{aligned} \tag{3}$$

2.2 Fusion Methods

Combining different and independent resources can increase the accuracy of biometric (or other) systems. Misclassification of some samples by a method or classifier can be compensated by combining different resources which in turn can be performed at different levels. There are various approaches for such fusion methods, of which we summarise the most common in the following.

Multi-instance systems use various sensors to capture samples. In multi-sensorial systems, samples from the same instance are captured using two or more different sensors (e.g., both visible light and infra-red cameras) are combined in a sensor level fusion approach to increase the robustness of the biometric system [6].

Combination at feature level can lead to improved performance as more information is available (compared to fusion at classification level, which is discussed below). Fusion of features is usually implemented by concatenating two or more feature vectors, i.e. if $f_1 = \{x_1, \ldots, x_n\}$ and $f_2 = \{y_1, \ldots, y_m\}$ are two feature vectors of lengths n and m, respectively, then the fused feature vector $f = \{x_1, \ldots, x_n, \ldots, y_1, \ldots, y_m\}$ is obtained by concatenation of f_1 and f_2.

Fusion at classification level, or classifier fusion, can improve recognition performance compared to simple individual classifiers. In general, we can distinguish three levels of fusion here, namely:

- *Abstract Level Fusion:* Abstract or decision level fusion can be seen as making a decision by combining the outputs of different classifiers for a test sample. It is the simplest fusion method and majority voting is the most commonly employed method here.
- *Rank Level Fusion:* Here, the outputs of each classifier (a subset of possible matches) are sorted in decreasing order of confidence so that each class has its own rank. Fusion can be performed by summing up the ranks of each class and the decision is given by the class of the highest rank.
- *Score (Measurement) Level Fusion:* Fusion rules on the vectors are derived to represent the distance between the test image and the training images. Thus, the output of each classifier is represented by scores or measurements. Fusion at this level combines the vectors of scores, and the decision is given by the class that has the minimum value. Assume that we want to classify an input pattern Z into one of m possible classes based on the evidence provided by R different classifiers. Let $\acute{x}_i$ be the score derived for Z from the i-th classifier, and let the outputs of the individual classifiers be $P(\omega_j | \acute{x}_i)$, i.e., the posterior probability of pattern Z belonging to class ω_j given the scores $\acute{x}_i$. If $c = \{1, 2, \ldots, m\}$ is the class to which Z is finally assigned, then this can be

done by the following rules [14]:

$$c = \arg \max_j \max_i P(\omega_j|x_i) \qquad (4)$$

$$c = \arg \max_j \min_i P(\omega_j|x_i)$$

$$c = \arg \max_j \mathrm{med}_i P(\omega_j|x_i)$$

$$c = \arg \max_j \mathrm{avg}_i P(\omega_j|x_i)$$

$$c = \arg \max_j \prod_i P(\omega_j|x_i)$$

3   Proposed Algorithm

In this paper, we propose a fast and accurate ear recognition system based on principal component analysis (PCA) and fusion at classification and feature levels. Figure 1 gives an overview of our proposed algorithm.

The first step in our approach is to divide each ear image into non-overlapping blocks. In each block, PCA features are extracted and each subimage is matched separately using a minimum distance classifier.

In the first model, the features of the blocks are combined at feature level. Here, the PCA algorithm is applied on each subimage and then the features of all blocks combined to form a feature vector. Classification is performed using a minimum distance classifier.

In the first model, the outputs of classifiers are combined at abstract, rank, or score levels. For decision level fusion, majority voting method is used to combine results from all blocks. For rank level fusion, the Borda count method is employed to combine the ranks obtained by the individual classifiers. Finally, for score level fusion, minimum, maximum, product, mean, and median rules are used to combine the scores.

Considering the computational complexity of our approach, the first step in our algorithm is to divide the ear images, which are of size $M \times N$, into Q

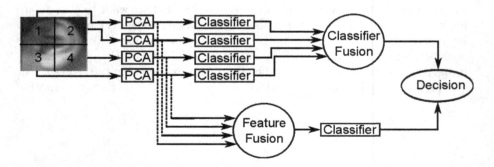

Fig. 1. Overview of our proposed ear recognition algorithm

equal-sized non-overlapping blocks. Consequently, the size of each image block will be $(M \times N)/Q$, and the computational complexity to perform PCA on an image block is thus of $O(K(MN/Q)^3)$. For all Q sub images, the computational complexity is thus of

$$O_{\text{blockPCA}} = O(QK(MN/Q)^3) = O(K(MN)^3/Q^2), \tag{5}$$

and hence significantly smaller, by a factor of Q^2, compared to the $O(K(MN)^3)$ of performing PCA on the full images. However, the recognition rate of each block will be lower compared to the recognition rate of the whole ear image. Thus, in our approach, the fusion techniques discussed above are applied to increase the recognition performance.

4   Experimental Results

In our experiments, we use a dataset of 102 ear images, 6 images for each of 17 subjects [2]. In particular, six views of the left profile from each subject were taken under uniform diffuse lighting conditions, while there are slight changes in the head position from image to image.

For our experiments, a minimum distance classifier is used based on three different metrics, namely Euclidean, city-block, and cosine metrics. In our first experiment, we applied classical PCA based on the whole ear image using 2 and 4 of the images for training, respectively. The results of this are summarised in Tables 1 and 2 in terms of recognition performance and computation time, respectively.

In the second experiment, the proposed PCA algorithm is applied. Here, the ear images are divided into into 4, 9, and 16 blocks, respectively to reduce the

Table 1. Recognition rates [%] using different minimum distance classifiers and PCA on whole ear images

Metric	2 Training images	4 Training images
Euclidean	89.7059	94.1176
City-block	88.2353	94.1176
Cosine	89.7059	94.1176

Table 2. Comparison of required CPU times of applying PCA on whole ear images and on image blocks

Blocks	2 Training images	4 Training images
1 (whole image)	607	615
$2 \times 2 = 4$	36	39
$3 \times 3 = 9$	9	10
$4 \times 4 = 16$	2	3

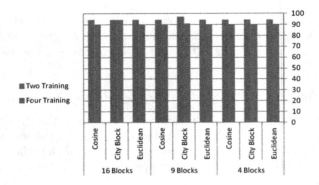

Fig. 2. Recognition rates [%] of the proposed feature fusion model at feature level

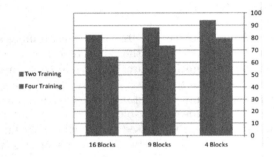

Fig. 3. Recognition rates [%] of the proposed classifier fusion model at abstract level

computational complexity of the proposed model. The resulting computation times are given in Table 2. As we can see from there, the time required for PCA calculation on blocks is very small compared to that of performing PCA on the whole images, while computation time decreased with decreasing block sizes.

After dividing the ear image into blocks, the features of all blocks are combined into one feature vector to perform feature level fusion, while matching is performed using minimum distance classifiers. The obtained results are given in Fig. 2. As shown there, feature fusion of image blocks can lead to better recognition performance compared to utilising the whole ear images. Moreover, the accuracies of different block sizes are approximately the same.

The final experiment is conducted to investigate the proposed classifier fusion models. Here, the features that result from each block are matched separately, while we combine the results of the classifiers at abstract level, rank level, and score level, respectively. The corresponding recognition results are shown in Figs. 3, 4, and 5, respectively.

From Figs. 3, 4, and 5, we can see that our proposed classifier fusion model achieves accuracies that exceed those obtained based on the whole ear images. Moreover, the accuracy of our model is in general inversely proportional to the number of blocks employed. When the number of blocks is increased, some of the resulting blocks will contain only background or small parts of the ear, which

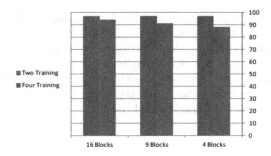

Fig. 4. Recognition rates [%] of the proposed classifier fusion model at rank level

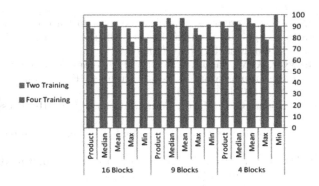

Fig. 5. Recognition rates [%] of the proposed classifier fusion model at score level

decreases the accuracy for these parts and hence affects the overall accuracy. This problem can in particular be observed for abstract level fusion, while for rank and score level fusion, using many ranks or scores may compensate the problem. Score level fusion leads to recognition performance that is significantly better compared to abstract and rank level fusion. Not surprisingly, abstract level fusion yields the lowest accuracy, as it is based purely on the decisions without further information.

Our proposed method also performs much better than some reported in the literature including that by Kumar and Zhang [7], which uses PCA to extract features from ear images and gives a recognition rate of 71.6 % on the same database, and [13], which combines PCA with linear discriminant analysis (LDA) and discrete cosine transform (DCT), respectively and yields a classification performance of 63.8 % on the database.

5   Conclusions

In this paper, we have presented an algorithm to identify persons using 2D ear images based on principle component analysis. Crucially, the computational complexity of PCA is addressed by partitioning the images into small blocks and performing PCA on the subimages separately. We then combine the blocks at feature and classification level, respectively, with the latter leading to the best results and significantly improved performance compared to performing PCA-based recognition based in the whole ear images. In addition to this increased classification accuracy, our approach also significantly reduces the computation time required, hence giving a fast and accurate ear recognition algorithm as demonstrated by a series of experimental results.

References

1. Candès, E.J., Li, X., Ma, Y., Wright, J.: Robust principal component analysis? J. ACM **58**(3), 11 (2011)
2. Carreira-Perpinan, M.: Compression neural networks for feature extraction: Application to human recognition from ear images. MS thesis, Faculty of Informatics, Technical University of Madrid, Spain (1995)
3. Chang, K., Bowyer, K.W., Sarkar, S., Victor, B.: Comparison and combination of ear and face images in appearance-based biometrics. IEEE Trans. Pattern Anal. Mach. Intell. **25**(9), 1160–1165 (2003)
4. Golub, G.H., van Loan, C.F.: Matrix Computations, vol. 3. JHU Press, Baltimore (2012)
5. Iannarelli, A.V.: Ear Identification. Paramont Publishing Company, Fremont (1989)
6. Jain, A., Nandakumar, K., Ross, A.: Score normalization in multimodal biometric systems. J. Pattern Recogn. **38**(12), 2270–2285 (2005)
7. Kumar, A., Zhang, D.: Ear authentication using log-Gabor wavelets. In: Defense and Security Symposium. pp. 65390A–65390A (2007)
8. Liu, L., Wang, Y., Wang, Q., Tan, T.: Fast principal component analysis using eigenspace merging. In: IEEE International Conference on Image Processing (ICIP), vol. 6, pp. VI-457 (2007)
9. Roweis, S.: EM algorithms for PCA and SPCA. In: Advances in Neural Information Processing Systems, vol. 10, pp. 626–632. MIT Press, Cambridge (1998)
10. Schilling, H.A., Harris, S.L.: Applied Numerical Methods for Engineers Using MATLAB. Brooks/Cole Publishing Co, Pacific Grove (1999)
11. Shlens, J.: A tutorial on principal component analysis. arXiv preprint arXiv:1404.1100 (2014)
12. Skarbek, W.: Merging subspace models for face recognition. In: Petkov, N., Westenberg, M.A. (eds.) CAIP 2003. LNCS, vol. 2756, pp. 606–613. Springer, Heidelberg (2003)
13. Tharwat, A., Hashad, A., Salama, G.: Human ear recognition based on parallel combination of feature extraction methods. Mediterr. J. Comput. Netw. **6**(4), 133–137 (2010)
14. Tharwat, A., Ibrahim, A.F., Ali, H.A.: Multimodal biometric authentication algorithm using ear and finger knuckle images. In: 7th International Conference on Computer Engineering & Systems, pp. 176–179 (2012)
15. Toygar, Ö., Acan, A.: Boosting face recognition speed with a novel divide-and-conquer approach. In: Aykanat, C., Dayar, T., Körpeoğlu, I. (eds.) ISCIS 2004. LNCS, vol. 3280, pp. 430–439. Springer, Heidelberg (2004)
16. Turk, M., Pentland, A.: Eigenfaces for recognition. J. Cogn. Neurosci. **3**(1), 71–86 (1991)

Data Mining Techniques
for Large Scale Data

Binarizing Change for Fast Trend Similarity Based Clustering of Time Series Data

Ibrahim K.A. Abughali$^{(\boxtimes)}$ and Sonajharia Minz

School of Computer and System Science,
Jawaharlal Nehru University, New Delhi, India
{barhoom.gali,sona.minz}@gmail.com

Abstract. It is observed that traditional clustering methods do not necessarily perform well on time-series data because of the temporal relationships in the observed values over a period of time. Another issue with time series is that databases contain bulk amount of data in terms of dimension and size. Clustering algorithms based on traditional measures of dissimilarity find trade-offs between efficiency and accuracy. In addition, time series analysis should be more concerned with the patterns in change and the points of change rather than the values of change. In this paper a new representation technique and similarity measure have been proposed for agglomerative hierarchical clustering.

Keywords: Time series representation · Similarity search · Clustering

1 Introduction

Today Time Series data management has become an interesting research topic for the data miners. Particularly, the clustering of time series has attracted the interest.

Clustering is the process of finding natural groups, called clusters, the grouping should maximize inter-cluster variance while minimizing intra-cluster variance [1], most of the clustering techniques can be two major categories, Partition-based clustering and Hierarchical Clustering [2]. Many of the traditional clustering algorithms use Euclidean distance or Pearson's correlation coefficient to measure the proximity between the data points. However, in case of time-series data these parameters involve the individual magnitudes at each time point therefore the traditional algorithms perform poorly with time-series expressions data, to overcome these limitations the proposed work aims to represent the variations in the measurements of the time-series for fast implementation of an efficient agglomerative nesting algorithm, the focus of this work is on fast whole sequence similarity search in the changes in respect to time rather than the values in the time series data.

The rest of the paper is organized as follows: Sect. 2 presents a brief review of related work. Sections 3 and 4 demonstrates the basic concept and presents the analysis of the proposed algorithm respectively. In Sects. 5 and 6 experimental and the conclusions and some future directions.

© Springer International Publishing Switzerland 2015
M. Kryszkiewicz et al. (Eds.): PReMI 2015, LNCS 9124, pp. 257–267, 2015.
DOI: 10.1007/978-3-319-19941-2_25

2 Related Work

Many clustering algorithms have been proposed such as k-means, DBSCAN, STING, p-cluster and COD [4–6]. One of the recently proposed algorithms is VCD algorithm [3] to analyze the trends of expressions based on their variation over time, using cosine similarity measure with two user inputs, it has been enhanced later in EVCD algorithm [2] for same purpose with one single user input and provides results in several levels which allows the user to select the most appropriate level by using different parameters such as the silhouette coefficient, number of clusters and clusters density. Both algorithms Enhanced Variation Co-expression Detection (EVCD) and (VCD) algorithms [2, 3] inferred that the cosine similarity measure was the most appropriate similarity measure for clustering the time varying microarray data.

3 Concepts and Definition

In order to determine the variation patterns in the time series based on the changes in the values observed at fixed time points binarization of the change has been proposed. Some related definitions are presented in this section.

3.1 Variation Vector

Given a sequence of $n + 1$ measurements observed at time periods $t_0, t_1, t_2...t_n$ to denote a univariate time series, say, $Y = \langle y_0, y_1, y_2...y_n \rangle \in \mathbb{R}^{n+1}$. A variation vector $Y_v \in \mathbb{R}^n$ of Y is a sequence of the differences denoted by, $Y_v = \langle d_1, d_2...d_n \rangle$, where $d_i = y_i - y_{i-1}$, for $1 \leq i \leq n$. The increase in the measurement $(y_i \geq y_{i-1})$ and its magnitude is represented by the difference $d_i \geq 0$. Similarly, the decrease $(y_i < y_{i-1})$ is computed as $d_i < 0$.

The trend is the tendency of a continuous process that is measured during a fixed time interval. The trend analysis may traditionally be carried out by plotting a trend curve or a trend line and by monitoring the increase (decrease) in the values. Thus trend analyses involve observation of the tendencies of the values by way of analyzing the changes that occur in terms of the quantum of the change and/or the nature of the changes. The pattern of increase or decrease in the values of the measurements may play a significant role in the trend analyses. Variation vectors quantify the difference in measurements at two consecutive time periods say t_i and t_{i+1} in terms. The directions of change, increase or decrease, may be captured by the positive or negative sign of the magnitude of difference d_i respectively. Therefore, a binary representation of the direction of change is suitable for computational efficiency. Binarization of the change for any time-series has been proposed by a direction vector. Further, the trend similarity based on the distance metric of the n-dimensional binary vectors has been defined.

3.2 Direction Vector

For a variation vector, $Y_v = \langle v_1, v_2, \ldots, v_n \rangle \in \mathbb{R}^n$, a direction vector $Y_d \in \{0, 1\}^n$ is defined as $Y_d = \langle b_1, b_2, \ldots, b_n \rangle$, where,

$$b_i = \begin{cases} 0 & \text{if } v_i \geq 0 \\ 1 & \text{if } v_i < 0 \end{cases}. \tag{1}$$

Example 1: Consider two time series $T_1 = \langle 3, 7, 2, 0, 4, 5, 9, 7, 2 \rangle$ and $T_2 = \langle 10, 15, 11, 5, 19, 25, 27, 24, 13 \rangle$. The corresponding variation vectors are, $V_1 = \langle 4, -5, -2, 4, 1, 4, -2, -5 \rangle$ and $V_2 = \langle 5, -4, -6, 14, 6, 2, -3, -11 \rangle$. The direction vectors of T_1 and T_2 are $D_1 = \langle 0, 1, 1, 0, 0, 0, 1, 1 \rangle$ and $D_1 = \langle 0, 1, 1, 0, 0, 0, 1, 1 \rangle$ respectively.

3.3 Trend Similarity

Let two time series $X = \langle x_0, x_1, x_2, \ldots, x_n \rangle$ and $Y = \langle y_0, y_1, y_2, \ldots, y_n \rangle$ be measured at the time $t_0, t_1, \ldots, t_n$. Let $X_v = \langle v_1, v_2, \ldots, v_n \rangle$ and $Y_v = \langle u_1, u_2, \ldots, u_n \rangle$ be the corresponding variation vectors and $X_d = \langle l_1, l_2, \ldots, l_n \rangle$ and $Y_d = \langle s_1, s_2, \ldots \ldots s_n \rangle$ be the corresponding direction vectors. Then X and Y are said to be similar in trend if and only if $l_i = s_i$ for $1 \leq i \leq n$.

Both direction vectors X_d and Y_d are n-bit binary vectors. For each i if $x_i \geq x_{i-1}$ in series X i.e. $v_i \geq 0$ then $l_i = 0$ and $l_i = 1$ for vice versa. In case of the time series Y the bit value of s_i would depict the increase if the value at t_i from the values at t_{i-i} as $u_i \geq 0$ and correspondingly, $s_i = 0$, and vice versa. If for each i, $l_i = s_i$ then Y is said to be trend similar to X. It may be noted that for the definition of similarity the magnitude of difference in the two time-series has not been considered. However, only the concept of direction of change i.e. increase or decrease, has been considered. The information in the direction vector may be utilized to determine the degree of similarity.

Example 2: Consider the direction vectors D_1 and D_2 in the above example corresponding the two time-series T_1 and T_2 each of length 9. The magnitude of the differences are represented by the variation vectors V_1 and V_2. It may be noted that for each i, $1 \leq i \leq 8$, $V_{1i} \neq V_{2i}$. However, D_1 and D_2 are bit-wise equal, i.e. $D_{1i} = D_{2i}$, for $1 \leq i \leq 8$, therefore, the two series T_1 and T_2 are observed to be similar in trend.

The following metric to measure the distance between two n-dimensional binary vectors has been considered in this work. Let $\beta = \{0, 1\}$ and $I_n = \{0, 1, 2 \ldots n\}$ then the binary function $d_{binary} : \beta \times \beta \to \beta$. For $b_1, b_2 \in \beta$,

$$d_{binary}(b_1, b_2) = \begin{cases} 0 & \text{if } b_1 = b_2 \\ 1 & \text{otherwise} \end{cases} \tag{2}$$

Then the distance function between a pair of n-dimensional binary vectors is $d_n : \beta^n \times \beta^n \to I_n$ Consider two n-dimensional binary vectors say $D_1, D_2 \in \beta^n$.

$$d_n(D_1, D_2) = \sum_{j=1}^{n} d_{binary}(b_{1j}, b_{2j}) \tag{3}$$

Let $d_n(D_1, D_2) = k$. Then $k = 0$ if $\sum_{i=1}^{n} d_{binary}(b_{1i}, b_{2i}) = 0$ and $k = n$ if $\sum_{i=1}^{n} d_{binary}(b_{1i}, b_{2i}) = n$. Therefore $0 \leq k \leq n$

Example 3: Consider the following two sequences as time series, $T_1 = \langle 3, 7, 2, 0, 4, 5, 9, 7, 2 \rangle$ and $T_3 = \langle 45, 80, 22, 10, 40, 63, 45, 90, 10 \rangle$, then variation vectors V_1 and V_3 of T_1 and T_3 are, $V_1 = \langle 4, -5, -2, 4, 1, 4, -2, -5 \rangle$ and $V_3 = \langle 35, -58, -12, 30, 23, -18, 45, -80 \rangle$, the direction vectors D_1 and D_3 are $D_1 = \langle 0, 1, 1, 0, 0, 0, 1, 1 \rangle$ and $D_3 = \langle 0, 1, 1, 0, 0, 1, 0, 1 \rangle$.

For $D_1, D_3 \in B^8$, the dissimilarity between D_1 and D_3 may be computed using the distance function d_8,

$$d_8(D_1, D_3) = 2 \tag{4}$$

where,

$$d_{binary}(b_{1i}, b_{2i}) = 1 \quad \text{for } i \in \{6, 7\} \tag{5}$$

and

$$d_{binary}(b_{1i}, b_{2i}) = 0 \quad \text{for } i \in \{1, 2, 3, 4, 5, 8\} \tag{6}$$

To allow difference in trends at the certain bits out of the n-bits, the concept of trend dissimilarity of degree-k has been considered where k ≤ n may be the number of bits at which the two n-dimensional direction vectors encounter bit-mismatch.

3.4   Trend Dissimilarity of Degree K

Given two n-dimensional time series T_i and T_j, and their respective direction vectors D_i and D_j, T_i and T_j are said to have dissimilarity of degree k, if $d_n(D_i, D_j) = k$, for $1 \leq k \leq n$.

The clusters at level-0 may contain identical objects. Consider any two arbitrary objects x and y, and the Euclidian distance function d, the traditional measure of dissimilarity. Then $d(x, y) = 0$, i.e. $\sqrt{\sum(x_i, y_i)^2} = 0$ if the two objects are identical. Therefore, the objects x and y must be grouped in the same cluster at level-0, say ith cluster denoted by, $C_{0,i}$. Let $C_{i,j}$ denote cluster-id j at level-i. Then the m clusters at level-0 are $C_{0,1}, C_{0,2}, C_{0,3}, \ldots, C_{0,m}$. Let a measure of dissimilarity at 1 bit represented by distance metric d_1 be associated to the clusters at level-1, dissimilarity at 2 bits represented by d_2 and so on. Then any two arbitrary objects x, y may be in the same cluster at level-1, $C_{1,j}$, only if, $0 < d(x, y) \leq d_1$. In this section the concept of Trend Cluster of level-k using the dissimilarity of degree-k is defined.

3.5   Trend Cluster of Level-K

For $\mathcal{T} = \{T_1, T_2, \ldots, T_m\}$, a set of n-dimensional time series of cardinality m, and the set of corresponding direction vectors $\Gamma = \{D_1, D_2, \ldots, D_m\}$, a trend cluster of level-k,

$C_{k,j}$ would include all time-series T_i and T_j in the same cluster if $d_n(D_i, D_j) = k$. However, if $d_n(D_i, D_j) \neq k'$ for all $k', 0 \leq k' < k$, then T_i and T_j will be allocated to distinct trend clusters of level-0, level-1, up to level-k', say $C_{k',i}$ and $C_{k',j}$, but would be grouped in the same trend clusters of level-k, say $C_{k,i}$.

Example 4: Consider time series T_1, T_2 and T_3 as in the Examples 1 and 3. The direction vectors of each is $D_1 = \langle 0,1,1,0,0,0,1,1 \rangle$, and $D_2 = \langle 0,1,1,0,0,0,1,1 \rangle$ $D_3 = \langle 0,1,1,0,0,1,0,1 \rangle$. Consider D_1 and D_2, $d_8(D_1, D_2) = 0$ therefore, T_1 and T_2 must be grouped in the same cluster of level-0. Consider D_1 and D_3, $d_8(D_1, D_3) = 2$. i.e. the series T_1 and T_3 have the trend dissimilarity of degree-2. Therefore, T_1 and T_3 must be grouped in different trend clusters of level-0 and level-1 say $C_{0,1}$ and $C_{0,3}$, and $C_{1,1}$ and $C_{1,3}$ respectively. However, the two must be grouped in the same trend cluster of level-2 say, $C_{2,1}$.

Example 5: Consider the 5-dimensional view of the four gene expressions a, b, c and d, as shown in Fig. 1. The direction vectors D_a and D_c are identical therefore genes a and c are trend similar. Even visually the vectors a and c are the most similar to each other than to the vectors b and d.

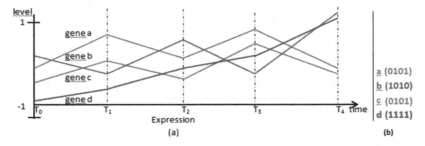

Fig. 1. Trend similarity in gene expressions

An advantage of this approach is the simplicity of representation of the objects of m-dimensional time series database, using only one bit to represent the change in value from time t_i to t_{i+1},

$$b_i = \begin{cases} 0 & x_{i+1} \geq x_i \\ 1 & x_{i+1} < x_i \end{cases}; 0 < i < m - 1 \tag{7}$$

The direction vectors are loss transformation of the original data from which no original values can be retrieved. Thus it is a novel representation from the perspective of security and privacy preservation of the original data.

4   Fast Trend Similarity-Based Clustering (FTSC) Algorithm

FTSC algorithm starts with generating the variation vectors, second is binarization of the variation vector, and third is direction vectors indicate similarity in trend in the time series thus forming the trend clusters of level-0 in the hierarchy of clusters. The higher level clusters may result from merging the closest clusters in the previous level starting by smaller clusters, each cluster is represented by a direction vector as medoid of the cluster. The distance between clusters is computed by the distance between the medoids of the two clusters.

Input: T, Time series database of n series;
         m+1, Dimension of each time series;
         m, maximum level of clustering;
Output: Hierarchical clusters up to level m;
Initialization:
$V = \emptyset;$ // set of Variation vectors.
$D = \emptyset;$ // set of Direction vectors.
$i = 0$;
$C_{i,j} = \emptyset,\ 1 \le j \le n;$ // i: cluster-level, j: cluster-id.
Begin
  I.   Variation Vector Generation
        $\forall t \in T,\ generate\ V(t);$
        $V = V \cup \{V(t)\};$
  II.   Binalization: Direction Vector Generation
                     $\forall X \in V, generate\ b(X);$
        $D = D \cup \{b(X)\};$
  III. Level-0 Cluster Generation
        $\forall D_t, D_l \in D$ if $d_m(D_t, D_l) = 0$
                     Then $C_{0,j} = merge(D_t, D_l)$ for $1 \le j \le n;$
  IV.   Higher Level Cluster Generation
        For i = 1 to m
                For (j= 1 to maximum-cluster at level i-1)
        and
                        (l= 1 to maximum-cluster at level i-1)
            For $\forall C_{i-1,j},\ C_{i-1,l}$ clusters of level i-1,
                If $d_m\left(C_{i-1,j}, C_{i-1,l}\right) \le i$ then construct a cluster
        of level-i
                $C_{i,s} = $ merge $(C_{i-1,j}, C_{i-1,l})$
End.

The FTSC algorithm is a nonparametric algorithm and it does not require any prior information related to data or number of clusters.

The asymptotic time complexity of the algorithm is quadratic on the product of the dimension of the time series and number of clusters level-i, $n_i < n$, therefore the complexity of the algorithm is $O((mn)^2)$. However, due to the binarization of the variation in the time series, the comparisons of the m bits and distance computation may be implemented using fast bit operators.

5 Experiments and Results

5.1 Data Sets

The experiments have been carried out to perform clustering on two microarray data sets and two financial data sets. Table 1 describes the data sets.

Table 1. Data set

Data set	Repository	Type	No of rows	No of dim
1	NCBI	Microarray/Affymetrix	12488	8
2	NCBI	Microarray /Drosophila genome	3456	8
3	PWT	Financial/exchange rates and PPPs over GDP	29	61
4	NSE	Financial/(NSE) India	1555	9

5.2 System Configuration

Windows 8 enterprise © 2012, 64-bit, with processor intel® core (TM) i7 CPU, U 640@1.20 GHz an. Dot Net platform has been used to implementation.

5.3 Design of Experiments

The experiments have been designed to assess the performance of FTSC algorithm in terms the efficiency and accuracy. Efficiency is mainly observed in terms of execution time. The accuracy of the algorithm is considered to be the consistency in cluster allocation to a time series irrespective of the number of re-executed, cluster allocation to multiple copies of the time series data, and the order of input of the time series to the algorithm. Second experiment compares both algorithms FTSC and EVCD.

5.4 Efficiency and Accuracy of FTSC

The first experiment has been designed to examine the speed of Fast Trend Similarity Clustering algorithm to cluster the four data sets. The experiment of running the program implementing the algorithm repeated five times, the average running time to yield the hierarchical clusters for each of the four data sets Affymetrix, Drosophila genome, Exchange Rates and PPPs over GDP and NSE with execution time 00:00:02.66, 00:00:01.72, 00:00:10.11 and 00:00:01.34 respectivly.

The outcomes of running the FTSC algorithm on Affymetrix are presented in Tables 2, 3 and 4. In Table 2 the 7-bit direction vector of gene Id 11251 is 0000001 which is in cluster $C_{0,0}$ while the two genes 11152 and 12182 in serial 7 and 8 have identical direction vectors 0000101. Therefore, $C_{0,3}$ includes two genes. The total clusters of level-0 is 115.

Table 2. Direction vectors, clusters of level-0 of AffyMetrix data

S.no.	GENE ID	Direction vector	Cluster no.
1	11251	0000001	0
2	6599	0000010	1
:	:	:	:
6	11278	0000011	2
7	11152	0000101	3
8	12182	0000101	3
:	:	:	:
13	8001	0001001	6
:	:	:	:
16	11668	0001001	6
:	:	:	:
12487	10226	1111011	114
12488	10461	1111011	114

Table 3. Level-3 cluster formation

Cluster id	Medoid	$C_{2,*}$	$C_{2,*}$	$C_{2,*}$	$C_{2,*}$	$C_{2,*}$	Cluster density
$C_{3,0}$	0	4	7	14	29	58	486
$C_{3,1}$	20	26	52	81	–	–	689
$C_{3,2}$	54	48	41	–	–	–	73
$C_{3,3}$	87	93	97	105	–	–	1657
$C_{3,4}$	107	77	–	–	–	–	103
$C_{3,5}$	9	70	–	–	–	–	48
:	:	:	:	:	:	:	:

Table 4. Level-4 cluster formation

Cluster id	Medoid	$C_{3,*}$	$C_{3,*}$	$C_{3,*}$	$C_{3,*}$	Cluster density
$C_{4,0}$	0	9	16	31	60	581
$C_{4,1}$	20	54	83	–	–	764
$C_{4,2}$	87	99	107	–	–	1880
:	:	:	:	:	:	:

Tables 3 and 4 present the clusters of level-2 and level-3 respectively. In the two tables the rows display all the clusters $C_{i,j}$, i denoting the cluster level and j the cluster ID. The cluster medoid has been presented in the second column by the identifier of the direction vector representing the cluster of level-0. In Table 3, the 3rd, 4th, 5th, 6th and 7th column display the clusters of level-2 that are merged to form the cluster of level-3. Thus the cluster id $C_{3,0}$ represented by the medoid 0 is formed by merging the clusters of level-2 represented by the medoids 4, 7, 14, 29 and 58 yielding the cluster with a total of 486 genes. The cluster $C_{3,1}$ is the outcome of merging three clusters of level-2 that are represented by the medoids 26, 52 and 81 to the cluster represented by medoid 20 at level-3 having a total of 689 genes. To obtain the clusters $C_{3,6}$ to $C_{3,15}$ no other clusters of level-2 were merged to the ones represented by the respective medoids indicated in column two. The blank '−' entries in the table indicate no clusters of level-2. Therefore, the row pertaining to the cluster $C_{3,6}$ with medoid 16 indicates that no cluster of level-2 satisfied the criterion for the merge operation although the total number of genes in the cluster $C_{3,6}$ is 2, where number of clusters of level-3 are 16.

The clusters from $C_{3,6}$ to $C_{3,15}$ in level-3 have not changed from the previous level with the same medoids and densities.

Similarly the Table 4 exhibits the details of the clusters of level-4. From both Tables 3 and 4 it may be observed that the cluster $C_{4,0}$ with medoid 0 has been formed by merging the clusters $C_{3,0}$, $C_{3,5}$, $C_{3,6}$, $C_{3,7}$ and $C_{3,11}$ referred to by the medoids 0, 9, 16, 31 and 60 respectively. It may also be observed that the density of $C_{4,0}$ is the sum of the densities of the $C_{3,0}$, $C_{3,5}$, $C_{3,6}$, $C_{3,7}$ and $C_{3,11}$. Similarly the cluster $C_{4,2}$ is formed by merging $C_{3,13}$, and $C_{3,4}$, to $C_{3,3}$ resulting in the density 1880.

As the FTSC algorithm is an agglomerative clustering algorithm yielding a hierarchical clustering of levels 0–7 for Affymetrix data. The cluster at the highest level $C_{6,0}$, represented by the medoid 0 includes all the 12488 genes (Figs. 2 and 3).

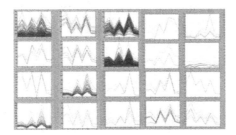

Fig. 2. Random clusters plot for DS 1 level 0

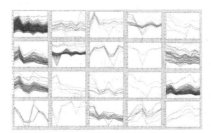

Fig. 3. Random clusters plot for DS 2 level 0

In order to estimate the efficiency, accuracy and sensitivity to order of data inputs, all the rows of the Affymetrix data set were duplicated four times and randomly shuffled. Therefore, the algorithm was executed with a total of $4 \times 12488 = 49952$ rows with 8 dimensions. The output of the program was a hierarchical clustering with levels

0–7 with same number of clusters at each level as before but the density of each cluster was four time the previous density. i.e. the cluster $C_{4,5}$ with inputs four time the first run was represented by a gene that had direction vector identical to the gene 9 and contained 192 genes. The same phenomenon was observed for all the clusters of each level from level-0 to level-7. Thus the accuracy of the algorithm has been assessed. The average running time of repeated execution of the four times the original data set was 00:00:10.714.

The repeated execution of the program after randomly shuffling the rows yielded the same number of clusters. However, each time the execution time was differed in the 3rd or the 4th decimal point with the mean being 00:00:02.6599 (Figs. 4 and 5).

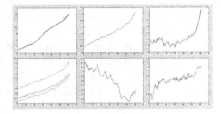

Fig. 4. Random clusters plot for DS 3 level 0

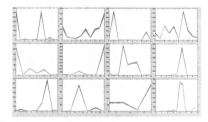

Fig. 5. Random clusters plot for DS 4 level 0

5.5   Comparison of FTSC and EVCD Algorithms

In this experiment the results of EVCD algorithm and FTSC algorithm have been compared. Two real world data sets Affymetrix and Drosophilia data sets as described in Table 1 are used in this experiment to assess the novelty of trend dissimilarity as the changes in the time series are represented by direction vectors. The EVCD algorithm is also a parametric algorithm while FTSC algorithm is not. EVCD algorithm requires one user input as the parameter ε. The experiment has been repeated for three values of ε, i.e. 0.01, 0.05 and 0.1 respectively. As EVCD performs a hierarchical clustering, for $\varepsilon = 0.01$, 10 clusters and 6 singletons were obtained at level 14, while for $\varepsilon = 0.05$, 10 clusters and 6 singletons were obtained at level 2, and finally 11 clusters and 2 Singleton were obtained at level 1 for $\varepsilon = 0.1$.

6 Conclusions

The experiments indicate that although the FTSC algorithm has the complexity $O((mn)^2)$ it is fast in terms of execution time due to the binarizing the change in the time-series. The binary representation in terms of the direction vector affect the distance computation implemented using bit level operators. The binarization also helps in privacy and security of the actual data. The nonparametric characteristic of the algorithm keeps the end user from exercise of parameter tuning. User also does not require any prior knowledge of the data or the clusters. The FTSC algorithm is time efficient and has the potential to yield accurate clusters of time-series data. The scalability of the algorithm in terms of multi-dimensions time-series and dealing with noise shall be investigated in future. To select a better medoid of the cluster of each higher level is also considered as future work.

References

1. Esling, P. Agon, C.: Time-series data mining. ACM Computing Surveys, **45**(1), 5 (2012)
2. Minz, S., Abughali, I.K.A.: Time-varying microarray data sets: co-expression detection. In: IEEE 2011 9th International Conference on ICT & Knowledge Engineering, IEEE Explore 978–1-4577-2162-5/11, pp. 43–46 (2011)
3. Yin, Z.-X., Chiang, J.-H.: Novel algorithm for coexpression detection in time-varying microarray data sets. IEEE/ACM Trans. Comput. Biol. Bioinf. **5**, 120–135 (2008)
4. Xu, R., Wunsch, D.: Survey of Clustering Algorithms. IEEE Trans. Neural Netw. **16**, 645–678 (2005)
5. G-Means Algorithm (2007). http://www.cs.utexas.edu/users/dml/Software/gmeans.html
6. Eisen, M.B., Spellman, P.T., Brown, P.O., Botstein, D.: Cluster analysis and display of genome-wide expression patterns. In: Proceedings Nat'l Academy of Sciences USA, vol. 95, issue no. 25, pp. 14863–14868 (1998)

Big Data Processing by Volunteer Computing Supported by Intelligent Agents

Jerzy Balicki[✉], Waldemar Korłub, and Jacek Paluszak

Faculty of Telecommunications, Electronics and Informatics, Gdańsk University of Technology, Narutowicza St. 11/12, 80-233 Gdańsk, Poland
balicki@eti.pg.gda.pl, waldemar.korlub@pg.gda.pl, jacekpaluszak@gmail.com

Abstract. In this paper, volunteer computing systems have been proposed for big data processing. Moreover, intelligent agents have been developed to efficiency improvement of a grid middleware layer. In consequence, an intelligent volunteer grid has been equipped with agents that belong to five sets. The first one consists of some user tasks. Furthermore, two kinds of semi-intelligent tasks have been introduced to implement a middleware layer. Finally, two agents based on genetic programming as well as harmony search have been applied to optimize big data processing.

1 Introduction

Big data (an acronym BD) can be very useful to achieve high-value information related to decision support, business intelligence or forecasting. Large volumes of data are published by many companies to the web, and also they deploy e-commerce applications that enable continuous self-service transactions via the web. We observe a migration of database capacities from terabytes to petabytes. Furthermore, we can expect that modern systems will require distributed databases with exabytes or even zetabytes. It is difficult to define big data, e.g. 10 terabytes is a large capacity for a banking transaction system, but small even to test a web search engine. However, we can treat this data as big if a data capacity is large enough to be uncooperative to work with some relational database management systems RDBMS like DB2, INGRES, MySQL, *Oracle, Sybase* or SQL *Server* [28, 31].

Some crucial difficulties with big data are related to: capture, storage, search, sharing, analytics, and visualizing. Few exabytes of data are captured per day from different sources: sensors, GPS, smartphones, microphones, cameras, tablets, computer simulations, satellites, radiotelescopes, and social networks via some wireless sensor networks. In result, the data store capacity has approximately doubled every three years for thirty years. Furthermore, *The Internet of Things* supports BD gathering. Currently, we can expect to store more or less zetabytes. data storage, their visualization, analysis and search are still considered as an open problem to solve, too [14, 24].

BD is not convenient to the most RDBMS because massively parallel software on thousands of servers is required. In applications of statistics and visualization, sizes of BD exceed the capability of commonly used tools. A BD size can increase to achieve

© Springer International Publishing Switzerland 2015
J. Wang and C. Yap (Eds.): FAW 2015, LNCS 9130, pp. 268–278, 2015.
DOI: 10.1007/978-3-319-19941-2_26

many petabytes for one volume. Fortunately, progress in speed of data communication can support BD processing. Another feature of BD is wide variety of them, what is related to a huge range of data types and sources. So, the 4Vs model can be described by: high *volume*, extraordinary *velocity*, great data *variety*, and *veracity* [30]. In BD, some regressions can be used to find predictions. On the other hand, some descriptive statistics can be developed for business intelligence.

Big data processing requires high performance architecture of distributed systems. In this paper, volunteer computing systems are proposed for big data processing. In the volunteer grids such as BOINC, *Folding@home*, and GIMPS, flat data sets are transformed into several millions of subsets that are processed parallel by volunteer computers. Performance of grid computing as *Folding@home* is estimated at 40 [PFLOPS]. For comparison, the fastest supercomputer performance *Tianhe*-2 reached 34 [PFLOPS] in 2014 [8]. Moreover, the BOINC system performance is 6 [PFLOPS] and computing power achieved for the most important projects using this software are: *SETI@Home* – 681 [TFLOPS] and *Einstein@Home* – 492 [TFLOPS]. GIMPS with 173 [TFLOPS] discovered the 48th Mersenne prime in 2013. The number of active volunteers can be estimated as 238,000 for BOINC [8].

Moreover, intelligent agents can be developed to efficiency improvement of a grid middleware layer. For example, an experimental volunteer and grid system called *Comcute* that is developed at *Gdańsk University of Technology* can be equipped with agents that belong to five sets [5, 11]. This grid was a virtual laboratory for experiments with big data and intelligent agents. Especially, some user tasks like the *Collatz* hypothesis verification or the 49th *Mersenne'* prime finding can move autonomously with a big amount of data from some source databases to some destination computers, and then outcomes are returned. If we consider above tasks, a reduction of databases can be done by dynamic memorizing the current period of Integers. However, a dilemma appears if we study some simulations of fire spread that is another *Comcute* project. In that case, some scenarios are analyzed, and several strategies are found [6].

Furthermore, two kinds of intelligent tasks have been considered to implement a middleware layer. Agents for data management send data from source databases to distribution agents. Then, distribution agents cooperate with web computers to calculate results and return them to management agents. Both types of agents can autonomously move from one host to another to improve quality of grid resource using.

Moreover, two group of agents based on genetic programming as well as harmony search have been introduced to optimize big data processing. A set of agents designed for local optimization are some harmony search schedulers. These schedulers can optimize resource using. They cooperate with distributors and managers to give them information about optimal workload in a grid. Finally, genetic programming has been applied for finding the compromise configurations of grid. These agents cooperate with harmony search schedulers to correct in local timetables.

In this paper, big data dilemmas are described in Sect. 2. Then, intelligent agents for big data are studied in Sect. 3. A description of agents based on genetic programming is included in Sect. 4. Moreover, some outcomes for numerical experiments are submitted.

2   Big Data System Architectures

Some current big data applications are based on tools such as *Hadoop* or *NoSQL* cluster. Scalability is their ability to handle an increasing amount of transactions and stored data at the same amount of time. *MongoDB* is the *NoSQL* database that supports the data stored to different nodes and has support for a number of programming languages.

The current solutions to big data dilemmas are based on parallel data processing. We can use a massively parallel cluster with lots of CPUs, GPUs, RAM and disks to obtain a high performance by data-based parallelism. It is important to deal with OLTP *online* transaction processing that is a class of information systems to manage transaction-oriented applications. Moreover, low response time for decision-support queries can be obtained for OLAP that is online analytical processing to answering multi-dimensional analytical queries rapidly. High reliability can be obtained through data replication. As a final point, we expect extensibility with the almost linear speed-up as well as linear scale-up. Performance can increase linearly for a constant database size, load, and proportional increase of the components (CPU, GPU, RAM, disk). On the other hand, linear scale-up means that performance is constant for proportional increase of CPUs and a linear growth of load and database size.

Figure 1 shows three cases of data-based parallelism. In the first case, the same operation is carried out on different data (Fig. 1a). This case can be considered for large query. For concurrent and different queries that operate on the same data, we can consider the second case on Fig. 1b.

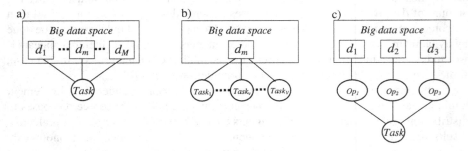

Fig. 1. Three cases of data-based parallelism

The third case (Fig. 1c) is related to complex query that is divided on some parallel operations acting on diverse data. Above three cases of data-based parallelism permit us to prepare two alternative architectures that support big data parallelism.

The main architecture that is convenient for write-intensive tasks is based on shared-memory computers like *Bull Novascale*, HP *Proliant* or IBM *Numascale*. Unfortunately, this architecture is based on NUMA *Non-Uniform Memory Access* server technology and it is not suited for big data. Several disks are shared by many processors via shared RAM. The architecture can support effectively applications, and it can support load balancing. On the other hand, the NUMA architecture is involved with interconnection limits and there are some difficulties with extensibility.

Architecture with shared-disk cluster is much more prepared for big data processing than NUMA architecture. *Storage Area Network* SAN interconnects private memory and disks that are shared by processors. Hosts like *Exadata, Oracle* RAC and IBM *PowerHA* are convenient for applications and some extensions can be made easily. However, a complex distributed lock manager is needed for cache coherence.

A crucial feature of BD is related to intensive reading from hard disks and then processing, instead of processing and then intensive writing. If we consider no sharing of memory or disks across nodes (Fig. 2), this system requires data partitioning of database like in servers: DB2 DPF, *MySQLcluster* or *Teradata*.

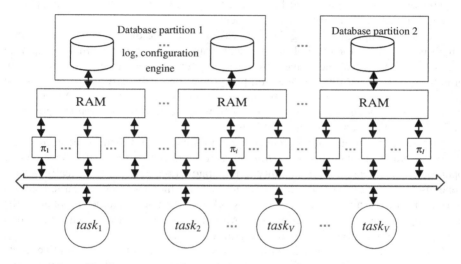

Fig. 2. Shared-nothing cluster architecture for big data [24]

Big data is spread over some partitions that run on one or some separate servers with own table spaces, logs, and configurations. A query is performed in parallel on all partitions. Such architecture can support *Google* search engine, *NoSQL* key-value stores (*Bigtable*). An advantage is the highest extensibility and low cost. In contrast, some updates and distributed transactions are not efficient.

3   Intelligent Agents for Big Data

Intelligent agents can improve efficiency of a grid middleware layer for big data processing. A volunteer grid can gather data from multiple sources, which may be heterogeneous and spread geographically across the world. Moreover, the collected data may be stored by a volunteer grid in multiple geographically spread facilities [9, 17].

Multiagent systems are well suited for BD acquisition because of mobility, which means the ability to move between different facilities. By doing that agents can get closer to the source of data or closer to the data they are about to process. It reduces bandwidth requirements and communication delays [9]. The ability to react upon sudden changes of the environment and to act proactively is important to provide foundation for handling changes in availability of data sources or collected data. An agent can make decision if move to another set of data or initiate communication with other agents [2, 7].

Other useful traits of agents include abilities to communicate and negotiate. In agent-based data mining system it is possible to distinguish different roles and groups of tasks that constitute the whole mining process. Individual roles can be assigned to agents. Through communication and negotiation working groups of agents can be established, each of them built of agents with a unified goal. Agents can improve efficiency of data mining compared to centralized approaches [37]. It was applied in different domains showing promising results for further research, e.g. banking and finance domain [21] or resource allocation in distributed environments [4, 9].

A common approach for big data processing is the use of *MapReduce* algorithm, which is optimized for parallel and asynchronous execution on multiple computing nodes [12]. Because of the proven usefulness of this approach in multiple areas, e.g. bioinformatics [15, 26], fraud detection [22], social network analysis [16, 25] – there are many software frameworks for performing this kind of computations. Among most popular is the *Apache Hadoop* software [12], which can utilize computing power of multiple machines in a distributed environment.

However, such tools often introduce certain limitations. Of them is the need to use internal storage mechanisms (e.g. *Hadoop* distributed file system [29]), for effective operation. It is an effect of an inability to integrate with external and typically hetero-geneous data sources. Data administrators are forced to move or duplicate large volumes of data from existing data stores to framework-specific ones. Another problem is the lack of support for online data and analysis. Moreover, many scenarios require extraction and merging of data to produce a meaningful result [32].

One more issue is that some frameworks may introduce architectural flaws like single point of failure (e.g. Hadoop before version 2.0 [33]) that have an influence on the whole system, in which they were deployed. Some of those issues can be addressed by using agent-oriented approach. Agents are designed for heterogeneous environments and are usually attributed with the ability to handle changes [36]. It translates to capability of integrating with different data sources. Reactive nature of agents enables them to work with online data streams with each new piece of information appearing in the stream interpreted as an event that requires agent response.

Another trait of agents is pro-activeness, which means that an agent can not only react to external events but also run actions on its own [34]. It is especially important in case of analysis, as there are often no clues about expected results known in advance. Because of that it is not possible to create the knowledge extraction algorithm *a priori*. Proactive behavior of agents can expose information that was not expected.

The problem of data acquisition can be solved be making use of another trait of agents – their mobility. It means that a software agent is not bound to any particular machine or execution container [20]. Individual agents can migrate between different nodes, to get closer to the sources of data. Instead of providing the data to the system, with multiagent approach the system acquires the information on its own so there is no need for data administrators to migrate or duplicate the data.

For real-time data analysis that takes into account both offline data and online streams Marz proposed the lambda architecture [23], which consists of three main components: batch layer, serving layer and speed layer. The batch layer is responsible for offline data processing and can be implemented using existing *MapReduce* frameworks like afore-mentioned *Hadoop*. This layer produces batch views of the data, which can be exposed to external applications. The serving layer serves prepared views to clients. The speed layer is responsible for real-time processing of data streams. It analyses data that was not yet processed by the batch layer. Speed layer produces real-time views that can be coupled with batch views to create complete representation of the extracted knowledge.

Twardowski and *Ryżko* further show that the lambda architecture can be defined in terms of a heterogeneous multiagent system [32]. There are several strong motivations for this approach. First of all, the lambda architecture gives only general guidelines. The actual realization requires integration of a few components: one for batch processing, another one for serving views, a different solution for real-time stream analysis and a component that merges real-time views with batch views. There are ready-made frame-works and libraries for each step that can be used to create a complete solution.

The use of multiagent environment will provide a common way for information exchange between different component and a common execution model [32]. The differences between individual components of the lambda architecture lead to inherently heterogeneous realizations so the ability to handle diversity in agent systems in another motivation for this approach. Figure 3 shows the lambda architecture employing multi-agent model.

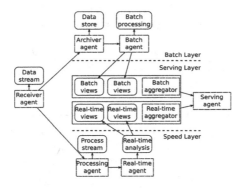

Fig. 3. Multiagent real-time processing utilizing lambda architecture [32]

Limitations of intelligent agents systems are related to higher complexity of software preparation that causes higher costs and higher probability of mistake appearing. What is more, some algorithms based on artificial intelligence have to be considered. These algorithms are probabilistic and it cause some unpredicted outcomes during some phases of their running. However, advantages of intelligent agents seem to be more important that their disadvantages. Among intelligent agents, agents based on genetic program-ming are worth to consideration regarding their capability of resource managing in grids applied for big data.

4 Agents Based on Genetic Programming

Intelligent agents based on genetic programming *AGPs* can optimize a grid resource management for big data queries [1, 18]. Especially, they have been dedicated to global optimization of middleware software module allocation in *Comcute* grid. In that system, they cooperate with intelligent agents based on harmony search *AHSs* that reconfigure some local parts of grids. *AGP* starts from a goal of load balancing to be achieved and then it creates a solver autonomously [19]. It is similar to deal with the dilemma from machine learning "How can computers be made to do what needs to be done, without being told exactly how to do it?" [27]. *AGP* uses the principle of selection, crossover and mutation to obtain a population of programs applied as a scheduler for efficient using big data by the *Comcute*. This scheduler optimizes some criteria related to load balancing and send a compromise solution to *AHSs* [35].

In *AGP*, a program is represented as a tree that consists on branches and nodes: a root node, a branch node, and a leaf node. A parent node is one which has at least one other node linked by a branch under it. The size of the parse tree is limited by the number of nodes or by the number of the tree levels. Nodes in the tree are divided on functional nodes and terminal ones. Mutation and crossover operate on trees, and chosen node from tree is a root of subtree that is modified regarding genetic operators.

Figure 4 shows architecture of the volunteer grid *Comcute* with one AGP agent that cooperates with two *AHS* agents. We divide the whole grid on several subgrids with at most 15 nodes.

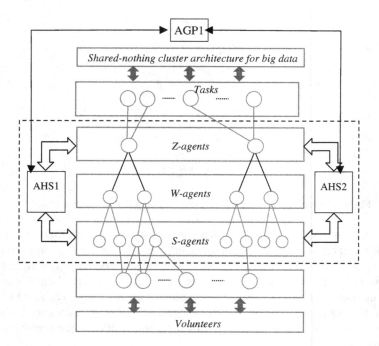

Fig. 4. Agent based on genetic programming in the *Comcute* for big data processing

AHS can find configuration for at most 15 nodes, 50 tasks and 15 alternatives of resource sets *ARS* on *PC/Windows 7/Intel i7*. Figure 5 shows an example of a compromise configuration for a subgrid in the *Comcute* that was found by *AHS1* for its area consisted of 15 nodes. Workload is characterized by two criteria [10, 13]. The first one is the CPU workload of the bottleneck computer (denoted as $\hat{Z}_{max}$), and the second one is the communication workload of the bottleneck server ($\tilde{Z}_{max}$).

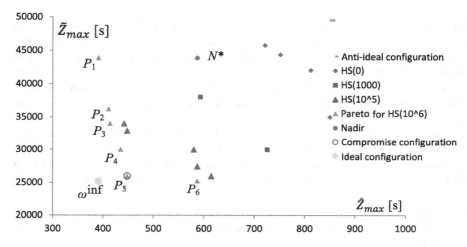

Fig. 5. A compromise configuration found by *AHS1*

The compromise configuration found by *AHS1* is specified in Table 1. The *W*-agent with number 5 as well as two *S*-agents (No. 6 and 27) should be moved to the node No. 1, where the third alternative of resource set *ARS* (*BizServer* E5-2660 v2) is assigned. Agent *AHS1* publishes this specification in the common global repository of grid and the other agents can read it to make decision related to moving to some recommended nodes. However, reconfiguration of resource requires a bit of time. So, a middleware agent reads the state of resources in the given grid node, and then it makes decision whether to go to that node or not.

Table 1. A specification of a compromise configuration from AHS1

Node i	1	2	3	4	5	6	7	8	9	10	11	12	13	14
ARS j	3	8	3	9	8	9	8	3	9	3	3	3	3	3
No. W	5		11	4	9		2,13	10		1,8	6,15	3,7	14	12
No. S	6,27	13,23,24,30	7,26	14,21,28	3,15,25	4,5,11,22	19	2	1,16,18,20	9	10	29	12,17	8

On the other hand, the *AGP* cooperates with several *AHSs*. It takes into account their recommendation for resource using by middleware agents. Moreover, the *AGP* optimizes the resource usage for the whole grid starting from the configuration obtained by set of *AHSs* and trying to improve it by multi-criteria genetic programing.

5   Concluding Remarks and Future Work

Shared-nothing cluster architecture for big data can be extended by cooperation with volunteer and grid computing. Moreover, intelligent agents in the middleware of grid can significantly support efficiency of proposed approach. Multi-objective genetic programming as relatively new paradigm of artificial intelligence can be used for finding Pareto-optimal configuration of the grid. Agents based on genetic programing can cooperate with harmony search agents to solve NP-hard optimization problem of grid resource using.

Our future works will focus on testing the other sets of procedures and terminals to find the compromise configurations for different criteria. Initial numerical experiments confirmed that sub-optimal in Pareto sense configurations can be found by *AGPs* and *AHSs*. Moreover, quantum-inspired algorithm can support big data, too [3].

Acknowledgements. This research is supported by Department of Computer Architecture, Faculty of Electronics, Telecommunications and Informatics, Gdańsk University of Technology under statutory activity grant.

References

1. Altameem, T., Amoon, M.: An agent-based approach for dynamic adjustment of scheduled jobs in computational grids. J. Comput. Syst. Sci. Int. **49**, 765–772 (2010)
2. Balicki, J.: Negative selection with ranking procedure in tabu-based multi-criterion evolutionary algorithm for task assignment. In: Alexandrov, V.N., van Albada, G.D., Sloot, P.M., Dongarra, J. (eds.) ICCS 2006. LNCS, vol. 3993, pp. 863–870. Springer, Heidelberg (2006)
3. Balicki, J.: An adaptive quantum-based multiobjective evolutionary algorithm for efficient task assignment in distributed systems. In: Mastorakis, N., et al. (eds.) Proceedings of the 13th WSEAS International Conference on Computers, Recent Advances in Computer Engineering, Rhodes, Greece, pp. 417–422 (2009)
4. Balicki, J., Kitowski, Z.: Multicriteria evolutionary algorithm with tabu search for task assignment. In: Zitzler, E., Deb, K., Thiele, L., Coello Coello, C.A., Corne, D.W. (eds.) EMO 2001. LNCS, vol. 1993, pp. 373–384. Springer, Heidelberg (2001)
5. Balicki, J., Korłub, W., Szymanski, J., Zakidalski, M.: Big data paradigm developed in volunteer grid system with genetic programming scheduler. In: Rutkowski, L., Korytkowski, M., Scherer, R., Tadeusiewicz, R., Zadeh, L.A., Zurada, J.M. (eds.) ICAISC 2014, Part I. LNCS, vol. 8467, pp. 771–782. Springer, Heidelberg (2014)
6. Balicki, J., Korlub, W., Krawczyk, H., et al.: Genetic programming with negative selection for volunteer computing system optimization. In: Paja, W.A., Wilamowski, B.M. (eds.) Proceedings of the 6th International Conference on Human System Interactions, Gdańsk, Poland, pp. 271–278 (2013)

7. Bernaschi, M., Castiglione, F., Succi, S.: A high performance simulator of the immune system. Future Gener. Comput. Syst. **15**, 333–342 (2006)
8. BOINC. http://boinc.berkeley.edu/, Accessed 25 Jan 2015
9. Cao, L., Gorodetsky, V., Mitkas, P.A.: Agent mining: the synergy of agents and data mining. IEEE Intell. Syst. **24**, 64–72 (2009)
10. Coello Coello, C.A., Van Veldhuizen, D.A., Lamont, G.B.: Evolutionary Algorithms for Solving Multi-Objective Problems. Kluwer Academic Publishers, New York (2002)
11. Comcute grid. http://comcute.eti.pg.gda.pl/, Accessed 25 Jan 2015
12. Dean, J., Ghemawat, S.: MapReduce: simplified data processing on large clusters. Commun. ACM **51**, 1–13 (2008)
13. Deb, K.: Multi-Objective Optimization Using Evolutionary Algorithms. Wiley, Chichester (2001)
14. Finkelstein, A., Gryce, C., Lewis-Bowen, J.: Relating requirements and architectures: a study of data-grids. J. Grid Comput. **2**, 207–222 (2004)
15. Gunarathne, T., et al.: Cloud computing paradigms for pleasingly parallel biomedical applications. In: Proceedings of the 19th ACM International Symposium on High Performance Distributed Computing, Chicago, pp. 460–469 (2010)
16. Guojun, L., Ming, Z., Fei, Y.: Large-scale social network analysis based on MapReduce. In: Proceedings of the International Conference on Computational Aspects of Social Networks, pp. 487–490 (2010)
17. Jennings, N.R., Wooldridge, M.: Applications of intelligent agents. In: Jennings, N.R., Wooldridge, M. (eds.) Intelligent Agents, pp. 3–28. Springer, New York (1998)
18. Kang, J., Sim, K.M.: A multiagent brokering protocol for supporting Grid resource discovery. Appl. Intell. **37**, 527–542 (2012)
19. Koza, J.R., et al.: Genetic Programming IV: Routine Human-Competitive Machine Intelligence. Kluwer Academic Publishers, New York (2003)
20. Leyton-Brown, K., Shoham, Y.: Multiagent Systems: Algorithmic: Game-theoretic and Logical Foundations. Cambridge University Press, Cambridge (2008)
21. Li, H.X., Chosler, R.: Application of multilayered multi-agent data mining architecture to bank domain. In: Proceedings of the International Conference on Wireless Communications and Mobile Computing, pp. 6721–6724 (2007)
22. Mardani, S., Akbari, M.K., Sharifian, S.: Fraud detection in process aware information systems using MapReduce. In: Proceedings on Information and Knowledge Technology, pp. 88–91(2014)
23. Marz, N., Warren, J.: Big Data - Principles and Best Practices of Scalable Realtime Data Systems. Manning Publications Co., USA (2014)
24. O'Leary, D.E.: Artificial intelligence and big data. IEEE Intell. Syst. **28**, 96–99 (2013)
25. Ostrowski, D.A.: MapReduce design patterns for social networking analysis. In: Proceedings of International Conference on Semantic Computing, pp. 316–319 (2014)
26. Qiu, X., et al.: Using MapReduce technologies in bioinformatics and medical informatics. In: Proceedings of the International Conference for High Performance Computing, Networking, Storage and Analysis, Portland (2009)
27. Samuel, A.L.: Programming computers to play games. Adv. Comput. **1**, 165–192 (1960)
28. Shibata, T., Choi, S., Taura, K.: File-access patterns of data-intensive workflow applications. In: Proceedings of the 10th IEEE/ACM International Conference on Cluster, Cloud and Grid Computing, pp. 522–525 (2010)
29. Shvachko, K., et al.: The Hadoop distributed file system. In: MSST, pp. 1–10 (2010)
30. Snijders, C., Matzat, U., Reips, U.-D.: 'Big Data': big gaps of knowledge in the field of Internet. Int. J. Internet Sci. **7**, 1–5 (2012)

31. Szabo, C., et al.: Science in the cloud: allocation and execution of data-intensive scientific workflows. J. Grid Comput. **12**, 223–233 (2013)
32. Twardowski, B., Ryzko, D.: Multi-agent architecture for real-time big data processing. In: Proceedings of the International Conference on Web Intelligence and Intelligent Agent Technologies, vol. 3, pp. 333–337 (2014)
33. Vavilapalli, V.K.: Apache Hadoop yarn: Yet another resource negotiator. In: Procedings of the 4th Annual Symposium on Cloud Computing, New York, USA, pp. 5:1–5:16 (2013)
34. Verbrugge, T., Dunin-Kęplicz, B.: Teamwork in Multi-Agent Systems: A Formal Approach. Wiley, Chichester (2010)
35. Węglarz, J., Błażewicz, J., Kovalyov, M.: Preemptable malleable task scheduling problem. IEEE Trans. Comput. **55**, 486–490 (2006)
36. Wooldridge, M.: Introduction to Multiagent Systems. Wiley, Chichester (2002)
37. Zhou, D., et al.: Multi-agent distributed data mining model based on algorithm analysis and task prediction. In: Proceedings of the 2nd International Conference on Information Engineering and Computer Science, pp. 1–4 (2010)

Two Stage SVM and kNN Text Documents Classifier

Marcin Kępa and Julian Szymański[✉]

Department of Computer Systems Architecture,
Gdańsk University of Technology, Gdańsk, Poland
markepa@pg.gda.pl, julian.szymanski@eti.pg.gda.pl

Abstract. The paper presents an approach to the large scale text documents classification problem in parallel environments. A two stage classifier is proposed, based on a combination of k-nearest neighbors and support vector machines classification methods. The details of the classifier and the parallelisation of classification, learning and prediction phases are described. The classifier makes use of our method named *one-vs-near*. It is an extension of the *one-vs-all* approach, typically used with binary classifiers in order to solve multiclass problems. The experiments were performed on a large scale dataset, with use of many parallel threads on a supercomputer. Results of the experiments show that the proposed classifier scales well and gives reasonable quality results. Finally, it is shown that the proposed method gives better performance compared to the traditional approach.

Keywords: SVM · k-nearest neighbor · Wikipedia · Documents categorization · Parallel classification

1 Introduction

Since the beginning of the Internet, its size and the amount of globally stored data has been growing. With every year, the estimated number of indexed web pages is increasing and today it is somewhere between 20 and 50 billion pages [1]. Because of the size of an average dataset, a need for automatic categorization arises. Repositories, such as Wikipedia, reaching 4.5 million articles, organized with hundreds of thousands of categories, could benefit from automatic categorization. There are many existing approaches to this problem, with different results both in terms of accuracy and performance [2–4], but there is still need for improvements in this area.

The aim of the work presented here is to propose a two stage classifier, capable of automatic categorization of text documents, from repositories containing over 100 k categories and millions of articles. The proposed classification is performed in two stages. The first one is a fast, initial classification stage, done by the k-nearest neighbours (kNN) classifier, where the dataset is limited to selected categories. The second stage is the final, accurate classification stage, done by

© Springer International Publishing Switzerland 2015
M. Kryszkiewicz et al. (Eds.): PReMI 2015, LNCS 9124, pp. 279–289, 2015.
DOI: 10.1007/978-3-319-19941-2_27

the support vector machines (SVM)classifier, trained on the limited dataset. The experiments designed to evaluate our approach are performed using Wikipedia data, processed with our application that allows us to construct its machine-processable representation [5]. The original contribution of this paper is the application of our method named *one-vs-near* in classification of large scale text documents repositories. This is done in order to improve the performance of a typical linear SVM in the *one-vs-all* setting.

The next section briefly describes SVM and kNN classifiers and the way they are incorporated to solve multiclass classification problems. Section 3 presents the details of the solution. Then, Sect. 4 briefly describes the Galera supercomputer, used to test the performance of our classifier in highly parallel environments and with big datasets. The experiments using our implementation, along with empirical results based on Wikipedia datasets, are given in Sect. 5. The last section summarizes the paper and gives ideas for future research in this area.

2   kNN and SVM Classifiers in a Typical Multiclass Setting

2.1   kNN in Multiclass Setting

One of the simplest, as well as the oldest machine classification techniques, is the approach called kNN [6]. In a typical setting each test object is assigned to a certain class, based on majority of its k nearest neighbors [6,7]. However, this approach can be computationally expensive for datasets containing millions of test objects. Some papers have shown [7,8] that kNN classifiers trained with the use of pre-labled examples can highly improve the quality of classification. Since a standard kNN approach can be very demanding performance-wise, modified solutions are introduced, eg. the centroid kNN [3]. The idea is to calculate a centroid for groups of feature vectors belonging to the same category and apply the similarity metric on these centroids, instead of on each feature vector individually. Given a set of S documents we can define the centroid vector as:

$$C = \frac{1}{|S|} \times \sum_{d \varepsilon S} d \tag{1}$$

where $|S|$ is the number of articles in a class S and d are the vectors representing the articles. After computing the centroids, we can use any similarity metric to compare them in the prediction phase. The complexity of prediction in such case (assigning labels to m_{test} test objects) is at most $O\left(m_{\text{test}} \cdot N\right)$, where N is the number of categories. The complexity of computing the model is at best only $O\left(m_{\text{train}}\right)$, where m_{train} is the number of training examples.

2.2   SVM in Multiclass Setting

SVM's are one of the most effective methods of text classification [9]. In its base form an SVM is a binary classifier that constructs a hyperplane $h()$ in a high dimensional feature space (examples are typically projected into that space by a kernel function), which is convex-optimized during training so that it separates the classes leaving maximal possible space (margins) between them. The prediction step can be summarized in a simple equation $a = h(x)$, where a is the activation of the hyperplane, and x is the feature vector (possibly transformed by a kernel) of a testing object. The sign of a decides which class is predicted, whereas the absolute value of a indicates the confidence of this decision. With advanced optimization algorithms used by an SVM, time complexity of training such hyperplane is $O\left(m_{\text{train}}\right)$. Although there are attempts to directly deal with multiclass problems using reformulated SVMs [10], most often such problems are divided into binary classifications and incorporate typical SVM classifiers summarized above.

In a popular *one–vs–all* scheme for each class a separate hyperplane is trained, by treating examples from that class as positives and all the remaining examples in the dataset as negatives. During prediction, a test object is assigned to a class which hyperplane's activation a is the highest (*winner takes all* strategy). Complexity of calculating the whole model in this setting is $O\left(m_{\text{train}} \cdot N\right)$. For such a classifier the prediction can be performed in a $O\left(m_{\text{test}} \cdot N\right)$ time, the same as for the kNN classifier. The comparative study of this multiclass SVM setting, as well as other less popular ones, can be found in [11]. It is important to note that the classification of Wikipedia belongs to a multi-label family of problems. In such cases, the *winner takes all* algorithm is replaced, each article is tested against every category in the dataset and the final result consists of categories with activation scores that exceed a specified threshold.

2.3   Hybrid Approaches

In order to improve the effectiveness of classification hybrid approaches of kNN and SVM are introduced. The approaches vary in the way the classification stages are combined and the types of datasets used. One of such methods is the HKNNSVM classifier, proposed in [12] that improves kNN classier's accuracy by limiting the dataset only to the support vectors of each category's hyperplane. It should be noticed the accuracy of the kNN classifier is slightly increased in this approach. Both the training and the prediction phase of the HKNNSVM require a bigger amount of computations than in our approach, which might be a problem when classifying big repositories such as Wikipedia.

Another approach proposed in [13] uses the kNN classifier to select the nearest neighbours for a given query. An SVM classifier (DAGSVM) is then employed in order to make the final decision. The classifier shows excellent accuracy in character recognition however, a similar approach wouldn't be as effective in the case of large scale text documents classification. The initial search for nearest neighbours amongst millions of articles, each containing thousands of features, would be very demanding performance-wise. Moreover, the fact that an SVM has to be trained for each query is also an issue in a dataset containing millions of examples. Because of that our method should prove to be more efficient at classifying sparse textual datasets. A similar approach to character recognition is also proposed in [14], however, just as the previous solution, it is not practical for a large and sparse dataset such as Wikipedia.

A different approach is also proposed in [15]. The solution uses the kNN rule in order to assign real value weights to the examples in the training dataset. This is unlike the standard SVM, where examples belonging to a class are assigned a 1 and all the others are assigned a -1. Just as in the previous examples this approach proves to be more accurate than a single SVM. Again, the computational complexity of this approach makes it impractical in case of large scale text documents classification. It is worth mentioning that unlike the other solutions our approach deals with multi-label problems.

3   Details of Our Approach

Our classification system consists of four modules: the data preparation module, the initial classification module, the final classification module, and the results evaluation module. The results of every stage are saved in the file system, which allows us to run the stages independently. The data preparation module is designed to filter the dataset in order to meet requirements of the classifier. The data evaluation module consists of programs designed to return quality scores of the classifier such as its precision and recall.

3.1   Two Stage Classification

The classifier uses the *one–vs–near* approach instead of the *one–vs–all* approach in order to limit the dataset during the learning phase [16]. For each category, for which the classifier is trained, the dataset is limited to its closest neighbours. The category neighbour list (used for limiting the dataset) is computed by the kNN classifier in the first stage. It is important for the initial stage to be lightweight, in order to minimize its impact on the overall performance of the classifier. The kNN computes the distances between every centroid in the form of an ordered list. This list is then saved in the filesystem and later used by the second stage classifier. The second stage SVM uses the saved neighbour list to limit the dataset used for training the classifier for a single category. Apart from this, the second stage classifier works as a standard *one–vs–all* approach for SVM. However, thanks to this difference it is possible to greatly limit the training dataset size for each binary classifier. Because of that, the training performance should be improved. Furthermore, the accuracy of the resulting classifier should be comparable to one trained on the entire dataset. This two stage approach is presented in Algorithm 1.

Algorithm 1. Two stage training

---
1. Get a category to train or end if no more categories exist
2. Get neighbours of that category from the neighbours list
3. Prepare the dataset containing only articles from neighbouring categories
4. Train the SVM classifier on that dataset and go to 1.

3.2   Parallelisation of the Computations

One of the main goals of our research is for the final classifier to be easily scalable. Both the training and the prediction tasks related to each SVM hyperplane and kNN centroid are intrinsically independent, therefore the job of dividing the problem between parallel compute nodes is straightforward. Each node is on its own responsible for downloading tasks from a task queue. Each task is in fact either a category to train (in the training phase) or an article for which classes are to be predicted (in the prediction phase). Each compute node picks up tasks from the task queue in batches.

In addition to machine level parallelisation, each node runs its computations in parallel threads. Managing to distribute the training and prediction procedures related to all classes over different compute nodes allows us to construct a scalable classifier. The classifier accesses its files through a Network File System (NFS) so that every machine works in the same directory and has access to the same files. As mentioned before, the jobs to be done are stored in a single file queue – the TODO file. The TODO file contains names of hyperplanes to train, in case of training and a list of objects to predict labels for, in case of prediction. Every node can obtain a certain number of jobs from the TODO file and run these jobs using available cores. Having done that it can receive new jobs and so on. Synchronization between nodes is obtained using Message Passing Interface (MPI) implementation Open MPI [17].

Because many parallel nodes need access to the TODO file, there is a need for some synchronization mechanism. The solution to this problem is to use the MPI in order to implement a master–slave scheme. The processes of the application are divided into a single master process and many slave processes. The master process is used to distribute the tasks between the slave processes and the slave processes are in turn used to conduct the computations. The access to the TODO file is granted to slave processes by the master process. Each slave has to request access to the queue from the master before downloading its tasks.

4   Test Environment

In order to test our approach on big datasets a parallel computations environment was needed. The classifier was tested on the Galera supercomputer, in Academic Computer Centre (CI TASK), part of Gdansk Univeristy of Technology. The cluster consists of 1344 2,33 GHz Intel Xeon QuadCore processors (5376 cores), 25 TB total system memory, 100 TB disk storage and Mellanox InfiniBand interconnect with 20 Gb/s bandwidth. The cluster is operated under a Linux family operating system. The total theoretical peak performance of the cluster is 50 TFLOPS. Upon its launch, the cluster performance was measured to be 38.2 TFLOPS.

The environment is configured to use the message passing interface (MPI) implementation for communication between different nodes of the cluster. The tasks are queued for execution with a portable batch system (PBS) based queue. For the purpose of this work only a fraction of the cluster was used, comprising of 500 cores. This was more than enough to test the classifier in a massively parallel environment. The results of these experiments can be seen in the next section.

5   Experiments

To evaluate the effectiveness of our approach a series of tests was performed. They were planned to check performance, scalability and F-score of the classifier. Initial tests have been conducted with smaller size data and without cross validation. The final tests have been performed using large scale datasets and with evaluation based on cross validation. The datasets used in this paper were created from the entire Wikipedia, based on 8th March 2013 dump [18]. This dump was processed using Matrix'u application [5] in order to create a bag of words [19] representation of the dataset. The dataset was then filtered, which among other things deleted administrative categories and merged small categories with their parents. Remaining very small categories (for small categories there is not enough examples in order to train an accurate classifier) were removed.

5.1   KNN and SVM Training Scalability

The first test was designed in order to test the scalability of the solution by comparing training phase performance of both classifiers (the initial kNN and the final SVM classifier). The dataset for this test was limited to 530 categories, containing 853 283 documents. Apart form testing the scalability in a highly parallel environment (between 8 and 160 logical processors) another important result of this test is the comparison of the fast initial kNN classifier and the final accurate SVM classifier. The results in this test represent only the time needed for creating the classification models. All additional time is subtracted form the results, they can be seen in Fig. 1. As expected the fast initial classifier is faster by an order of magnitude. This result shows that it is indeed feasible to use the kNN classifier in order to conduct a fast initial classification and then to use that data to improve performance of the SVM classifier. Moreover, the scalability of both classifiers is very good.

5.2   Classification Quality for Big Data

After performing the scaling tests for small data, the next step was to run the classifier on the entire Wikipedia. After filtering, the dataset for this test consisted of 2 992 212 articles grouped in 18 335 categories. The quality of the classifier was validated in 10-fold cross validation. The calculated values are the precision, recall and the F-score. In order to compare the quality of the stage classifier with the standard one–vs–all approach, as well as the centroid kNN classifier, all three classifiers were trained on the same dataset. Together all the models trained for a single classifier took around 50 gigabytes of disk space.

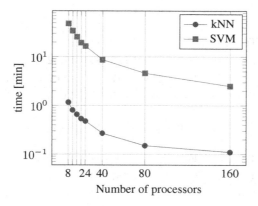

Fig. 1. SVM training performance on Galera cluster

First, the one–vs-all SVM classifier was run. In order to find best results (modify the recall of the classifier), different acceptance thresholds were tested. This means that the SVM classification rule was altered, by an additional parameter t, changing the hyperplanes $h(x)$ position:

$$Category(x) = sign(h(x) - t) \qquad (2)$$

Additional explanation of this threshold, as well as more advanced approaches to its optimization can be found in [20]. The results of this test can be seen in Table 1. As we can see, the optimal results were achieved for 0.05 acceptance threshold, further increase of the threshold gave better precision but the recall suffered greatly. On the other hand, decreasing the threshold gave better recall, but the precision deteriorated quickly.

As mentioned before, the same dataset was classified using the centroid kNN classifier, with cosine similarity used as the distance metric. The F-score of this classification is much lower however, it is still acceptable. For the kNN classifier the acceptance threshold is the minimal level of cosine similarity, between a category and the article feature vector, used to determine whether it belongs to that category. The recall (depending on the acceptance threshold) was as high as 50 %, which is still usable. On the other hand, limiting the results even more, by increasing the acceptance threshold, gave good precision of over 50 % with small

Table 1. SVM classifier precission

Accept thresh.	True pos.	False pos.	False neg.	Precision	Recall	F-score
0.30	3 681 502	1 700 894	7 033 603	68.39 %	34.35 %	45.74 %
0.05	4 239 491	2 905 837	6 475 614	59.33 %	39.56 %	**47.47 %**
0	4 365 491	3 405 395	6 349 614	56.17 %	40.74 %	47.23 %

Table 2. kNN classifier precission

Accept thresh.	True pos.	False pos.	False neg.	Precision	Recall	F-score
0.2	4 831 716	22 737 151	5 883 389	17.52 %	45.09 %	25.06 %
0.3	3 542 955	12 079 258	7 172 150	22.67 %	33.06 %	**26.90 %**
0.9	1 644 218	1 431 920	9 070 887	53.45 %	15.34 %	23.84 %

Table 3. Two stage classifier precission

Accept thresh.	True pos.	False pos.	False neg.	Precision	Recall	F-score
0.1	4 127 571	6 161 031	6 587 534	40.11 %	38.52 %	39.30 %
0.3	3 676 520	2 819 521	7 038 585	56.59 %	34.31 %	**42.72 %**
0.35	3 572 710	2 441 656	7 142 395	59.40 %	33.34 %	42.71 %

recall of 15 %. Some example results in relation to the acceptance threshold can be seen in Table 2.

Finally the two stage classifier was tested on the same dataset. Based on the results from small data tests, the amount of neighbours was set to 30 % of categories. The results of this test can be seen in Table 3. The F-score is considerably better than for the kNN classifier and slightly lower than for a one–vs–all solution (but with much better performance). Furthermore, it is worth noting that the two stage classifier achieved F-score comparable to classifiers that took part in the Pascal Large Scale Hierarchical Text Classification Challenge (LSHTC3) [21]. Comparing results form this paper to the ones from Wikipedia based tasks in LSHTC3, we can see that these values are very similar. For example, the best F-score for Wikipedia datasets in LSHTC3 was 49 % for the medium and 45 % for the large dataset.

5.3 Performance for Big Data

Another important experiment was to test the performance of the training and the prediction phase for all three classifiers. The results presented in Fig. 2 are for the same dataset as before, with 18 335 categories and 2 992 212 articles. The tests were conducted with 62 compute nodes, each with 8 logical processors, giving 496 processors in total. The results are averaged in 10-fold cross validation. We can clearly see that in the training phase the two stage approach is considerably faster, than the traditional one–vs–all classifier. Although the centroid kNN approach presents poor precision, the training time is shorter by an order of magnitude. This means that this approach could still be useful in certain cases where lower precision is not an issue. All in all, the multi stage approach presents itself as a good way to increase performance of SVM classification, while maintaining high F-score.

As mentioned in the previous sections all presented classifiers are comparable when it comes to computational complexity of the prediction phase. The

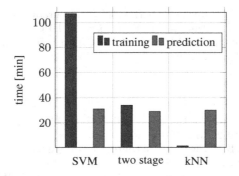

Fig. 2. Performance comparison of different solutions for big data

prediction times for the entire dataset are therefore very similar for the three approaches. The calculated times are as follows: 29 min for the two stage, 31 min for the SVM and 30 min for the kNN classifier.

6   Conclusions and Future Work

The aim of the research presented in this paper was to develop and evaluate a parallel multi stage approach to classification of text documents with SVM. Our solution was designed to be used with large scale text document repositories in mind, such as Wikipedia. The results of the experiments show that our approach scales up well and gives good F-score. The two stage approach based on *one–vs–near* scheme was tested on big datasets created from the entire Wikipedia. Precision and recall of our solution proved to be comparable to the typical *one–vs–all* scheme, while significantly improving the time needed for classifiers construction. Additionally it was proven that the simple centroid kNN classifier can also produce useful quality results, with classifier creation time shorter by an order of magnitude. Although the problem of text documents classification was extensively tested in many works (eg. [2,4,22]), there is still some room for further research and improvements in this area.

There are many yet untested approaches to this problem that are worth pursuing. It would be interesting to verify how substituting the initial kNN classifier with different approaches would impact the quality and performance of the classifier as well as allow to mining the relations between categories [23] . It would be also interesting to test the two stage approach with different kinds of SVM solvers and their parameters. The performance of the classifier could also be improved using an array data DBMS (such as SciDB) to store the feature vectors, instead of plain text files. This would further improve the performance of the classifier and possibly decrease the memory requirements. Also improvement of managing the threads distribution using BeesyCluster [24] can lead to results improvement.

Acknowledgments. The work was performed within grant "Modeling efficiency, reliability and power consumption of multilevel parallel HPC systems using CPUs and GPUs" sponsored by and covered by funds from the National Science Center in Poland based on decision no DEC-2012/07/B/ST6/01516.

References

1. de Kunder, M: The size of the world wide web (2014). http://www.worldwideweb size.com/. Accessed 22 May 2014
2. Gantner, Z., Lars, S.-T.: Automatic content-based categorization of wikipedia articles. In: Proceedings of the 2009 Workshop on The People's Web Meets NLP: Collaboratively Constructed Semantic Resources, People's Web 2009, pp. 32–37. Association for Computational Linguistics, Stroudsburg (2009)
3. Han, E.-H.S., Karypis, G.: Centroid-based document classification: analysis and experimental results. In: Zighed, D.A., Komorowski, J., Żytkow, J.M. (eds.) PKDD 2000. LNCS (LNAI), vol. 1910, pp. 424–431. Springer, Heidelberg (2000)
4. Fan, R.-E., Chang, K.-W., Hsieh, C.-J., Wang, X.-R., Lin, C.-J.: Liblinear: a library for large linear classification. J. Mach. Learn. Res. **9**, 1871–1874 (2008)
5. Szymański, J.: Wikipedia articles representation with matrix'u. In: Hota, C., Srimani, P.K. (eds.) ICDCIT 2013. LNCS, vol. 7753, pp. 500–510. Springer, Heidelberg (2013)
6. Cover, T.M., Hart, P.E.: Nearest neighbor pattern classification. IEEE Trans. Inf. Theory **13**, 21–27 (1967)
7. Weinberger, K.Q., Blitzer, J., Saul, L.K.: Distance metric learning for large margin nearest neighbor classification. In: Advances in Neural Information Processing Systems 18, pp. 1473–1480. MIT Press, Cambridge (2005)
8. Draszawka, K., Szymanski, J.: Thresholding strategies for large scale multi-label text classifier. In: IEEE 2013 the 6th International Conference on Human System Interaction (HSI), pp. 350–355 (2013)
9. Joachims, T.: Text categorization with support vector machines: learning with many relevant features. In: Nédellec, C., Rouveirol, C. (eds.) ECML. LNCS, vol. 1398, pp. 137–142. Springer, Heidelberg (1998)
10. Crammer, K., Singer, Y.: On the algorithmic implementation of multiclass kernel-based vector machines. J. Mach. Learn. Res. **2**, 265–292 (2002)
11. Duan, K.-B., Keerthi, S.S.: Which is the best multiclass SVM method? An empirical study. In: Oza, N.C., Polikar, R., Kittler, J., Roli, F. (eds.) MCS 2005. LNCS, vol. 3541, pp. 278–285. Springer, Heidelberg (2005)
12. Vinoth, R., Jayachandran, A., Balaji, M., Srinivasan, R.: A hybrid text classification approach using KNN and SVM. Int. J. Adv. Found. Res. Comput. (IJAFRC) **1**(3), 20–26 (2014)
13. Zhang, H., Berg, A., Maire, M., Malik, J.: SVM-KNN: discriminative nearest neighbor classification for visual category recognition. In: Proceedinngs of 2006 IEEE Computer Society Conference on Computer Vision and Pattern Recognition, vol. 2, pp. 2126–2136 (2006)
14. Shih, Y., Wei, D.: Machine learning final project: Handwritten sanskrit recognition using a multi-class SVM with K-NN guidance (2011)
15. Hsu, C.-C., Yang, C.-Y., Yang, J.-S.: Associating kNN and SVM for higher classification accuracy. In: Hao, Y., Liu, J., Wang, Y.-P., Cheung, Y., Yin, H., Jiao, L., Ma, J., Jiao, Y.-C. (eds.) CIS 2005. LNCS (LNAI), vol. 3801, pp. 550–555. Springer, Heidelberg (2005)

16. Balicki, J., Szymanski, J., Kępa, M., Draszawka, K., Korlub, W.: Improving effectiveness of svm classifier for large scale data. In: Proceeedings of the 14th International Conference on Artificial Intelligence and Soft Computing (in print). Springer (2015)
17. Gabriel, E., et al.: Open MPI: goals, concept, and design of a next generation mpi implementation. In: Kranzlmüller, D., Kacsuk, P., Dongarra, J. (eds.) EuroPVM/MPI 2004. LNCS, vol. 3241, pp. 97–104. Springer, Heidelberg (2004)
18. Wikipedia: Wikipedia database dump (2014). http://dumps.wikimedia.org/enwiki/20140102/. Accessed 25 January 2014
19. Szymanski, J.: Comparative analysis of text representation methods using classification. Cybern. Syst. **45**, 180–199 (2014)
20. Shanahan, J.G., Roma, N.: Improving SVM text classification performance through threshold adjustment. In: Lavrač, N., Gamberger, D., Todorovski, L., Blockeel, H. (eds.) ECML 2003. LNCS (LNAI), vol. 2837, pp. 361–372. Springer, Heidelberg (2003)
21. Institute of Informatics and Telecommunications - NCSR Demokritos in Greece: Large scale hierarchical text classification challenge (2015). http://lshtc.iit.demokritos.gr/. Accessed 18 January 2015
22. Hsu, C.-W., Chang, C.-C., Lin, C.-J.: A practical guide to support vector classification. Technical report, Department of Computer Science, National Taiwan University (2003)
23. Szymański, J.: Mining relations between wikipedia categories. In: Zavoral, F., Yaghob, J., Pichappan, P., El-Qawasmeh, E. (eds.) NDT 2010. CCIS, vol. 88, pp. 248–255. Springer, Heidelberg (2010)
24. Czarnul, P.: Modeling, run-time optimization and execution of distributed workflow applications in the JEE-based beesycluster environment. J. Supercomput. **63**(1), 1–26 (2010)

Task Allocation and Scalability Evaluation for Real-Time Multimedia Processing in a Cluster Environment

Jerzy Proficz[✉] and Henryk Krawczyk[✉]

Gdansk University of Technology, Gdansk, Poland
jerp@task.gda.pl, hkrawk@eti.pg.gda.pl

Abstract. An allocation algorithm for stream processing tasks is proposed (Modified Best Fit Descendent, MBFD). A comparison with another solution (BFD) is provided. Tests of the algorithms in an HPC environment are described and the results are presented. A proper scalability metric is proposed and used for the evaluation of the allocation algorithm.

Keywords: Scalability · Task allocation · HPC · KASKADA platform

1 Introduction

The pressure of the society demanding secure public environment enforced the governments and local authorities to deploy special measurements related to monitoring of cities and the surrounding areas. The ubiquitous recording devices, installed in the public space, enable recording and *post mortem* analysis of dangerous events. Currently, however, the main goal is to perform such analysis in real-time, mainly to prevent the dangerous situations from happening at all.

The Mayday Euro 2012 project was conceived to provide the proper means for introducing the online multimedia stream processing services into the world of supercomputer datacenters. KASKADA platform [4], an HPC system realizing the above objectives, including execution of soft real-time multimedia applications and underlying services, was developed at the TASK Academic Computer Center of the Gdansk University of Technology in Poland. The platform was deployed on a computer cluster Galera, consisting of 672 compute nodes and 4032 processor cores.

KASKADA platform supports two main groups of functionalities, the first one is related to development of multimedia applications and services, while the second one enables their execution in the supercomputer environment. In this article we focus on

The experiments described in this article were performed at the Academic Computer Center (TASK) of Gdansk University of Technology. The work was realized as a part of MAYDAY EURO 2012 and CD NIWA projects, Operational Programme Innovative Economy 2007-2013, Priority 2 "Infrastructure area R&D".

© Springer International Publishing Switzerland 2015
M. Kryszkiewicz et al. (Eds.): PReMI 2015, LNCS 9124, pp. 290–300, 2015.
DOI: 10.1007/978-3-319-19941-2_28

the latter, specifically on task management issues related to resource allocation and their influence on the system scalability. We propose a new heuristic approach supporting task allocation on a compute cluster with the soft real-time constraints.

Our contribution in this paper is as follows:

- a new allocation algorithm for real-time multimedia tasks specifically tailored for non-linear resource increase,
- a scalability metric adopted for real-time multimedia processing,
- an empirical evaluation of the proposed algorithm in a real computer cluster environment, showing its superior scalability in comparison with the typical approach.

The following section provides background information about the described problem, including main definitions and related works. The next section is related to the scalability concept and its meaning for real-time multimedia stream processing systems. Then we present the proposed allocation algorithm and its evaluation in a real data center environment. Finally, we provide conclusions and plans for future works.

2 Problem Description and Related Works

KASKADA platform provides facilities for acquisition, archiving and processing of large numbers of multimedia streams, recorded by different types of devices, including video cameras (PAL and HD), thermographic (infrared) cameras, and microphones. Such massive data flow requires well established network connections: an external fiber network, used for connecting the recording devices, as well as interconnect one, used for the exchange of partially processed (intermediate) data. Moreover, this flowing data needs to be processed by tasks realizing complex algorithms, which usually requires a high performance computer [4].

During our earlier research, we observed a certain phenomenon related to the allocation of more than one multimedia stream processing task on a single compute node c. In such conditions, the total processor utilization of a set of tasks T increases more than the sum of each task processor utilization (γ_i), i.e.:

$$\gamma(c, T) \geq \sum_{i=1}^{|T|} \gamma_i \tag{1}$$

Moreover, in [7] we proposed a method for estimating the total utilization for a given set of tasks T and a specific node c. We proposed the following formula:

$$\gamma(c, T) \approx \eta_h \left(\sum_{i=1}^{|T|} |SI_i| \right) \sum_{i=1}^{|T|} \gamma_i \tag{2}$$

where η_h is a correction function determined experimentally for a specific type of the processed stream h, and SI_i is a set of input streams for task i.

The typical process of service execution in KASKADA platform is performed as follows: the user using his/her workstation invokes a feature in the web application, which forwards this request trough the web service interface. The parallel processing management server detects the type of the call, and if a complex services is involved, it decomposes the complex service structure into simpler ones, then selects the proper tasks and finally performs an allocation of the resources in the cluster. Finally, the threads and processes of the tasks are started and the initial notification delivered to the client. Afterwards, during the task execution, the procedure of monitoring is performed and the results (output multimedia streams and event messages) are delivered and archived.

A resource allocation algorithm accepts a set of tasks to be allocated as the input and assigns these tasks to the cluster nodes. The algorithm considers two constraints. The former is related to the real-time nature of the allocated tasks, the chosen nodes need to perform proper provisioning by providing enough computing resources, in our case the processor time. The latter constraint, which is not obligatory, is related to general cluster management, specifically deciding if the cluster is to be loaded uniformly or should the tasks be concentrated on a minimal number of nodes. Because of the possibility of blocking new computationally-intensive tasks when all nodes are partially utilized, as well as the opportunity for utilizing power saving strategies, we decided to use the concentration approach.

In the related works, the scalability of a real-time system was considered in [1], where a set of five scheduling algorithms was implemented in LITMUS (LInux Testbed for MUltiprocessor Scheduling in Real-Time systems) environment and used for testing a Sun Niagara multicore platform. The scalability was analyzed for a fixed number of logical processing cores (32) in comparison to the number of executed tasks. The results were visualized as schedulability, the factor measuring ratio of the non-delayed computations as a function the utilization cap, i.e. the load of the tested processors.

Another example of scalability assessment can be found in [2]. It describes a real-time cyberinfrastructure, including a hardware and software platform for Real-time Online Interactive Applications (ROIA) – mainly Massively Multiplayer Online Games (MMOG). They assume the platform is scalable when it is able to maintain the real-time constraints as the number of users increases, by using parallel and distributed mechanisms in application design while increasing the server number. The results were presented as the average CPU utilization for a given server number, in opposition to the served clients number.

In [6] Yu Tang presented the Pull-Based Distributed Scheduling (PBDS) algorithm for a cluster architecture of real-time servers. The described scalability tests were performed for the number of cluster nodes increasing from 10 to 30, and two load levels: light and heavy, with varying transfer rates expressed in Kb/s. The results were presented as the average response time of the system (in seconds) and the overall throughput (in Kb/s) in comparison to the number of nodes, separately for the light and heavy loads.

The common characteristic of the described scalability evaluations is the usage of some performance parameter like response time, throughput or generated frames per second, as a speedup substitute. Then, like in the formal definition [5], these measurements are presented as a function of either the computation units number (e.g. nodes, GPUs) or the problem size (utilization cap or data size in the definition [5]). The scalability itself is not directly evaluated, except from the approximating statements like it "is nearly linear" in [2].

3   Scalability Definition

Scalability is intuitively defined as a capability to hold performance up across machine sizes and problem sizes. This definition can be formally formulated using the concept of speedup, which is a two-dimensional function of a given system [5]:

$$\varepsilon(p, n) = \frac{\tau(1, n)}{\tau(p, n)} \tag{3}$$

where p is the number of the installed computation units (e.g. processors, cluster nodes), n is the input data size, and $\tau(p, n)$ is the measured time of computation for the given computation units number and data size. For a fixed data size, the ideally scalable system speedup is defined as $Sp(p) = p$, the high performance scalability assumes $p/2 < Sp(p) < p$ for a given p range. Similarly, for a fixed computation unit size, $p = P$ and a given data size range, the system highly scalable when $P/2 < Sp(n) < P$. Kuck in [5] defines other ranges of speedup for intermediate cases, as well as unacceptable scalability levels.

KASKADA platform is a system realizing computations on continuously flowing data – streams. The processing is performed in real-time, with possible delays, and even with acceptable low level data loss. However, the computation times are fixed and they are not dependent on the data size (e.g. video resolution or fps) nor the number of computation units (nodes); otherwise the results returned to the user would be unacceptable. Thus the formula (3) cannot be used directly, because the speedup would always equal 1 ($\varepsilon(p, n) \approx 1$).

The more appropriate scalability metrics, concerning quality of processing, was proposed by Jogalekar in [3]. It is based on the productivity defined as:

$$\iota(p) = \frac{\lambda(p)f(p)}{\kappa(p)} \tag{4}$$

where p is the scale of the system (e.g. the number of installed computation units, processors, nodes etc.), $f(p)$ is the average value of each response, directly related to the quality provided by the system, and $\lambda(p)$ is the throughput of the system, $\kappa(p)$ is the cost of the system. The scalability metric itself is defined as the ratio of the productivity of two scales: p_1 and p_2:

$$\psi(p_1, p_2) = \frac{\iota(p_2)}{\iota(p_1)} \tag{5}$$

Usually the scale p_1 is fixed, so the above metric can be written as $\psi(p_2)$. In the case of KASKADA platform, we decided to use normalized output data loss as a quality metric:

$$f(p) = \frac{1}{1 + \frac{\varphi(p)}{L}} \tag{6}$$

where $\varphi(p)$ is the measured fraction of output data lost and L is the acceptable level of data loss. We arbitrary set its value to 1 %, thus the above equation can be simplified as:

$$f(p) = \frac{1}{1 + 100\varphi(p)} \tag{7}$$

For the platform at a given scale p, we can measure the throughput $\lambda(p)$ as a number of video frames (or other stream elements) entering the system, which is directly proportional to the number of processed streams. On the other hand, we assume the cost of the system is directly proportional to the number of used nodes, thus we can denote $\kappa(p)$ as the number of used nodes. However, the allocation algorithm can also allocate a node only partially, thus this value can be fractional. Moreover, we also assume the value for fixed referential productivity $\iota(p_l)$ is to be measured for a minimum scale (number of used nodes), but with the data loss being below the threshold $\varphi(p) < L$.

The above scalability metric has a specific feature: its values can be greater than 1.0. It is caused by the granularity of the performed services and discretion of a single node, e.g. for one node the system can allocate only one service, however for two nodes it can provide resources for 3 services, due to the possibility of allocating the tasks of the third service on both nodes.

4   Allocation Algorithm

We can assume two main allocation strategies: load balancing and load concentration. The former is widely used for uniform workload distribution, especially when the goal is the maximum resource usage and the computations scale well. The latter causes the resources to be used partially, which can be important for energy savings when the unused hardware (e.g. compute nodes) can be switched off.

One of the most important elements of computation management is proper task provisioning. This is a critical issue for real-time tasks, when resource starvation can cause indivertible loss of processed data and unacceptable decrease of reliability. Thus a large pool of unused resources is necessary for execution of large number of such real-time tasks, which makes the load concentration strategy more appropriate. Moreover, in our case the fact that two or more cooperating tasks are located on the same compute node can be used for network traffic minimization, as well as for decreasing the processor and memory load in the case when data encoding and packaging can be skipped.

In the case of KASKADA platform, load concentration can be realized using an allocation algorithm resolving the well known bin packing problem with variable size of the bins. We assume the task (processor) load is the size of the packed element, and the free resources – (partially) unused computation nodes are the bins. The maximum additional loads which can be placed on the nodes correspond to the capacity of the bins. The above problem is NP-hard, we proposed and analyzed a set of heuristics in [8], where the simulation showed that for the given environment BFD (Best Fit Descending) algorithm is the most suitable.

KASKADA platform is responsible for proper allocation and provisioning of multimedia processing tasks, executed with the soft real-time constraints. We denote task i as:

$$t_i = (a_i, SI_i, SO_i, \gamma_i) \tag{8}$$

where a_i is the task algorithm, SI_i is the set of input streams, SO_i is the set of output streams, and finally γ_i is processor utilization generated by the task in the case it is executed on a node exclusively. Because of the computation specifics, we assume the memory and network bandwidth are not a bottle-neck, thus they can be properly provisioned.

KASKADA platform realizes a service call by executing a set of interconnected computation tasks. If the allocation algorithm allocates more than one task to a specific node, its resources, especially processor cores and memory bus must be shared. Because of the specific type of computation, i.e. multimedia processing, a nonlinear processor load increase occurs [7], namely a task executed exclusively on a node causes lower processor utilization than the same task executed together with other tasks.

To provide the proper allocation, i.e. without the possibility of task starvation, the allocation algorithm requires a prediction function for processor utilization, which should accept the task descriptions (denoted in Eq. 1) as input, and provide as a result the total processor utilization of all tasks executed together on a node. We assume the function is non-decreasing with the increasing task set. We proposed the realization of such a function in [7], it consists of a product of two components:

$$\gamma(c, T) \approx \left(\sum\nolimits_{i=1}^{|T|} \gamma_i \right) \cdot \eta_h \left(\sum\nolimits_{i=1}^{|T|} |SI_i| \right) \tag{9}$$

the former is a plain sum of all processor utilizations caused by the tasks (the set T) executed on a node exclusively, and the latter is the correction function η_h, determined experimentally for a specific media type h. The above method is realized in the PRED function in listing in Fig. 1. The linear utilization $util$ is computed in the loop (lines 5–7) and the correction function $\eta_{pal}(str)$ (used in line 8) is kept as an array in memory. The function $\omega(c)$ returns the array of the tasks already allocated to the node c, $|t.SI|$ and $t.\gamma$ are respectively the number of input streams and the initial utilization (measured on a node exclusively) of the task t.

```
1.  function PRED(c, t)
2.     Tc := ω(c)
3.     util := t.γ
4.     str := |t.SI|
5.     for j := 1 to |Tc|
6.         util := util + Tc[j].γ
7.         str := str + |Tc[j].SI|
8.     return η_pal(str) · util
```

Fig. 1. The prediction function formula based on the correction function (2) from [7]

The listing in Fig. 2 presents the pseudo-code of the proposed allocation algorithm MBFD (Modified Best Fit Descending). The provided set T, consisting of tasks to be allocated on the nodes of the cluster C, is used to create the list T*, ordered descendingly by the task processor utilizations (γ_i). The algorithm divides the cluster (cases in lines 4, 16 and 24) to be used by three types of the multimedia streams: video PAL (set C_{PAL}), video HD (set C_{HD}), audio/non-multimedia (set C_{NIL}), with possible extension for additional types implemented in the function $\nu(t)$, which detects the stream type of the task t (line 3). Because of the high non-linearity of tasks processing HD streams, only one such task can be executed on a single node.

```
1. create T* using set T, sorted decreasing with γᵢ
2. for i := 1 to |T|
3.    switch ν(T*[i])
4.    case PAL:
5.       find c∈CPAL where PRED(c, T*[i]) is max and less-equal 1
6.       if c<>null then
7.          μ(T*[i]) := c
8.       else
9.          find c∈CNIL where PRED(c, T*[i]) is max and less-equal 1
10.         if c<>null then
11.            μ(T*[i]) := c
12.            CNIL := CNIL \ {c}
13.            CPAL := CPAL ∪ {c}
14.         else
15.            FAIL("allocation unavailable")
16.   case HD:
17.      find c∈CNIL where PRED(c, T*[i]) is max and less-equal 1
18.      if c<>null then
19.         μ(T*[i]) := c
20.         CNIL := CNIL \ {c}
21.         CHD := CHD ∪ {c}
22.      else
23.         FAIL("allocation unavailable")
24.   case AUD/NIL:
25.      find c∈CHD where PRED(c, T*[i]) is max and less-equal 1
26.      if c<>null then
27.         μ(T*[i]) := c
28.      else
29.         find c∈CPAL gdzie PRED(c, T*[i]) is max and less-equal 1
30.         if c<>null then
31.            μ(T*[i]) := c
32.         else
33.            find c∈CNIL where PRED(c, T*[i]) is max and less-equal 1
34.            if c <> null then
35.               μ(T*[i]) := c
36.            else
37.               FAIL("allocation unavailable")
```

Fig. 2. The allocation algorithm MBFD: Modified Best Fit Descending

The main loop of the algorithm iterates trough the list of tasks (T*). Every consecutive task is matched to the to the best node, i.e. the one which utilization, after the task allocation, will be maximal but not greater than 1. The low influence of tasks with audio or without streams implies that the algorithm can group them with both HD and PAL streams (lines 25, 29, 33), however tasks processing PAL and HD streams should not be grouped together on the same node, thus they can use the audio/no stream (from set C_{NIL}) and free nodes (lines 5, 9 for PAL and line 17 for HD). In case there are no matching nodes in the cluster for the task being allocated, an error (failure) is reported and the algorithm is finished. The outcome allocation is stored as an associative array $\mu(t)$, with the key being the allocated task t, and the value the best matching node of the cluster c.

5 Experimental Evaluation

The experiments were performed using a Galera computer cluster, placed at the TASK Academic Computer Center of the Gdansk University of Technology. For the algorithm evaluation we used an increasing set of compute nodes, starting with one and up to one hundred. Each node consists of two Intel Xeon E5345 2.33 GHz (8 cores) processors, 8 GB RAM, and 20 Gbps Infiniband network interface connected in the fat tree structure. The evaluation was performed against four benchmark services. All of them processed PAL video streams, and consist of simple services executed as connected computation tasks. Table 1 presents the description of the services. The motivation to select these services was to examine the most popular algorithms used by KASKADA platform users, as well as to check the allocation algorithm against various workloads.

Table 1. The benchmark services

ID	Functionality	Input stream no	Output stream no	Task no	Min node no (p_1)
fd4	Detection of human faces, every 4th frame, saving face image to a file.	1	1	4	1
fdt	Detection of human faces using data decomposition to 32 sub-tasks.	1	1	35	4
fd32	Detection of human faces every 32nd frame, and notification generation.	1	1	5	1
mrg	Merging of two streams and providing time mark in the output stream.	2	1	3	1

Two algorithms were to be under examination: MBFD, the proposed solution to providing accurate computation resources, with the correction related to the non-linear processor utilization increase, as well as its simplified version: BFD, which simply adds the task loads without the correction. The hypothesis was the former provides better scalability than the latter, thus its usage is justified for parallel, real-time multimedia stream processing.

During the test execution the consecutive benchmark services were started with an increasing number of compute nodes in the cluster. The minimum node number $p_l(s)$ depends on the service, and is directly related to the fixed reference productivity $\iota(p_l)$ used during scalability metric computation (see Eq. 5). Every experiment was repeated 5 times and the average value was considered. The measurements were continued until the maximum possible number of cluster nodes were used, in our case we could use 100 nodes at most, so $P_{MAX} = 100$. Then the next service was tested, and after all tests and allocation algorithms are measured the procedure was finished. The experiments were performed for the services processing PAL video streams.

Figures 3 and 4 present the resulting scalability metric measured during the execution of the test services. The first chart consists of the results related to the BFD algorithm, the one based solely on the bin packing problem heuristic, and the second is related to

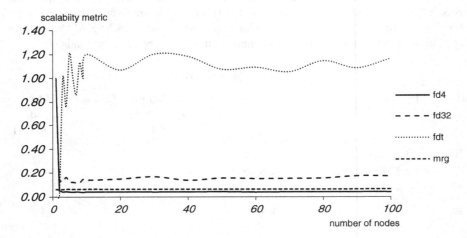

Fig. 3. The scalability metric measures for the BFD allocation algorithm

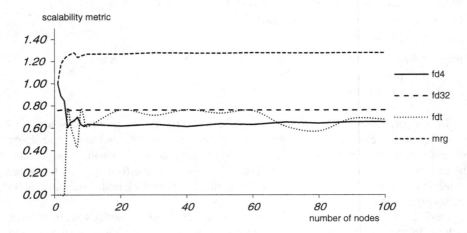

Fig. 4. The scalability metric measures for the MBFD allocation algorithm

the MBFD algorithm, supported with the non-linear processor utilization prediction function. In general the smaller services (e.g. mrg with MBFD) have better scalability, however in other tests we observed the dependence of scalability value on the algorithms used in the services rather than the service complexity.

We can observe that in most cases the MBFD heuristic provides better results (higher scalability values), which is consistent with the assumption of non-linear load increase of the executed tasks. The only exception is related to the fdt service (which dropped down for MBFD), where the distribution of stream data to 32 tasks causes disappearance of the non-linear load increase effect – in this case, they behave like normal compute tasks. It is worth to mention the fluctuation of measurement for the larger services (consisting of a higher number of simple services, e.g. fdt), which is especially visible for a smaller number of nodes (but not only, see fdt for 60 + nodes using MBFD). It is caused by the service complexity, because quite often allocating an additional node does not increase the number of services executed in the cluster, e.g. for 4 and 5 nodes only one fdt service can be executed.

6 Conclusions and Future Works

In the article we proposed a new solution for stream processing task allocation in an HPC environment. The proposed heuristic is based on the observation of non-linear load increase for real-time streaming tasks and is related to the well-known BFD algorithm. The experimental results confirmed our first assumptions, providing the basis for the implementation of the allocation component for KASKADA platform [4].

In the future works, we are considering migration of the whole KASKADA platform [4] to the cloud environment, where the harder constraints related to more loose SLA in comparison to a typical HPC cluster can affect the performance of the allocation algorithm. On the other hand, we want to adapt the above solution to a heterogeneous environment, especially with the consideration of GPGPU devices.

References

1. Brandenburg, B.B., Calandrino, J.M., Anderson, J.H.: On the scalability of real-time scheduling algorithms on multicore platforms: a case study. In: Real-Time Systems Symposium, pp.157–169, 30 November–3 December (2008)
2. Gorlatch, S., Glinka, F., Ploss, A.: Towards a scalable real-time cyberinfrastructure for online computer games. In: 2009 15th International Conference on Parallel and Distributed Systems (ICPADS), pp.722–727, 8–11 December 2009
3. Jogalekar, P., Woodside, M.: Evaluating the scalability of distributed systems. IEEE Trans. Parallel Distrib. Syst. **11**(6), 589–603 (2000)
4. Krawczyk, H., Proficz, J.: KASKADA - multimedia processing platform architecture. In: Proceedings of the 2010 International Conference on Signal Processing and Multimedia Applications (SIGMAP), pp.26–31, 26–28 July 2010
5. Kuck, D.J.: High Performance Computing, Challenges for Future Systems. Oxford University Press, New York (1996)

6. Tang, Y., Changqin, B., Zhenghua, W., Sheng, S.: A scalable approach to distributed scheduling scheme on cluster architecture of real-time server. In: 2010 2nd International Conference on Computer Engineering and Technology (ICCET), vol. 3, pp. V3-123–V3-127, 16–18 April 2010
7. Krawczyk, H., Proficz, J., Daca, B.: Prediction of Processor Utilization for Real-Time Multimedia Stream Processing Tasks, Distributed Computing and Internet Technology, pp. 278–289. Springer, Heidelberg (2013)
8. Krawczyk, H., Proficz, J.: The task graph assignment for KASKADA platform. In: Proceedings of the 2010 International Conference on Software and Data Technologies (ICSOFT), pp. 192–197, 22–24 July 2010

Fuzzy Computing

Concept Synthesis Using Logic of Prototypes and Counterexamples: A Graded Consequence Approach

Soma Dutta[1]([✉]) and Piotr Wasilewski[2]

[1] Institute of Mathematics, University of Warsaw, Warszawa, Poland
somadutta9@gmail.com
[2] Institute of Informatics, University of Warsaw, Warszawa, Poland
piotr@mimuw.edu.pl

Abstract. This paper is a preliminary step towards proposing a scheme for synthesis of a concept out of a set of concepts focusing on the following aspects. The first is that the semantics of a set of simple (or independent) concepts would be understood in terms of its prototypes and counterexamples, where these instances of positive and negative cases may vary with the change of the context, i.e., a set of situations which works as a precursor of an information system. Secondly, based on the classification of a concept in terms of the situations where it strictly applies and where not, a degree of application of the concept to some new situation/world would be determined. This layer of reasoning is named as logic of prototypes and counterexamples. In the next layer the method of concept synthesis would be designed as a graded concept based on the already developed degree based approach for logic of prototypes and counterexamples.

Keywords: Fuzzy approximation space · Similarity relation · Graded consequence · Concept synthesization

1 Introduction

In our everyday decision making we face a few challenges. Some of them are, selecting relevant interpretations of the vague concepts involved in our daily communications, synthesis or analysis of new concepts in terms of the available (component) concepts, and deciding the applicability of a concept to a newly appeared situation based on some initial situations or worlds where the applicability of the concept of concern is known. So, while proposing the theory we would focus on the following issues; one is the semantics of the vague concepts, second is a general concept synthesis mechanism to generate compound (or dependent) concepts based on the simple (or independent) concepts, and the third is to allow open world scenario so that based on the available database decisions about the new situations/worlds/cases can be induced. Plenty of approaches concerning fuzzy sets, rough sets and their combinations are available [1,6,7,13–20] to deal

© Springer International Publishing Switzerland 2015
M. Kryszkiewicz et al. (Eds.): PReMI 2015, LNCS 9124, pp. 303–313, 2015.
DOI: 10.1007/978-3-319-19941-2_29

with the above mentioned aspects of human reasoning. Some of the theories [15, 19, 20] are used and verified against the real practical needs of automated decision support system.

We, in this paper, set a preliminary step to develop a theoretical background for a logic of prototypes and counterexamples, where a vague concept would be understood by clear positive and negative instances of its application. While developing the theory we would keep a room open for appearance of new cases, for which decisions regarding the applicability of a concept need to be taken. We would take a degree based approach to design the logic of prototypes and counterexamples based on the already developed notion of fuzzy approximation space [6]. Given this ground set-up, where concepts are understood in terms of their prototypes and counterexamples, interrelation between a set of (sub)concepts and a concept is defined as an application of graded consequence relation [2, 3]. Thus synthesis of a concept based on the interpretations of its component concepts is done in a graded approach. Drawing analogy from the real needs of a computer-aided decision support system, particularly in the field of healthcare, we would try to explain why the theory, proposed in this paper, is worth exploring further.

2   Logic of Prototypes and Counterexamples

In this section we start with the notion of fuzzy approximation space as given in [6].

Definition 1 [6]. A fuzzy approximation space is defined as a pair (U, R) where R is a fuzzy binary relation on U, i.e., $R : U \times U \mapsto [0, 1]$. The fuzzy lower and upper approximations $\underline{R}F, \overline{R}F : U \mapsto [0, 1]$ of a crisp or fuzzy subset F of U is defined as follows. $\underline{R}F(u) = \inf_{v \in U} \{R(u, v) \rightarrow F(v)\}$ and $\overline{R}F(u) = \sup_{v \in U} \{R(u, v) * F(v)\}$, where $* : [0, 1] \times [0, 1] \mapsto [0, 1]$ is a t-norm [12], and $\rightarrow : [0, 1] \times [0, 1] \mapsto [0, 1]$ is the S-implication [12] with respect to $*$ such that $a \rightarrow b = 1 - (a * (1 - b))$.

Let us start with an overview of the problem, a mathematical model of which would be our target to achieve under the name of 'logic of prototypes and counterexamples'. Let we have a clinical record of n number of patients' details with respect to some m number of parameters/attributes. These parameters might be some objective values of some clinical tests, called signs, or some subjective features experienced by the patients, called symptoms. With respect to the state of each patient, the values corresponding to all these parameters are converted, by some mean, to the values over a common scale, say $[0, 1]$. That is, if an m-tuple $\langle a_1, a_2, \ldots, a_m \rangle$ from $[0, 1]^m$ represents the rates of the m parameters corresponding to a patient, then we say $\langle a_1, a_2, \ldots, a_m \rangle$ describes the state of a patient. Based on the rates assigned to all the parameters by each patient, i.e. a m-tuple of values $\langle a_1, a_2, \ldots, a_m \rangle$ from $[0, 1]^m$, which cases representing the states of the patients are how much similar or dissimilar may be anticipated. Now, one task is to make a tentative diagnosis about a patient whose measurement concerning the m-tuple of parameters appears to be new with respect to the database of the n patients.

So, in the general set-up we start with a set S of finitely many situations, say $\{s_1, s_2, \ldots, s_n\}$, and P of finitely many parameters $\{p_1, p_2, \ldots, p_m\}$. Each s_i, $i = 1, 2, \ldots n$, is considered to be a function $s_i : P \mapsto [0, 1]$. Let the consolidated data of each situation is stored in the form of a set $\{\langle s_i(p_1), s_i(p_2), \ldots, s_i(p_m)\rangle : s_i \in S\}$, which is a subset of $[0, 1]^m$. Let $W \subseteq [0, 1]^m$ and $\{\langle s_i(p_1), s_i(p_2), \ldots, s_i(p_m)\rangle : s_i \in S\} \subseteq W$. Each member of W may be called a world. We now consider a fuzzy approximation space $\langle W, Sim\rangle$, where Sim is a fuzzy similarity relation between worlds of W. That is, $Sim : W \times W \mapsto [0, 1]$, and we assume Sim to satisfy the following properties.

(i) $Sim(\omega, \omega) = 1$ (reflexivity)
(ii) $Sim(\omega, \omega') = Sim(\omega', \omega)$ (symmetry)
(iii) $Sim(\omega, \omega') * Sim(\omega', \omega'') \leq Sim(\omega, \omega'')$ (transitivity)

Following [6], the fuzzy approximation space $\langle W, Sim\rangle$ is based on the unit interval $[0, 1]$ endowed with a t-norm $*$ and a S-implication operation $\rightarrow$, as mentioned in Definition 1. In Sect. 3, apart from $\rightarrow$ we would also require $\rightarrow_*$, the residuum operation with respect to $*$ in $[0, 1]$.

We now propose to represent any (vague) concept α by a pair (α^+, α^-) consisting of the positive instances (prototypes) and negative instances (counterexamples) of α respectively, where $\alpha^+, \alpha^- \subseteq W$ and $\alpha^+ \cap \alpha^- = \phi$.

Definition 2. Given the fuzzy approximation space $\langle W, Sim\rangle$, and a concept α represented by (α^+, α^-), the degree to which α applies to a world $\omega \in W$, denoted by $gr(\omega \models \alpha)$, is given by:

$$gr(\omega \models \alpha) = 1 \text{ if } \omega \in \alpha^+,$$
$$= 0, \text{ if } \omega \in \alpha^-, \text{ and}$$
$$= \overline{Sim}\alpha^+(\omega) * \neg\overline{Sim}\alpha^-(\omega), \text{ otherwise.}$$

The fuzzy upper approximations $\overline{Sim}\,\alpha^+$ and $\overline{Sim}\,\alpha^-$ are defined following the Definition 1, i.e., $\overline{Sim}\,\alpha^+(\omega) = \sup_{u \in W} Sim(\omega, u) * \alpha^+(u) = \sup_{u \in \alpha^+} Sim(\omega, u)$. Similar is the case for $\overline{Sim}\,\alpha^-$. For further references we should also note that the fuzzy lower approximation is defined as $\underline{Sim}\,\alpha^-(\omega) = \inf_{u \in W}\{Sim(\omega, u) \rightarrow \alpha^-(u)\} = 1 - \inf_{u \notin \alpha^-} Sim(\omega, u)$. The simplifications of $\overline{Sim}\,\alpha^+(\omega)$ and $\underline{Sim}\,\alpha^-(\omega)$ are done based on the properties of $*$, $\rightarrow$, and $\neg$, where $a \rightarrow 1 = 1$, $a \rightarrow 0 = \neg a = 1 - a$.

The third case given in the Definition 2 formalizes the idea that *a concept α applies to a newly appeared world ω if ω is similar to a world which lies in the possible zone of the positive instances of α, and it is not that ω is similar to a world which lies in the possible zone of the negative instances of α.*

In the definition of $\models$, we have not specified the definition for Sim. Sim may be interpreted in different ways. Sometimes Sim is considered to be inversely proportional to a notion of distance. Here, we may think that given a set of prototypes and counterexamples of a property α, whether at a new situation/world

α applies or not, would depend on the arguments which go in favour of *that the world qualifies* α, as well as the arguments which go against; checking the above could be equivalent to check the degree of strength of the arguments which go in favour of considering the new world ω to be similar to the set of positive instances of α, and the degree of strength of the arguments which go against for considering ω similar to the worlds of α^+. The arguments which go against considering ω to be similar to the worlds of α^+ can be counted as the same as the arguments which go in favour of considering ω similar to the worlds of α^-. So, Sim would have some relations with the *argument in favour* and the *argument against*. For this paper we do not need to specify what do we really mean by the *argument in favour* and the *argument against*; let us just equate $\overline{Sim}\,\alpha^+(\omega)$ $= D_{af}(\omega, \alpha)$, the *degree of arguments in favour of* ω qualifies α and $\overline{Sim}\,\alpha^-(\omega)$ $= D_{ag}(\omega, \alpha)$, the *degree of arguments against* ω qualifies α. So, given a concept α and world $\omega \notin \alpha^+, \alpha^-$, $gr(\omega \models \alpha) = D_{af}(\omega, \alpha) * \neg D_{ag}(\omega, \alpha)$. That is, the underlying meaning is, a concept α applies to a world ω if we have arguments in favour of ω *qualifies* α and we do not have arguments against ω *qualifies* α.

It is also to be noted, that when $D_{af}(\omega, \alpha) = \sup_{u \in \alpha^+} Sim(\omega, u) = Sim(\omega, \omega')$ for some $\omega' \in \alpha^+$, we say $D_{af}(\omega, \alpha)$ attains at ω'; i.e., the (highest) degree of argument in favour of ω qualifies α attains at ω'.

We now would like to explore the properties of the relation $\models$.

Theorem 1. (i) $gr(\omega \models \alpha) = 1$ if $\omega \in \alpha^+$.
(ii) If $\alpha^+ \subseteq \beta^+$ and $\beta^- \subseteq \alpha^-$, then $gr(\omega \models \alpha) \leq gr(\omega \models \beta)$.
(iii) For $\omega \in \alpha^+$ and $\omega' \in \beta^+$ if $D_{af}(\omega, \beta)$ attains at ω' and $D_{ag}(\omega, \gamma) = 0$, then $gr(\omega \models \beta) * gr(\omega' \models \gamma) \leq gr(\omega \models \gamma)$.

Proof. (i) is obvious from the definition of $\models$.
(ii) $gr(\omega \models \alpha) = \sup_{u \in \alpha^+} Sim(\omega, u) * \neg \sup_{u \in \alpha^-} Sim(\omega, u)$
$\qquad\qquad \leq \sup_{u \in \beta^+} Sim(\omega, u) * \neg \sup_{u \in \alpha^-} Sim(\omega, u)$ [since $\alpha^+ \subseteq \beta^+$]
...(a)
Since $\beta^- \subseteq \alpha^-$, $\neg \sup_{u \in \alpha^-} Sim(\omega, u) \leq \neg \sup_{u \in \beta^-} Sim(\omega, u)$ ...(b)
Combining (a) and (b) we have, $gr(\omega \models \alpha) \leq gr(\omega \models \beta)$.
(iii) Let for two concepts α, β, $\omega \in \alpha^+$ and $\omega' \in \beta^+$. Let us consider another concept γ.
$gr(\omega \models \beta) * gr(\omega' \models \gamma)$
$= [\sup_{u \in \beta^+} Sim(\omega, u) * \neg \sup_{u \in \beta^-} Sim(\omega, u)]$
$\quad * [\sup_{v \in \gamma^+} Sim(\omega', v) * \neg \sup_{v \in \gamma^-} Sim(\omega', v)]$
$= [\sup_{u \in \beta^+} Sim(\omega, u) * \sup_{v \in \gamma^+} Sim(\omega', v)]$
$\quad * [\neg \sup_{u \in \beta^-} Sim(\omega, u) * \neg \sup_{v \in \gamma^-} Sim(\omega', v)]$
$\leq [\sup_{u \in \beta^+} Sim(\omega, u) * \sup_{v \in \gamma^+} Sim(\omega', v)]$
$= [Sim(\omega, \omega') * \sup_{v \in \gamma^+} Sim(\omega', v)]$ [since $D_{af}(\omega, \beta)$ attains at ω']
$= \sup_{v \in \gamma^+} [Sim(\omega, \omega') * Sim(\omega', v)]$ [by $a * \sup_i b_i = \sup_i (a * b_i)$]
$\leq \sup_{v \in \gamma^+} Sim(\omega, v) * \neg \sup_{v \in \alpha^-} Sim(\omega, v)$ [as Sim is transitive and $D_{ag}(\omega, \gamma)$
$= 0$]
$= gr(\omega \models \gamma)$

A notion of logical consequence relation is usually supposed to be a relation between a set of sentences/propositions and a sentence. In model theoretic terms, we also talk about a model or a structure, which is a tuple consisting of the interpretations of the logical constants, predicates, and function symbols, satisfies a sentence. Here, the relation $\models$ is considered to be a binary relation between a world of W and a concept. In that sense one may be skeptic about calling $=$ a consequence relation or satisfaction relation. Nevertheless, it is to be noted that the above three properties of $\models$ seem to have some connection with our standard notion of logical consequence.

The first property of Theorem 1 ensures that any positive instance of a concept α qualifies α; this clearly has similarity with the condition overlap of a logical consequence. Property (ii) of $\models$ ensures that if two concepts α and β are such that the positive instances of α are the positive instances of β, and negative instances of β are the negative instances of α, then if ω qualifies α, ω qualifies β too. In ordinary context, where a concept applies and does not apply are complementary to each other we do only talk about positive instances of the concept. But in the context of vague concepts where the prototypes and the counterexamples of the concept do not exhaust the whole universe, (ii) may be considered similar to the monotonicity property of a logical consequence. The property (iii) is a variant version of the condition cut of a logical consequence. In the context of logic of prototypes and counterexamples, in order to claim that a world qualifies a concept we need to take care of both for and against arguments for such a claim. So, (iii) ensures that, *if ω, a positive instance of α, qualifies β by attaining the argument in favour at the world ω' belonging to the positive instances of β, then if ω' qualifies γ and we do not have any argument against ω qualifies γ, then ω qualifies γ.*

Thus, the name 'logic of prototypes and counterexamples', for this particular system of reasoning about whether a concept applies to a world, justifies itself.

3 Concept Synthesis: An Application of Graded Consequence

In the previous section we have proposed a way to calculate the degree of applicability of a concept to a world from the available prototypes and counterexamples of the concept where it strictly applies and does not apply respectively. We also have observed that the relation $\models$ between a world and a concept satisfies some properties which might be viewed as properties similar to a notion of consequence. In this section we would make a step towards synthesis of compound concepts from a set of concepts. More specifically we would be interested in obtaining a grade to which a (compound) concept can be derived from a set of (sub)concepts. The idea is built as follows.

We start with $\mathcal{C}_B$, a set of basic concepts relevant for an arbitrarily fixed domain. We restrict $\mathcal{C}_B$ to be a finite set as in the field of application we indeed deal with a finite set of basic concepts in order to understand a bigger set of concepts, which again needs not be infinite for meeting our everyday purpose. Now for each $c_b \in \mathcal{C}_B$, we have (c_b^+, c_b^-) such that $c_b^+, c_b^- \subseteq W$ and $c_b^+ \cap c_b^- = \phi$.

Given the classification of concepts in terms of their prototypes and counterexamples, we induce a fuzzy relation $\models$ over $W \times C_B$ mapping each pair from $W \times C_B$ to $[0,1]$. Let C be the set of all possible concepts over the same domain specified for C_B, and $C_B \subsetneq C$. As discussed in Sect. 2, the concepts are classified in terms of its prototypes and counterexamples based on a set of situations S. So, if S changes the classification of the same concept may vary. That is, the subjective view of a concept is incorporated in the notion of classification of a concept in terms of its prototypes and counterexamples. Although interpretation of a vague concept and its interrelation with other concept are very much subjective in nature, there might be a universally accepted common view as well. So, we assume that C is endowed with an order relation $\subseteq_k$, may be called *universal knowledge ordering* of concepts, such that $(C, \subseteq_k)$ forms a poset. Informally the order relation $\subseteq_k$ is such that if it is commonly agreed that *stomach infection* is a subconcept of *food poisoning*, then the former is included in the latter under the relation $\subseteq_k$. We skip the definition of $\subseteq_k$ for the time being, as to proceed in this paper a formal definition of $\subseteq_k$ is not necessary. Each concept $c_b \in C_B$ is independent in the sense that there is no $c_b' \in C_B$ such that $c_b' \subseteq_k c_b$. We now extend the definition of $\models$ on $W \times C$ in the following way.

Definition 3. $\models_e$ is a fuzzy relation from $W \times C$ to $[0,1]$ such that
$gr(\omega \models_e c) = gr(\omega \models c)$ if $c \in C_B$,
$\qquad = \dfrac{\Sigma_{c' \in C_B'} gr(\omega \models c')}{|C_B'|}$, otherwise where $C_B' = \{c' \in C_B : c' \subseteq_k c\}$.

The value of $gr(\omega \models_e c)$ for $c \notin C_B$ can be zero in two possible cases. One is when there is no concept in C_B which is a subconcept of c. The other case is when the concepts from C_B, which are included in c, do not hold at the world ω, i.e., they hold to the degree 0 at ω.

Now we propose a method of derivation of a concept from a set of concepts, where the derivation is a matter of grade.

Definition 4. Given any collection of worlds $\{\omega_i\}_{i \in I}$, taken from W, the fuzzy relation $|\approx$ over $P(C) \times C$, assuming values from $[0,1]$, is defined as below.
$gr(\{c_j : j = 1, 2, \ldots, n\} |\approx c) = \inf_{i \in I}[\inf_{j=1}^{n} gr(\omega_i \models_e c_j) \to_* gr(\omega_i \models_e c)]$,
where $\to_*$ is the residuum of $*$ in $[0,1]$.

It is now easy to check that $|\approx$ is a graded consequence relation [3,4], i.e., $|\approx$ satisfies the following properties.

(GC1) If $c \in \{c_j : j = 1, 2, \ldots, n\}$ then $gr(\{c_j : j = 1, 2, \ldots, n\} |\approx c) = 1$.
(GC2) If $C_1 \subseteq C_2(\in P(C))$, then $gr(C_1 |\approx c) \leq gr(C_2 |\approx c)$.
(GC3) $\inf_{c' \in C'} gr(C |\approx c') * gr(C \cup C' |\approx c) \leq gr(C |\approx c)$.

As in the beginning of this section we have mentioned that for our daily purpose we do not need to deal with an infinite set of concepts over a fixed domain, we have proposed the definition of $|\approx$ for synthesis of a concept from a set of finite number of concepts. The definition can be easily considered for infinite case, for which one may be referred to the work of the theory of graded consequence [2,3]. At this stage, again a similar point, as discussed at the end of Sect. 2, arises. The notion of graded consequence relation [2,3] originally has been defined as a fuzzy relation between $P(F)$, the power set of all sentences over a language, to F, the set of all sentences. That is, according to the definition of graded consequence [3] $|\approx$ is a function from $P(F) \times F$ to some set of values; in particular the value set could be $[0,1]$. Here, $|\approx$ maps every pair from $P(\mathcal{C}) \times \mathcal{C}$ to $[0,1]$. Matching both the context from the mathematical perspective sounds well, though conceptually $|\approx$ in the present context is different from that of the context of [2–5].

Now question arises why at all we choose this way of synthesis of a concept from a set of concepts. First of all, this allows the concept synthesis method to exploit the so-far developed logical background of the theory of graded consequence [2–5,8–11]. Theory of graded consequence gives a set-up where *object language entities*, which are here concepts, as well as the notion of *deriving a concept from a set of concepts* are of matter of grade. Besides, the theory of graded consequence allows the languages and the interpretations of different levels, like object, meta, and meta-meta levels, of a logical discourse to be distinctive. As a result, object level and meta-level of a logic may have different reasoning patterns and necessities. Like, here in the object level we have a graded relation $\models$ which has its own characteristics as mentioned in Theorem 1; the base algebraic structure for $\models$ is $([0,1], \wedge, \vee, *, \rightarrow)$. In the next level, we have another relation $|\approx$ with its own characteristics, with respect to the algebraic structure $([0,1], \wedge, \vee, *, \rightarrow_*)$. Dealing with vague concepts, it is quite natural to think that our mental process of synthesis of a vague concept from a set of vague concepts may not be crisp always, and it may need different layers of learning and synthesizing. So, the background of the theory of graded consequence might be advantageous for the theory of concept synthesis.

Let us check what else this definition of $|\approx$ can yield. We would try to explain a few natural expectations to a concept synthesis operation in the context of this definition.

(1) Let $\{c_j : j = 1, 2, \ldots, n\}$ be such that $c_j \in \mathcal{C}_B$ and $c_j \subseteq_k c$ for $j = 1, \ldots, n$. So, in accordance with the universal ordering $\subseteq_k$ between c_j's and c, it is expected that c can be derived from its subconcepts should have an affirmative answer. Below we present that the definition of $|\approx$ respects this natural demand in the following way.

Let ω_i be a world from the collection $\{\omega_i\}_{i \in I}$, and $gr(\omega_i \models_e c_j) = a_{ij}$ for $j = 1, 2, \ldots, n$. Then by Definition 3, $gr(\omega_i \models_e c) = \frac{\sum_{j=1}^n a_{ij}}{n}$. Now $\inf_j a_{ij} \leq a_{ij}$ for $j = 1, 2, \ldots, n$. So, $n(\inf_j a_j) \leq \sum_{j=1}^n a_{ij}$, i.e. $\inf_j a_{ij} \leq \frac{\sum_{j=1}^n a_{ij}}{n}$. Hence using property of $\rightarrow_*$ we have $gr(\{c_j : j = 1, 2, \ldots, n\} |\approx c) = 1$.

(2) Let us refer to the following quotation by Leslie Valiant [21], the 2010 Turing award winner. *"A specific challenge is to build on the success of machine learning so as to cover broader issues in intelligence. This requires, in particular a reconcillation between two contradictory characteristics - the apparent logical nature of reasoning and the statistical nature of learning".*

When at a world certain basic constituent properties of a compound concept hold, we often take a statistically average point of view to decide the applicability of the compound concept at that world. The Definition 3 in that way justifies itself. This may be referred to as *statistical nature of learning* of the applicability of a concept to a world. On the other way, when we need to check that whether a set of concepts C yields a complex concept c, we need to be logical in our way of deriving conclusion. That is, we need to check that whether each of the available worlds satisfying each concept of C, also satisfies the concept c. The Definition 4 takes care of this logical reasoning part. Hence, this definition of concept synthesis is an attempt towards reconcillation of statistical nature of learning and logical nature of reasoning.

(3) Let us think about some practical problems, and check how this proposed theory is dealing with those problems. In this regard, we would like to illucidate the idea by considering examples from the field of health-care system. Let $\{d_1, d_2, \ldots, d_l\}$ be the subconcepts of a disease d, and among them $d_{i_1}, \ldots, d_{i_k}$ for $1 \leq i_1 \leq i_k \leq l$ get only positive values at some world ω_i. Then the degree to which d applies to the world must get decreased compare to the case when all of $d_1, \ldots d_l$ are tested positive. The Definition 3 shows that in the former case the value becomes $\frac{\Sigma_{j=1}^{k} gr(\omega_i \models_e d_{i_j})}{l}$, which is less than the value $\frac{\Sigma_{j=1}^{l} gr(\omega_i \models_e d_{i_j})}{l}$ for the latter case.

Let us assume that a new problem d has emerged, and the medical domain knowledge does not have the information that which of the independent concepts of C_B are contained in d, in the sense of $\subseteq_k$. So, following the Definition 3 at all worlds the degree of applicability of d would be 0. This reflects the problem that *for a new concept there might not be any relevant information available according to the domain knowledge base; and there might not be also any statistical record available in favour of the concept as at each world the applicability of the concept gets nullified.* In such a situation, based on experiences, one may still try to relate some known problems with the problem of concern, and want to check the relationship of a set of concepts $\{c_1, \ldots, c_r\}$ with the problem d. As $a \rightarrow_* 0$ is not necessarily 0, that none of the worlds qualifies d does not imply $gr(\{c_1, \ldots, c_r\} \mid\approx d)$ to be zero.

The example below gives an overview of the application of the proposed theory. We start with a database of finitely many patients, and a fragment of the concept ontology from the medical domain specifying a set of independent concepts (C_B), and their relations with other compound concepts of $C(\supseteq C_B)$. From the database, positive and negative cases of each disease from C_B are determined, and then diagnosis of a particular disease at some particular case/situation/world, which might be new, is determined following the Definitions 2 and 3; using this at the next level the degree of relatedness of a set of diseases from C with a disease, say d, is computed following the Definition 4.

Example 1. Let we have a clinical database of a set of situations, $S = \{s_1, s_2, s_3, \ldots, s_9\}$ with respect to a set P of parameters/attributes, consisting of *temperature, blood-pressure, blood-tests, ecg, headache, sneezing, convulsion,* *vomiting, skin-rash, dizziness, stomach-upset, stomach pain.* Sequentially let us call these parameters as $p_1, p_2, p_3, \ldots, p_{12}$. As $p_1, \ldots, p_4$ are determined by some objective values of some tests these are called signs; the rest are symptoms, determined by some subjective values as experienced by particular patients. Let $C_B = \{Fev, Allergy, Stomach_{inf}, HBP, LBP, Vertigo, Unconsciousness\}$, and C is the union of C_B and $\{Fev_c, Fev_v, Stroke, Food\text{-}poisoning, Viral_{inf}, Peptic\text{-}ulcer\}$. The relations among the dependent and the independent concepts of C are as follows. $Fev \subseteq_k Fev_c$; Fev, $Allergy \subseteq_k Fev_v$; Fev, HBP, $Vertigo$, $Unconsciousness \subseteq_k Stroke$; Fev, $Stomach_{inf} \subseteq_k Food\text{-}poisoning$; Fev, $Stomach_{inf}$, $Allergy \subseteq_k Viral_{inf}$; and $Stomach_{inf} \subseteq_k Peptic\text{-}ulcer$. Fev_c and Fev_v respectively stand for *fever due to cold* and *viral fever*. HBP, LBP, $Stomach_{inf}$, and $Viral_{inf}$ respectively stand for *high blood pressure, low blood pressure, stomach infection* and *viral infection*. Each s_i is identified with its state given by $\langle s_i(p_1), s_i(p_2), s_i(p_3), \ldots, s_i(p_{12}) \rangle \in [0,1]^{12}$, and $W(\subseteq [0,1]^{12})$ contains $\langle s_i(p_1), s_i(p_2), s_i(p_3), \ldots, s_i(p_{12}) \rangle$ for each s_i. The tuple of values corresponding to each s_i, and the positive and negative cases of each disease from C_B are given in the following table (Table 1). Let us use d_1 for Fev, d_2 for $Allergy$, d_3 for $Stomach_{inf}$, d_4 for HBP, d_5 for LBP, d_6 for $Vertigo$, and d_7 for $Unconsciousness$. For each s_i if d_j receives $+$, then s_i is the positive case for d_j, and if it receives $-$, then s_i is the negative case of d_j. Now as $s_2 \in HBP^+, Vertigo^+, Unconsciousness^+$, following the concept ontology we may need to check $gr(s_2 \models_e Stroke)$. Following the Definitions 2 and 3, $gr(s_2 \models_e Stroke) = \frac{3}{4} + \frac{1}{4}[\overline{Sim}_{Fev^+}(s_2) * \neg\overline{Sim}_{Fev^-}(s_2)]$. Also, as $s_8 \in Vertigo^+, Unconsciousness^+$ verifying $gr(s_8 \models_e Stroke)$ is natural too. In this case, as $s_8 \in Fev^-, HBP^-$ $gr(s_8 \models_e Stroke)$ turns out to be $\frac{1}{2}$.

Table 1. Patients data table

	p_1	p_2	p_3	p_4	p_5	p_6	p_7	p_8	p_9	p_{10}	p_{11}	p_{12}	d_1	d_2	d_3	d_4	d_5	d_6	d_7
s_1	.8	.3	.5	0	.7	.7	0	0	0	0	0	0	+		-	-		-	-
s_2	.5	.8	.7	.7	.7	0	.7	0	.3	.9	0	0			-	+	-	+	+
s_3	.9	.5	.7	0	.7	.8	0	0	0	0	0	0	+		-	-		-	-
s_4	.4	.3	.6	0	0	0	0	.5	0	0	.7	.6	-	-	+	-		-	-
s_5	.7	.1	.7	0	.3	0	0	.7	0	0	.9	.7		-	+	-		-	-
s_6	.7	.2	.8	0	.3	0	0	.5	.7	0	.7	.7		+	+	-		-	-
s_7	.9	.5	.8	0	.8	.8	0	0	.3	.2	0	0	+		-	-		-	-
s_8	.3	.1	.3	.8	.7	0	.7	0	0	.8	0	0	-	-	-	-	+	+	+
s_9	.6	.7	.7	.5	.5	0	0	.9	0	.2	.9	1		-	+			-	-

Let a new situation s_{10}, with the tuple $\langle .5, .5, .5, .5, .7, 0, .8, .5, .7, .8, .5, 0 \rangle$ of values corresponding to the respective parameters, appear. The task is to make a diagnosis for s_{10} based on the *for* and *against* arguments that which of the situations from S are closely similar to s_{10}. Though for the progress of this paper it is not required to specify the exact formula for computing Sim, let us follow an intuitive reasoning for justifying the case for this example; formulation of the exact definition for Sim is one of our future agendas. The tuple for s_{10} indicates that both p_7 and p_{10} are having the highest values. Also we can notice that p_7, p_{10} have greater values at s_2 and s_8, both of which are the positive cases of $Vertigo$ and $Unconsciousness$. So, let us assume that s_{10} is very similar to s_2 and s_8. The next prominent feature for s_{10} is p_9. One can notice p_9 is most aggravated in the case of s_6, and so a close similarity between s_6 and s_{10} may also be assumed. The diagnosis for the known cases shows, s_6 is the only positive case of $Allergy$. Now as from the concept ontology one can observe that $Vertigo, Unconsciousness \subseteq_k Stroke$ and $Allergy \subseteq_k Viral_{inf}$, finding out the degree of relationship among $\{Viral_{inf}, Vertigo, Unconsciousness\}$ and $Stroke$, i.e., $gr(\{Viral_{inf}, Vertigo, Unconsciousness\} \mid \approx Stroke)$ might be suggestive.

4   Future Directions

The study made in this paper is an initial step towards obtaining a concept synthesis method using finitely many observations about the prototypical cases and counterexamples of its component concepts. The newness of this approach, perhaps, lies in its way of proposing both *satisfiability of a concept at a world*, and *determinning that a concept is synthesized out of a set of concepts* as matters of grade, and allowing both the levels of $\models$ and $\mid\approx$ to have different logical characteristics. A few directions of further research are as follows. (i) How the theory of argumentation in favour of a claim, and against a claim can be incorporated in order to come to a decision that a situation ω is similar to another situation ω', or if some concept applies to ω, that also applies to ω'? (ii) What would be the properties of $\models_e$ when it would be extended on $P(W) \times C$ with the standard set theoretic operations on $P(W)$? (iii) How to incorporate the views of a set of sets of situations, say $\{\omega_i\}_{i \in I}$, $\{\omega_j\}_{j \in J}$, ..., $\{\omega_l\}_{l \in L}$, allowing exchange of views among different sets of situations in the process of concept synthesis?

Acknowledgement. Authors of this paper are thankful to Professor Andrzej Skowron for his valuable suggestions regarding the development of this work. This work has been carried out during the tenure of ERCIM Alain Bensoussan fellowship of the first author, and this joint work is partially supported by the Polish National Science Centre (NCN) grants DEC-2011/01/D/ST6/06981.

References

1. Bazan, J.G., Skowron, A., Swiniarski, R.: Layered learning for concept synthesis. Trans. Rough Sets V: J. Subline LNCS **4100**, 39–62 (2006)
2. Chakraborty, M.K.: Use of fuzzy set theory in introducing graded consequence in multiple valued logic. In: Gupta, M.M., Yamakawa, T. (eds.) Fuzzy Logic in Knowledge-Based Systems, Decision and Control, pp. 247–257. Elsevier Science Publishers, B.V. (North Holland) (1988)
3. Chakraborty, M.K.: Graded consequence: further studies. J. Appl. Non-Classical Logics **5**(2), 127–137 (1995)

4. Chakraborty, M.K., Basu, S.: Graded consequence and some metalogical notions generalized. Fundamenta Informaticae **32**, 299–311 (1997)
5. Chakraborty, M.K., Dutta, S.: Graded consequence revisited. Fuzzy Sets Syst. **161**, 1885–1905 (2010)
6. Dubois, D., Prade, H.: Rough fuzzy sets and fuzzy rough sets. Int. J. Gen. Syst. **17**, 191–200 (1990)
7. Dubois, D., Prade, H.: Putting rough sets and fuzzy sets together. In: Słowñiski, R. (ed.) Intelligent Decision Support, vol. 11, pp. 203–232. Springer, Dordrecht (1992)
8. Dutta, S., Chakraborty M.K.: Grade in metalogical notions: a comparative study of fuzzy logics. Accepted in Mathware and Soft Computing
9. Dutta, S., Basu, S., Chakraborty, M.K.: Many-valued logics, fuzzy logics and graded consequence: a comparative appraisal. In: Lodaya, K. (ed.) Logic and Its Applications. LNCS, vol. 7750, pp. 197–209. Springer, Heidelberg (2013)
10. Dutta, S., Chakraborty, M.K.: Graded consequence with fuzzy set of premises. Fundamenta Informaticae **133**, 1–18 (2014)
11. Dutta, S., Bedregal, B.R.C., Chakraborty, M.K.: Some instances of graded consequence in the context of interval-valued semantics. In: Banerjee, M., Krishna, S.N. (eds.) ICLA. LNCS, vol. 8923, pp. 74–87. Springer, Heidelberg (2015)
12. Klir, G.J., Yuan, B.: Fuzzy Sets and Fuzzy Logic: Theory and Applications. Prentice Hall of India, New Delhi (1995)
13. Polkowski, L., Skowron, A.: Rough mereological approach to knowledge-based distributed AI. In: Lee, J.K., Liebowitz, J., Chae, J.M. (eds.) Critical Technology: Proceeding of the Third World Congress on Expert Systems, pp. 774–781. Cognizant Communication Corporation, New York (1996)
14. Nguyen, S.H., Bazan, J., Skowron, A., Nguyen, H.S.: Layered Learning for Concept Synthesis. In: Peters, J.F., Skowron, A., Grzymała-Busse, J.W., Kostek, B., Świniarski, R.W., Szczuka, M.S. (eds.) Transactions on Rough Sets I. LNCS, vol. 3100, pp. 187–208. Springer, Heidelberg (2004)
15. Sanchez, E. (ed.): Fuzzy Logic and The Semantic Web. Elsevier, Amsterdam (2006)
16. Skowron, A., Stepaniuk, J.: Information granules and rough-neural computing. In: Pal, S.K., Polkowski, L., Skowron, A. (eds.) Rough-Neural Computing: Techniques for Computing with Words, Series: Cognitive Technologies, pp. 43–84. Springer, Heidelberg (2004)
17. Skowron, A., Stepaniuk, J.: Hierarchical modelling in searching for complex patterns: constrained sums of information systems. J. Exp. Theor. Artif. Intell. **17**(1–2), 83–102 (2005)
18. Skowron, A., Jankowski, A., Wasilewski, P.: Interactive computational systems: rough granular approach. In: Popowa-Zeugmann, L. (ed.) Proceedings of the Workshop on Concurrency, Specification, and Programming (CS&P 2012). Informatik-Bericht, vol. 225, pp. 358–369. Humboldt University, Berlin (2012)
19. Skowron, A., Jankowski, A., Wasilewski, P.: Risk management and interactive computational systems. J. Adv. Math. Appl. **1**, 61–73 (2012)
20. Vetterlein, T., Cibattoni, A.: On the (fuzzy) logical content of CADIAG-2. Fuzzy Sets Syst. **161**(14), 1941–1958 (2010)
21. Web page of Professor Valiant. http://people.seas.harvard.edu/valiant/research interests.htm

Fuzzy Rough Sets Theory Reducts for Quantitative Decisions – Approach for Spatial Data Generalization

Anna Fiedukowicz[✉]

Faculty of Geodesy and Cartography, Department of Cartography,
Warsaw University of Technology, Warsaw, Poland
a.fiedukowicz@gik.pw.edu.pl

Abstract. One of the most important objectives within the scope of current cartography is the creation of system controlling the process of geographical data generalisation. Firstly, it requires selection of the features crucial from the point of view of the decision making process. Such tools as reducts and fuzzy reducts, though useful, are still insufficient for the quantitative decisions, common in cartographical generalization. Thus the author proposed a modification in fuzzy reducts calculating, which can allow to calculate them with regard to a continuous decision variable. The proposed method is based on the t-norm of fuzzy indiscernibility based on attribute value and fuzzy indiscernibility based on decision, which is calculated for each pair of objects. The solution seems to be more intuitive than the ones established previously.

Keywords: Fuzzy rough sets · Feature selection · Generalisation of geographic information · Cartography

1 Introduction

One of the most important challenges of modern cartography is the automation of the geographical information generalisation processes [1, 4, 9, 10]. It requires acquisition of data's crucial information, general patterns, and tendencies and their subsequent retaining on the lower levels of detail (LoD), which corresponds to lower map scales. This task up to now has been tackled manually by the skilled cartographers and it seems to be difficult to algorithmize. Even though there is an array of algorithms of generalisation, which address particular generalisation operators (such as: objects selection, simplification, smoothing, aggregation, amplification, etc.), what still poses the problem is the control of the entire process of generalisation – starting from the decision, "whether to generalise at all", through the choice of the appropriate operator and algorithm, up to the final selection of parameters of the latter [1, 4, 9, 10].

The facts described above point to the conclusion that, apart from the particular tools employed in the process of generalisation, the decision-making system is required in order to manage the operations on many different levels. Such system should possess and utilise the skilled cartographer's knowledge. However, such skills usually result from years of practice and experience along with refined aesthetic taste, and therefore they are not available explicitly. Taking the above into consideration, according to the

© Springer International Publishing Switzerland 2015
M. Kryszkiewicz et al. (Eds.): PReMI 2015, LNCS 9124, pp. 314–324, 2015.
DOI: 10.1007/978-3-319-19941-2_30

author, the methods of knowledge data discovery (KDD) might be used to convert this hidden knowledge into an explicit form [5, 10].

Therefore the author's ultimate aim is to create a base of fuzzy rules governing the process of generalisation. The step prior to that, then, will be to choose the features crucial for the process of generalisation itself, which will lead to simplification of the decision system. The following paper addresses the problem of attribute choice methodology (using the reducts) which takes into consideration the specific data features. While in the previous work [10] the classical rough set approach for categorical data was used, this paper focuses on the problem of numeric feature selection based on numeric decisions – it discusses and provides some new extensions for fuzzy rough set (FRST) approach. The rough set approaches are chosen for the feature selection as they are easily understandable and give good intuition about why certain attribute is selected.

2 Data Specifics

The currently developed geospatial databases provide an array of information - in the form of attributes (projected in database structure), as well as more implicit features (connected with objects' geometry and their topography) - which can be used in the process of generalisation of geographical information. In that way the data is specified by a number of attributes that can potentially be used in the further generalisation process management.

What is worth emphasising, is the fact that the attributes can be expressed in different measurement scales: qualitative (ranging from binary scale, through classifying scale, to ordinal scale) as well as quantitative. Thus, the decision attribute can also be represented in different measurement scales.

In this article selection generalisation operator is considered, with decision expressed in two measurement scales (Table 1):

1. Binary – for the systems created on one particular data level of detail (corresponding with the scale 1:20 000): 1 - the object is selected, 0 - object is not selected;
2. Quantitative – for the system with universal character allowing to choose objects on any map scale (within assumed range) – the attribute's value is a corresponding scale denominator.

However, the second above is strongly preferred. Firstly it does not require designing separate systems for each of the desired scales. In the past, when analogue maps prevailed, it was possible to distinguish the scales in which the data were to be generated (they corresponded to the scales in which the maps were printed). However, nowadays most maps are accessible interactively via the Internet and the end user can choose any scale, thus the generalisation to all levels of detail is useful.

The test dataset corresponds to the topographical data collected at the 1:10 0000 level of detail, which is available for the whole Poland's area in National Cartographical Database (pl. *Państwowy Zasób Kartograficzny*) and known as BDOT10 k (pl. *Baza Danych Obiektów Topograficznych* - Topographical Objects Database). However the data are strongly simplified (Fig. 1).

Table 1. Attributes values of test dataset (buildings): a_1 – building function (r – residential, o – office, s – shops & services, g – religious), a_2 – public function (1 – yes, 0 – no), a_3 – area (in square meters), a_4 – shortest distance to the river, a_5 – shortest distance to the road, a_6 – shortest distance to another building, a_7 – shortest distance to the forest, a_8 – shortest distance to built-up area; attributes a_3 to a_8 are calculated basing on objects geometry, a_4 to a_8 are expressed in meters; decision attributes (established by an expert) in different scales: dec_1 – quantitative scale, dec_2 – binary scale (for the LoD 1: 20 000 – $dec_2 = 1$ for $dec_1 \geq 20 000$)

ID	a_1	a_2	a_3	a_4	a_5	a_6	a_7	a_8	dec_1	dec_2
1	r	0	101	304	35	43	126	0	0	15000
2	r	0	149	275	33	43	156	0	0	15000
3	r	0	93	252	37	41	200	0	0	15000
4	r	0	284	222	40	41	238	0	0	15000
8	r	0	892	216	124	76	212	0	0	17000
5	g	1	1721	32	53	420	537	0	1	50000
6	s	0	796	114	26	26	327	0	1	25000
7	s	0	585	122	26	26	321	0	1	25000
9	s	1	5174	70	94	143	234	0	1	40000
10	o	1	2840	69	54	95	140	0	1	35000
11	o	0	2015	19	19	46	394	0	1	27000
12	r	0	130	855	327	581	20	542	1	23000

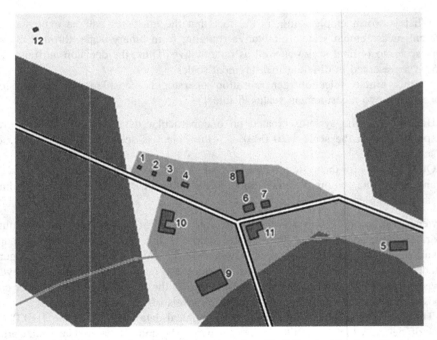

Fig. 1. Graphical representation of test dataset (numbered buildings) with other objects assuring spatial context: forests, roads, river, built-up area

3   Rough Set Based Feature Selection

3.1   Rough Sets

The rough sets theory allows to reduce the complexity of a system by searching of reduct B – the subset of the entire attribute A set [6–8, 11]. The following search is based on the discernibility relation, which can be defined as:

$$R_B = \{(x, y) \in X^2 \ and \ (\forall a \in B)(a(x) = a(y))\}$$

The so called decision reduct ensure the preservation of the original discernibility towards the decision: If the objects from different decision classes are discernible on the attribute set A, they are also discernible on its subset $B \in A$, being a reduct. The reduct has a minimal character, which means that none of the reduct's attributes can be omitted without losing of discernibility mentioned above [6–8].

The approach described is connected with particular constraint: attributes as well as the decisions should be expressed in the classification (not orderly) scale. Other ways, a discretisation is required, what entails a partial loss of information (including e.g. the order of distinguished classes).

One of the extensions for this approach considers graded indiscernibility between objects. Thus, the classes of attributes can be more or less similar to each other [8]. Established dissimilarities between attribute classes – degrees of discernibility can be expressed in the form of a matrix (example – Table 2).

Table 2.   Different discernibility degrees for classes of attribute a_1

a_1	r	s	o	g
r	0	0.8	1	1
s	0.8	0	0.5	1
o	1	0.5	0	1
g	1	1	1	0

3.2   Dominance-Based Rough Set Approach

The dominance based rough set (DBRS) approach, which is an extention of rough set theory, enables, without the loss of information, the use of attributes as well as decisions expressed in the ordinal scale. The theory postulates the apporoximation (and consequently reducts' calculation) for the union of the subsequent decision classes. The theory is insufficient, however, in the case of the attributes expressed in quantitative scales, as it indeed assumes the monotonous relation between the attributes, but does not establish the distance between the subsequent classes [3].

3.3   Fuzzy Rough Sets

The hybrid of the fuzzy sets theory and rough sets theory, enabling to create fuzzy reducts, employs the attributes in quantitative scale. The discernibility relation based on the equality of the attributes was replaced with the measure expressing the closeness of the objects represented by a fuzzy discernibility relation R [2, 11, 12].

The discernibility matrix then, existing also in the traditional fuzzy sets theory, is filed with the measure of closeness (based on each attribute) for each pair of objects with different decision value. In this paper the value of the fuzzy indiscernibility is calculated as follows:

$$R_a(x, y) = \frac{|a(x) - a(y)|}{l(a)} \qquad (1)$$

Subsequently, on the basis of the discernibility matrix quality of the reduct can be calculated by finding best fuzzy discernibility R_b for each pair of objects and finding its' minimal value out of all pairs (instead of max any co-norm $\perp$ can be used):

$$q(B) = \min(\max(R_{b \in B})) \qquad (2)$$

This approach is based on the original RST assumption that each reduct is as good as its weakest component, meaning it is as good as the least discernible pair of objects belonging to separate decision classes. Therefore, in the original approach minimum operator is used, however some authors find it overly restrictive and allow the use of an average instead [2]:

$$q(B) = \text{mean}(\max(R_{b \in B})) \qquad (3)$$

Such approach was used in the following work. The quality $q(B)$ can be then compared with the quality of the whole attribute set A (where ε is the acceptable tolerance of the quality loose) [2]:

$$q(B) \geq (1 - \varepsilon)q(A) \qquad (4)$$

Another approaches for reduct evaluation is to punish pairs of objects (x, y) which belong to different decision classes but are nearly indiscernible using reduct's attributes. It can be expressed as [12]:

$$P_B = \mathcal{T}_{b \in B}(1 - R_b(x, y)) \qquad (5)$$

The FRST approaches are the first enabling the calculation of the reducts for the data presented at the beginning, without discretization. It is possible, however, to establish only the reduct for the first variant (1) – binary decision scale.

4 Fuzzy Rough Reducts for Quantitative Decision

4.1 Adaptations of Existing FRST Methods

Few among the authors directly address the problem of reducts for the decisions in the quantitative scale (dec_2), however some of the solutions form Sect. 3.3 can be adapted for this case.

In the formula (2) or (3) similarity of objects by decision can be added (now all the objects' pair are compared):

$$q(B) = \text{mean}(\max(R_{b \in B}, 1 - R_{dec})) \tag{6}$$

However, this solution disadvantage is that it promotes too much the pairs of objects which are indiscernible according to the decision ($R_{dec} \cong 0$) which is in fact not interesting when looking for decisive reducts.

Also formula (5) can be adapted by adding fuzzy indiscernibility relation to the t-norm [12]:

$$\mathcal{T}(P_B, R_{dec}) \tag{7}$$

The final punishment is calculated as a sum:

$$Sim(d/B) = \sum_{x,y:d(x) \neq d(y)b \in B} R_{dec} \prod_{b \in B}(1 - R_b) \tag{8}$$

The disadvantage of this approach (formulas 7 and 8) is that it does not seem intuitive as it includes indiscernibility by decision for data already aggregated by t-norm.

4.2 Proposed Solutions

The proposed solution intends to be more intuitive for non-mathematical expert. It is based on the necessity to establish the value of the relative relation R_{a_rel} considering the objects' relation R both on attribute (R_a) and decision (R_{dec}). Therefore, it is proposed to calculate the relative tolerance relation for each pair of objects as a t-norm of R_a and R_{dec}:

$$R_{a\_rel} = \mathcal{T}(R_a, R_{dec}) \tag{9}$$

The most interesting from the point of view of applications described in introduction, seem to be such t-norms as:

1. product(R_a, R_{dec})
2. Hamacher product(R_a, R_{dec}): $T_{H_0}(a,b) = \begin{cases} 0 & if\ a = b = 0 \\ \frac{ab}{a+b-ab} & otherwise \end{cases}$
3. min(R_a, R_{dec})

The further proceedings are identical as in the classic FRST method, though all possible pairs of objects are compiled and Rb_rel is used instead of Rb:

$$q(B) = \text{mean}\left(\max\left(R_{a\_rel\ (b \in B)}\right)\right) \tag{10}$$

Such an approach allows to follow the significance of each attribute (in relation to the decision) for each pair of objects. This can be valuable from the point of view of expert using decision system as it allows to intuitionally understand the importance of attributes.

5   Experiments on Test Data

5.1   Some General Assumptions

The calculation of fuzzy discernibility indicator R_a for the pair of objects depends on the scale in which the attribute a was depicted. Therefore:

- For the attribute a_2 (expressed in binary scale) the classical discernibility approach, based on the equivalence relation was employed,
- For the attribute a_1 (expressed in the classifying scale) similarity matrix was employed (Table 2),
- For the attributes a_3 to a_8 (in quantitative scale) tolerance relation R basing on formula (1) was employed.

Due to the specificity of the described problem establishing of all possible reducts of the set was not necessary. In practice, for the purpose of the further application only 1, sometimes few, reducts will be used. The accessibility to the attributes necessary for the calculation is usually high as they are available in the databases either as the descriptive attributes designed in database structure, or are easy to calculate on the basis of objects geometry. Therefore, the reducts were calculated with Johannson's heuristic. It operates as follows: every time such attribute is added to the reduct, which results in the biggest increase of the quality q (understood as in formula 3). This steps may be repeated until:

- Obtaining the quality q fulfils the condition (4) assumed by the user or
- The point when adding another attributes results in the increase of quality q lower then estabilished Δq.

In this work the second method (with $\Delta q = 0.02$) was employed, due to the necessity of maintaining a low system complexity (and consequently not overly numerous reducts), if its higher complexity did not increase the overall quality significantly.

Similarly, the other reducts can be calculated (starting from the subsequent attribute), however this work limited itself to only 1 reduct in each example.

Table 3. Subsequent steps of fuzzy reduct creation for the binary decision, including the corresponding reducts qualities (the fields with the highest accuracy are highlighted, while the elements added to the reduct are in bold)

No	current reduct	quality	quality of a reduct after adding new attribute:							
			a_1	a_2	a_3	a_4	a_5	a_6	a_7	a_8
1	{}	0,00	**0,77**	0,43	0,33	0,29	0,22	0,28	0,30	0,14
2	{a_1}	0,77	0,77	0,80	0,80	0,87	0,90	**0,91**	0,82	0,91
3	{a_1, a_6}	0,91	0,91	**0,94**	0,93	0,91	0,91	0,91	0,91	0,91
4	{a_1, a_2, a_6}	0,94	0,94	0,94	0,94	0,94	0,94	0,94	0,94	0,94
5	{a_1, a_2, a_6}	**0,94**								

5.2 Fuzzy Reducts for Binary Decision

Determination of the reduct started with combining the objects belonging to different decision classes (dec_1) into pairs and calculating the discernibility matrix of relations R (according to the rules described in Sect. 5.1). Then the consecutive elements of the reduct were established with use of greedy heuristic basing on the q value – Table 3.

Consequently reduct {a_1, a_2, a_8} have been established. Adding another attribute would not increase the quality of the reduct so no other attribute was added. What is worth mentioning is the high quality of the decision reduct, as the ability to distinguish the object with different decisions, is high.

Table 4. Subsequent steps of fuzzy reduct creation for the quantitative decision, including the corresponding qualities (highlights as in Table 3); the method used: according to the formula 6

No	current reduct	quality	quality of a reduct after adding new attribute:							
			a_1	a_2	a_3	a_4	a_5	a_6	a_7	a_8
1	{}	0,00	**0,90**	0,90	0,73	0,68	0,67	0,73	0,71	0,68
2	{a_1}	0,90	0,90	**0,94**	0,92	0,90	0,91	0,91	0,90	0,92
3	{a_1, a_2}	0,94	0,94	0,94	0,94	0,94	0,96	**0,96**	0,94	0,96
4	{a_1, a_2, a_6}	0,96	0,96	0,96	0,96	0,96	0,96	0,96	0,96	0,96
5	{a_1, a_2, a_6}	**0,96**	or {a1, a2, a5} or {a1, a2, a8}							

5.3 Fuzzy Reducts for Quantitative Decisions

First the method using formula 6 was used. Results of following steps are presented in Table 4. One of the possible decision reduct is exactly the same that the one described in Sect. 5.2, although it allows to distinguish a number of object types. However accuracies seem to be over-optimistic since the goal was to distinguish between more exact decisions and the qualities here are higher than for the binary decision.

The next step was to test the method proposed in Sect. 4.2. In this case all possible pairs of objects were juxtaposed. For each of them R_a, R_{dec1} and R_{a_rel} were calculated (in two variants: with use of T-norm product and Hamacher product). Then the quality

Table 5. Subsequent steps of fuzzy reduct creation for the quantitative decision, including the corresponding qualities (highlights as in Table 3); the t-norm used: $R_{rel} = $ product(R_a, R_{dec})

No	current reduct	quality	quality of a reduct after adding new attribute:							
			a_1	a_2	a_3	a_4	a_5	a_6	a_7	a_8
1	{}	0,00	**0,30**	0,26	0,15	0,10	0,07	0,13	0,13	0,04
2	$\{a_1\}$	0,30	0,30	**0,33**	0,32	0,31	0,32	0,32	0,31	0,31
3	$\{a_1, a_2\}$	0,32	0,33	0,33	0,33	0,34	0,35	**0,35**	0,34	0,35
4	$\{a_1, a_2, a_6\}$	0,35	0,35	0,35	0,35	0,35	0,35	0,35	0,35	0,35
5	$\{a_1, a_2, a_6\}$	**0,35**	or $\{a_1, a_2, a_5\}$ or $\{a_1, a_2, a_8\}$							

Table 6. Subsequent steps of fuzzy reduct creation for the quantitative decision, including the corresponding qualities (highlights as in Table 3); the t-norm used: $R_{rel} = $ Hamacher product(R_a, R_{dec})

No	current reduct	quality	quality of a reduct after adding new attribute:							
			a_1	a_2	a_3	a_4	a_5	a_6	a_7	a_8
1	{}	0,00	**0,31**	0,26	0,19	0,13	0,10	0,16	0,18	0,04
2	$\{a_1\}$	0,31	0,31	**0,34**	0,34	0,33	0,34	0,33	0,33	0,33
3	$\{a_1, a_2\}$	0,34	0,33	0,33	0,33	0,34	0,35	**0,35**	0,34	0,35
4	$\{a_1, a_2, a_6\}$	0,35	0,36	0,36	0,36	0,36	0,36	0,36	0,36	0,36
5	$\{a_1, a_2, a_6\}$	**0,35**	or $\{a_1, a_2, a_5\}$ or $\{a_1, a_2, a_8\}$							

Table 7. Subsequent steps of fuzzy reduct creation for the quantitative decision (decision artificially brought to the binary scale), including the corresponding accuracies (highlights as in Table 3); the t-norm used: product and Hamacher product (results for both are identical)

No	current reduct	quality	quality of a reduct after adding new attribute:							
			a_1	a_2	a_3	a_4	a_5	a_6	a_7	a_8
1	{}	0,00	**0,41**	0,23	0,17	0,15	0,12	0,15	0,16	0,08
2	$\{a_1\}$	0,41	0,41	0,42	0,42	0,46	0,48	**0,48**	0,43	0,48
3	$\{a_1, a_6\}$	0,48	0,48	**0,50**	0,49	0,48	0,48	0,48	0,48	0,48
4	$\{a_1, a_2, a_6\}$	0,50	0,50	0,50	0,50	0,50	0,50	0,50	0,50	0,50
5	$\{a_1, a_2, a_6\}$	**0,50**								

based on R_{a_rel} was calculated for particular attributes – the subsequent steps of the heuristics are illustrated in the tables below (Tables 5 and 6).

The methods results in the same reducts as the previous ones, irrespectively which of the t-norm is used. What is more, even though qualities values differ depending on the used t-norm (and differ even more form the corresponding ones in Table 4), there seem to be noticeable tendencies in its distribution over the attributes.

In the next stage, in order to test the universality of the method, the same steps were taken, though in this case on binary decision (dec_1) – Table 7. In this case the results, irrespective of the chosen t-norm, were alike ($R_{dec} \in \{0, 1\}$, so all t-norms have the same value in the R_b function). It should be noted, that the result (calculated reduct) and the relation between the qualities of particular reducts are identical to those calculated with the classic method described in Sect. 5.2. The ratio of corresponding accuracies in the Tables 3 and 7 equals c.a. 0.53, which reflects the proportion of the number of pairs in different decision classes and the total number of pairs (35/66), or in other words the average discernibility of all pair of objects towards the decision.

6 Conclusions

The work above addressed the problem of reduct calculation in the case of decisions in quantitative scale. In the process a tolerance relation (R_{rel}), understood as t-norm of tolerance relations of attribute (R_a) and decision (R_{dec}), was employed.

The conducted tests using the two types of t-norm – product and Hamacher product – gave similar results, meaning they both allowed to achieve the same reducts irrespective of differences in the quality value of the reduct. Generally the reducts' qualities values calculated with this method were lower than for binary decision, what seems justifiable taking into consideration the necessity to distinguish the objects according to their continuous decision value. They are also lower than the qualities calculated by one of the existing methods (formula 6), however those qualities seem to be over-optimistic as they grow unreasonably thanks to the objects having the same or similar decision values.

The methodology employed in the case of artificially binary decision allowed to achieve the same reduct as it was the case in the original FRST method. However, even though the relation of quality values between particular attributes were maintained the absolute values of quality were different. It is a result of calculating the discernibility relation for all pair of objects.

The tests indicate that proposed method can be applied in the process of general-isation of geographical data mentioned above. However there is also a potential for other application areas. Depending on the application type other t-norm for calculating R_{rel} can be used.

The main advantage over the other method is that it is intuitively understandable and prevents black-box solution, allowing user to follow the importance of each attribute on every single step of the reduct computation. Therefore the method can be employed among others in the creation of the system of generalisation control, on the first stage of its development – attribute selection.

Acknowledgments. Author would like to thank Prof. Dominik Ślęzak for consultations and benevolent intelectual support.

References

1. Burghardt, D., Duchene, C., Mackaness, W. (eds.): Abstracting Geographic Information in a Data Rich World. Lecture Notes in Geoinformation and Cartography Series. Springer, Berlin (2014)
2. Cornelis, Ch., Jensen, R., Martín, G.H., Slezak, D.: Attribute selection with fuzzy decision reducts. Inf. Sci. **180**(2), 209–224 (2010)

3. Greco, S., Matarazzo, B., Słowiński, R.: Multicriteria classification by dominance-based rough set approach. In: Kloesgen, W., Zytkow, J. (eds.) Handbook of Data Mining and Knowledge Discovery. Oxford University Press, New York (2002)
4. Mackaness W.: Understanding geographic space. In: Mackaness, w., Ruas, A., Sarjakoski, T. (eds.) Generalisation of Geographic Information: Cartographic Modelling and Application. Elsevier, Oxford (2007)
5. Miller, H.J., Han, J.: Geographic Data Mining and Knowledge Discovery. Taylor & Francis, London (2001)
6. Pawlak, Z.: Rough sets. Int. J. Parallel Prog. 11(5), 341–356 (1982)
7. Pawlak, Z.: Rough Sets: Theoretical Aspects of Reasoning about Data. Kluwer Academic Publishing, Dordrecht (1991)
8. Pawlak, Z., Skowron, A.: Rough sets: some extensions. Inf. Sci. 177, 28–40 (2007)
9. Olszewski R., Kartograficzne modelowanie rzeźby terenu metodami inteligencji obliczeniowej, Prace Naukowe - Geodezja, z. 46, Oficyna Wydawnicza Politechniki Warszawskiej, Warszawa (2009)
10. Olszewski, R., Fiedukowicz, A.: Supporting the process of monument classification based on reducts, decision rules and neural networks. In: Kryszkiewicz, M., Cornelis, C., Ciucci, D., Medina-Moreno, J., Motoda, H., Raś, Z.W. (eds.) RSEISP 2014. LNCS, vol. 8537, pp. 327–334. Springer, Heidelberg (2014)
11. Riza, L.S., Janusz, A., Bergmeir, C., Cornelis, C., Herrera, F., Ślęzak, D., Benítez, J.M.: Implementing algorithms of rough set theory and fuzzy rough set theory in the R package "RoughSets". Inf. Sci. 287, 68–89 (2014)
12. Ślęzak, D., Betliński, P.: A role of (not) crisp discernibility in rough set approach to numeric feature selection. In: Hassanien, A.E., Kim, T.-H., Ramadan, R., Salem, A.-B.M. (eds.) AMLTA 2012. CCIS, vol. 322, pp. 13–23. Springer, Heidelberg (2012)

Fuzzy Rough Sets Theory Applied to Parameters of Eye Movements Can Help to Predict Effects of Different Treatments in Parkinson's Patients

Anna Kubis[1]([⊠]), Artur Szymański[2],
and Andrzej W. Przybyszewski[2,3]

[1] Faculty of Applied Mathematics, AGH University of Science and Technology,
A. Mickiewicza 30, 30-059 Kraków, Poland
anna.kubis@fis.agh.edu.pl
[2] Polish-Japanese Institute of Information Technology,
Koszykowa 86, 02-008 Warsaw, Poland
artur.szymanski@pja.edu.pl
[3] Department of Neurology, University of Massachusetts Medical School,
65 Lake Av., Worcester, MA 01655, USA
andrzej.przybyszewski@umassmed.edu

Abstract. Parkinson (PD) is the second most common neurodegenerative disease (ND) with characteristic movement disorders. There are well defined standard procedures to measure disease stage (Hohen Yahr scale), progression and effects of treatments (UPDRS – unified Parkinson Disease Rate Scale). But these procedures can only be performed by experienced neurologist and they are partly subjective. The purpose of our work was to test objective and non-invasive method that may help to estimate disease stage by measuring fast and slow eye movements (EM). It was demonstrated earlier that EM changes in PD. We have measured reflexive saccades (RS) and slow pursuit ocular movements (POM) in four sessions related to different treatments. With help of fuzzy rough sets theory (FRST) we have related measurements with expert's opinion by generalizing experimental finding by *fuzzy rules*. In order to test our approach, we have divided our measurements into training and testing sets. In the second test, we have removed expert's decisions and predicted them from the training set in two situations: on the basis of only classical neurological measurements and on the basis of EM measurements. We have observed, on 12 PD patients basis, an increase in predictions accuracy when eye movements were included as condition attributes. Our results with help of the FRST suggest that EM measurements may become an important diagnostic tool in PD.

Keywords: Deep brain stimulation (DBS) · Reducts · Information table · Decision rules · Rough sets

© Springer International Publishing Switzerland 2015
M. Kryszkiewicz et al. (Eds.): PReMI 2015, LNCS 9124, pp. 325–334, 2015.
DOI: 10.1007/978-3-319-19941-2_31

1 Introduction

The brain's deeper computational properties are still not well understood. We are even not sure if brain computations are more powerful than the Turing machine and such models as ARNN (analog recurrent neural networks) or coupled nonlinear oscillators are appropriate [1, 2]. For example, we do not know exactly how brain processes are affected by nerve cell deaths in the neurodegenerative diseases (ND) such as Parkinson or Alzheimer. It is well documented, however, that the disease starts long before the observed first symptoms and individual pathological mechanisms have a large spectrum. In Parkinson's, for example, the first motor symptoms are observed when 70 −80 % cells in responsible structure (substantia nigra) are dead and once cells are dead there is no chance for their recovery.

We can register symptoms of ND such as motor and/or mental disorders (dementias) and even provide symptomatic relief, though the structural effects of these are in most cases not yet understood. Fortunately, with early diagnosis there are often many years of disease progression with symptoms that, when they are precisely monitored, may result in improved therapies.

One of the purposes of this work is to try to extract knowledge from symptoms in order to model possible mechanisms of disease progression and adjust therapies in timely precise matter.

The majority of neurologists use the standard statistical methods to analyze the results of PD patients' treatment. As every patient suffers from PD in a different way and reacts differently to the treatment, averaging methods can lead to the confusing results. Therefore, in continuation of [3, 4], we propose to extend statistical analysis to data mining techniques in order to adjust PD treatment to an individual patient. Our method is based on fuzzy rough sets theory application as this approach should better fit to predict partly noisy and continuous medical measurements than previously proposed rough sets theory [3, 4].

As PD progression biomarker we have used measurements of eye movements. It is well established on the animal experimental basis that the basal ganglia are involved in the eye movement's control (see review [5]). It was also demonstrated on human subjects that fast (saccades) and also slow (pursuit ocular movements) eye movements are affected in Parkinson's diseases [6, 7].

Generally, different treatments are based on the UPDRS (Unified Parkinson's Disease Rating Scale) measurements, in particular on UPDRS II (activity of daily living), UPDRS III (examination of motor symptoms), UPDRS V (modifies Hoehn and Yahr staging – stage of the disease) and UPDRS VI (Schwab and England activities of daily living scale). As these measurements are strongly doctor dependent and partly subjective, we propose to use the eye movement (EM) as an individual doctor independent measure to improve diagnosis and objectivity. In the consequence, in our analysis in addition to standard neurological measurements, we have added EM parameters as condition attributes, doctors' expertise as the decision attribute and placed them in the decision table [8]. As the data in the table were related to different treatments, our purpose was to use the data mining techniques to estimate and to predict effectiveness of different therapies related to individual patients.

2   Methods

We have performed our analysis on PD data used earlier in [3, 4]. All of the 12 patients had implanted electrodes in the subthalamic nucleus that is a standard procedure in advanced Parkinson's. As number of PD patients is relatively small, results of this study are preliminary. The measurements were conducted in four sessions (S1–S4): in the first session (S1) patients were off medications (L-Dopa) and DBS stimulators was OFF; in the second session (S2) patient were off medication, but the stimulator was ON; in the third session (S3) patients were after his/her doses of L-Dopa and stimulator was OFF, and in the fourth session (S4) patients were on medication with the stimulator ON. The data set consisted of the estimation of the disease advancement made during the medical interview (expressed by Unified Parkinson Disease Rate Scale - UPDRS) related to changes in motor performance, behavioral dysfunction, cognitive impairment and functional disability, and EM measurements. We have evaluates saccadic and slow pursuit eye movements. The EM were recorded by head-mounted saccadometer (Ober Consulting, Poland). We have used an infrared eye track system coupled with a head tracking system (JAZZ-pursuit – Ober Consulting, Poland). In the EM measurements patient was sitting at the distance of 60–70 cm from the monitor with head supported by a headrest in order to minimize head motion. We measured fast eye movements in response to a light spot switched on and off, which moved horizontally from the straight eye fixation position (0 °) to 10 ° to the left or 10 ° to the right after arbitrary time ranging between 0.5–1.5 s. When the patient fixated eyes on the spot in the middle marker (0 °) the spot then changed color from white to green, indicating a signal for performance of RS (reflexive saccades); or from white to red meaning a signal for performing AS (antisaccades) – not evaluated in this study. Then the central spot was switched off and one of the two peripheral targets, selected at random with equal probability, was illuminated instead (non-overlapping test). Patients had to look at the targets and follow them as they moved in the RS task. After making a saccade to the peripheral target, the target remained on for 0.1 s after which another trial was initiated. In each test the subject had to perform 10 RS and 10 AS in a row in Med-off (medication off) within two situations: with DBS off (S1) and DBS on (S2). In the next step the patient took medication and had a break for one half to one hour, and then the same experiments were performed, with DBS off (S3) and DBS on (S4). Slow EM – pursuit ocular movements (POM) were measured in response to a light spot with horizontal sinusoidal movements (with slow (0.125 HZ), medium (0.25 Hz) and fast (0.5 Hz) frequencies), placed from 10 ° to the left to 10 ° to the right. POM measurements were performed in four different sessions in similar procedures as described above for RS measurements.

In this work we have analyzed only RS data using the following parameters: averaged for both eyes: delay (RS latency), amplitude (RS amplitude), duration (RS duration), velocity (RS velocity). We have analyzed POM data using the following parameters averaged for both eyes: gain (eye movement amplitude/sinus amplitude) and accuracy (difference between sinusoid and eye positions) for three different frequencies. More details can be found in [3, 4].

3   Theoretical Basis

Our data were represented as a decision table. In the rows we put the measurements' values for respective patients during each single session. As columns we use patient's number, patient's age, session number, estimations of UPDRS, Schwab and England and Hoehn and Yahr scales and EM measurements: RS parameters and slow, medium and fast sinus parameters for POM.

As fuzzy rough set theory (FRST) is an extension of rough set theory (RST) [8] we define here a similarity or tolerance relation [9, 10] instead of crisp equivalence. The tolerance relation $R_a(x, y)$ determines the discernibility between the values of the specific attribute for a pair of observation. There are several means to describe this relation $R_a(x, y)$ as presented below after [8–12]:

$$R_a(x, y) = 1 - \frac{|a(y) - a(x)|}{|a_{min} - a_{max}|} \tag{1}$$

In this way, the value of tolerance relations is directly proportional to the absolute value of the difference between the attribute's values for the two observations.

$$R_a(x, y) = e^{\frac{-(a(y)-a(x))^2}{2\sigma_a}} \tag{2}$$

where σ_a stands for standard deviation for the given attribute a. This equation includes standard deviation of the data, therefore in most cases it is more sensible for the behavior of the data than Eq. (1) mentioned above.

$$R_a(x, y) = e^{\frac{-\|a(y)-a(x)\|^2}{d}} \tag{3}$$

where d is a positive number. In our case, we take absolute value as a norm and variance σ_a in place of d.

In the next step, we have normalized the differences between each pair of conditional attributes' values. For this purpose, we have used a t-norm, marked τ. For a given pair of attributes a and b we get $R_{\{a,b\}}(x, y) = \tau(R_a(x, y), R_b(x, y))$. In order to get the value of the relations for the whole set of conditional attributes B, it is enough to normalize the difference between the first pair and the successive element and then by recurrence the difference between the value for the set got at the preceding step and a successive added element: $R_{\{a,b,c\}}(x, y) = \tau(R_{\{a,b\}}(x, y), R_c(x, y))$. The two most commonly used t-norms are: t.cos and Łukasiewicz t-norm, described respectively by equations:

$$R_{\{a,b\}}(x, y) = \max\{0, R_a(x, y) \cdot R_b(x, y) - \sqrt{1 - R_a(x, y)} \cdot \sqrt{1 - R_b(x, y)}\}, \tag{4}$$

$$R_{\{a,b\}}(x, y) = \max\{0, R_a(x, y) + R_b(x, y) - 1\}. \tag{5}$$

Tolerance relation defined by (1) is transitive in both t-norms, while tolerance relations defined with (2, 3) are transitive only with t-norm [9].

In case of modelling the difference between values for the decision attribute, we usually use the relation of identity: $R_d(x, y) = \begin{cases} 0, x = y \\ 1, x \neq y \end{cases}$.

In FRS concept, for the sets *(U, B)* of observations and condition attributes we define *B*-lower and *B*-upper approximations separately for every observation *x*. For each of the observations *x* we define *B*-lower approximation as: $(R_B \downarrow X)(x) = \inf_{y \in U} I(R_B(x, y), X(y))$, where *I* is an implicator [9]. The *B*-lower approximation for the observation *x* is then the set of observations which are the most similar to observation *x* and it can predict the decision attribute with the highest confidence, based on conditional attributes *B*.

The *B*-upper approximation is defined by $(R_B \uparrow X)(x) = \sup_{y \in U} \tau(R_B(x, y), X(y))$. Then, in fuzzy rough sets approach, the *B*-upper approximation is a set of observations for which the prediction of decision attribute has the smallest confidence.

Another term used in further explanations is positive region for an element y. The fuzzy B-positive region is a fuzzy set in the set *U* that contains each observation x to the extent that all objects with approximately equal values for the set of conditional attributes *B* have equal values for decision attribute. Formally after [9]: $POS_B(y) = \bigcup (R_B \downarrow R_d x)(y)$.

The predictive ability for d of the set of conditional attributes *B* is reflected in the degree of dependency defined as $\gamma_B = \frac{|POS_B|}{|POS_{A \setminus \{d\}}|} = \frac{\sum_{x \in U} POS_B(x)}{\sum_{x \in U} POS_{A \setminus \{d\}}(x)}$. If there does not exist any other subset B' of B such that $POS_{B'} = POS_{A \setminus \{d\}}$, B is called a decision reduct.

The rules in FRST approach are constructed from tolerance classes and corresponding decision concepts. A ready fuzzy rule will be a triple *(B, C, D)*, where *B* is a set of conditional attributes that appear in the rule, *C* stands for fuzzy tolerance class of object and *D* stands for decision class of object.

Apparently, many terms are defined differently in rough sets (applied and described in [3, 4]) and fuzzy rough sets approach, e.g. in RST upper approximation is a global term, defined for the whole data set while in FRST we define upper approximation separately for each element. Those sets are also larger in fuzzy method, as they contain observations that are not necessarily identical with the observation for which we define the upper approximation. As a consequence of this difference, in most cases we get close to 100 % of class coverage for predictors in FRST approach, while the coverage in RST is usually much lower.

4 Results

Below are examples of decision tables that include fast eye movements - reflexive saccades: RS (Table 1) and slow, pursuit eye movements - POM (Table 2) parameters.

Table 1. A part of the decision table for the first experiment including RS

Pat	Age	Sess	UPDRS III	Schwab EngSc	Scc Durat	SccLat t	SccAmp	SccVel	UPDRS Total
11	58	1	45	60	52	259	15	552,3	75
11	58	2	17	90	52	204	14	549,8	32
11	58	3	32	60	46	315	12	514	62
11	58	4	10	90	54	239	11	371,7	25

Pat - patient's number, age - patient's age, Sess - session number, UPDRS III - motor tests, SchwabEngSc –Schwab & England activity, SccDurat - RS duration, SccLat - RS latency, SccAmp - RS amplitude, SccVel - RS velocity, UPDRS Total - sum of a UPDRS I to VI.

On the basis of Table 1 we understand rules as (the last column is the decision attribute) for the first row:

$$('Pat' = 11)\&('age' = 58)\&('Sess' = 1)\&\ldots = > ('UPDRS III' = 45) \quad (6)$$

The rule should be read as follows: if for patient #11 and his/her age 58 and session S1 and value equal to 45 and … then his/her UPDRS III value is 45. We get such rules separately for each of the rows of the decision table. The main purpose of our analysis is to reduce number and increase universality of these rules.

In order to create fuzzy rules, we have used the algorithm called Hybrid Fuzzy-Rough Rule Induction and Feature Selection and described in detail in [13–15]. In the mentioned algorithm feature selection (a process of finding a subset of attributes which represent the same information as the complete feature set) and rule induction are performed simultaneously.

Table 2. A part of the decision table including POM

Pat	Age	Sess	HY scale	Schwab EngSc	gxss	gxms	gxfs	accss	accms	accfs
11	58	1	4	60	1.0583	0.9327	0.8487	0.8863	0.8151	0.7434
11	58	2	1	90	1.0077	0.9145	0.8823	0.5768	0.658	0.6929
11	58	3	4	60	1.0336	1.0046	0.9498	0.6864	0.652	0.4858
11	58	4	1	90	1.0588	1.0262	0.9408	0.7838	0.8019	0.8384

Pat- patient's number, age - patient's age, Sess - session number, HYscale - Hoehn and Yahr's scale, SchwabEngSc- as above, gxss/gxms/gxfs - gain for slow/medium/fast sinus, accss/accms/accfs - accuracy for slow/medium/fast sinus.

The rules determining UPDRS III are important in prediction of PD symptoms while the rules for session numbers are crucial in measuring the effects of different treatments. In order to predict results from new patients, we have performed the *test-and-train scenario* (e.g. [8]). For this purpose we divide the data set into two parts: training set, containing 75 % of the data and testing set, composed of the remaining 25 % that we have tested. We have removed decision attributes from the test set and compared them with attributes values obtained from our rules.

As the test-and-train scenario strongly depends on which part of our measurements was taken as training and which part was tested. In order to make the result possibly generalized, we have divided our experimental set into 4 subsets (4-fold-test). Then we treated each of them separately as a testing set, using the sum of other sets as a training set. The mean of four predictions gave the final measure of the accuracy.

In order to measure the effects of the treatment, we performed the prediction of the session numbers as a decision attribute. In the first step, as other attributes we used patient's number, age, results of UPDRS III, UPDRS IV and UPDRS Total, result in Hoehn and Yahr's scale and in Schwab and England's scale.

To make the prediction, we have used the RoughSets package in R environment. We checked the results of the prediction using different tolerance and t-norm definitions (Table 3).

Table 3. Global accuracy for different parameters chosen for the prediction of session numbers (S1−S4) without EM attributes

tolerance relation	t-norm	global accuracy
eq.(1)	eq.(5)	0.33
eq.(1)	eq.(4)	0.38
eq.(2)	eq.(4)	0.29
eq.(3)	eq.(4)	0.29

We chose then Eqs. 1 and 4 as parameters for our prediction. Its results are presented in the confusion matrix below (Table 4).

Table 4. Confusion matrix for different session numbers (S1−S4) without EM attributes

Predicted

		1	2	3	4	ACC
	1	0.5	0.5	0.25	0	0.4
	2	0	2	0.25	0	0.89
Actual	3	0.75	0.75	0	0	0
	4	0.25	0.75	0	0	0
	TPR	0.33	0.5	0	NaN	

TPR: True positive rates for decision classes, ACC: Accuracy for decision classes. Class coverage for predictors: 1 and global coverage = 1, and **global accuracy = 0.42**.

In order to compare how eyes movements parameters change the diagnostic ability of the data set, in the second step we have used the parameters of eyes movements: POM gain and accuracy for medium sinus (Table 5).

Table 5. Global accuracy for prediction of session numbers (S1–S4) including POM attributes

tolerance relation	t-norm	global accuracy
eq.(1)	eq.(5)	0.46
eq.(1)	eq.(4)	0.55
eq.(2)	eq.(4)	0.46
eq.(3)	eq.(4)	0.5

We chose then Eqs. (1) and (4) as parameters for our prediction. Its results are presented in the confusion matrix below (Table 6).

Table 6. Confusion matrix for different session numbers (S1–S4) including POM

Predicted

	1	2	3	4	ACC
1	1	0.5	0	0	0.67
2	0.25	0.75	0	0.25	0.6
3	0	0.25	0.75	0.5	0.5
4	0.5	0	0.5	0.75	0.42
TPR	0.57	0.5	0.6	0.5	

Actual

TPR: True positive rates for decision classes, ACC: Accuracy for decision classes. Class coverage for predictors: 1 and global coverage = 1, and **global accuracy = 0.55**.

In order to predict individual patient's symptoms related to different treatments, we made prediction of the UPDRS III. To estimate the global accuracy for the predictions of UPDRS attributes, we decided to recognize the prediction as accurate if it does not differ from the actual values from more than 20 % of values range.

As our purpose was to find if RS (saccade) attributes are significant, we began with prediction of UPDRS III using classical neurological measures but without UPDRS total and without EM parameters. Below, in Table 7 we gave results of global accuracies using different parameters of tolerance and t-norm. The best result gave Eqs. 1 and 6: the global accuracy was 46 %.

In the next step, we have tested results of UPDRS III prediction using in addition to standard neurological data (without UPDRS total) also RS duration and amplitude. The best result - global accuracy of 63 % - was obtained for Eqs. 2 and 5 (Table 8).

Table 7. Global accuracy for different parameters chosen for the prediction of UPDRS values, not including any eye movements parameters

tolerance relation	t-norm	global accuracy
eq.(1)	eq.(5)	0.46
eq.(1)	eq.(4)	0.33
eq.(2)	eq.(4)	0.38
eq.(3)	eq.(4)	0.38

Table 8. Global accuracy for the UPDRS III prediction including RS

tolerance relation	t-norm	global accuracy
eq.(1)	eq.(5)	0.58
eq.(1)	eq.(4)	0.54
eq.(2)	eq.(4)	0.63
eq.(3)	eq.(4)	0.54

5   Conclusions

We have presented a comparison of several tolerance and t-norm equations in prediction results of different treatments in Parkinson patients using fuzzy rough set theory (FRST). We have performed similar calculations for the symptoms predictions also using FRST. Our results demonstrated that attribute related to the eye movements are important and gave better predictions than only classical neurological measurements. This work is continuation of our previous papers where rough set theory (RST) was used. The global coverage results were better when FRST was used, however the global accuracy was higher with RST, but number of measurement is relatively small. A big advantage of the eye movement (EM) measurements is that they might be perform without doctor help, objectively with high precision and in the near future at patient's home. With help of the data mining methods such as RST or FRST these data can be automatically evaluated in order to give instant, objective advice to individual patient – it is the future method related to the tele-medicine. However, in order to be able to use the analyzed methods in practical applications, we need to perform measurements and confirm our results on larger group of patients that is actually in work in-progress.

Acknowledgements. This work was partly supported by projects Dec-2011/03/B/ST6/03816 from the Polish National Science Centre.

References

1. Siegelmann, H.T.: Neural and super-turing computing. Mind. Mach. **13**, 103–114 (2003)
2. Przybyszewski, A.W., Linsay, P.S., Gaudiano, P., Wilson, C.M.: Basic difference between brain and computer: integration of asynchronous processes implemented as hardware model of the retina. IEEE Trans Neural Netw. **18**, 70–85 (2007)
3. Przybyszewski, A.W., Szlufik, S., Dutkiewicz, J., Habela, P., Koziorowski, D.M.: Machine learning on the video basis of slow pursuit eye movements can predict symptom development in Parkinson's patients. In: Nguyen, N.T., Trawiński, B., Kosala, R. (eds.) ACIIDS 2015. LNCS, vol. 9012, pp. 268–276. Springer, Heidelberg (2015)
4. Przybyszewski, A.W., Kon, M., Szlufik, S., Dutkiewicz, J., Habela, P., Koziorowski, D.M.: Data mining and machine learning on the basis from reflexive eye movements can predict symptom development in individual Parkinson's patients. In: Gelbukh, A., Espinoza, F.C., Galicia-Haro, S.N. (eds.) MICAI 2014, Part II. LNCS, vol. 8857, pp. 499–509. Springer, Heidelberg (2014)
5. Hikosaka, O., Takikawa, Y., Kawagoe, R.: Role of the basal ganglia in the control of purposive saccadic eye movements. Physiol. Rev. **80**, 953–978 (2000)
6. Jones, G.M., DeJong, J.D.: Dynamic characteristics of saccadic eye movements in Parkinson's disease. Exp. Neurol. **31**, 17–31 (1971)
7. Ladda, J., Valkovič, P., Eggert, T., Straube, A.: Parkinsonian patients show impaired predictive smooth pursuit. J. Neurol. **255**, 1071–1078 (2008)
8. Pawlak, Z.: Rough Sets: Theoretical Aspects of Reasoning about Data. Kluwer, Dordrecht (1991)
9. Radzikowska, A.M., Kerre, E.E.: A comparative study of fuzzy rough sets. Fuzzy Sets Syst. **126**, 137–156 (2002)
10. Riza, L.S., Janusz, A., Bergmeir, C., Cornelis, C., Herrera, F., Ślęzak, D., Benítez, J.M.: Implementing algorithms of rough set theory and fuzzy rough set theory in the R package, "RoughSets". Inf. Sci. **287**, 68–69 (2014)
11. Hu, Q., Yu, D., Pedrycz, W., Chen, D.: Kernalized fuzzy rough sets and their applications. IEEE Trans. Knowl. Data Eng. **23**, 1471–1649 (2011)
12. Cornelis, C., De Cock, M., Radzikowska, A.: Fuzzy rough sets: from theory into practice. In: Pedrycz, W., Skowron, A., Kreinovich, V. (eds.) Handbook of Gradual Computing, pp. 533–552. Wiley, Chichester (2008)
13. Jensen, R., Cornelis, C., Shen, Q.: Hybrid fuzzy-rough rule induction and feature selection. In: IEEE International Conference on Fuzzy Systems (FUZZ-IEEE), pp. 1151–1156 (2009)
14. Jensen, R., Shen, Q.: New approaches for fuzzy-rough feature selection. In: Proceedings of the 19th International Conference on Fuzzy Systems, vol. 149, issue no. 1, pp. 5–20 (2005)
15. Bazan, J., Nguyen, H.S., Nguyen, T.T., Skowron, A., Stepaniuk, J.: Decision rules synthesis for object classification. In: Orłowska, E. (ed.) Incomplete Information: Rough Set Analysis, pp. 23–57. Physica, Verlag (1998)

Determining OWA Operator Weights
by Maximum Deviation Minimization

Wlodzimierz Ogryczak$^{(\boxtimes)}$ and Jaroslaw Hurkala

Institute of Control and Computation Engineering,
Warsaw University of Technology, Warsaw, Poland
{w.ogryczak,j.hurkala}@elka.pw.edu.pl

Abstract. The ordered weighted averaging (OWA) operator uses the weights assigned to the ordered values of the attributes. This allows one to model various aggregation preferences characterized by the so-called orness measure. The determination of the OWA operator weights is a crucial issue of applying the operator for decision making. In this paper, for a given orness value, monotonic weights of the OWA operator are determined by minimization of the maximum absolute deviation inequality measure. This leads to a linear programming model which can also be solved analytically.

1 Introduction

The problem of aggregating numerical attributes to form an overall measure is of considerable importance in many disciplines. The most commonly used aggregation is based on the weighted sum. The preference weights can be effectively introduced with the so-called Ordered Weighted Averaging (OWA) aggregation developed by Yager [18]. In the OWA aggregation the weights are assigned to the ordered values (i.e. to the smallest value, the second smallest and so on) rather than to the specific criteria. Since its introduction, the OWA aggregation has been successfully applied to many fields of decision making including also ones modeling risk averse preferences in decisions under uncertainty [9] as well as those requiring equity and fairness while aggregating several agents gains [10,11]. The OWA operator allows us to model various aggregation functions from the maximum through the arithmetic mean to the minimum. Thus, it enables modeling of various preferences from the optimistic to the pessimistic one.

Several approaches have been introduced for obtaining the OWA weights with a predefined degree of orness [2,17]. O'Hagan [7] proposed a maximum entropy approach, which involved a constrained nonlinear optimization problem with a predefined degree of orness as its constraint and the entropy as the objective function. Actually, the maximum entropy model can be transformed into a polynomial equation and then solved analytically [3]. A minimum variance approach to obtain the minimal variability OWA operator weights was also considered [4]. The minimax disparity approach proposed by Wang and Parkan [15] was the first method of finding OWA operator using Linear Programming (LP). This method determines the OWA operator weights by minimizing the

© Springer International Publishing Switzerland 2015
M. Kryszkiewicz et al. (Eds.): PReMI 2015, LNCS 9124, pp. 335–344, 2015.
DOI: 10.1007/978-3-319-19941-2_32

maximum difference between two adjacent weights under a given level of orness. The minimax disparity approach was further extended [1,14] and related to the minimum variance approaches [6]. The maximum entropy approach has been generalized for various Minkowski metrics [20,21] in some cases expressed with LP models [16]. The LP model of the mean absolute deviation has been also considered [8]. In this paper we analyze a possibility to use another LP solvable models. In particular, we develop the LP model to determine the OWA operator weights by minimizing the Maximum Absolute Deviation inequality measure. In addition to the LP model an analytical formula is also derived.

2   Orness and Inequality Measures

The OWA aggregation with weights $\mathbf{w} = (w_1, \ldots, w_m)$ of vector $\mathbf{y} = (y_1, \ldots, y_m)$ is mathematically formalized as follows [18]. First, we introduce the ordering map $\Theta : R^m \to R^m$ such that $\Theta(\mathbf{y}) = (\theta_1(\mathbf{y}), \theta_2(\mathbf{y}), \ldots, \theta_m(\mathbf{y}))$, where $\theta_1(\mathbf{y}) \geq \theta_2(\mathbf{y}) \geq \cdots \geq \theta_m(\mathbf{y})$ and there exists a permutation τ of set I such that $\theta_i(\mathbf{y}) = y_{\tau(i)}$ for $i = 1, \ldots, m$. Next, we apply the weighted sum aggregation to ordered vectors $\Theta(\mathbf{y})$, i.e. the OWA aggregation takes the following form:

$$A_\mathbf{w}(\mathbf{y}) = \sum_{i=1}^m w_i \theta_i(\mathbf{y}). \tag{1}$$

The OWA aggregation may model various preferences from the optimistic (max) to the pessimistic (min) Yager [18] introduced a well appealing concept of the orness measure to characterize the OWA operators. The degree of orness associated with the OWA operator $A_\mathbf{w}(\mathbf{y})$ is defined as

$$\text{orness}(\mathbf{w}) = \sum_{i=1}^m \frac{m-i}{m-1} w_i \tag{2}$$

For the max aggregation representing the fuzzy 'or' operator with weights $\mathbf{w} = (1, 0, \ldots, 0)$ one gets $\text{orness}(\mathbf{w}) = 1$ while for the min aggregation representing the fuzzy 'and' operator with weights $\mathbf{w} = (0, \ldots, 0, 1)$ one has $\text{orness}(\mathbf{w}) = 0$. For the average (arithmetic mean) one gets $\text{orness}((1/m, 1/m, \ldots, 1/m)) = 1/2$. A complementary measure of andness defined as $\text{andness}(\mathbf{w}) = 1 - \text{orness}(\mathbf{w})$ may be considered. OWA aggregations with orness greater or equal 0.5 are considered or-like whereas the aggregations with orness smaller or equal 0.5 are treated as and-like. The former corresponds to rather optimistic preferences while the latter represents rather pessimistic (risk-averse) preferences.

The OWA aggregations with monotonic weights are either or-like or and-like. Exactly, decreasing weights $w_1 \geq w_2 \geq \ldots \geq w_m$ define an or-like OWA operator, while increasing weights $w_1 \leq w_2 \leq \ldots \leq w_m$ define an and-like OWA operator. Actually, the orness and the andness properties of the OWA operators with monotonic weights are total in the sense that they remain valid for any subaggregations defined by subsequences of their weights. Such OWA aggregations allow one to model equitable or fair preferences [10,11], as well as risk aversion in decisions under uncertainty [13].

Yager [19] proposed to define the OWA weighting vectors via the regular increasing monotone (RIM) quantifiers, which provide a dimension independent description of the aggregation. A fuzzy subset Q of the real line is called a RIM quantifier if Q is (weakly) increasing with $Q(0) = 0$ and $Q(1) = 1$. The OWA weights can be defined with a RIM quantifier Q as $w_i = Q(i/m) - Q((i-1)/m)$. and the orness measure can be extended to a RIM quantifier (according to $m \to \infty$) as follows [19]

$$\text{orness}(Q) = \int_0^1 Q(\alpha) \, d\alpha \tag{3}$$

Thus, the orness of a RIM quantifier is equal to the area under it.

Monotonic weights can be uniquely defined by their distribution. First, we introduce the right-continuous cumulative distribution function (cdf):

$$F_{\mathbf{w}}(d) = \sum_{i=1}^m \frac{1}{m} \delta_i(d) \quad \text{where} \quad \delta_i(d) = \begin{cases} 1 & \text{if } w_i \le d \\ 0 & \text{otherwise} \end{cases} \tag{4}$$

which for any real value d provides the measure of weights smaller or equal to d. Alternatively one may use the left-continuous right tail cumulative distribution function $\overline{F}_{\mathbf{w}}(d) = 1 - F_{\mathbf{w}}(d)$ which for any real value d provides the measure of weights greater or equal to d.

Next, we introduce the quantile function $F_{\mathbf{w}}^{(-1)} = \inf \{\eta : F_{\mathbf{y}}(\eta) \ge \xi\}$ for $0 < \xi \le 1$ as the left-continuous inverse of the cumulative distribution function $F_{\mathbf{w}}$, ie., $F_{\mathbf{w}}^{(-1)}(\xi) = \inf \{\eta : F_{\mathbf{w}}(\eta) \ge \xi\}$ for $0 < \xi \le 1$. Similarly, we introduce the right tail quantile function $\overline{F}_{\mathbf{w}}^{(-1)}$ as the right-continuous inverse of the cumulative distribution function $\overline{F}_{\mathbf{w}}$, i.e., $\overline{F}_{\mathbf{w}}^{(-1)}(\xi) = \sup \{\eta : \overline{F}_{\mathbf{w}}(\eta) \ge \xi\}$ for $0 < \xi \le 1$. Actually, $\overline{F}_{\mathbf{w}}^{(-1)}(\xi) = F_{\mathbf{w}}^{(-1)}(1 - \xi)$. It is the stepwise function $\overline{F}_{\mathbf{w}}^{(-1)}(\xi) = \theta_i(\mathbf{w})$ for $\frac{i-1}{m} < \xi \le \frac{i}{m}$.

Dispersion of the weights distribution can be described with the Lorenz curves and related inequality measures. Classical Lorenz curve used in income economics as a cumulative population versus income curve to compare equity of income distributions. Although, the Lorenz curve for any distribution may be viewed [5] as a normalized integrated quantile function. In particular, for distribution of weights $\mathbf{w}$ one gets

$$L_{\mathbf{w}}(\xi) = \frac{1}{\mu(\mathbf{w})} \int_0^\xi F_{\mathbf{w}}^{(-1)}(\alpha) d\alpha = m \int_0^\xi F_{\mathbf{w}}^{(-1)}(\alpha) d\alpha \tag{5}$$

where while dealing with normalized weights w_i we have always $\mu(\mathbf{w}) = 1/m$. Graphs of functions $L_{\mathbf{w}}(\xi)$ are piecewise linear convex curves. They are nondecreasing, due to nonnegative weights w_i. A perfectly equal distribution weights ($w_i = 1/m$ for all $i = 1, \ldots, m$) has the diagonal line as the Lorenz curve.

Alternatively, the upper Lorenz curve may be used which integrates the right tail quantile function. For distribution of weights $\mathbf{w}$ one gets

$$\overline{L}_{\mathbf{w}}(\xi) = \frac{1}{\mu(\mathbf{w})} \int_0^\xi \overline{F}_{\mathbf{w}}^{(-1)}(\alpha)d\alpha = m \int_0^\xi \overline{F}_{\mathbf{w}}^{(-1)}(\alpha)d\alpha \qquad (6)$$

Graphs of functions $\overline{L}_{\mathbf{w}}(\xi)$ are piecewise linear concave curves. They are nondecreasing, due to nonnegative weights w_i. Similar to $L_{\mathbf{w}}$, the vector of perfectly equal weights has the diagonal line as the upper Lorenz curve. Actually, both the classical (lower) and the upper Lorenz curves are symmetric with respect to the diagonal line in the sense that the differences

$$\bar{d}_{\mathbf{w}}(\xi) = \overline{L}_{\mathbf{w}}(\xi) - \xi \quad \text{and} \quad d_{\mathbf{w}}(\xi) = \xi - L_{\mathbf{w}}(\xi) \qquad (7)$$

are equal for symmetric arguments: $\bar{d}_{\mathbf{w}}(\xi) = d_{\mathbf{w}}(1 - \xi)$. Hence,

$$\overline{L}_{\mathbf{w}}(\xi) + L_{\mathbf{w}}(1 - \xi) = 1 \quad \text{for any } 0 \leq \xi \leq 1 \qquad (8)$$

Note that in the case of nondecreasing OWA weights $0 \leq w_1 \leq \ldots \leq w_m \leq 1$ the corresponding Lorenz curve $L_{\mathbf{w}}(\xi)$ is (weakly) increasing with $L_{\mathbf{w}}(0) = 0$ and $L_{\mathbf{w}}(1) = 1$ as well as the OWA weights can be defined with L as $w_i = L_{\mathbf{w}}(i/m) - L_{\mathbf{w}}((i-1)/m)$. Hence, $L_{\mathbf{w}}$ may be considered then as a RIM quantifier generating weights $\mathbf{w}$ [13]. Following Eq. (3), the orness measure of RIM quantifier is given as $\text{orness}(L) = \int_0^1 L(\alpha)\, d\alpha$, thus equal to the area under $L_{\mathbf{w}}$. Certainly, for any finite m the RIM orness $\text{orness}(L_{\mathbf{w}})$ differs form $\text{orness}(\mathbf{w})$, but the difference depends only on the value of m, Exactly,

$$\text{orness}(L_{\mathbf{w}}) = \sum_{i=1}^m \frac{m-i}{m} w_i + \sum_{i=1}^m \frac{1}{2m} w_i = \frac{m-1}{m}\text{orness}(\mathbf{w}) + \frac{1}{2m} \qquad (9)$$

In the case of nonincreasing OWA weights $1 \geq w_1 \geq \ldots \geq w_m \geq 0$ the corresponding upper Lorenz curve $\overline{L}_{\mathbf{w}}(\xi)$ is (weakly) increasing with $\overline{L}_{\mathbf{w}}(0) = 0$ and $\overline{L}_{\mathbf{w}}(1) = 1$ as well as the OWA weights can be defined with $\overline{L}$ as $w_i = \overline{L}_{\mathbf{w}}(i/m) - \overline{L}_{\mathbf{w}}((i-1)/m)$. Hence, $\overline{L}_{\mathbf{w}}$ may be considered then as a RIM quantifier generating weights $\mathbf{w}$. Similar to (9) the difference between the RIM orness $\text{orness}(L_{\mathbf{w}})$ and $\text{orness}(\mathbf{w})$ depends only on the value of m.

Typical inequality measures are some deviation type dispersion characteristics. They are *inequality relevant* which means that they are equal to 0 in the case of perfectly equal outcomes while taking positive values for unequal ones.

The simplest inequality measures are based on the absolute measurement of the spread of outcomes, like the *(Gini's) mean absolute difference*

$$\Gamma(\mathbf{w}) = \frac{1}{2m^2} \sum_{i=1}^m \sum_{j=1}^m |w_i - w_j| \qquad (10)$$

or the *maximum absolute difference*

$$D(\mathbf{w}) = \max_{i,j=1,\ldots,m} |w_i - w_j|. \qquad (11)$$

In most application frameworks better intuitive appeal may have inequality measures related to deviations from the mean outcome like the *maximum absolute deviation*

$$\Delta(\mathbf{w}) = \max_{i \in I} |w_i - \mu(\mathbf{w})|. \tag{12}$$

Note that the *standard deviation* σ (or the *variance* σ^2) represents both the deviations and the spread measurement.

Deviational measures may be focused on the downside semideviations or the upper ones. One may define the *maximum downside semideviation* $\Delta^d(\mathbf{w})$ and the *maximum upside semideviation* $\Delta^u(\mathbf{w})$, respectively

$$\Delta^d(\mathbf{w}) = \max_{i \in I}(\mu(\mathbf{w}) - w_i) \quad \text{and} \quad \Delta^u(\mathbf{w}) = \max_{i \in I}(w_i - \mu(\mathbf{w})). \tag{13}$$

In economics one usually considers relative inequality measures normalized by the mean. Among many inequality measures perhaps the most commonly accepted by economists is the *Gini index*, which is the relative mean difference

$$G(\mathbf{w}) = \Gamma(\mathbf{w})/\mu(\mathbf{w}) = m\Gamma(\mathbf{w}). \tag{14}$$

Similarly, one may consider the relative maximum deviation

$$R(\mathbf{w}) = \Delta(\mathbf{w})/\mu(\mathbf{w}) = m\Delta(\mathbf{w}). \tag{15}$$

Note that due to $\mu(\mathbf{w}) = 1/m$, the relative inequality measures are proportional to their absolute counterparts and any comparison of the relative measures is equivalent to comparison of the corresponding absolute measures.

The above inequality measures are closely related to the Lorenz curve [10] and its differences from the diagonal (equity) line (7). First of all

$$G(\mathbf{w}) = 2 \int_0^1 \bar{d}_\mathbf{w}(\alpha)d\alpha = 2 \int_0^1 d_\mathbf{w}(\alpha)d\alpha \tag{16}$$

thus

$$G(\mathbf{w}) = 2 \int_0^1 \overline{L}_\mathbf{w}(\alpha)d\alpha - 1 = 1 - 2 \int_0^1 L_\mathbf{w}(\alpha)d\alpha. \tag{17}$$

Recall that in the case of nondecreasing OWA weights $0 \leq w_1 \leq \ldots \leq w_m \leq 1$ the corresponding Lorenz curve $L_\mathbf{w}(\xi)$ may be considered as a RIM quantifier generating weights $\mathbf{w}$. Following Eq. (9), one gets

$$G(\mathbf{w}) = 1 - 2\text{orness}(L_\mathbf{w}) = \frac{m-1}{m}(1 - 2\text{orness}(\mathbf{w})) \tag{18}$$

enabling easy recalculation of the orness measure into the Gini index and vice versa. Similarly, in the case of nonincreasing OWA weights $1 \geq w_1 \geq \ldots \geq w_m \geq 0$, one gets

$$G(\mathbf{w}) = 2\text{orness}(\overline{L}_\mathbf{w}) - 1 = \frac{m-1}{m}(2\text{orness}(\mathbf{w}) - 1). \tag{19}$$

3 Maximum Deviation Minimization

We focus on the case of monotonic weights. Following Eqs. (18) and (19), the Gini index is then uniquely defined by a given orness value. Nevertheless, one may still select various weights by minimizing the Maximum Deviation (MD) measure. Although related to the Lorenz curve it is not uniquely defined by the Gini index and the orness measure. Actually, the MD minimization approach may be viewed as the generalized entropy maximization based on the infinity Minkowski metric [16].

Let us define differences

$$\bar{d}_i(\mathbf{w}) = \bar{L}_{\mathbf{w}}(\frac{i}{m}) - \frac{i}{m} \quad \text{and} \quad d_i(\mathbf{w}) = \frac{i}{m} - L_{\mathbf{w}}(\frac{i}{m}) \qquad \text{for } i = 1, \ldots, m \quad (20)$$

where due to nonnegativity of weights, for all $i = 1, \ldots, m-1$

$$\bar{d}_i(\mathbf{w}) \leq \frac{1}{m} + \bar{d}_{i+1}(\mathbf{w}) \quad \text{and} \quad d_i(\mathbf{w}) \leq \frac{1}{m} + d_{i-1}(\mathbf{w}) \qquad (21)$$

with $d_0(\mathbf{w}) = \bar{d}_0(\mathbf{w}) = 0$ and $d_m(\mathbf{w}) = \bar{d}_m(\mathbf{w}) = 0$. Thus

$$\bar{d}_{m-i}(\mathbf{w}) \leq \frac{i}{m} \quad \text{and} \quad d_i(\mathbf{w}) \leq \frac{i}{m} \qquad \text{for } i = 1, \ldots, m-1 \quad (22)$$

The Gini index represents the area defined by $\bar{d}_i(\mathbf{w})$ or $d_i(\mathbf{w})$, respectively,

$$G(\mathbf{w}) = \frac{2}{m} \sum_{i=1}^{m-1} \bar{d}_i(\mathbf{w}) = \frac{2}{m} \sum_{i=1}^{m-1} d_i(\mathbf{w}) \qquad (23)$$

while the relative maximum deviation may be represented as [10]

$$R(\mathbf{w}) = m\Delta(\mathbf{w}) = \max\{m\Delta^d(\mathbf{w}), m\Delta^u(\mathbf{w})\} = \max\{md_1(\mathbf{w}), m\bar{d}_1(\mathbf{w})\}$$
$$= \max\{md_1(\mathbf{w}), md_{m-1}(\mathbf{w})\} = \max\{m\bar{d}_1(\mathbf{w}), m\bar{d}_{m-1}(\mathbf{w})\} \qquad (24)$$

Note that due to (22) $m\Delta^d(\mathbf{w}) \leq 1$.

Assume there is given orness value $0.5 \leq \alpha \leq 1$ and we are looking for monotonic weights $1 \geq w_1 \geq \ldots \geq w_m \geq 0$ such that orness$(\mathbf{w}) = \alpha$ and the (relative) maximum deviation $R(\mathbf{w})$ is minimal. Following Eqs. (19), (23) and (24), it leads us to the problem

$$\min \max\{m\bar{d}_1(\mathbf{w}), m\bar{d}_{m-1}(\mathbf{w})\}$$
$$\text{s.t.} \frac{2}{m} \sum_{i=1}^{m-1} \bar{d}_i(\mathbf{w}) = \frac{m-1}{m}(2\alpha - 1) \qquad (25)$$

with additional (22) constraints. This allows us to form the following LP model

$$\min \quad md \qquad (26)$$
$$\text{s.t.} \quad \bar{d}_1 \leq d, \quad \bar{d}_{m-1} \leq d \qquad (27)$$
$$\bar{d}_1 + \ldots + \bar{d}_{m-1} = (m-1)(\alpha - 0.5) \qquad (28)$$
$$0 \leq \bar{d}_i \leq \frac{1}{m} + \bar{d}_{i+1} \qquad \text{for } i = 1, \ldots, m-1 \qquad (29)$$

with variables $\bar{d}_i$ for $i = 1, \ldots, m - 1$, auxiliary variable d and constant $\bar{d}_m = 0$. Having solved the above LP problem, the corresponding weights can be simply calculated according to the following formula (with $\bar{d}_0 = \bar{d}_m = 0$):

$$w_i = \bar{d}_i - \bar{d}_{i-1} + \frac{1}{m} \qquad \text{for } i = 1, \ldots, m \tag{30}$$

Symmetrically, assume there is given orness value $0 \le \alpha \le 0.5$ and we are looking for monotonic weights $0 \le w_1 \le \ldots \le w_m \le 1$ such that orness$(\mathbf{w}) = \alpha$ and the (relative) maximum deviation $R(\mathbf{w})$ is minimal. Following Eqs. (18), (23) and (24), it leads us to the problem

$$\min \max\{md_1(\mathbf{w}), md_{m-1}(\mathbf{w})\}$$
$$\text{s.t. } \frac{2}{m} \sum_{i=1}^{m-1} d_i(\mathbf{w}) = \frac{m-1}{m}(1 - 2\alpha) \tag{31}$$

with additional (22) constraints. Thus leading to the LP problem

$$\min md$$
$$\text{s.t. } d_1 \le d, \quad d_{m-1} \le d$$
$$d_1 + \ldots + d_{m-1} = (m - 1)(0.5 - \alpha) \tag{32}$$
$$0 \le d_i \le \tfrac{1}{m} + d_{i-1} \qquad \qquad \text{for } i = 1, \ldots, m - 1$$

with variables d_i for $i = 1, \ldots, m - 1$, auxiliary variable d and constant $d_0 = 0$. The corresponding weights can be found according to the formula

$$w_i = d_{i-1} - d_i + \frac{1}{m} \qquad \text{for } i = 1, \ldots, m \tag{33}$$

where $d_0 = d_m = 0$.

LP models (26)–(29) and (32) allow for application standard optimization techniques to solve them. However, their structure is so simple that the problem of maximum deviation minimization can also be solved analytically. We will show this in details for the case of $0.5 \le \text{orness}(\mathbf{w}) \le 1$ and the corresponding model (26)–(29) (Fig. 1).

One may take advantage of the fact that an optimal solution to the minimax problem $\min\{\max\{y_1, y_2\} : \mathbf{y} \in Q\}$ are perfectly equal values $y_1 = y_2$ or one of them, say y_2, reaches its upper bound $U_2 = \max\{y_2 : \mathbf{y} \in Q\}$ while the other takes the larger value $y_1 > U_2$. Hence, when the required orness level is small enough (still not below 0.5), then the optimal solution is defined by

$$\bar{d}_1 = \bar{d}_{m-1} = \bar{h}$$

where $\bar{h}$ is defined by the orness Eq. (28) while leaving inequalities (29) inactive. The optimal solution is then defined by

$$\frac{m}{i}\bar{d}_i = \frac{m}{i}\bar{d}_{m-i} = \bar{h} \quad \text{for } 1 \le i \le \frac{m}{2}$$

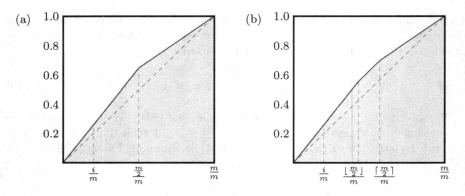

Fig. 1. Areas under Lorenz curve for minimal maximum deviation: even (a) vs. odd (b) number of weights

In the case of odd $m = 2n + 1$ one has

$$\bar{d}_i = \frac{i}{m}\bar{h} \quad \text{and} \quad \bar{d}_{m-i} = \frac{i}{m}\bar{h} \quad \text{for } 1 \leq i \leq n$$

thus leading to the equation

$$\bar{d}_1 + \ldots + \bar{d}_{m-1} = 2 \sum_{i=1}^{n} n\frac{i}{m}\bar{h} = \frac{n(n+1)}{m}\bar{h} = (m-1)(\alpha - 0.5)$$

and $\bar{h} = \frac{4m(\alpha - 0.5)}{m+1}$. Note that following Eq. (30) such a solution is generated by weights:

$$w_i = \frac{1}{m} + \frac{4(\alpha - 0.5)}{m + 1} \quad \text{for } i = 1, \ldots, n$$

$$w_{n+1} = \frac{1}{m}$$

$$w_i = \frac{1}{m} - \frac{4(\alpha - 0.5)}{m + 1} \quad \text{for } i = n + 2, \ldots, m$$

In the case of even $m = 2n$ one has

$$\bar{d}_i = \frac{i}{m}\bar{h} \quad \text{and} \quad \bar{d}_{m-i} = \frac{i}{m}\bar{h} \quad \text{for } 1 \leq i \leq n$$

although $\bar{d}_n$ and $\bar{d}_{m-n}$ is the same variable. This leads to the equation

$$\bar{d}_1 + \ldots + \bar{d}_{m-1} = \sum_{i=1}^{n} \frac{i}{m}\bar{h} + \sum_{i=1}^{n-1} \frac{i}{m}\bar{h} = \frac{n^2}{m}\bar{h} = (m-1)(\alpha - 0.5)$$

and $\bar{h} = \frac{4(m-1)(\alpha-0.5)}{m}$. Note that following Eq. (30) such a solution is generated by weights:

$$w_i = \frac{1}{m} + \frac{4(m-1)(\alpha-0.5)}{m^2} \quad \text{for } i = 1, \ldots, n$$

$$w_i = \frac{1}{m} - \frac{4(m-1)(\alpha-0.5)}{m^2} \quad \text{for } i = n+1, \ldots, m$$

The above analytical formulae for weights are valid as long as the required orness level α is small enough (still not below 0.5) allowing constraint (22) to remain inactive. This is equivalent to the restriction $\bar{h} \leq 1$ thus

$$\alpha \leq \frac{m+1}{4m} + 0.5 \quad \text{and} \quad \alpha \leq \frac{m}{4(m-1)} + 0.5$$

for odd and even m, respectively.

When the required orness level is higher, then constraint (22) becomes active, thus enforcing zero weights within the second part of the sequence. Exactly, there exists $1 \leq \kappa \leq m/2$ such that

$$w_1 = \ldots = w_{\kappa-1} \geq w_\kappa \geq w_{\kappa+1} = \ldots = w_m = 0$$

where $\frac{m}{i}\bar{d}_{m-i}(\mathbf{w}) = 1$ for $i < m - \kappa$.

4   Conclusion

The determination of ordered weighted averaging (OWA) operator weights is a crucial issue of applying the OWA operator for decision making. We have considered determining monotonic weights of the OWA operator by minimization of the maximum (absolute) deviation inequality measure. This leads us to a linear programming model which can also be solved analytically. The analytic approach results in simple direct formulas. The LP models allow us to find weights by the use of efficient LP optimization techniques and they enable easy enhancement of the preference model with additional requirements on the weights properties. The latter is the main advantage over the standard method of entropy minimization. Both the standard method and the proposed one do have their analytical solutions. However, if we try to elaborate them further by adding some auxiliary (linear) constraints on the OWA weights, then the entropy minimization model forms a difficult nonlinear optimization task while the maximum deviation minimization is still easily LP-solvable.

Acknowledgements. The research was partially supported by the National Science Centre (Poland) under the grant DEC-2012/07/B/HS4/03076.

References

1. Amin, G.R., Emrouznejad, A.: An extended minimax disparity to determine the OWA operator weights. Comput. Ind. Eng. **50**, 312–316 (2006)

2. Fuller, R.: On obtaining OWA operator weights: a short survey of recent developments. In: Proceedings 5th IEEE International Conference on Computer Cybernetics. Gammarth, Tunisia, pp. 241–244(2007)
3. Fuller, R., Majlender, P.: An analytic approach for obtaining maximal entropy OWA operator weights. Fuzzy Sets Syst. **124**, 53–57 (2001)
4. Fuller, R., Majlender, P.: On obtaining minimal variability OWA operator weights. Fuzzy Sets Syst. **136**, 203–215 (2003)
5. Gastwirth, J.L.: A general definition of the Lorenz curve. Econometrica **39**, 1037–1039 (1971)
6. Liu, X.: The solution equivalence of minimax disparity and minimum variance problems for OWA operators. Int. J. Approx. Reasoning **45**, 68–81 (2007)
7. O'Hagan, M.: Aggregating template or rule antecedents in real-time expert systems with fuzzy set logic. In: Proceedings 22nd Annual IEEE Asilomar Conference Signals, Systems, Computer. Pacific Grove, CA, pp. 681–689 (1988)
8. Majdan, M., Ogryczak, W.: Determining OWA operator weights by mean absolute deviation minimization. In: Rutkowski, L., Korytkowski, M., Scherer, R., Tadeusiewicz, R., Zadeh, L.A., Zurada, J.M. (eds.) ICAISC 2012, Part I. LNCS, vol. 7267, pp. 283–291. Springer, Heidelberg (2012)
9. Ogryczak, W.: Multiple criteria linear programming model for portfolio selection. Ann. Oprs. Res. **97**, 143–162 (2000)
10. Ogryczak, W.: Inequality measures and equitable locations. Ann. Oprs. Res. **167**, 61–86 (2009)
11. Ogryczak, W., Luss, H., Pióro, M., Nace, D., Tomaszewski, A.: Fair optimization and networks: a survey. J. Appl. Math. **2014**, 1–25 (2014). Art. ID 612018
12. Ogryczak, W., Śliwiński, T.: On solving linear programs with the ordered weighted averaging objective. Eur. J. Opnl. Res. **148**, 80–91 (2003)
13. Ogryczak, W., Śliwiński, T.: On efficient WOWA optimization for decision support under risk. Int. J. Approx. Reason. **50**, 915–928 (2009)
14. Wang, Y.-M., Luo, Y., Liu, X.: Two new models for determining OWA operator weights. Comput. Ind. Eng. **52**, 203–209 (2007)
15. Wang, Y.-M., Parkan, C.: A minimax disparity approach for obtaining OWA operator weights. Inf. Sci. **175**, 20–29 (2005)
16. Wu, J., Sun, B.-L., Liang, C.-Y., Yang, S.-L.: A linear programming model for determining ordered weighted averaging operator weights with maximal Yagers entropy. Comput. Ind. Eng. **57**, 742–747 (2009)
17. Xu, Z.: An overview of methods for determining OWA weights. Int. J. Intell. Syst. **20**, 843–865 (2005)
18. Yager, R.R.: On ordered weighted averaging aggregation operators in multicriteria decision making. IEEE Trans. Syst., Man Cyber. **18**, 183–190 (1988)
19. Yager, R.R.: Quantifier guided aggregation using OWA operators. Int. J. Intell. Syst. **11**, 49–73 (1996)
20. Yager, R.R.: Measures of entropy and fuzziness related to aggregation operators. Inf. Sci. **82**, 147–166 (1995)
21. Yager, R.R.: On the dispersion measure of OWA operators. Inf. Sci. **179**, 3908–3919 (2009)

Fuzzy Set Interpretation of Comparator Networks

Łukasz Sosnowski[1,2] and Dominik Ślęzak[3,4]([✉])

[1] Systems Research Institute, Polish Academy of Sciences,
Ul. Newelska 6, 01-447 Warsaw, Poland
[2] Dituel Sp. z o.o., Ul. Ostrobramska 101 Lok. 206, 04-041 Warsaw, Poland
l.sosnowski@dituel.pl
[3] Institute of Mathematics, University of Warsaw,
Ul. Banacha 2, 02-097 Warsaw, Poland
[4] Infobright Inc., Ul. Krzywickiego 34 Lok. 219, 02-078 Warsaw, Poland
slezak@mimuw.edu.pl

Abstract. We discuss how to model similarities between compound objects by utilizing networks of comparators. The framework is used to construct identification and classification systems. Comparing to our previous research, we pay a special attention to fuzzy-set-inspired foundations of how compound signals are processed through the network. We also reconsider some of already-known examples of applications of comparator networks, now using the proposed fuzzy-set-based terminology.

Keywords: Networks of comparators · Compound object similarities · Fuzzy sets and relations · Semantic parsing of bibliography items

1 Introduction

Similarity is one of fundamental aspects of reasoning in artificial intelligence [1]. In this paper, we show a similarity-based approach to constructing classification and identification models for compound objects. By an object we mean an element of real world, which can be stored as a data object represented using ontology specified in alignment with domain knowledge about a given problem [2]. By a compound object we mean an object, which combines a plurality of objects within ontology-definable structure. Such structure can be further used to synthesize similarities basing on analysis of object components [3].

The proposed framework is based on hierarchical comparisons of investigated objects with reference sets reflecting different levels of object structures. As a case study, we consider a task of identifying components in texts representing bibliography items [4]. The process of assigning dynamically derived text

This work was partly supported by Polish National Centre for Research and Development (NCBiR) grants O ROB/0010/03/001 in frame of Defence and Security Programmes and Projects and PBS2/B9/20/2013 in frame of Applied Research Programmes, as well as by Polish National Science Centre (NCN) grants DEC-2012/05/B/ST6/03215 and DEC-2013/09/B/ST6/01568.

M. Kryszkiewicz et al. (Eds.): PReMI 2015, LNCS 9124, pp. 345–353, 2015.
DOI: 10.1007/978-3-319-19941-2_33

fragments to particular categories relies on comparing them with a reference database of publications treated as compound objects [5]. It can be envisioned as a resemblance-based recognition method, where similarity to labeled objects enables us to extrapolate assignments onto new items.

In our previous research, we investigated a number of applications of compound object comparators going beyond typical classification tasks [6]. We also attempted to provide possibly complete description of how to construct networks of comparators in practice. However, a deeper analysis of theoretical foundations of our approach has been still missing. Thus, the main focus in this paper is on mathematical interpretation of transmitting comparisons through a network, using mainly terminology of fuzzy sets and relations [7].

The paper is organized as follows: Sect. 2 introduces basic notions corresponding to a single comparator of compound objects. Section 3 establishes foundations for similarity-related operations, which occur inside multilayered networks of comparators. Section 4 recalls and clarifies already-mentioned experiments with analysis of bibliography items. Section 5 summarizes our work and specifies some research directions for nearest future.

One can think about our approach as analogous to feedforward neural networks [8]. However, comparators work with two kinds of information: about an input object described by its possible structural characteristics and attribute values, and about its similarities to reference objects produced as outputs of previous network layers. Coexistence of these two components makes our model unique. On the other hand, it is certainly useful to compare it with other frameworks, such as those developed using already-mentioned fuzzy sets [9], those based on rough sets and rough mereology [10] and others.

2   Basics of Comparators

Comparator *com* computes a vector of similarities of an input object $u \in U$ to elements of a subset $X(u)$ of reference set *ref*. The aim of *com* is to narrow down the space of reference objects comparable to a given u. Such output can be used as final result of a comparison module embedded into a bigger application, it can be also combined with outputs of other comparators or transmitted to comparators in the next layer of a more compound network.

Comparator *com* receives a value of u on an attribute a, denoted as $a(u)$, and compares it to values $a(x)$ for reference objects $x \in X(u)$. The choice of a and other parameters inside *com* are based on domain knowledge about a given problem. A content of $X(u) \subseteq ref$ may depend on outcomes of other comparators in a network. At the start of computations, we assume $X(u) = ref$. Given input set U, we will represent *com* as function $\mu_{com} : U \times 2^{ref} \to [0,1]^{ref}$, where $[0,1]^{ref}$ denotes all fuzzy sets over discrete domain *ref*.

Fuzzy sets $\mu_{com}(u, X(u))$ are computed in several steps. First, $u \in U$ is compared to each $x \in X(u)$ separately. The result of comparison can be represented as fuzzy relation $\mu_a(u, x)$ between values (representations) of a for u and x. Quantities of $\mu_a(u, x)$ are then filtered in two stages. First, we check through predefined exception rules to exclude reference objects, which should not be compared with u based on available information. Secondly, we check whether the remaining quantities are not lower than an activation threshold $p > 0$. Overall, we can use a formula modifying initial μ_a as follows:

$$\mu_a^*(u, x) = \begin{cases} \mu_a(u, x) \text{ if } x \in X(u) \text{ and } \mu_a(u, x) \geq p \\ \qquad \text{and there are no rules which} \\ \qquad \text{disallow comparing } u \text{ with } x \\ 0 \qquad \text{otherwise} \end{cases} \tag{1}$$

The next two steps, represented as filtering function $f : [0, 1]^{ref} \rightarrow [0, 1]^{ref}$ and sharpening function $s : [0, 1]^{ref} \rightarrow [0, 1]^{ref}$, aim at further filtration of similarity coefficients by their mutual comparisons and strengthening the highest remaining weights. For this purpose, it is more convenient to think about vector $\mu_a^*(u)$ with coordinates defined as $\mu_a^*(u)[i] = \mu_a^*(u, x_i)$, $i = 1, ..., |ref|$.

The role of f is to increase the amount of zero coordinates of $\mu_a^*(u)$. For example, one can set to 0 all elements, which are not among n highest similarity scores, for $n \geq 1$. Calculation of most of filtering functions considered in our previous research can be optimized by splitting coordinates onto blocks, deriving f concurrently and merging results. It is particularly important for applications, which require operating with large cardinalities of ref.

The role of s is to introduce non-linearity, whose benefits can be compared to a usage of exponential functions in feedforward neural networks. Let us put $\mu_a^f(u) = f(\mu_a^*(u))$. The following formula for s works only with non-zero coefficients and keeps maximal values of $\mu_a^f(u)$ unchanged. These properties are important for both the speed and accuracy of computing.

$$s(\mu_a^f(u))[i] = \begin{cases} \max_u \cdot e^{\mu_a^f(u)[i] - \max_u} \text{ if } \mu_a^f(u)[i] > 0 \\ 0 \qquad \text{otherwise} \end{cases} \tag{2}$$

where $\max_u = \max_i \mu_a^f(u)[i]$. Derivation of $s(f(\mu_a^*(u)))$ can be also expressed using operations on fuzzy sets and relations, where similarities between u and reference objects correspond to fuzzy membership degrees.

Vector $s(f(\mu_a^*(u)))$ can be treated as output of single comparator. In a larger network, there can exist several interrelated comparators looking at the same ref by means of different attributes. Let us denote by $com_1, ..., com_l$ a set of such comparators, working with $a_1, ..., a_l$ respectively. We put

$$\mu_{com}(u, X(u)) = \overline{(s(f(\mu_{a_1}^*(u))), ..., s(f(\mu_{a_l}^*(u))))} \tag{3}$$

as output of composite comparator com containing $com_1, ..., com_l$ as its parts. The role of function $\overline{(...)}$ is to synthesize local outcomes in order to send further a unique signal related to ref. Such synthesis can be based on fuzzy t-norms and s-norms, statistical tools, election algorithms and so on [11].

3 Comparator Networks

Let us denote a network of comparators by net. Performance of net can be characterized analogously to a single comparator, by a function $\mu_{net} : U \rightarrow$

$[0,1]^{ref}$. Overall outcome can be utilized directly in object identification process or, e.g., as an input to a similarity-based classifier, which checks sums of weights of reference objects dropping into particular decision classes.

Structure of *net* is similar to multilayered feedforward neural networks, although transmitted signals and calculations inside nodes are different. Each layer of *net* contains a set of comparators and a specific translating/aggregating mechanism. Comparators run in parallel, usually basing on different attributes. Thus, from computational perspective, we can see that concurrency can be achieved both at the level of single comparators and their larger groups. The role of translator is to convert comparator outputs to information about reference objects that would be useful for the next layer. The role of aggregator is to choose the most likely outputs of the translator, in case there was any non-uniqueness in assigning information about input objects to comparators. We will see that those roles can be interpreted using fuzzy t-norms and fuzzy s-norms.

Each object is described using ontologies defined by concepts and relationships between them. Given a hierarchy of concepts, one can consider relationships of generalization and decomposition. Generalization is a relationship of being a sub-object of another object, while decomposition is a relationship of being a parent (super-object) of a set of sub-objects. Particular layers of *net* usually correspond to hierarchy levels, so transitions between them correspond to generalization of decomposition. This affects the way of handling both input and reference objects, as well as modeling similarities between them.

Consequently, in a single *net*, different comparators can refer to different types and levels of reference (sub-)objects, using different attributes and parameters. Thus, the first task is to extract for a given u its structural representation, i.e., all its parts and their corresponding attribute values. Moreover, it is not always obvious which parts of u should be compared to particular reference sets. In such cases, a single u can yield multiple possible combinations of assignments of its parts to particular comparators. All such alternative representations, denoted as u', should be processed through the first layers and, later, the most probable assignments of u's parts to particular categories of reference objects can be derived. One can think about collections of possible representations u' as information granules $g(u)$ created around input objects $u \in U$ [12].

Inputs to each layer are determined by values of attributes for $u \in U$ or its sub-objects. However, subsets of reference (sub-)objects, which u is going to be compared to, are induced dynamically by comparators in previous layers. In the simplest scenario, comparators in preliminary layers aim at reducing subsets of potentially comparable reference objects using relatively easily-computable attributes, leaving more complex calculations to further layers, where the number of reference items to be compared is already decreased. In other cases, initial layers work with attributes specified for sub-objects, producing vectors of similarities that need translation to the level of similarities between more compound objects, whose attributes are analyzed later. However, the complexity does not need to grow with consecutive layers. In some applications, the first layers can work with relatively basic attributes of compound objects, whose similarities are then translated to lower structural levels for detailed processing.

Types of reference objects can vary from layer to layer or even within a single layer. Comparators in a given layer usually refer to entities at the same level of ontology-based hierarchy of considered objects. However, a given hierarchy level can include multiple types of entities. Let us denote by $\mu_{net}^k(u)$ an outcome of the k-th layer for input object $u \in U$, after applying above-mentioned operations of translation and aggregation. Denote by $ref_1^{k+1}, ..., ref_{m(k+1)}^{k+1}$ reference sets used by comparators in the $(k+1)$-th layer. Our goal in this section is to specify function $\mu_{net}^k : U \rightarrow [0,1]^{ref_1^{k+1}} \times ... \times [0,1]^{ref_{m(k+1)}^{k+1}}$, which takes into account similarity vectors obtained from comparators in the k-th layer. Once we have $\mu_{net}^k(u)$, we can forward it as a signal granule and prepare subsets $X(u)_1^{k+1} \subseteq ref_1^{k+1}, ..., X(u)_{m(k+1)}^{k+1} \subseteq ref_{m(k+1)}^{k+1}$ to be utilized by next comparators. Those two types of granules – the above signal granule and previously-mentioned information granule $g(u)$ – illustrate a twofold way of operating with information about objects throughout networks of comparators.

The central part of μ_{net}^k is matrix M_{net}^k with dimensions $|ref_1^k|+...+|ref_{m(k)}^k|$ and $|ref_1^{k+1}| + ... + |ref_{m(k+1)}^{k+1}|$, which links the k-th and the $(k+1)$-th layers of *net*. In its simplest implementation, it is a sparse boolean matrix encoding these of combinations of reference (sub-)objects in sets $ref_1^k, ..., ref_{m(k)}^k$ and $ref_1^{k+1}, ..., ref_{m(k+1)}^{k+1}$, which structurally correspond to each other. Matrices are created during the process of defining reference sets, whose elements are decomposed due to their ontology-based specifications. Connections can be also additionally weighted with degrees expressing, e.g., to what extent particular sub-objects should influence similarities between their parents.

Translation can be executed as a product of M_{net}^k with concatenated vectors of similarities obtained as outputs of comparators $com_1^k, ..., com_{m(k)}^k$ in the k-th layer, for each of possible representations of u gathered in information granule $g(u)$. Let us enumerate all such representations as $u_1', ..., u_{|g(u)|}'$ and denote by $G_{net}^k(u)$ the matrix of all possible output combinations, that is:

$$G_{net}^k(u) = \begin{bmatrix} \mu_{com_1^k}(u_1')[1] & \cdots & \mu_{com_1^k}(u_{|g(u)|}')[1] \\ \vdots & \ddots & \vdots \\ \mu_{com_1^k}(u_1')\left[|ref_1^k|\right] & \cdots & \mu_{com_1^k}(u_{|g(u)|}')\left[|ref_1^k|\right] \\ \mu_{com_2^k}(u_1')[1] & \cdots & \mu_{com_2^k}(u_{|g(u)|}')[1] \\ \vdots & \ddots & \vdots \\ \mu_{com_{m(k)}^k}(u_1')\left[|ref_{m(k)}^k|\right] & \cdots & \mu_{com_{m(k)}^k}(u_{|g(u)|}')\left[|ref_{m(k)}^k|\right] \end{bmatrix} \quad (4)$$

We can represent the mechanism for computing $\mu_{net}^k(u)$ as follows:

$$\mu_{net}^k(u)[i] = \max_j \min\left((M_{net}^k G_{net}^k(u))[i][j], 1\right) \quad (5)$$

where $[i][j]$ denotes coordinates of matrix $M_{net}^k G_{net}^k(u)$. Surely, specification of required operations in terms of matrices and vectors helps in efficient implementation. On the other hand, we can see below that these calculations can be indeed interpreted by means of well-known t-norms and s-norms.

Firstly, for a given $u'_j \in g(u)$, column $(M^k_{net}G^k_{net}(u))[j]$ represents possible similarities of u to reference objects in the $(k+1)$-th layer. Each of those similarities is computed as a sum of similarities between components of u (distributed among comparators according to combination u'_j) and reference objects in the k-th layer. If it exceeds 1, then of course we cut it down. Thus, similarities between objects at the $(k+1)$-th layer are computed as Łukasiewicz's t-norm of similarities between the corresponding objects at the k-th layer.

Secondly, in order to finally assess similarity of u to a given reference object in the $(k+1)$-th layer, we look at all combinations in $g(u)$ and choose the maximum possible score. Thus, we follow Zadeh's s-norm. Intuitively, our usage of t-norm corresponds to taking a conjunction of component similarities in order to judge similarity between compound objects, while our usage of s-norm reflects a disjunction of all alternative ways of obtaining that similarity. From this perspective, our current implementation reflects one of possible specifications and other settings of t-norm and s-norm could be considered as well.

Surely, the above layout is still a kind of simplification. As noted in Sect. 2, some comparators can comprise of multiple sub-units referring to different attributes or even different types of objects. However, function-based interpretation of network performance enables to look at such composite cases as a recursive specification of how information is flowing. Moreover, it lets better understand how to adapt existing data-based learning approaches, such as error backpropagation in neural networks transmitting compound signals [13], which might be utilized, e.g., to adjust weights in translation matrices.

Actually, the topic of learning comparator networks is far wider. For example, parameters responsible for synthesis of partial outcomes of composite comparators, usage of layer outcomes to specify reference subsets for next layers, as well as aggregation of final network results can be all tuned by basing on, e.g., evolutionary algorithms [14]. Moreover, some attribute and object selection methods developed within already-mentioned framework of rough sets could be utilized to optimize configuration of comparators and reference sets [15].

4   Illustrative Example

Methods outlined in previous sections have been used in a number of academic and commercial projects. As a case study, let us discuss the task of analysis of bibliography items, described in more detail in [5]. The goal here is to determine structural patterns of references represented as unstructured texts, so their fragments get identified as members of classes such as author names, paper titles, publication dates and so on. The comparator-network-based solution aimed at this kind of text processing was designed as a component of the system responsible for indexing articles stored in scientific repositories [4].

As an example, the text "*Sosnowski, Ł.: Framework of Compound Object Comparators. Intelligent Decision Technologies (2015)*" should be recognized as aligned with structural pattern ATJY, where A, T, J and Y stand for authors, title, journal and year, respectively. Also, "*Sosnowski, Ł.*" should be identified as existing or added as new element of reference set of authors etc.

Such recognition process can be divided into preprocessing, parsing and classification. The first stage is responsible for filtering out completely useless characters (e.g.: exclamation marks). The second stage splits text onto potentially meaningful parts, basing on appropriate interpretation of punctuation and additional rules aiming at merging some of produced parts together and final cleaning. As a result, we obtain components for further usage.

For the third stage, we employ the network with input layer containing comparators corresponding to the following categories of reference objects acquired from the considered repository [4]: Authors (A), Book (B), Country (C), Doi (D), Journal (J), Pages (P), Proceedings (R), Series (S), Title (T), Volume (V), Year (Y). Different comparators work with different attributes. We assume that elements of reference sets are already correctly classified.

Comparator dedicated to authors includes sub-comparators looking at sorted initials (si), longest lengths of text fragments (ll) and full strings representing authors (au). Their similarity measures are as follows:

$$\mu_{si}(u, x) = 1 - \frac{d_L(u, x)}{max\{n(u), n(x)\}}$$

$$\mu_{ll}(u, x) = 1 - \frac{|n(u) - n(x)|}{max\{n(u), n(x)\}}$$

$$\mu_{au}(u, x) = \frac{1 + pos(u, x) - neg(u, x)}{2 + pen(u, x)} \tag{6}$$

where $n(x)$ denotes the length of x (if x and u are empty, then we put $\mu_{si}(u, x) = \mu_{ll}(u, x) = 1$), $d_L(u, x)$ denotes Levenshtein's edit distance, $pos(u, x)$ is the average similarity between tokens occurring within u and their corresponding best-matching tokens within x, $neg(u, x)$ is the ratio of tokens within u, for which we could not find any sufficiently similar tokens within x (please note that both pos and neg cannot exceed 1), and $pen(u, x)$ is the number of tokens within x, which were not chosen as best-matching counterparts for any tokens within u. Similarities used in other comparators are defined analogously, sometimes also involving comparisons of regular expression patterns.

For experiments, for training and testing, we use data sets with 132 and 268 texts, respectively. Training data set is used to fill in reference set of structural patterns. For each of 132 texts, we manually detect and classify their parts to A/B/C/D/J/P/R/S/T/V/Y categories and treat obtained sequences of codes (such as ATJY above) as structural reference objects.

Network is initiated with default activation thresholds $p = 0.5$ and uniform aggregation/translation weights. For each comparator and each text used for training, there is a dedicated unit test, which checks whether comparator's output includes correct answer. If not, then – depending on specific situation – reference set is enriched with a new object, which covers a given case, or comparator's activation threshold is set to be less rigorous.

Each of parsed test texts is processed in two stages. Firstly, our network completes part classification and produces the sets of candidate structural patterns. Then, the network conducts structure classification based on comparing those candidates with patterns in structural reference set. Final result is an ordered subset of reference structural patterns.

Table 1. Upper left/right: best/worst results obtained when using part classification. Lower left/right: best/worst results obtained using complete process (part classification + structure classification). P_* (where $*$ is p or m), R_* and F_* stand for precision, recall and F_1-score, respectively. $_p$ and $_m$ stand for measurements related to outcomes of part classification and structure classification, respectively.

Pattern	P_p	R_p	F_p	P_m	R_m	F_m	Pattern	P_p	R_p	F_p	P_m	R_m	F_m
ATR	1.00	1.00	1.00	0.75	1.00	0.86	ATC	0.33	0.67	0.44	0.00	0.00	0.00
RY	1.00	1.00	1.00	1.00	1.00	1.00	ATYATYP	0.60	0.43	0.50	1.00	0.43	0.60
ATJVY	1.00	1.00	1.00	1.00	0.80	0.89	AYTRY	0.43	0.60	0.50	1.00	0.60	0.75
AT	1.00	0.98	0.99	0.61	0.92	0.73	ATPYC	0.54	0.80	0.64	0.60	0.60	0.60
ATRYP	1.00	0.93	0.96	1.00	0.73	0.83	ATJVYP	1.00	0.50	0.65	0.50	0.25	0.33
ATVPYD	0.91	0.98	0.95	0.92	0.84	0.86	ATVY	0.60	0.75	0.67	1.00	0.75	0.86
ATJVPYD	1.00	0.90	0.94	0.98	0.71	0.81	ATVYPR	0.56	0.83	0.67	1.00	0.50	0.67
ATJVPY	1.00	0.86	0.92	1.00	0.58	0.72	ATJPY	0.94	0.58	0.70	0.89	0.60	0.71
AJVPYD	0.86	1.00	0.92	0.86	1.00	0.92	ATRPYC	0.71	0.77	0.73	1.00	0.77	0.86
ATVPY	1.00	0.85	0.91	1.00	0.60	0.75	ATAT	0.57	1.00	0.73	0.00	0.00	0.00
Pattern	P_p	R_p	F_p	P_m	R_m	F_m	Pattern	P_p	R_p	F_p	P_m	R_m	F_m
RY	1.00	1.00	1.00	1.00	1.00	1.00	AYTJP	1.00	0.70	0.82	0.00	0.00	0.00
ATRY	0.89	0.88	0.86	1.00	0.88	0.93	ATC	0.33	0.67	0.44	0.00	0.00	0.00
ATRPY	0.93	0.90	0.90	0.99	0.88	0.92	ATAT	0.57	1.00	0.73	0.00	0.00	0.00
AJVPYD	0.86	1.00	0.92	0.86	1.00	0.92	ATJYR	1.00	0.60	0.75	0.00	0.00	0.00
ATRJPY	0.83	0.83	0.83	1.00	0.83	0.91	AYT	0.94	0.87	0.87	0.13	0.13	0.13
ATRPYCD	0.84	0.85	0.84	0.97	0.85	0.91	ATJVYP	1.00	0.50	0.65	0.50	0.25	0.33
ATPYD	0.83	1.00	0.91	0.83	1.00	0.91	AYTP	0.73	0.85	0.76	0.37	0.30	0.33
ATY	0.88	0.91	0.87	0.90	0.90	0.90	ATYR	1.00	0.75	0.86	0.50	0.50	0.50
ATJVY	1.00	1.00	1.00	1.00	0.80	0.89	ATYP	0.83	0.69	0.75	0.61	0.47	0.53
ATRP	1.00	0.75	0.86	0.88	0.88	0.88	ATPYC	0.54	0.80	0.64	0.60	0.60	0.60

Table 1 includes results in terms of standard evaluation measures, such as precision, recall and F_1-score [16]. It shows the best and the worst results for part classification (the first stage only) and complete solution (both stages mentioned above). Global average values of F_1-score are equal to 0.86 and 0.78 for the first case and the second case, respectively.

The reason for lower F_1-score in the second case is that some structural patterns obtained for test texts may not be present in structural reference sets, so performing structure classification is actually a harder task. It can also happen that inputs are corrupted or wrongly created, which is a bigger problem for entire texts than for their parts. Still, obtained results make it possible to use this solution in practice, if applied together with incremental methods for cleaning, unifying and extending reference sets.

5 Conclusion

Networks of comparators are useful for solving decision problems requiring similarity modelling. They are characterized by a common modular approach to various tasks, such as classification, identification, etc., based on comparator units and their corresponding reference sets. In this paper, we showed to what extent networks of comparators can be described using fuzzy set terminology and operations. We hope that reported mathematical foundations will lead toward new areas of applications of our methodology.

References

1. Tversky, A., Shafir, E.: Preference, Belief, and Similarity - Selected Writings. Bradford Books. MIT Press, Cambridge (2004)
2. Staab, S., Maedche, A.: Knowledge portals - ontologies at work. AI Mag. **22**(2), 63–75 (2001)
3. Schickel-Zuber, V., Faltings, B.: OSS - a semantic similarity function based on hierarchical ontologies. In: Proceedings of IJCAI, pp. 551–556 (2007)
4. Ślęzak, D., Stencel, K., Nguyen, H.S.: (No)SQL platform for scalable semantic processing of fast growing document repositories. ERCIM News **2012**(90), 50–51 (2012)
5. Sosnowski, Ł.: Framework of Compound Object Comparators. Intelligent Decision Technologies (2015). doi:10.3233/IDT-140229
6. Szczuka, M.S., Sosnowski, Ł., Krasuski, A., Kreński, K.: Using domain knowledge in initial stages of KDD - optimization of compound object processing. Fundam. Inform. **129**(4), 341–364 (2014)
7. Kacprzyk, J.: Multistage Fuzzy Control- A Model-based Approach to Fuzzy Control and Decision Making. Wiley, Chichester (2012)
8. Rutkowski, L.: Computational Intelligence - Methods and Techniques. Springer, Heidelberg (2008)
9. Marín, N., Medina, J.M., Pons, O., Sánchez, D., Vila, M.A.: Complex object comparison in a fuzzy context. Inf. Softw. Technol. **45**, 431–444 (2003)
10. Pawlak, Z., Skowron, A.: Rough sets - some extensions. Inf. Sci. **177**(1), 28–40 (2007)
11. Sosnowski, Ł., Pietruszka, A., Krasuski, A., Janusz, A.: A resemblance based approach for recognition of risks at a fire ground. In: Ślęzak, D., Schaefer, G., Vuong, S.T., Kim, Y.-S. (eds.) AMT 2014. LNCS, vol. 8610, pp. 559–570. Springer, Heidelberg (2014)
12. Zadeh, L.A.: Computing with Words - Principal Concepts and Ideas. Volume 277 of Studies in Fuzziness and Soft Computing. Springer, Heidelberg (2012)
13. Szczuka, M.S., Ślęzak, D.: Feedforward neural networks for compound signals. Theoret. Comput. Sci. **412**(42), 5960–5973 (2011)
14. Stahl, A., Gabel, T.: Using evolution programs to learn local similarity measures. In: Ashley, K.D., Bridge, D.G. (eds.) ICCBR 2003. LNCS, vol. 2689. Springer, Heidelberg (2003)
15. Janusz, A., Ślęzak, D., Nguyen, H.S.: Unsupervised similarity learning from textual data. Fundam. Inf. **119**(3–4), 319–336 (2012)
16. Tjong Kim Sang, E.F.: Introduction to the CoNLL-2002 shared task: language-independent named entity recognition. In: Procedings of CoNLL, pp. 155–158 (2002)

Inverted Fuzzy Implications in Backward Reasoning

Zbigniew Suraj[1] and Agnieszka Lasek[2]([mail])

[1] Chair of Computer Science, Faculty of Mathematics and Natural Sciences,
University of Rzeszów, Prof. S. Pigonia Str. 1, 35-310 Rzeszów. Poland
zbigniew.suraj@ur.edu.pl
[2] Department of Electrical Engineering and Computer Science,
Lassonde School of Engineering, York University, 4700 Keele Street, Toronto, Canada
alasek@cse.yorku.ca

Abstract. Fuzzy inference systems generate inference results based on fuzzy IF-THEN rules. Fuzzy implications are mostly used as a way of interpretation of the IF-THEN rules with fuzzy antecedent and fuzzy consequent. From over eight decades a number of different fuzzy implications have been described, e.g. [6–10]. This leads to the following question: how to choose the proper function among basic fuzzy implications. In our paper, we propose a new method for choosing implication. Our method allows to compare two fuzzy implications. If the truth value of the consequent and the truth value of the implication are given, by means of inverse fuzzy implications we can easily optimize the truth value of the implication antecedent. In other words, we can choose the fuzzy implication, which has the highest or the lowest truth value of the implication antecedent or which has higher or lower truth value than another implication.

Keywords: Fuzzy logic · Fuzzy implications · Inverted fuzzy implications · Backward reasoning

1 Introduction

There were proposed various methods of knowledge representation and reasoning. One of the most popular approaches to knowledge representation are the fuzzy rules. However, reasoning is mainly classified into two types: forward reasoning and backward reasoning. The inference mechanism of forward reasoning is based on a data-derived way, and has a powerful prediction ability, which is capable of alarming latent hazards, forthcoming accidents, and faults. By contrast, backward reasoning is based on a goal-derived manner, has explicit objectives, which are generally to search the most possible causes related to an existing fact. Backward reasoning plays an essential role in fault diagnosis, accident analysis, and defect detection. This kind of reasoning uses fuzzy logic [4] to reason about data in the inference mechanism instead of a variety of other logics, including Boolean logic, (non-fuzzy) many-valued logics, nonmonotonic logics, etc.

© Springer International Publishing Switzerland 2015
M. Kryszkiewicz et al. (Eds.): PReMI 2015, LNCS 9124, pp. 354–364, 2015.
DOI: 10.1007/978-3-319-19941-2_34

From imprecise inputs and fuzzy rules imprecise conclusions are obtained. Approximate reasoning with fuzzy sets encompasses a wide variety of inference schemes.

Paper [5] discusses different representations of rules in a nonfuzzy setting and extends these representations to rules with a fuzzy conclusion part. It introduces the different types of fuzzy rules and put them in the framework of fuzzy sets and possibility theory.

In [2] a new fuzzy reasoning method by optimizing the similarity of truth-tables is presented. Its basic idea is to find a fuzzy set such that the truth-tables generated by the antecedent rule and the consequent rule are as similar as possible.

Fuzzy rules are often presented in the form of implications. A typology of fuzzy rules and the problem of multiple-valued implications are discussed in paper [4]. It reviews the problem of representing fuzzy knowledge, and ranges from linguistic variables to conditional if-then rules and qualified statements.

Fuzzy implications can be represented in many ways. One of them is the functional representation (e.g. [12,13,17]). The definition of fuzzy implications and their mathematical properties were studied e.g. in [1] and [16]. One of basic problems in building an inference system is choosing the relevant fuzzy implication (e.g. [2,11]). In [11] authors proposed a method allowing to choose the most suitable fuzzy implication in an inference system application. They introduced an algorithm that calculates the distance between two fuzzy implications and which is based on generalized modus ponens.

In [14], we have presented a fuzzy forward reasoning methodology for rule-based systems using the functional representation of fuzzy rules (fuzzy implications). In this paper, we extend a methodology for selecting relevant fuzzy implications from [14] in backward reasoning. The proposed methodology takes full advantage of the functional representation of fuzzy implications and the algebraic properties of the family of all fuzzy implications. It allows to compare two fuzzy implications. If the truth value of the conclusion and the truth value of the implication are given, we can easily optimize the truth value of the implication premise. In particular, in this paper we introduced an algorithm of finding the fuzzy implication which has the highest truth value of the antecedent when the truth value of the consequent and the truth value of the implication are given. This methodology can be useful for the design of inference engine based on the rule knowledge for a given rule-based system.

In the solution in this paper to divide the domain of fuzzy implications into areas, in which it will be possible to select appropriate fuzzy implication we had to use the Lambert W function, also called product logarithm. Lambert W function is a special function used when solving equations containing unknown to both the base and the exponent power. It is defined as the inverse of $f(z) = ze^z$, where z belongs to the set of complex numbers. It is marked $W(z)$. Thus, for each complex number z holds: $z = W(z)e^{W(z)}$.

The rest of this paper is organized as follows. Section 2 contains basic information on fuzzy implications. In Sect. 3 the research problem is formulated.

Section 4 presents the solution of the given research problem. Section 5 is devoted to the pseudo-code of an algorithm for determining a basic fuzzy implication which has the highest truth value of the antecedent when the truth value of the consequent and the truth value of the implication are given. Section 6 includes summarizing of our research and some remarks.

2   Preliminaries

In this section we recall a definition of a fuzzy implication and we list a few of basic fuzzy implications known from the subject literature [1].

A function $I : [0,1]^2 \to [0,1]$ is called *a fuzzy implication* if it satisfies, for all $x, x_1, x_2, y, y_1, y_2 \in [0,1]$, the following conditions:

- if $x_1 \leq x_2$, then $I(x_1, y) \geq I(x_2, y)$, i.e., $I(., y)$ is decreasing;
- if $y_1 \leq y_2$, then $I(x, y_1) \leq I(x, y_2)$, i.e., $I(x, .)$ is increasing;
- $I(0,0) = 1$;
- $I(1,1) = 1$;
- $I(1,0) = 0$.

There exist uncountably many fuzzy implications. The following Table 1 contains a few examples of basic fuzzy implications. Figure 1 gives us some plots of these functions.

Table 1. Examples of basic fuzzy implications

Name	Year	Formula of basic fuzzy implication
Łukasiewicz	1923, [10]	$I_{LK}(x,y) = min(1, 1 - x + y)$
Gödel	1932, [4]	$I_{GD}(x,y) = \begin{cases} 1 \ if \ x \leq y \\ y \ if \ x > y \end{cases}$
Reichenbach	1935, [12]	$I_{RC}(x,y) = 1 - x + xy$
Kleene-Dienes	1938, [9]; 1949, [3]	$I_{KD}(x,y) = max(1 - x, y)$
Goguen	1969, [8]	$I_{GG}(x,y) = \begin{cases} 1 \ if \ x \leq y \\ \frac{y}{x} \ if \ x > y \end{cases}$
Rescher	1969, [13]	$I_{RS}(x,y) = \begin{cases} 1 \ if \ x \leq y \\ 0 \ if \ x > y \end{cases}$
Yager	1980, [18]	$I_{YG}(x,y) = \begin{cases} 1 \quad if \ x = 0 \ and \ y = 0 \\ y^x \ if \ x > 0 \ or \ y > 0 \end{cases}$
Weber	1983, [17]	$I_{WB}(x,y) = \begin{cases} 1 \ if \ x < 1 \\ y \ if \ x = 1 \end{cases}$
Fodor	1993, [3]	$I_{FD}(x,y) = \begin{cases} 1 \qquad\qquad if \ x \leq y \\ max(1 - x, y) \ if \ x > y \end{cases}$

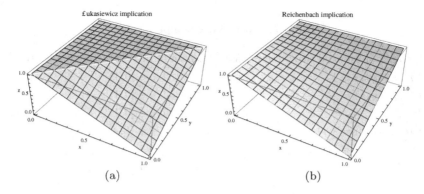

Fig. 1. Plots of I_{LK} and I_{RC} fuzzy implications

3   Problem Statement

Our goal is to elaborate an algorithm to find a method of selecting fuzzy implication in view of the value of the implication antecedent.

Assume that there is given a basic fuzzy implication $z = I(x, y)$, where x,
y belong to $[0,1]$. y is the truth value of the consequent and is known. z is
the truth value of the implication and is also known. In order to determine the
value of the truth of the implication antecedent x it is needed to compute the
inverse function $InvI(y, z)$. In other words, the inverse function $InvI(y, z)$ has
to be determined. Not every of basic implications can be inverted. The function
can be inverted only when it is injective.

Choosing implications in the opposite situation, where the truth value of the
antecedent and the truth value of the implication are given are described in the
paper [14] and primary results regarding this problem in forward reasoning are
included in the paper [15].

4   Results

Table 2 lists inverse fuzzy implications and their domains and in Fig. 2 there are
some plots of them.

The domains of every considered inverted fuzzy implications are included in
a half of the unit square, where $y \leq z < 1$ and $y \in (0, 1)$. Only one inverted
fuzzy implication has a domain which is smaller than this area. This is inverted
Fodor implication and in the whole its domain ($y \leq z < 1 - y, y \in [0, 1]$) this
function is equal to inverted Kleene-Dienes implication.

For $y \leq z < 1 - y$ there are the following inequalities: $InvI_{FD} = InvI_{KD} <
InvI_{RC} < InvI_{LK}$, $InvI_{YG} < InvI_{RC} < InvI_{LK}$, $InvI_{GG} < InvI_{LK}$. A graph-
ical representation of the ordering of inverted basic fuzzy implications is given
in Fig. 3.

For $1 - y \leq z < 1$ and $y \leq z$ there are the same inequalities, but without
inverted Fodor implication, because this function does not exist in this area.

Table 2. Inverted fuzzy implications

Formula of inverted fuzzy implication	Domain of inverted fuzzy implication
$InvI_{LK}(y,z) = 1 - z + y$	$y \leq z < 1, y \in [0,1)$
$InvI_{RC}(y,z) = \frac{1-z}{1-y}$	$y \leq z \leq 1, y \in [0,1)$
$InvI_{KD}(y,z) = 1 - z$	$y < z \leq 1, y \in [0,1)$
$InvI_{GG}(y,z) = \frac{y}{z}$	$y \leq z < 1, y \in (0,1)$
$InvI_{YG}(y,z) = \log_y z$	$y \leq z < 1, y \in (0,1)$
$InvI_{FD}(y,z) = 1 - z$	$y < z < 1 - y, y \in [0,1)$

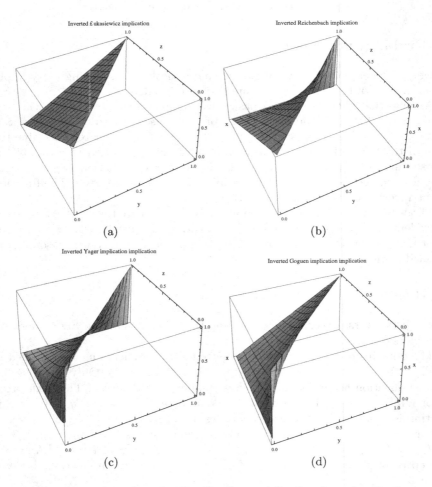

Fig. 2. Plots of $InvI_{LK}$, $InvI_{RC}$, $InvI_{YG}$ and $InvI_{GG}$ fuzzy implications

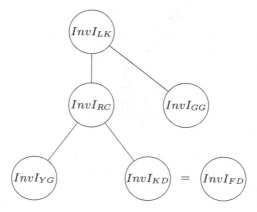

Fig. 3. A graphical representation of the ordering of inverted basic fuzzy implications for $y \leq z < 1 - y$

The resulting inverse functions can be compared with each other so that it is possible to order them. However, some of those functions are incomparable in the whole domain. Nevertheless, by dividing the domain into separable areas (see Fig. 4), we obtained 19 inequalities between inverted fuzzy implications for any $y \leq z < 1$ and $y \in (0,1)$. A few from this inequalities are given below and their graphical representation is in Table 3. This is only a part of the analysis. Full description of the analysis will be in the full version of this work.

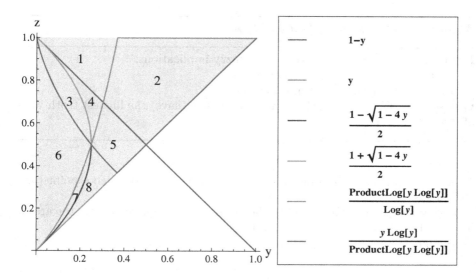

Fig. 4. The unit square $[0,1]^2$ divided into separable areas

1. For $z > 1 - y$ and $z > \frac{ProductLog(yLog(y))}{Log(y)}$
   $InvI_{YG} < InvI_{KD} < InvI_{RC} < InvI_{GG} < InvI_{LK}$
2. For $z > y$ and $z > 1 - y$ and $z < \frac{ProductLog(yLog(y))}{Log(y)}$
   $InvI_{KD} < InvI_{YG} < InvI_{RC} < InvI_{GG} < InvI_{LK}$
3. For $z < \frac{1+\sqrt{1-4y}}{2}$ and $z > \frac{yLog(y)}{ProductLog(yLog(y))}$
   $InvI_{YG} < InvI_{GG} < InvI_{KD} < InvI_{RC} < InvI_{LK}$
4. For $z > \frac{1+\sqrt{1-4y}}{2}$ and $z > \frac{ProductLog(yLog(y))}{Log(y)}$ and $z < 1 - y$
   $InvI_{YG} < InvI_{KD} < InvI_{GG} < InvI_{RC} < InvI_{LK}$

All inequalities given in Table 3 can be proven in a similar way. As examples, we will consider one of inequalities. Let $y \in (0, 1)$ and $z \in (y, 1)$. $y < z$, so obviously $y^2 < yz$. By adding and subtracting $1 - z + y$ to the equation we obtained $1 - z < 1 - z + y - y + yz - y^2$. And therefore, $\frac{1-z}{1-y} < 1 - z + y$. This completes the proof of the inequality: $InvI_{RC} < InvI_{LK}$ in domains of these functions.

5   Algorithm

Below we present the pseudo-code of the algorithm ($DetermineImplicationGTVA$) for determining a basic fuzzy implication which has the highest truth value of the antecedent whereas the truth value of the consequent and the truth value of the implication are given.

The algorithm uses the results of our research presented in Table 3. The first step in the algorithm determines to which area $(1) - (19)$ from Table 3 point (y, z) belongs to.

Algorithm. $DetermineImplicationGTVA$

Input: W - a given subset of the basic fuzzy implications;
    y - the truth value of the consequent;
    z - the truth value of the implication
Output: $I \in W$ - fuzzy implication(s) which has (have) the highest truth value of the antecedent

---

1. $a \leftarrow area(y, z)$ //determines the area from $(1) - (19)$ to which a point (y, z) belongs to;
2. **order** the set W with respect to the graph G_a of inequalities from the area a;
3. $I \leftarrow$ the maximal element(s) from the ordered set W;
4. **return** I;

Table 3. A part of the table of inequalities (4 out of 19 possible cases)

No	Area and inequality	Chart of area	Graph of inequalities
1	For $z > 1 - y$ and $z > \frac{ProductLog(yLog(y))}{Log(y)}$ $InvI_{YG} < InvI_{KD} <$ $InvI_{RC} < InvI_{GG} <$ $InvI_{LK}$		
2	For $z > y$ and $z > 1 - y$ and $z < \frac{ProductLog(yLog(y))}{Log(y)}$ $InvI_{KD} < InvI_{YG} <$ $InvI_{RC} < InvI_{GG} <$ $InvI_{LK}$		

(Continued)

Table 3. (*Continued*)

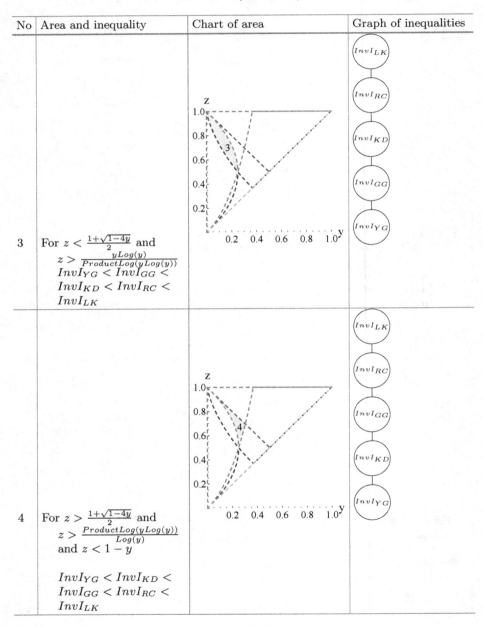

No	Area and inequality	Chart of area	Graph of inequalities
3	For $z < \frac{1+\sqrt{1-4y}}{2}$ and $z > \frac{yLog(y)}{ProductLog(yLog(y))}$ $InvI_{YG} < InvI_{GG} < InvI_{KD} < InvI_{RC} < InvI_{LK}$		
4	For $z > \frac{1+\sqrt{1-4y}}{2}$ and $z > \frac{ProductLog(yLog(y))}{Log(y)}$ and $z < 1 - y$ $InvI_{YG} < InvI_{KD} < InvI_{GG} < InvI_{RC} < InvI_{LK}$		

6   Concluding Remarks

In this paper, we introduced an algorithm of finding the fuzzy implication which has the highest truth value of the antecedent from a given subset of the basic fuzzy implications, when the truth value of the consequent and the truth value of the implication are given.

Also it turned out, that an implication which has the largest truth value of the antecedent is always the inverted Łukasiewicz implication.

In the solution in this paper we used Lambert W function. Lambert W function cannot be expressed in terms of elementary functions. For this reason it is impossible to calculate the coordinates of a point (y_0, z_0) in an analytical way (area number 19 in Table 3).

It is possible to avoid this problem if we skip Yager fuzzy implication in our analysis. This is one of problems which we would like to investigate applying the approach presented in the paper.

Acknowledgments. This work was partially supported by the Center for Innovation and Transfer of Natural Sciences and Engineering Knowledge at the University of Rzeszów. We would like to thank the anonymous referees for critical remarks and useful suggestions to improve the presentation of the paper.

References

1. Baczyński, M., Jayaram, B.: Fuzzy Implications. STUDFUZZ, vol. 231. Springer, Heidelberg (2008)
2. Deng, G., Jiang, Y.: Fuzzy reasoning method by optimizing the similarity of truth-tables. Inf. Sci. **288**(20), 290–313 (2014)
3. Dienes, Z.P.: On an implication function in many-valued systems of logic. J. Symb. Logic **14**, 95–97 (1949)
4. Dubois, D., Prade, H.: Fuzzy sets in approximate reasoning, part 1: inference with possibility distributions. Fuzzy Sets Syst. **40**(1), 143–202 (1991)
5. Dubois, D., Prade, H.: What are fuzzy rules and how to use them. Fuzzy Sets Syst. **84**(2), 169–185 (1996)
6. Fodor, J.C.: On contrapositive symmetry of implications in fuzzy logic. In: Proceedings of the 1st European Congress on Fuzzy and Intelligent Technologies (EUFIT 1993), pp. 1342–1348. Verlag der Augustinus Buchhandlung, Aachen (1993)
7. Gödel, K.: Zum intuitionistischen Aussagenkalkul. Auzeiger der Akademie der Wissenschaften in Wien. Math. Naturwiss. Klasse **69**, 65–66 (1932)
8. Goguen, J.A.: The logic of inexact concepts. Synthese **19**, 325–373 (1969)
9. Kleene, S.C.: On a notation for ordinal numbers. J. Symb. Logic **3**, 150–155 (1938)
10. Łukasiewicz, J.: Interpretacja liczbowa teorii zdań. Ruch Filozoficzny **7**, 92–93 (1923)
11. Papadopoulos, B., Trasanides, G., Hatzimichailidis, A.: Optimization method for the selection of the appropriate fuzzy implication. J. Optim. Theory Appl. **134**(1), 135–141 (2007)
12. Reichenbach, H.: Wahrscheinlichkeitslogik. Erkenntnis **5**, 37–43 (1935)
13. Rescher, N.: Many-Valued Logic. McGraw-Hill, New York (1969)

14. Suraj, Z., Lasek, A., Lasek, P.: Inverted fuzzy implications in approximate reasoning. In: 23th International Workshop on Concurrency, Specification and Programming, Chemnitz, Germany, 29 September– 01 October 2014

15. Suraj, Z., Lasek, A.: Toward optimization of approximate reasoning based on rule knowledge. In: Proceedings of the International Conference on Systems and Informatics, Shanghai, China, 15–17 November 2014

16. Tick, J., Fodor, J.: Fuzzy implications and inference processes. Comput. Inf. **24**, 591–602 (2005)

17. Weber, S.: A general concept of fuzzy connectives, negations and implications based on t-norms and t-conorms. Fuzzy Sets Syst. **11**, 115–134 (1983)

18. Yager, R.R.: An approach to inference in approximate reasoning. Int. J. Man-Mach. Stud. **13**, 323–338 (1980)

Rough Sets

Generating Core Based on Discernibility Measure and MapReduce

Michal Czolombitko$^{(\boxtimes)}$ and Jaroslaw Stepaniuk

Faculty of Computer Science, Bialystok University of Technology,
Wiejska 45A, 15-351 Bialystok, Poland
{m.czolombitko,j.stepaniuk}@pb.edu.pl
http://www.wi.pb.edu.pl

Abstract. In this paper we propose a parallel method for generating attribute core based on distributed programming model MapReduce and rough set theory. The results of the experiments on real dataset show that the proposed method is effective for big data.

Keywords: Rough sets · MapReduce · Core

1 Introduction

Since the massive data could be stored in cloud platforms, data mining for the large datasets is hot topic. Parallel methods of computing are alternative for large datasets processing and knowledge discovery for large data. MapReduce is a distributed programming model, proposed by Google for processing large datasets, so called Big Data. Users specify the required functions Map and Reduce and optional function Combine. Every step of computation takes as input pairs $< key, values >$ and produces another output pairs $< key', values' >$. In the first step, the Map function reads the input as a set $< key, values >$ pairs and applies user defined function to each pair. The result is a second set of the intermediate pairs $< key', values' >$, sent to Combine or Reduce function. Combine function is a local Reduce, which can help to reduce final computation. It applies second user defined function to each intermediate key with all its associated values to merge and group data. Results are sorted, shuffled and sent to the Reduce function. Reduce function merges and groups all values to each key and produces zero or more outputs.

Rough set theory is mathematical tool for dealing with incomplete and uncertain information [4]. The notion of core is very important in the rough set theory. In decision table some of the condition attributes may be superfluous (redundant in other words). This means that their removal cannot worsen the classification. The set of all indispensable condition attributes is called the core. One can also observe that the core is the intersection of all decision reducts – each element of the core belongs to every reduct. Thus, in a sense, the core is the most important subset of condition attributes. None of its elements can be removed without affecting the classification power of all condition attributes. A much

© Springer International Publishing Switzerland 2015
M. Kryszkiewicz et al. (Eds.): PReMI 2015, LNCS 9124, pp. 367–376, 2015.
DOI: 10.1007/978-3-319-19941-2_35

more detailed description of the concept of the core can be found, for example, in the book [6] or in the article [4].

There are some research works combining MapReduce and rough set theory. In [10] parallel method for computing rough set approximations was proposed. In [9] method for computing core based on finding positive region was proposed. They also presented parallel algorithm of attribute reduction in [8]. In [2] is proposed a design of a Patient-customized Healthcare System based on the Hadoop with Text Mining for an efficient Disease Management and Prediction.

In this paper we propose parallel method for generating attribute core based on distributed programming model MapReduce and rough set theory. This paper is organized as follows. Section 1 includes background introduction to rough sets and two algorithms of generating core. Parallel algorithm based on discernibility measure of a set and MapReduce is proposed in Sect. 3. Results of experiments and analysis is presented in Sect. 4. Conclusions and future work are drawn in the last Section.

2 Pseudocode for Generating Core

2.1 Pseudocode for Generating Core Using Discernibility Matrix

In order to compute the core we can use discernibility matrix introduced by Prof. Andrzej Skowron (see e.g. [4,6]). The core is the set of all single element entries of the discernibility matrix.

Let $DT = (U, A \cup \{d\})$ be a decision table, where U is a set of objects, A is a set of condition attributes and d is a decision attribute. Below one can find pseudocode for an algorithm of calculating core $C \subseteq A$ using discernibility matrix $[DM(x,y)]_{x,y \in U}$, where $DM(x,y) = \{a \in A : a(x) \neq a(y) \text{ and } d(x) \neq d(y)\}$.

INPUT: discernibility matrix $[DM(x,y)]_{x,y \in U}$
OUTPUT: core $C \subseteq A$
1: $C \leftarrow \emptyset$
2: **for** $x \in U$ **do**
3:     **for** $y \in U$ **do**
4:         **if** $cardinality(DM(x,y)) = 1$ **and** $DM(x,y) \not\subseteq C$ **then**
5:             $C \leftarrow C \cup DM(x,y)$
6:         **end if**
7:     **end for**
8: **end for**

The main concept of this algorithm is based on a property of singletons i.e. cells from discernibility matrix consisted of the only one attribute. This property tells that any singleton cannot be removed without affecting the classification power. Input for the algorithm is the discernibility matrix DM. Output is core C consisting of a subset of condition attributes set denoted as A. Core is initialized as empty set in line 1. Two loops in lines 2 and 3 are responsible for iteration over all objects in discernibility matrix. In the condition instruction in line 4 it is checked if a matrix cell contains only one attribute. If so, then this attribute is added to the core C.

The main disadvantage of using discernibility matrix for big datasets is its size. Memory complexity of creating this type of square matrix is equal to $(cardinality(U))^2 * cardinality(A)$. This makes the algorithm showed in this subsection infeasible for big data. In the next subsection we present an approach more feasible for big data.

2.2 Pseudocode for Generating Core Based on Discernibility Measure of a Set of Attributes

Generating core based on discernibility measure was discussed in [3]. Counting table CT is a two-dimensional array indexed by values of information vectors (vector of all values of an attribute set $B \subseteq A$) and decisions values, where

$$CT(i,j) = cardinality(\{x \in U : \vec{x}_B = i \text{ and } d(x) = j\})$$

Pessimistic memory complexity of creating this type of matrix is equal to $(caridinality(U) * cardinality(V_d))$, where V_d is a set of all decisions. The discernibility measure $disc(B)$ of set of attributes $B \subseteq A$ can be calculated from the counting table as follows:

$$disc(B) = \frac{1}{2} \sum_{i,j} \sum_{k,l} CT(i,j) \cdot CT(k,l), \text{if } i \neq k \text{ and } j \neq l$$

Below is the pseudocode for this algorithm:

INPUT: decision table $DT = (U, A \cup \{d\})$
OUTPUT: core $C \subseteq A$
1: $C \leftarrow \emptyset$ {C is equal to empty set}
2: $CT \leftarrow 0$ {All values in counting table CT are equal to 0}
3: **for** $x \in U$ **do**
4:     $CT(\vec{x}_A, d(x)) \leftarrow CT(\vec{x}_A, d(x)) + 1$
5: **end for**
6: $disc(A) \leftarrow 0$
7: **for** $[x]_A \in U/IND(A)$ {$U/IND(A)$ is partition of U defined by A} **do**
8:     **for** $[y]_A \in U/IND(A)$ **do**
9:         **if** $\vec{x}_A \neq \vec{y}_A$ and $d(x) \neq d(y)$ **then**
10:             $disc(A) \leftarrow disc(A) + CT(\vec{x}_A, d(x)) \cdot CT(\vec{y}_A, d(y))$
11:         **end if**
12:     **end for**
13: **end for**
14: $CT \leftarrow 0$
15: **for** $a \in A$ **do**
16:     $B \leftarrow A - \{a\}$
17:     $disc(B) \leftarrow 0$
18:     **for** $x \in U$ **do**

19:    $CT(\vec{x}_B, d(x)) \leftarrow CT(\vec{x}_B, d(x)) + 1$

20:    **end for**

21:    **for** $[x]_B \in U/IND(A)$ **do**

22:      **for** $[y]_B \in U/IND(A)$ **do**

23:        **if** $\vec{x}_B \neq \vec{y}_B$ **and** $d(x) \neq d(y)$ **then**

24:          $disc(B) \leftarrow disc(B) + CT(\vec{x}_B, d(x)) \cdot CT(\vec{y}_B, d(y))$

25:        **end if**

26:      **end for**

27:    **end for**

28:    **if** $disc(B) < disc(A)$ **then**

29:      $C \leftarrow C \cup \{a\}$

30:    **end if**

31: **end for**

Input to the algorithm is a decision table DT, and output is the core C of DT. In the beginning core C is initialized as empty set and all values in the counting table are set to zero. First loop in line 3 generate counting table for set of all conditional attributes. For each object in decision table, value in array is increased by one. Indexes of value are information vector of object and its value of decision. In the line 6 value of discernibility measure of set of all condition attributes, $disc(A)$, is initialized as zero. Next two loops in lines 7 and 8 take subsequent equivalence classes to comparison. If information vectors of these classes are not equal, $disc(A)$ is increased by product of two values from the counting tables where indexes are values information vectors and different decisions. Next step is computing the discernibility measure of set of attributes after removing one of them. In line 14 all values in the counting table are set to zero. Loop in line 15 takes attribute a from set of all condition attributes A. Set B is initialized as set of all condition attributes after removing this attribute. Similarly as above, in lines 15–27 is calculated $disc(B)$. Finally, values of the discernibility measure of set of all attributes and after removing attribute a are compared in the line 28. In case of difference, this attribute is added to the core.

3   MapReduce Implementation

The main concept of the proposed algorithm is parallel computation of counting tables. The proposed algorithm consists of the four steps: Map, Combine, Reduce and Compute Core.

Step 1. Map

INPUT: key : subtable id, $value$: decision subtable $DT_i = (U_i, A \cup \{d\})$

OUTPUT: $< key', value' >$ pair where key' : $(\vec{x}_B, d(x))$ and $value'$ is id of the object x

1: **for** $x \in U_i$ **do**

2:    $key' \leftarrow (\vec{x}_A, d(x))$

3:    $value' \leftarrow$ id of the object x

4:    $emit(< key', value' >)$

5:    **for** $a \in A$ **do**
6:        $B \leftarrow A - \{a\}$
7:        $key' \leftarrow (\vec{x}_B, d(x))$
8:        $value' \leftarrow$ id of the object x
9:        $emit(< key', value' >)$
10:   **end for**
11: **end for**

Input to function Map are: key is a subtable id stored in HDFS, and $value$ is decision subtable $DT_i = (U_i, A \cup \{d\})$. First loop in line 1 takes an object x from decision subtable DT_i and emits pair $< key', value' >$, where key' is information vector with respect to set of all condition attributes and decision of the object x, and $value$ is id of the this object. Loop in line 5 takes attribute a from set of all condition attributes A. Set B is initialized as set A after removing this attribute. Next step is emitting pair $< key', value' >$, where key' is information vector with respect to set B and decision of the object x, and $value'$ is id of the this object.

Step 2. Combine
INPUT: $< key, value >$ where $key : (\vec{x}_B, d(x))$ and $value$ is id of the object x.
OUTPUT: $< key', value' >$ where $key : (\vec{x}_B, d(x))$ and $value'$ is the number of objects belonging to equivalence class $[x]_B$ with equal decision values, from decision subtable DT_i.
1: $key' \leftarrow key$
2: $value' \leftarrow 0$
3: **for** each value **do**
4:    $value' \leftarrow value' + 1$
5: **end for**
6: $emit(< key', value' >)$

Input to function Combine are $< key, value >$ pairs where $key : (\vec{x}_B, d(x))$ and $value$ is id of the object. This function accepts a key and a set of values associated with this key from the local Map. Function emits pairs $< key', value' >$, where key' is $(\vec{x}_B, d(x))$ and $value'$ is the number of objects associated with this key from decision subtable DT_i.

Step 3. Reduce
INPUT: $< key, value >$ where $key : (\vec{x}_B, d(x))$ and $value$ is the number of objects belonging to equivalence class $[x]_B$ with equal decision values from decision subtable DT_i.
OUTPUT: the files containing pairs $< key', value' >$ where $key : (\vec{x}_B, d(x))$ and $value'$ is the number of objects belonging to equivalence class $[x]_B$ with equal decision values from decision table DT.
1: $key' \leftarrow key$
2: $value' \leftarrow 0$
3: **for** each value **do**

4:     $value' \leftarrow value' + value$
5: **end for**
6: $saveToFile('A - B'.dat, < key', value' >)$

Input to function Reduce are $< key, value >$ pairs where $key : (\vec{x}_B, d(x))$ and $value$ is the number of objects belonging to equivalence class $[x]_B$ with the same decisions from decision subtable DT_i. Function emits pairs $< key', value' >$, where key' is $(\vec{x}_B, d(x))$ and $value'$ is a number of the objects associated with this key from decision table DT. Each pair is saved to file, which name based on index removed attribute from the set of all condition attributes A.

Step 4. Compute Core

INPUT: the files containing the pairs $< key, value >$ where $key : (\vec{x}_B, d(x))$ and $value$ is number of the objects belong to $[x]_B$ with the same decision value.

OUTPUT: core $C \subseteq A$
1: $C \leftarrow \emptyset$
2: $disc(A) \leftarrow 0$
3: **for** $x \in \emptyset.dat$ **do**
4:     **for** $y \in \emptyset.dat$ **do**
5:         **if** $\vec{x}_B \neq \vec{y}_B$ **and** $d(x) \neq d(y)$ **then**
6:             $disc(A) \leftarrow disc(A) + value(x) \cdot value(y)$
7:         **end if**
8:     **end for**
9: **end for**
10: **for** $a \in A$ **do**
11:     $disc(A - \{a\}) \leftarrow 0$
12:     **for** $x \in a.dat$ **do**
13:         **for** $y \in a.dat$ **do**
14:             **if** $\vec{x}_B \neq \vec{y}_B$ **and** $d(x) \neq d(y)$ **then**
15:                 $disc(A - \{a\}) \leftarrow disc(A - \{a\}) + value(x) \cdot value(y)$
16:             **end if**
17:         **end for**
18:     **end for**
19:     **if** $disc(A - \{a\}) < disc(A)$ **then**
20:         $C \leftarrow C \cup \{a\}$
21:     **end if**
22: **end for**

Input to the function Compute Core is directory contains files, which names based on index of removed attribute from set of all condition attributes, and output is core. In the beginning core C is initialized as empty set and discernibility measure of set of all attributes is set to zero. First two loops in lines 3 and 4 take subsequent lines from file with counting table for all condition attributes to comparison. If these two lines contains information about two different information vectors and decisions, $disc(A)$ is increased by product of two values these lines.

Similarly operations are repeated for each file contains $< key, value >$ pairs. Name of these files based on removed attribute from original set of condition attributes. If computed discernibility measure of set of attributes without this attribute is less than for all attributes, this attribute is added to the core.

4   Experimental Results

The algorithm was running on the Hadoop MapReduce platform [1], where Hadoop MapReduce is a YARN-based system for parallel processing of big datasets. In the experiments, Hadoop 2.5.1 version was used. Cluster of compute nodes contains 19 PC's. All the PC's has four 3.4 GHz cores and 8 GB of memory.

In this paper, we present the results of the conducted experiments using data about children with insulin-dependent diabetes mellitus (type 1). Diabetes mellitus is a chronic disease of the body's metabolism characterized by an inability to produce enough insulin to process carbohydrates, fat, and protein efficiently. Treatment requires injections of insulin. Twelve condition attributes, which include the results of physical and laboratory examinations and one decision attribute (microalbuminuria) describe the database used in our experiments. The data collection so far consists of 107 cases. The database is shown at the end of the paper [5]. A detailed analysis of the above data is in Chap. 6 of the book [6].

This database was used for generating bigger datasets consisting of $0.5 \cdot 10^6$ to $30 \cdot 10^6$ of objects. New datasets were created by randomly multiplying the rows of original dataset. Numerical values were discretized and each attributes value was encoded using two digits.

4.1   Speedup

In speedup tests, the dataset size is constant and the number of nodes grows in each step of experiment. To measure speedup, we used dataset contains $30 \cdot 10^6$ objects. The speedup given by the n-times larger system is measured as [7]:

$$Speedup(n) = \frac{t_n}{t_1}$$

where n is number of the nodes in cluster, t_1 is the computation time on one node, and t_n is the computation time on n nodes.

The ideal parallel system with n nodes provides n times speedup. The linear speedup is difficult to achieve because of the I/O operations data from HDFS and the communications cost between nodes. Table 1 shows the computational time of generating core from dataset containing $30 \cdot 10^6$ objects. The number of nodes varied from 1 to 19. Figure 1 shows, the proposed algorithm has a good performance until the number of nodes is less than 14. Mapper doesn't always work on node where is stored block of file in HDFS. In this case block of file with data is sending from node, where is stored to another, where is processing. The main reason why speedup isn't linear is overhead read and write operations. Generally, if the number of node is bigger, the speed performs better.

Table 1. Speedup experiment results

Number of the nodes	1	2	3	4	5	6	7	8	9	10	11	12	13	14	15	16	17	18	19
Time (s)	415	300	191	131	104	89	77	73	68	62	60	54	52	52	54	47	56	46	45

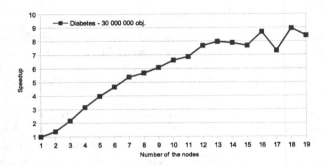

Fig. 1. Speedup

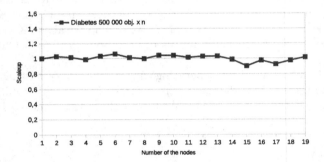

Fig. 2. Scaleup

4.2 Scaleup

Scaleup analysis measures stability system when system and dataset size grow in each step of experiment. The scaleup coefficient is defined as follows [7]:

$$Scaleup(DT, n) = \frac{t_{DT_{1,1}}}{t_{DT_{n},n}}$$

where $t_{DT_{1,1}}$ is the computational time for dataset DT on one node, and $t_{DT_{n},n}$ is the computational time for n-larger dataset DT on n nodes.

We demonstrate how good the proposed parallel algorithm handles larger datasets when more nodes are available. In scaleup experiments there is linear relationship between number of nodes and dataset size. Core is generated for the dataset consisting of $0.5 \cdot 10^6$ objects on one node and for $9.5 \cdot 10^6$ objects on 19 nodes. Clearly, Fig. 2 shows that the proposed algorithm is scalable.

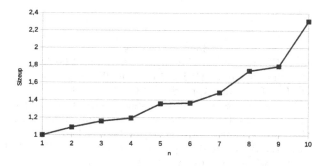

Fig. 3. Sizeup

4.3 Sizeup

In sizeup tests, the number of nodes is constant, and the dataset size grows in each step of experiment. Sizeup measures how much time is needed for calculations when the size of dataset n-larger than the original dataset. Sizeup is defined as follows [7]:

$$Sizeup(DT) = \frac{t_{DT_n}}{t_{DT_1}}$$

where t_{DT_1} is execution time for a given dataset DT, and t_{DT_n} is execution time n-larger dataset than DT. Figure 3 shows the sizeup experiments results on ten nodes. Results shows that the proposed algorithm has a very good sizeup performance. When dataset grows ten times, the computational time grows 2.3 times.

5 Conclusions and Future Research

In this paper, generating core for large datasets based on rough set theory is studied. A parallel attribute reduction algorithm is proposed, which is based on the distributed programming model of MapReduce and the core computation algorithm using discernibility measure of a set of attributes. It is worth noting that a very interesting element of the paper is an usage of a counting table instead of a discernibility matrix. The results of the experiments show that the proposed method is efficient for large data, and it is a useful method for data mining and knowledge discovery for big datasets. Our future research work will focus on optimizing data placement in Hadoop to improve efficiency and on applications of distributed in-memory computing for attribute reduction.

Acknowledgements. The research is supported by the Polish National Science Centre under the grant 2012/07/B/ST6/01504 (Jaroslaw Stepaniuk) and by the scientific grant S/WI/3/2013 (Michal Czolombitko).

The experiments were performed using resources co-financed with the European Union funds as a part of the "Centre for Modern Education of the Bialystok University of Technology" project (Operational Programme Development of Eastern Poland)

References

1. Hadoop MapReduce. http://hadoop.apache.org
2. Lee, B.K., Jeong, E.H.: A design of a patient-customized healthcare system based on the hadoop with text mining (PHSHT) for an efficient disease management and prediction. Int. J. Softw. Eng. Appl. **8**(8), 131–150 (2014)
3. Nguyen, H.S.: Approximate boolean reasoning: foundations and applications in data mining. In: Peters, J.F., Skowron, A. (eds.) Transactions on Rough Sets V. LNCS, vol. 4100, pp. 334–506. Springer, Heidelberg (2006)
4. Pawlak, Z., Skowron, A.: Rudiments of rough sets. Inf. Sci. **177**(1), 3–27 (2007)
5. Stepaniuk, J.: Knowledge discovery by application of rough set models. In: Polkowski, L., Tsumoto, S., Lin, T.Y. (eds.) Rough Set Methods and Applications. STUDFUZZ, vol. 56, pp. 137–233. Springer, Heidelberg (2000)
6. Stepaniuk, J.: Rough-Granular Computing in Knowledge Discovery and Data Mining. SCI, vol. 152. Springer, Heidelberg (2008)
7. Xu, X., Jäger, J., Kriegel, H.P.: A fast parallel clustering algorithm for large spatial databases. Data Min. Knowl. Discov. **3**, 263–290 (1999)
8. Yang, Y., Chen, Z., Liang, Z., Wang, G.: Attribute reduction for massive data based on rough set theory and MapReduce. In: Yu, J., Greco, S., Lingras, P., Wang, G., Skowron, A. (eds.) RSKT 2010. LNCS, vol. 6401, pp. 672–678. Springer, Heidelberg (2010)
9. Yang, Y., Chen, Z.: Parallelized computing of attribute core based on rough set theory and MapReduce. In: Li, T., Nguyen, H.S., Wang, G., Grzymala-Busse, J., Janicki, R., Hassanien, A.E., Yu, H. (eds.) RSKT 2012. LNCS, vol. 7414, pp. 155–160. Springer, Heidelberg (2012)
10. Zhang, J., Li, T., Ruan, D., Gao, Z., Zhao, C.: A parallel method for computing rough set approximations. Inf. Sci. **194**, 209–223 (2012)

Music Genre Recognition in the Rough Set-Based Environment

Piotr Hoffmann[✉] and Bożena Kostek

Faculty of Electronics, Telecommunications and Informatics, Audio Acoustics Laboratory,
Gdańsk University of Technology, 80-233 Gdańsk, Poland
phoff@sound.eti.pg.gda.pl, bokostek@audioacoustics.org

Abstract. The aim of this paper is to investigate music genre recognition in the rough set-based environment. Experiments involve a parameterized music database containing 1100 music excerpts. The database is divided into 11 classes corresponding to music genres. Tests are conducted using the Rough Set Exploration System (RSES), a toolset for analyzing data with the use of methods based on the rough set theory. Classification effectiveness employing rough sets is compared against k-Nearest Neighbors (k-NN) and Local Transfer function classifiers (LTF-C). Results obtained are analyzed in terms of global class recognition and also per genre.

Keywords: Music processing · Rough sets · Genre recognition · k-Nearest Neighbors · RSES system · LTF-C

1 Introduction

There exist many methods that can be used for data storage, analysis and classification. The main feature of these methods should be universality and efficiency. With regard to universality a system should allow for collecting and storing various data sets, regardless of the processes and phenomena described in them. The effectiveness of the system should enable users to make data analysis and classification easily, and to control these processes. Handling data efficiently requires storing them in tables with objects (rows) and attributes (columns), describing single instances. Data in their nature could be imprecise, uncertain and/or incomplete, thus this requires special preparation so that they can be processed, mined and classified. Another important issue is to select significant components present in the tables to provide their discernibility within the classes. For the analysis of data with characteristics described above the rough set-based methods are very useful, as they generate interpretable results in the form of reducts and rules. One of the well-known systems for the analysis and classification of data is the Rough Set Exploration (RSES) system that returns extracted rules and reducts acquired from the rough set-based analysis [3, 4]. It is worth to emphasize that the RSES system is a software tool that enables to carry out large-scale computational experiments related to the analysis of array data using the rough set theory [13, 16].

© Springer International Publishing Switzerland 2015
M. Kryszkiewicz et al. (Eds.): PReMI 2015, LNCS 9124, pp. 377–386, 2015.
DOI: 10.1007/978-3-319-19941-2_36

In this publication, the authors illustrate the process of recognizing music genres using the rough set theory. One of the main objectives of the data analysis shown in this paper is to uncover underlying causes or factors and to determine the relationship between objects (audio tracks belonging to music genres) in the case study related to the Music Information Retrieval (MIR) domain [5]. The classification is carried out on a set of music descriptors using the methods available in the RSES software. In addition, data are to undergo a pre-processing of feature vector parameters employing the PCA method [1, 15] and parameter weighting. Lastly, a comparison of genre classification results employing two sets of music excerpts is provided.

2   Data Preprocessing

The theory of rough sets was created in the early 80s of the twentieth century. Its main use is for synthesizing and analyzing data sets efficiently. Methods based on the rough set theory have been used, among others, in data mining and knowledge discovery in complex tasks of classification and computer decision support systems [13]. Currently, it is one of the fastest growing methods within the artificial intelligence domain. In the rough set theory a requirement that a data set needs to have a clearly defined boundaries is discarded [2, 17]. The scope of rough sets is defined by the lower and upper approximations of tabular data, obtained experimentally.

The difference between the upper and lower approximation is the border area, which includes all cases, that cannot be seamlessly classified on the basis of the current knowledge. The lower approximation set contains all objects for which there is no doubt that they are representatives of this set in view of knowledge. Objects that cannot be excluded that they are representatives of this set belong to the upper approximation. Boundary of the set are all of the objects for which it is not known whether or not they are representatives of a given set. The larger area border set, the more objects in it are less precise. The theory of rough sets allows the processing of both quantitative and qualitative data tables, called decision tables.

The basic structure of the data in the information systems using rough set theory is a table. All data are grouped in tables according to the principle that the rows of tables are objects, and attributes are columns. The formula (1) presents the information system [17].

$$SI = \ < U, A, V, f >$$ (1)

where:

U - non-empty, finite set of objects,
A - non-empty finite set of attributes,
V - set of attribute values,
f - function of information, which is the Cartesian product of a set of objects and a set of attribute values.

In Music Information Retrieval systems tables filled with music descriptors constitute the information system. In such systems, each track is parameterized and then stored in a table [5, 8]. A special case of information systems are decision tables, which describes cases (also called examples or objects) using conditional attribute values and a decision. Attributes are independent variables while the decision is a dependent variable, which means that the conditional attributes determine the value of a decision.

Based on the decision table only, it is impossible to directly know the relationship between the conditional attributes and decision describing objects. Therefore, it is necessary to further process them to extract dependencies. Attribute reduction is very crucial in the rough set-based data analysis because it is used to induce decision rules without reducing the classification accuracy [19]. In the rough set theory the reduct is generally defined as a minimal subset of attributes that can classify the same domain of objects as unambiguously as the original set of attributes, which means that the reduct is a minimal subset of attributes having the characteristics of the whole collection. For a given information system may have multiple reducts, consisting of a variable number of arguments. The major problem is, however, the identification and removal of attributes that are unnecessary. The process of determining reducts is considered as a bottleneck in the inference systems based on rough sets. On the other hand, reducts and decision making may be acquired for large systems of dozens or hundreds of attributes using genetic algorithms. After preparing reducts it is possible to generate data available in the decision-making system in the form of logical rules used in the classification process. The rules have the conditional form, and their number is equal to the maximum number of objects multiplied by the number of distinguishable reducts. However, not all rules are needed to use or to be implemented in the decision-making process. The reduction method applied in the RSES system was very thoroughly described by its authors [2], thus it will not be recalled here.

The main element of music genre recognition systems is the optimized parameter input. The extracted feature vector should have a detailed description of parameterized samples and preserve a very good separability. Taking into account these assumptions feature vector containing 173 elements has been prepared. The vector includes parameters associated with the MPEG 7 standard [12] and the melcepstral parameters [8–10]. The list of parameters include: Spectral flatness Measure Spread Spectrum Audio, Audio Spectrum Envelope, Spectral Centroid, Temporal centroid. Full list of parameters was shown in the study [18]. Frequency band used for the parameter is in the range from 63 to 8000 Hz. The prepared feature vector is used to describe each signal frame.

173-element vector generates a very large amount of information describing a song. Consequently, this leads to an extensive amount of data undergoing classification, which in the context of the usage of e.g. the k-Nearest Neighbors classifier is important. It was therefore decided to apply Principal Component Analysis (PCA) to reduce data redundancy [1]. This is to identify patterns in the data and present them in such a way as to indicate their similarities and differences. The PCA method uses the variance of the data to prepare a new database parameters. The new descriptors are linear combinations of parameters that carry much information about the test set. It was experimentally checked that the PCA method can shorten the given feature vector of 173 descriptors to 19, which significantly reduces the computation time. In addition, the use of the described analysis improves the classification efficiency, which was presented in an earlier paper by the authors [7].

3 Experiments

This Section describes the results of the experiments in which the rough set theory was applied to music genre classification. For this purpose rough set classification learning algorithm provided by RSES was used. Experiments aimed at a comparison between standard classification algorithms - k-Nearest Neighbors and Local Transfer function classifiers [10, 11]. The rule set decision algorithm based on conditional rules calculates the attributes of the new object, which is essential for decisions related to the content of reducts. Then, it looks for rules that match attribute values, if there are no matching rules, the result is the most common decision, or the least expensive decision. In the case that multiple rules match attribute values, they may indicate a number of decisions, then the vote should be taken that selects the answer that appears most often [6].

The k-Nearest Neighbors algorithm is the simplest one, and as such is very commonly used for classification. We utilized the minimal distance that uses the Euclidean distance function. Its aim is to predict the class membership of objects. The decision is based on the k-closest objects. An object is classified by the majority vote [14].

LTF-C algorithm is a neural network employed for classification tasks with the architecture similar to the radial network (RBF), but different training algorithms. It consists of two layers of neurons. The first layer - hidden - contains neurons with Gaussian transfer function, that detects cluster of patterns of the same class in the training data. Each neuron of this layer is assigned a class that tries to detect the cluster [10]. The second layer consist of linear neurons that segregate responses of hidden neurons

Table 1. Number of excerpts in the Synat and GZTAN databases

Genre:	Synat	GZTAN
Pop	100	100
Rock	100	100
Country	100	100
R&B	100	–
Rap & Hip-Hop	100	100
Classical	100	100
Jazz	100	100
Dance & Dj	100	100
NewAge	100	–
Blues	100	100
Hard Rock & Metal	100	100
Reggae	–	100

according to the assigned classes and add them by formulating a final answer network structure [10].

All classification tests were carried out in the RSES environment. Tests were performed on two data sets. The first one, called "Synat", contained 1100 audio excerpts divided into 11 most popular music genres. The second one GZTAN [17], a commercial data set, contained 1000 audio files divided into 10 music genres. Sizes and content of these two sets are presented in Table 1 [7, 8]. The length of each music excerpt is 30 s. Both datasets were created in a similar way to reflect a variety of music genres and contain audio files belonging to the most popular music genres.

In preparing parameters for the classification the data volume was reduced using the method of PCA. The number of parameters after employing PCA was 33, which accounted for 80 % of information retained from the entire feature vector. Figure 1 presents a block diagram of the proposed processing path of music genre according to rough set-based analysis.

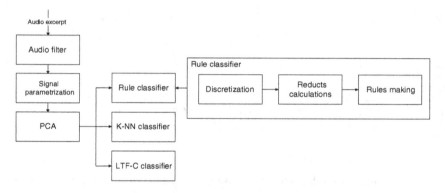

Fig. 1. Block diagram of processing in the experiment

Classification in the testing phase was performed with the default settings of the algorithms used. The RSES system automatically chooses the optimum values of parameters for the most effective results. This confirms the validity of the additional data preparation prior to processing information using the rough set theory. In Table 2 the results of tests conducted are shown. The table shows the results for the feature vectors without the PCA method and with the use of PCA. At the same time, it should be noted that feature vectors not reduced by the PCA method were processed by algorithms for much longer.

The effectiveness of the classification algorithms reached 70 % when the PCA method was not used and 85 % when it was employed. Analyzing the results obtained for individual genres, it can be noticed that classical music genre is distinguishable among genres as it has got a very good 90 % classification effectiveness for each test set. Similarly, very good results were achieved for rock and hard rock genres with the use of the PCA method.

Table 2. Classification effectiveness of music genres in Synat dataset.

Genre [%]	Rule classification		k-NN		LTF-C	
	No PCA	PCA	No PCA	PCA	No PCA	PCA
Blues	0.844	0.917	0.744	0.818	0.724	1
Classical	0.909	1	0.889	1	0.849	1
Country	0.786	1	0.886	0.9	0.862	0.9
DanceDj	0.84	0.778	0.8	0.727	0.72	0.697
HardRock	0.65	1	0.75	0.905	0.65	0.857
Jazz	0.736	0.778	0.636	0.909	0.656	0.727
NewAge	0.867	1	0.767	0.909	0.747	0.879
Pop	0.788	0.87	0.688	0.861	0.608	0.778
RB	0.739	0.767	0.639	0.757	0.717	0.595
Rap	0.585	0.783	0.485	0.871	0.635	0.806
Rock	0.749	1	0.649	0.958	0.529	0.958
Σ	0.772	0.899	0.721	0.874	0.699	0.836

Table 2 clearly shows the gain after applying the method using the rough set theory. On average it is about 5 % better than while employing two other methods, which should be considered a very good result despite a much longer data processing. Longer data processing in classification systems based on the theory of rough sets is due to the discretization step and generation of reducts. In particular, the step of reduct generating is a very demanding for the available resources. In Fig. 2 the results obtained in the experiments in the form of graphs were summarized. The dashed line presents the average value of the individual results.

The RSES system occurred to be the most effective classification algorithm, the weakest algorithm in ranking was the neural network. It was also the least balanced in its indications, i.e. recognizing individual music genres with a large discrepancy. The k-Nearest Neighbors algorithm was already investigated by the authors in the classification of music genres. The publication cited in here showed that the k-NN algorithm achieved the best results on music parameterized data [7]. However, the application of rough sets in the classification process of music genres with respect to existing algorithms that were used by the authors resulted in about 5 % better performance, which should be considered as a very good result, because this is due to only the change of the classification method without additional processing.

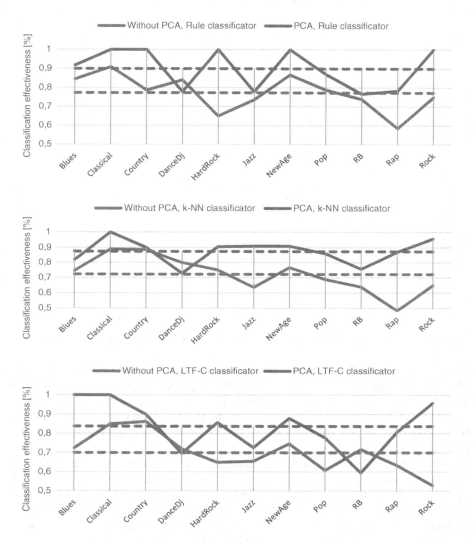

Fig. 2. Classification effectiveness of the algorithms employed

To confirm the results obtained by the authors another experiment was conducted using a commercially available GZTAN database [17]. In the experiment, eight music genres common for both databases were compared in the context of the classification efficiency. The experiment uses the k-NN algorithm. In Fig. 3 results obtained from the experiments for various music genres are shown.

The results obtained for the GZTAN database compared to the SYNAT database are 7 % lower. Similarly to the database SYNAT in the case of database GZTAN an increase in the effectiveness of the recognition genres after PCA use can be noticed. Significant differences in the recognition of music genres were found in the case of DanceDj genre,

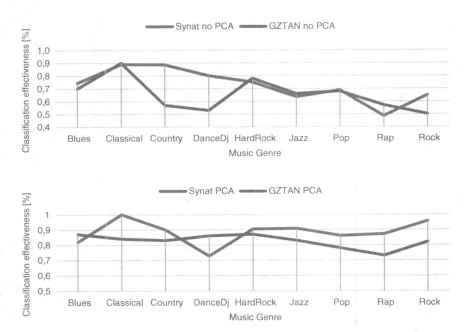

Fig. 3. Classification effectiveness of the k-NN algorithm using two different data sets.

which can be caused by a different description of DanceDj genre in GZTAN database. The reason for the lower classification efficiency is probably a greater variety of songs in the GZTAN database. Tracks occurring in the database fully describe the variety of music genres. Furthermore the database contains also recordings with reduced quality, which may have an adverse effect on the effectiveness of the parameterization of the proposed solution.

In order to confirm the statistical significance of the obtained results the T-Student test was performed. Values were calculated for each analyzed genre. To reject the null hypothesis of statistical insignificance of the results T parameter should be higher than 2.228. The critical value is based on the value from the T-Student distribution table for 10 degrees of freedom. In Table 3 the exact values of the parameter T for all experiments are shown. Seven of eight results of experiments can be considered as statistically significant. Small variations obtained in these results show the correctness of the results in the statistical sense. The only experiment that did not show statistical significance of the results was carried out based on the GZTAN database without PCA.

Table 3. Summary of the results of T-Student Test

	Rule classification		k-NN		LTF-C		k-NN GZTAN	
	No PCA	PCA	No PCA	PCA	No PCA	PCA	No PCA	PCA
T-value	2.422	2.825	2.463	2.387	2.439	2.557	2.261	2.446

To summarize, it may be concluded that rule decision algorithm gives in most cases better results than minimal distance or neural algorithms. But this is associated with a significantly longer duration of data processing. Moreover, in the case of a trial experiment conducted by the authors when a very large database with more than 30000 parameterized music excerpts was used, the system was not able to calculate reducts due to lack of the application memory resources. The test confirms earlier findings on the computational complexity of the process of generating reducts. The additional use of the PCA method in the decision process further improved the effectiveness of the classification decision by reducing the number of attributes.

4   Conclusions

In the experiments conducted, the authors used the RSES application for testing the effectiveness of recognizing music genres using the rough set theory. The application enables to efficiently carry out data analysis and generates classification tables. The authors examined also the effect of using the PCA method in data processing according to the theory of rough sets. Experiments were conducted on two different data sets.

The classification effectiveness achieved in the experiments is very good (above 85 %) for the whole data set. In other publications related to music genre classification [10, 11, 20] when k-NN and rule classification algorithms were used, the results were worse. The main reason for such good results obtained by the authors may be a unique parameterization module which was applied along with the PCA dimensionality reduction. Very good results have been achieved through the effective parameterization, accurately describing the test set. The data after analysis with the PCA method were very well separable and were described by a much smaller number of data without losing much information. Calculating reducts from the PCA-reduced data improves the classification effectiveness up to 10 %. Performed statistical analysis confirmed the statistical significance of results. Most of experiments can be defined as statistically significant.

Performed test show how important in decision making process is the data preprocessing step. Discretization of tabular data, reducts and rules calculation require operation time and are resource-absorbing, but for smaller sets data processing based on rough sets can be more effective, regardless of the resources involved.

Acknowledgements. This work was partially supported by the grant no. PBS1/B3/16/2012 entitled "Multimodal system supporting acoustic communication with computers" financed by the Polish National Centre for Research and Development and the company Intel Technology Poland.

References

1. Abdi, H., Williams, L.J.: Principal component analysis. Wiley Interdisc. Rev. Comput. Stat. **2**, 433–459 (2010)

2. Bazan, J.G., Nguyen, H.S., Nguyen, T.T., Skowron, A., Stepaniuk, J.: Decision rule synthesis for object classification. In: Orłowska, E. (ed.) Incomplete Information: Rough Set Analysis, vol. 13, pp. 23–57. Physica - Verlag, Heidelberg (1998)
3. Bazan, J., Szczuka, M.S., Wróblewski, J.: A new version of rough set exploration system. In: Alpigini, J.J., Peters, J.F., Skowron, A., Zhong, N. (eds.) RSCTC 2002. LNCS (LNAI), vol. 2475, pp. 397–404. Springer, Heidelberg (2002)
4. Kostek, B.: Music Information Retrieval in Music Repositories. In: Skowron, A., Suraj, Z. (eds.) Rough Sets and Intelligent Systems - Professor Zdzisław Pawlak in Memoriam. ISRL, vol. 42, pp. 463–489. Springer, Heidelberg (2013)
5. Kostek, B.: Perception-Based Data Processing in Acoustics, Applications to Music Information Retrieval and Psychophysiology of Hearing. Cognitive Technologies. Springer, Heidelberg (2005)
6. Kostek, B.: Soft Computing in Acoustics, Applications of Neural Networks, Fuzzy Logic and Rough Sets to Musical Acoustics. Studies in Fuzziness and Soft Computing. Physica Verlag, Heidelberg (1999)
7. Hoffmann, P., Kostek, B.: Music data processing and mining in large databases for active media. In: Ślęzak, D., Schaefer, G., Vuong, S.T., Kim, Y.-S. (eds.) AMT 2014. LNCS, vol. 8610, pp. 85–95. Springer, Heidelberg (2014)
8. Kostek, B., Hoffmann, P., Kaczmarek, A., Spaleniak, P.: Creating a reliable music discovery and recommendation system. In: Bembenik, R., Skonieczny, L., Rybiński, H., Kryszkiewicz, M., Niezgódka, M. (eds.) Intelligent Tools for Building a Scientific Information Platform: From Research to Implementation. SCI, vol. 541. Springer, Switzerland (2013)
9. Kostek, B., Kupryjanow, A., Zwan, P., Jiang, W., Raś, Z.W., Wojnarski, M., Swietlicka, J.: Report of the ISMIS 2011 contest: music information retrieval. In: Kryszkiewicz, M., Rybinski, H., Skowron, A., Raś, Z.W. (eds.) ISMIS 2011. LNCS, vol. 6804, pp. 715–724. Springer, Heidelberg (2011)
10. Kotropoulos, C., Benetos, E., Panagakis, E.: Music genre classification: a multilinear approach. In: ISMIR (2008)
11. Mlynek, D., Zoia, G., Scaringella, N.: Automatic genre classification of music content. IEEE Signal Process. Mag. **23**(2), 133–141 (2006)
12. MPEG 7 standard. http://mpeg.chiariglione.org/standards/mpeg-7
13. Pawlak, Z.: Rough sets. Int. J. Comput. Inform. Sci. **11**(5), 341–356 (1982)
14. Schnitzer, D., Flexer, A., Widmer, G.: A fast audio similarity retrieval method for millions of music tracks. Multimedia Tools Appl. **58**, 23–40 (2012)
15. Shlens, J.: A Tutorial on Principal Component Analysis, Version 2, 10 December 2005
16. Skowron, A., Polkowski, L. (ed.): Rough Sets in Knowledge Discovery, vols. 1 and 2, Physica Verlag, Heidelberg (1998)
17. Tzanetakis, G., Cook, P.: Musical genre classification of audio signals. IEEE Trans. Speech Audio Process. **10**(5), 293–302 (2002)
18. Tzacheva, A.A., Bell, K.J.: Music information retrieval with temporal features and timbre. In: An, A., Lingras, P., Petty, S., Huang, R. (eds.) AMT 2010. LNCS, vol. 6335, pp. 212–219. Springer, Heidelberg (2010)
19. Wróblewski, J.: Covering with reducts - a fast algorithm for rule generation. In: Polkowski, L., Skowron, A. (eds.) RSCTC 1998. LNCS (LNAI), vol. 1424, pp. 402–407. Springer, Heidelberg (1998)
20. Zheng, J., Oussalah, M.: Automatic System for Music Genre Classification, University of Birmingham, Electronics, Electricial and Computer Engineering (2006)

Scalability of Data Decomposition Based Algorithms: Attribute Reduction Problem

Piotr Hońko[✉]

Faculty of Computer Science, Bialystok University of Technology,
Wiejska 45A, 15-351 Bialystok, Poland
p.honko@pb.edu.pl

Abstract. This paper studies the issue of scalability of data decomposed based algorithms that are intended for attribute reduction. Two approaches that decompose a decision table and use the relative discernibility matrix method to compute all reducts are investigated. The experiments results reported in this paper show that application of the approaches makes it possible to gain a better scalability compared with the standard algorithm based on the relative discernibility matrix method.

Keywords: Rough sets · Attribute reduction · Data decomposition · Scalability

1 Introduction

Attribute reduction is a challenging task in areas such as data mining and pattern recognition. Rough set theory [9] as a mathematical tool to deal with inconsistent data has commonly been used to investigate this issue from a theoretical and practical viewpoint (e.g. [6,10,11]). Much research has been devoted to finding a reduct, especially a minimal one. Although one reduct is sufficient to reduce the attribute set, the problem of finding all reducts still has its justification. A deeper analysis of the data can be conducted when all reducts are known.

The method proposed in [10] for finding all reducts is based on a discernibility matrix and its alternative representation in the form of discernibility function. The idea of relative discernibility matrix/function for attribute reduction in decision tables has been intensively studied by many researchers (e.g. [2,5,7]).

The main problem to face when developing a method for finding reducts is the computationally complexity of the attribute reduction task. Finding all reducts is proven to be an NP-hard problem [10]. Much effort has therefore been made to accelerate the attribute reduction process (e.g. [1,12,14]).

Another direction for making attribute reduction methods more efficient for large databases is to divide the attribute reduction problem into subproblems. It can be done by applying data decomposition based attribute reduction

The project was funded by the National Science Center awarded on the basis of the decision number DEC-2012/07/B/ST6/01504.

M. Kryszkiewicz et al. (Eds.): PReMI 2015, LNCS 9124, pp. 387–396, 2015.
DOI: 10.1007/978-3-319-19941-2_37

approaches. Such a solution makes it possible to considerably decrease the space complexity. This limitation is essential since approaches for computing all reducts are mainly based on the discernibility matrix method which leads to quadratic complexity with respect to the data size.

One can encounter a few studies on the use of data decomposition for finding all reducts of a decision table. In [3], the discernibility matrix of a decision table is divided into submatrices. The reduct set is computed based on those obtained from the submatrices. In [4], a general data decomposition based approach for computing all reducts of an information system and decision table is proposed. It can be treated as a generalization of the approach from [3]. A data decomposition based method proposed in [13] uses the core attribute to generate all minimal reducts.

All the above approaches were verified theoretically; however, no experimental research has been reported yet. A practical verification is needed to evaluate an important property of a data decomposition based approach, i.e. its scalability compared with the approach that operates on the whole data.

The paper's contribution is to define the notion of scalability in the context of data decomposition based algorithms. The paper also presents experimental research that shows the scalability of data decomposition based algorithms for attribute reduction. The algorithms use the relative discernibility matrix method and are constructed based on the approach introduced in [4] and on its modification proposed in this paper.

Section 2 restates basic notions related to attribute reduction in rough set theory. Section 3 introduces a data decomposition based approach for attribute reduction and proposes its modification using dual reducts. The problem of scalability of data decomposition based algorithms is investigated in Sect. 4 from the theoretical and practical viewpoints. Section 5 provides concluding remarks.

2   Basic Notions

This section restates basic definitions from rough set theory related to attribute reduction.

Definition 1. *[9] (decision table) A decision table is a pair $DT = (U, A \cup \{d\})$, where U is a non-empty finite set of objects, called the universe, A is a non-empty finite set of condition attributes, and $d \notin A$ is the decision attribute.*

Each attribute $a \in A \cup \{d\}$ is treated as a function $a : U \to V_a$, where V_a is the value set of a.

For a decision table a relative indiscernibility relation and relative reduct of the attribute set are defined.

Definition 2. *[8,9] (relative indiscernibility relation) A relative indiscernibility relation $IND(B, d)$ generated by $B \subseteq A$ on U is defined by*

$$IND(B, d) = \{(x, y) \in U \times U : (x, y) \in IND(B) \vee d(x) = d(y)\}, \qquad (1)$$

where $IND(B) = \{(x, y) \in U \times U : \forall_{a \in B} a(x) = a(y)\}$.

Definition 3. *[8, 9] (relative reduct of attribute set) A subset B of A is a relative reduct of A if and only if*

1. $IND(B, d) = IND(A, d)$,
2. $\forall_{\emptyset \neq C \subset B} IND(C, d) \neq IND(B, d)$.

The set of all relative reducts of A on U is denoted by $RED(A, d)$.

The relative reduct set of a decision table can be computed using a relative discernibility function.

Definition 4. *[10] (relative discernibility function) A relative discernibility function f_{DT} of a decision table $DT = (U, A \cup \{d\})$ is a Boolean function of k Boolean variables $a_1^*, \ldots, a_k^*$ that correspond, respectively, to attributes $a_1, \ldots, a_k \in A$ and is defined by*

$$f_{DT}(a_1^*, \ldots, a_k^*) = \bigwedge_{c_{x,y}^d \neq \emptyset} \bigvee_{a \in c_{x,y}^d} a^* \tag{2}$$

where $(c_{x,y}^d)$ is the relative discernibility matrix of DT such that $\forall_{x,y \in U} c_{x,y}^d = \{a \in A : a(x) \neq a(y), d(x) \neq d(y)\}$.

A prime implicant[1] $a_{i_1}^* \wedge \cdots \wedge a_{i_k}^*$ of f_{DT} is equivalent to a relative reduct $\{a_{i_1}, \ldots, a_{i_k}\}$ of DT. For details, see e.g. [4,10].

3   Decomposition of Decision Table

This section introduces two data decomposition based approaches for attribute reduction.

Let $DT = (U, A \cup \{d\})$ be a decision table.

3.1   Reduct Based Approach

In this approach partial results are reduct sets of subtables of the decision table.

A relative indiscernibility relation and relative reduct of attribute set on a universe subset are defined as follows.

Definition 5. *[4](relative indiscernibility relation on universe subset) A relative indiscernibility relation $IND_X(B, d)$ generated by $B \subseteq A$ on $X \subseteq U$ is defined by*

$$IND_X(B, d) = IND(B, d) \cap X \times X \tag{3}$$

Definition 6. *[4] (relative reduct of attribute set on universe subset) A subset $B \subseteq A$ is a relative reduct of A on $X \subseteq U$ if and only if*

---

[1] An implicant of a Boolean function is any conjunction of literals (variables or their negations) such that, if the values of these literals are true under an arbitrary valuation of variables, then the value of the function under the valuation is also true. A prime implicant is a minimal implicant (with respect to the number of literals).

1. $IND_X(B, d) = IND_X(A, d)$,
2. $\forall_{\emptyset \neq C \subset B} IND_X(C, d) \neq IND_X(B, d)$.

The set of all relative reducts of A on $X \subseteq U$ is denoted by $RED_X(A, d)$.

To decompose a decision table (see Fig. 1), each its decision class (i.e., the set $X_v = \{x \in U : d(x) = v\}$, where $v \in V_d$) is divided into subsets (middle subtables), then each pair of subsets of different classes is merged into one set (final subtables). To compute relative reduct sets of a decision table, the subreduct sets of all the final subtables are joined using the following operation.

Definition 7. *[4] (operation $\dot{\cup}$) An operation $\dot{\cup}$ on families of sets is defined by*

1. $\mathcal{S} \dot{\cup} \emptyset = \emptyset \dot{\cup} \mathcal{S} = \mathcal{S}$;
2. $\mathcal{S} \dot{\cup} \mathcal{S}' = \{S \cup S' : S \in \mathcal{S}, S' \in \mathcal{S}'\}$;
3. $\dot{\bigcup}_{i=1}^{k} \mathcal{S}_i = \mathcal{S}_1 \dot{\cup} \mathcal{S}_2 \dot{\cup} \cdots \dot{\cup} \mathcal{S}_k$, *where* $k > 1$.

The family of attribute subsets created by the above operation includes, in general, not only reducts but also supersets of them. To remove unnecessary sets, the following operation is used. Let $min(\mathcal{S})$ be the set of minimal elements of a family $\mathcal{S}$ of sets partially ordered by the relation $\subseteq$.

Theorem 1. *[4] Let* $U = \bigcup_{i=1}^{k} X_{v_i}$, *where* X_{v_i} *is a decision class,* $v_i \in V_d$ *and* $k > 1$ *is the number of different classes in DT. Let* $\mathcal{X}_{v_i}$ *be a covering of* X_{v_i} *$(1 \leq i \leq k)$. The following holds*

$$RED(A, d) = min \left(\dot{\bigcup}_{\substack{1 \leq i < j \leq k, \\ X \in \mathcal{X}_{v_i}, Y \in \mathcal{X}_{v_j}}} RED_{X \cup Y}(A, d) \right) \tag{4}$$

The approach based on Theorem 1 can use any attribute reduction algorithm for computing all reducts in subtables.

3.2   Dual Reduct Based Approach

This subsection proposes an approach that uses dual reducts of subtables of the decision table.

Proposition 1. *The following holds*

$$RED(A, d) = min \left(\dot{\bigcup}_{c_{x,y}^d \neq \emptyset} \{\{a\} : a \in c_{x,y}^d\} \right) \tag{5}$$

Proof. (sketch) We have $RED(A, d) = PI(\bigwedge_{c_{x,y}^d \neq \emptyset} \bigvee_{a \in c_{x,y}^d} a^*)$, where $PI(p)$ is the set of all prime implicants of a Boolean expression p. We obtain $\dot{\bigcup}_{c_{x,y}^d \neq \emptyset} \{\{a\} : a \in c_{x,y}^d\} = \bigwedge_{c_{x,y}^d \neq \emptyset} \bigvee_{a \in c_{x,y}^d} a^*$ and $PI(f_{IS}(a_1^*, \ldots, a_m^*)) \Leftrightarrow min(\mathcal{S}_{f_{IS}})$ where $\mathcal{S}_{f_{IS}}$ is the family of sets corresponding to f_{IS} (for details, see [4]).

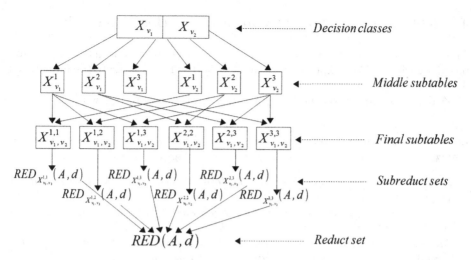

Fig. 1. Data decomposition based attribute reduction of $DT = (U, A \cup \{d\})$, where $U = X_{v_1} \cup X_{v_2}, V_d = \{v_1, v_2\}$, and each class is divided into three subtables

Definition 8. *The dual reduct set to $RED(A, d)$ is defined by*

$$RED^{-1}(A, d) = min\left(\dot{\bigcup}_{S \in RED(A,d)} \{\{a\} : a \in S\}\right) \qquad (6)$$

Proposition 2. *The following holds*

$$RED^{-1}(A, d) = min(\{c_{x,y}^d \neq \emptyset : x, y \in U\}) \qquad (7)$$

This can be proved analogously to Proposition 1.

Theorem 2. *Let $U = \bigcup_{i=1}^{k} X_{v_i}$, where X_{v_i} is a decision class, $v_i \in V_d$ and $k > 1$ is the number of different classes in DT. Let $\mathcal{X}_{v_i}$ be a covering of X_{v_i} $(1 \leq i \leq k)$. The following holds*

$$RED(A, d) = min\left(\dot{\bigcup}_{S \in \mathcal{S}} \{\{a\} : a \in S\}\right) \qquad (8)$$

where $\mathcal{S} = min\left(\bigcup_{\substack{1 \leq i < j \leq k, \\ X \in \mathcal{X}_{v_i}, Y \in \mathcal{X}_{v_j}}} RED_{X \cup Y}^{-1}(A, d)\right).$

Proof. (sketch) By Proposition 1 and $\mathcal{S} = min(\{c_{x,y}^d \neq \emptyset : x, y \in U\})$.

The approach based on Theorem 2 can use any dual attribute reduction algorithm for computing all dual reducts in subtables.

4   Scalability of Data Decomposition Based Algorithms

This section defines the notion of scalability in the context of data decomposition based algorithms. The experiments results reported here illustrate the introduced definitions.

A data decomposition based algorithm is understood in this work as follows.

Definition 9. *(data decomposition based algorithm) A data decomposition based algorithm (dd-algorithm for short) is defined by the following properties*

1. *Decomposition of the database into certain number of portions such that the union of them is the whole database.*
2. *The use of an additional algorithm, called embedded algorithm, to compute partial results on portions or their combinations, e.g. union.*
3. *Merging the partial results to obtain the final result that coincides with that computed by the embedded algorithm on the whole database.*

4.1   Scalability with Respect to the Number of Data Portions

To evaluate the scalability of a dd-algorithm the following definition is proposed. Let n be the data size and p be the number of portions the data is divided into.

Definition 10. *(scalability with respect to the number of data portions) A dd-algorithm is scalable with respect to the number of portions the data is divided into if its run-time is constant as p is increased and n is constant.*

Theoretically, p can be increased up to n, i.e. each data portion includes one object. Normally, such a "dense" data decomposition is unnecessary or even undesirable, e.g. due to the large number of files, each including one object. In practice, the number of data portions can be indirectly defined by the maximal allowed size of the data portion, e.g. memory capacity limitation.

From practical viewpoint, scalability to some extent is sufficient.

Definition 11. *(scalability to extent p' with respect to the number of data portions) A dd-algorithm is scalable to extent p' with respect to the number of portions the data is divided into if its run-time is constant as p is increased up to p' and n is constant.*

To verify scalability of approaches introduced in Sect. 3, ten databases (see Table 1) taken from the UCI Repository (http://archive.ics.uci.edu/ml/) were tested. The discernibility matrix method was employed to compute reducts from decomposed and non-decomposed data for performance comparison. The approach was implemented in C++ and tested using a laptop (Intel Core i5, 2.3 GHz, 4 GB RAM, Windows 7). Tables 2, 3 and 4 show algorithms run-time in seconds. Each test was repeated at least once to eliminate the influence of other running processes on the time performance.

The most interesting observation derivable from Table 2 is that the dd-algorithms are about twice faster compared to the standard algorithm. The latter

Table 1. Characteristics of databases

sym[a]	db	attr	obj	cls	red	sym	db	attr	obj	cls	red
D1	electricity brd	5	45781	31	2	D2	kingrook vs king (krk)	7	28056	18	1
D3	pima ind. diab	9	768	2	28	D4	nursery	9	12960	5	1
D5	shuttle	10	43500	7	19	D6	australian credit appr	15	690	2	44
D7	adult	15	32561	2	2	D8	mushroom	23	8124	7	4
D9	trains	33	10	2	333	D10	sonar, mines vs. rocks	61	208	2	1314

[a]sym – the symbol of a database used in Tables 2, 3 and 4; db – database's name; attr, obj, cls, red – the number of attributes, objects, classes, and reducts, respectively.

is more time consuming because of operations such as loading the whole data into the memory, constructing all pairs of objects to check which of them are to be used to compute the discernibility matrix cells. The only exception is the last database, i.e. *sonar, mines vs. rock*, where the dd-algorithms need (a little) more time than the standard one. This phenomena can be caused by the large number of computations due to a big number of reducts (1314).

The result of the RA version are comparable with those of the DRA one for databases with small number of attributes or reducts. In the remaining cases, the former version is more time consuming. The main reason is the $\dot{\cup}$ operation (see Definition 7 and Theorem 1) is used directly to compute the final reduct set based on subreduct ones. This solution is not efficient since for each two subreduct sets to be joined we construct all possible combinations of their reducts and then check these sets to find the minimal ones. Joining of two subreduct sets in such way that the minimal sets are directly obtained could considerably speed up finding the reduct set. This issue is to be the direction of future work.

Table 2. Attribute reduction with varying number of data portions

db	RDM [a]	RA				DRA			
	1*	1	2	5	10	1	2	5	10
D1	649.87	277.28	278.41	279.52	277.63	276.84	277.70	278.55	277.63
D2	249.63	116.14	116.60	117.09	118.18	116.63	117.27	117.98	118.91
D3	000.14	000.06	000.06	000.09	000.14	000.06	000.06	000.06	000.06
D4	050.06	021.07	021.29	021.91	021.22	021.21	021.04	021.27	021.15
D5	377.57	144.43	145.62	143.79	145.58	144.71	145.78	144.65	146.13
D6	000.22	000.11	000.19	000.97	002.39	000.11	000.11	000.11	000.13
D7	299.46	121.14	120.05	119.92	122.44	121.76	120.64	119.85	121.80
D8	050.98	026.84	027.50	030.69	049.52	025.86	026.40	026.28	026.77
D9	000.27	000.25	000.52	000.48	—	000.02	000.25	000.25	—
D10	090.01	090.53	102.18	181.51	343.16	091.36	100.36	096.17	092.27

[a]RDM – the relative discernibility matrix used on non-decomposed data; RA (DRA) – (dual) reduct based approach; 1*, 1, 2, 5, 10 – the number of portions each class of the databases was decomposed into (1* – no decomposition).

Summing up, the DRA version is more scalable (at least to do degree 10) that the standard one w.r.t. to the number of data portions. The scalability of RA version depends on the sizes of attribute set and reduct set.

4.2 Scalability with Respect to the Data Size

The scalability of a dd-algorithm with respect to the data size is defined as follows.

Definition 12. *(scalability w.r.t. the data size) A dd-algorithm with the fixed $p > 1$ is scalable with respect to the data size if its run-time is constant in comparison with the run-time of the algorithm with $p = 1$ as m is increased*[2].

Note that the scalability of a dd-algorithm w.r.t. the data size does not coincide with the general scalability of an algorithm w.r.t. the data size[3]. The general scalability of a dd-algorithm depends mainly on that of the embedded algorithm.

To investigate this type of scalability, each of three selected databases (i.e., *electricity board, shuttle,* and *sonar, mines vs. rocks*) was divided into ten samples. Both versions of the data decomposition based algorithm (DRA and RA) were tested for $p = 2$.

As previously, one can observe in Table 3 that the dd-algorithms are about twice faster compared to the standard one (the *sonar, mines vs. rocks* database is an exception). Therefore, one can conclude that the scalability of the dd-algorithms w.r.t. the data size (at least for ten times bigger databases) is comparable with that of the standard one.

Table 3. Attribute reduction with growing data size

db	alg	1[a]	2	3	4	5	6	7	8	9	10
D1	RDM	006.34	025.51	058.57	102.60	159.83	229.84	313.17	408.66	525.81	649.87
	RA	002.92	011.63	025.67	046.64	071.33	102.74	139.87	182.61	230.99	283.66
	DRA	002.87	011.76	025.82	046.70	071.50	102.87	139.61	182.84	231.19	283.44
D5	RDM	003.45	014.13	034.22	061.47	096.75	139.22	188.43	243.68	308.62	377.57
	RA	001.48	005.84	013.12	023.30	036.05	052.11	070.82	092.87	118.02	145.62
	DRA	001.51	005.90	013.21	023.43	036.58	052.25	071.45	092.98	117.76	145.78
D10	RDM	000.27	000.93	004.21	005.53	010.17	015.67	027.15	053.47	088.75	090.01
	RA	000.27	000.84	003.63	007.78	014.71	030.37	038.17	069.79	101.03	102.18
	DRA	000.27	000.89	004.31	005.54	009.74	015.30	026.65	051.50	088.90	100.36

[a]1,...,10 mean how many samples of the original database were taken.

4.3 Full Scalability

Using Definition 10 or 12 one can define the full scalability of a data decomposition based algorithm.

---

[2] An algorithm with $p = 1$ is understood as one that is run on non-decomposed data.
[3] An algorithm is scalable w.r.t. the data size if its run-time grows linearly in proportion to the data size.

Definition 13. *(full scalability based on scalability w.r.t. p) A dd-algorithm is fully scalable if its scalable with respect to p as n is increased.*

Definition 14. *(full scalability based on scalability w.r.t. n) A dd-algorithm is fully scalable if its scalable with respect to n as p is increased.*

The above scalability, like that from Definition 10, is not necessary in practice. Therefore, a less strict version is proposed.

Definition 15. *(quasi full scalability) A dd-algorithm is quasi fully scalable if it is scalable with respect to the data size as $p = n/n_p$, where n_p is the fixed data portion size.*

The quasi full scalability is desirable when the data size grows over time and the data partition size is limited in advance. In such a case, we have to increase the number of data portions at any time the data size grows respectively.

This type of scalability was studied using the same databases as in Table 3[4]. For the first two databases the results reported in Table 4 are coincided with those from Table 3. Namely, the increase of p does not influence the run-time of the dd-algorithms. For the third database slower run-time can be observed only for the DRA version. The other version is not efficient due to the same reason as in Sect. 4.1. Therefore, DRA version can be treated as quasi fully scalable.

Table 4. Attribute reduction with growing data size and number of data portions

db	alg	1	2	3	4	5	6	7	8	9	10
D1	RDM	006.34	025.51	058.57	102.60	159.83	229.84	313.17	408.66	525.81	649.87
	RA	002.87	011.55	025.80	045.84	071.81	103.62	140.24	186.92	232.42	283.94
	DRA	002.89	011.50	025.79	045.99	071.64	102.96	140.42	185.94	232.07	284.26
D5	RDM	003.45	014.13	034.22	061.47	096.75	139.22	188.43	243.68	308.62	377.57
	RA	001.50	005.93	013.32	023.66	036.98	053.35	072.92	095.66	119.67	148.15
	DRA	001.48	005.94	013.34	023.64	037.09	053.19	072.64	096.46	120.11	147.50
D10	RDM	000.27	000.93	004.21	005.53	010.17	015.67	027.15	053.47	088.75	090.01
	RA	000.25	000.83	003.99	010.37	024.66	058.27	098.81	172.99	267.15	344.14
	DRA	000.27	000.90	004.23	005.48	009.93	015.88	026.52	046.70	082.40	092.70

5 Conclusion

This paper studied the problem of scalability of data decomposition based algorithms. General definitions devoted to investigating the scalability of such algorithms were proposed. They were applied to evaluate data decomposition based algorithms using the relative discernibility matrix method for computing all reducts of a decision table.

---

[4] The number of portions each class of the database is divided into grows proportionally to the number of taken samples.

The experimental research done under this paper showed that it is possible to obtain the same or better scalability of an attribute reduction algorithm using a data decomposition based approach. The version using dual reducts is more likely to be scalable than that using reducts themselves. In the latter case, the main reason for the increase of the number of computations is that the cardinalities of subreduct sets can be big, often considerably bigger than that of the reduct set. However, the reduct based approach can be improved by applying a more efficient method for computing the final reduct set based on subreduct ones.

References

1. Chen, D., Zhao, S., Zhang, L., Yang, Y., Zhang, X.: Sample pair selection for attribute reduction with rough set. IEEE Trans. Knowl. Data Eng. **24**(11), 2080–2093 (2012)
2. Degang, C., Changzhong, W., Qinghua, H.: A new approach to attribute reduction of consistent and inconsistent covering decision systems with covering rough sets. Inf. Sci. **177**(17), 3500–3518 (2007)
3. Deng, D., Huang, H.-K.: A new discernibility matrix and function. In: Wang, G.-Y., Peters, J.F., Skowron, A., Yao, Y. (eds.) RSKT 2006. LNCS (LNAI), vol. 4062, pp. 114–121. Springer, Heidelberg (2006)
4. Hońko, P.: Attribute reduction: A horizontal data decomposition approach. Soft Comput. (2014). doi:10.1007/s00500-014-1554-8
5. Hu, X., Cercone, N.: Learning in relational databases: a rough set approach. Comput. Intell. **11**(2), 323–338 (1995)
6. Kryszkiewicz, M.: Comparative study of alternative type of knowledge reduction in inconsistent systems. Int. J. Intell. Syst. **16**, 105–120 (2001)
7. Kryszkiewicz, M.: Rough set approach to incomplete information systems. Inf. Sci. **112**(1–4), 39–49 (1998)
8. Miao, D., Zhao, Y., Yao, Y., Li, H.X., Xu, F.: Relative reducts in consistent and inconsistent decision tables of the Pawlak rough set model. Inf. Sci. **179**(24), 4140–4150 (2009)
9. Pawlak, Z.: Rough Sets. Theoretical Aspects of Reasoning about Data. Kluwer Academic, Dordrecht (1991)
10. Skowron, A., Rauszer, C.: The discernibility matrices and functions in information systems. In: Słowiński, R. (ed.) Intelligent Decision Support. Springer, Amsterdam (1992)
11. Swiniarski, R.: Rough sets methods in feature reduction and classification. Int. J. Appl. Math. Comput. Sci. **11**(3), 565–582 (2001)
12. Thi, V.D., Giang, N.L.: A method for extracting knowledge from decision tables in terms of functional dependencies. Cybern. Inf. Technol. **13**(1), 73–82 (2013)
13. Ye, M., Wu, C.: Decision table decomposition using core attributes partition for attribute reduction. In: ICCSE. vol. 23, pp. 23–26. IEEE (2010)
14. Zhang, X., Mei, C., Chen, D., Li, J.: Multi-confidence rule acquisition oriented attribute reduction of covering decision systems via combinatorial optimization. Knowl.-Based Syst. **50**, 187–197 (2013)

Application of Fuzzy Rough Sets to Financial Time Series Forecasting

Mariusz Podsiadło[1] and Henryk Rybinski[2]([✉])

[1] Misys Plc, London, UK
mariusz.podsiadlo@misys.com
[2] Warsaw University of Technology, Warsaw, Poland
hrb@ii.pw.edu.pl

Abstract. This paper investigates experimentally the feasibility of Fuzzy Rough Sets in building trend prediction models for financial time series, as related research is scarce. Aside of the standard classification accuracy measures, financial profit and loss backtesting using a sample market timing strategy was performed, and profit related quality of the tested methods compared against that of buy&hold strategy applied to the used market indices. The experiments show that Fuzzy Rough Sets models present a viable basis for forecasting market movement direction and thus can support profitable market timing strategies.

Keywords: Artificial intelligence · Rough sets · Fuzzy rough sets · Financial time series prediction

1 Introduction

Forecast of financial markets' conditions is crucial not only for involved institutions and individuals but also for economical well-being of entire nations. On the other hand, financial markets are an inexhaustible source of noisy and incomplete information in a form of financial time series. The number of dimensions, many of which have hidden correlations, and amount of data to be analysed make delivery of good models with sufficient predictive quality very challenging[1]. In this context, a correct forecast of market direction was shown to be sufficient to generate profitable trading strategies [10].

Attempts to employ data discovery and soft computing methods to deliver predictive models have a long track record, including a growing set of reported work based on Rough Sets [13]. Rough Sets and Fuzzy Rough Sets models [14] deliver a way to infer knowledge from noisy and incomplete data and to automatically select significant data features (reducts). This forms a good basis to cope with the challenges of financial time series like data inconsistency and number of dimensions. Rough Sets were shown to be applicable to any time scale, down

---

[1] It is enough to look at the size of quantitative financial engineering teams in any major financial institution.

© Springer International Publishing Switzerland 2015
M. Kryszkiewicz et al. (Eds.): PReMI 2015, LNCS 9124, pp. 397–406, 2015.
DOI: 10.1007/978-3-319-19941-2_38

to intraday trading [9]. However, the recent advances in the field of fuzzy rough sets [14,16] have not been verified in the area of financial time series forecasting yet, although the flexible definition of a similarity relation and ability to work directly with numeric variables make fuzzy rough sets an interesting candidate. Furthermore, related experiments using soft computing models often reported good classification accuracy but financial backtesting and the actual financial performance of the evaluated models were seldom shown. This prompted some reports, where several soft computing models were tested and deemed to deliver performance at best on par with statistical methods and simple buy and hold strategy [7].

Consequently, this work attempts to contribute new research data with regards to the feasibility of rough sets and fuzzy rough sets models to the task of financial time series forecasting. Both, the classification accuracy and financial performance of the examined models were tested using large real life data sets of several well known market-neutral indices. This work extends also on research described in [12] by considering Fuzzy Rough Sets.

This paper is organized as follows: Sect. 2 describes the experiment setup, used data, preprocessing and prediction performance analysis framework. Section 3 presents and discusses the experiment results. Section 4 concludes the paper.

2   Experiment Setup

In this study, Rough Sets (RS), Variable Precision Rough Sets (VPRS) and Fuzzy Rough Sets [14] combined with the k-nearest neighbour algorithm [16] were used to predict direction of movements of US S&P500, German DAX and Hong Kong Hang Seng (HSI) stock market indices[2]. The implementation of the positive region based fuzzy-rough nearest neighbor algorithm [16] provided in [15] was used, with two definitions of the lower/upper set approximation based on:

1. VQRS: Vaguely Quantified Rough Sets [3].
2. Lukasiewicz triangular norm operators, with tolerance relation [5]:

$$R_a(x, y) = 1 - \frac{(|a(x) - a(y)|)}{|a_{max} - a_{min}|} \tag{1}$$

Predictive models based on Variable Precision Rough Sets [17] were used as the reference. A detailed review of roughs sets, fuzzy rough sets, their extensions and their applications in finance is provided in [4,8,13].

All the models were applied to time series data samples of the used stock market indices in order to generate movement predictions using a walk-forward rolling time window cross-validation procedure [6]. Subsequently, using a back-testing procedure applied to Exchange Traded Funds (ETFs) associated with

---

[2] Beating a buy and hold strategy in these efficient markets ought to be challenging.

respective indices, the financial profit and loss of a sample market timing strategy and the *buy and hold* benchmark were compared. The experiment environment was built using R package RoughSets [15], ROSETTA Rough Sets system [11] and an SQL database.

2.1   Input Data

Historical daily Open, Low, High, Close prices, and Volume (OHLC&V) time series of the S&P500, DAX and HSI stock market indices were used to train the models and generate index movement predictions. Additional input was provided by daily closing values of the S&P 500 near-term volatility index VIX[3], which was used to gauge market confidence and covered the below-given periods. Data was divided into training, calibration and test samples, whereas the training sample was set to be 3333 trading days, and the testing period was fixed between 29th of January 2010 and 31st of December 2013 for all indices. This caused a slight difference in the calendar coverage and total sample lengths due to differing holiday calendars of the used indices, i.e. for S&P500 from 1st of August 1996 to 30th of December 2013, DAX from 4th of September 1996 to 27th of December 2013, and HSI from 20th of November 1996 to 30th of December 2013. The associated ETFs, i.e. SPDR S&P500, iShares Core DAX UCITS, and Tracker Fund of Hong Kong, designed to track the performance of the respective indices, were subsequently used to backtest the financial performance of predictions generated for the test sample. The data was adjusted for dividends, splits and mergers. All the used indices and ETF data covered the same time period. Figure 1 displays relative performance of the used index time series over the applied time period.

The input data included 16 conditional variables and one decision variable. The following time series attributes were used as conditional variables: *Open, Close, Volume, VIX Close*. The following derived technical indicators were used as conditional variables: *Acceleration* ($n = 5$), *Average Directional Index* ($n = 50$), *Commodity Channel Index* ($n = 20$), *Chaikin Oscillator* ($n = 3, m = 10$), *Exponential Moving Average* ($n = 50$), *MACD* ($n = 12, m = 26$), *Momentum* ($n = 5$), *Price Oscillator* ($n = 10, m = 5$), *RSI* ($n = 14$), *Price Rate of Change* ($n = 1$), *Williams A/D*, *Williams %R*. See [1] for detailed description of the indicators[4].

Simple daily return R_i of the given index for a future period i served as the basis for the decision variable d_i (prediction target) defined to be the change direction of the daily simple return R_i as follows:

$$d_i = \begin{cases} down & \text{if } R_i \in (-\infty, 0) \\ up & \text{otherwise} \end{cases} . \tag{2}$$

---

[3] http://www.cboe.com/VIX.
[4] The Internet also provides abundance of related information.

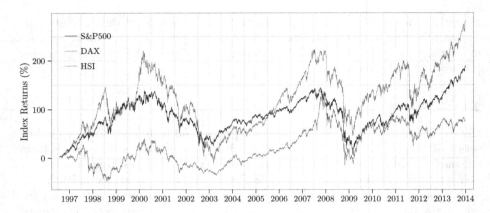

Fig. 1. Cumulated return of used indices over the applied time period (Color figure online)

where:

- d_i - the decision variable for i-th sample,
- $R_i = \left(\frac{p_i}{p_{i-1}} - 1\right) * 100$ - simple daily return of the Close price p_i for i-th period (trading day).

The discrete decision variable d_i was used for both, rough sets and fuzzy rough sets classifier models.

2.2 Market Timing Strategy

The financial performance of the tested models was verified using a long-only market timing strategy defined as follows:

$$\text{if } d_i = \begin{cases} down & \text{then } sell \text{ at } Open_{i+1} \\ up & \text{then } buy \text{ at } Open_{i+1} \\ otherwise & \text{then no action} \end{cases} \tag{3}$$

where:

- d_i - decision variable for i-th sample,
- $Open_{i+1}$ - the open price at the next period ($i + 1$-th sample).

Aside of handling the up and down forecast, the strategy accommodated for the case where a prediction model generated no prediction for the forecasted period. In this case the strategy generated no trading signal resulting in no action for the period. The strategy was defined as long only, i.e. no short position was allowed. The *buy and hold* strategy was used as the benchmark strategy. For the purpose of financial performance comparison the ability to buy/sell fractional shares was enabled. This allowed to fully utilize the available equity (initial equity was set to 10'000) in both, buy&sell and market timing strategies. The transaction costs (commissions, slippage, etc.) were ignored in this experiment, as the focus was on the prediction performance.

2.3   Setup and Testing of Classifier Models

All the models were tested with each of selected indices (see Fig. 1) using the walk-forward method with a rolling time window [6]. The walk-forward method derives from the standard cross-validation but also observes the time order of time series and so prevents sample and *look-ahead bias*. For each iteration of the walk-forward cross-validation a consecutively aligned training, calibration and testing samples were selected. The training sample was fixed at 3333 trading days. Multiple combinations of calibration and testing samples were defined drawing their lengths from the sets {21,42,63} and {3,5,21} trading days, respectively.

For each combination of training-calibration-test sample set in the walk-forward cycle, predictive models were created using the training sample. The parameter-driven models, i.e. the fuzzy rough positive region based nearest neighbour model with VQRS approximation ($FPOSNN_{VQRS}$) and the VPRS based model, were then tuned using a grid search on the respective calibration sample, so the optimal (i.e. resulting in the highest classification accuracy/lowest error) model parameter set for the given calibration sample and iteration of the walk forward procedure was found. For the reference VPRS model the tuned parameter was the VPRS $\beta \in \{0.0, 0.25, 0.49\}$. For the $FPOSNN_{VQRS}$ model, the optimal boundaries (α, β) of the *most* quantifier were searched in the set $(\alpha, \beta)_{most} \in \{(0.2, 1), (0.4, 1), (0.6, 1)\}$. The *some* quantifier had fixed boundary parameters set at $(\alpha, \beta)_{some} = (0.1, 0.6)$. The FPOSNN model using the Lukasiewicz $\mathcal{T}$-norm operators ($FPOSNN_{\mathcal{T}_L}$) did not require a specific tuning. The FPOSNN models used the fixed number of neighbours $k = 10$ in the middle of the range suggested in [16].

For VPRS-based analysis the input variables were preprocessed as follows:

1. A subset of technical indicators and the VIX Close values, for which a common consensus on interpretation of their values exists, were discretized using manually defined intervals.
2. The remaining conditional variables were first normalized using the z-score method and subsequently discretized using the equal binning algorithm with 3 bins. The mean and standard deviation of the learning sample as well as its equal binning cuts were used to normalize and discretize the validation and testing samples.

For the fuzzy rough sets-based processing all conditional variables were normalized, like in case of the rough sets model. Fuzzy rough sets do not require discretization of conditional variables. The decision variable d (Eq. 2) was encoded as 1 for an *up* movement, and -1 for a *down* movement. The VPRS model set the decision variable d to 0 in case, the ruleset was unable to predict its value (i.e. no matching rule was found, as per Eq. 3).

3   Experiment Results

The walk-forward procedure was executed using rough sets and fuzzy rough sets models generated for all combinations of test sample lengths and model specific

Table 1. Prediction performance of tested models

Test Sample Length	Test Sample Hit Ratio (%)								
	$VPRS$			$FPOSNN_{VQRS}$			$FPOSNN_{T_L}$		
	S&P500	HSI	DAX	S&P500	HSI	DAX	S&P500	HSI	DAX
Calibration period = 21									
3	53.09	**50.51**	51.64	*54.20*	50.10	51.44	53.90	50.20	50.65
5	52.29	**50.51**	**51.74**	53.60	50.00	51.54	53.19	49.80	50.94
21	52.58	**50.51**	49.95	53.90	49.70	51.64	52.89	50.10	51.44
Calibration period = 42									
3	51.87	50.61	49.85	52.89	50.81	52.14	53.18	51.11	52.04
5	51.25	50.71	50.55	**53.09**	50.61	52.04	51.87	51.42	**52.23**
21	50.30	**51.62**	48.96	51.17	50.20	51.84	50.25	50.30	52.14
Calibration period = 63									
3	50.56	52.13	49.35	**52.89**	50.00	52.83	52.28	50.10	52.23
5	51.47	51.52	49.85	52.28	50.30	52.83	52.08	50.71	52.53
21	49.85	*52.43*	49.85	52.48	49.80	*52.93*	52.58	50.10	52.63

parameters as defined in Sect. 2.3. Both, the predictive and financial performance were tested with results described below.

3.1   Prediction Performance

Table 1 presents the total classification performance of the generated models. The prediction models based on $FPOSNN_{VQRS}$ delivered classification performance with the highest accuracy for S&P500 and DAX indices. i.e. 54.21 % (calibration sample/forecast period=21/3 trading days) and 52.93 % (calibration sample/forecast period=63/21 trading days), respectively. The VPRS rule based model delivered the best accuracy for HSI with 52.43 % vs. the worst performing $FPOSNN_{VQRS}$ model with classification accuracy of 49.80 % for the calibration sample/forecast period=63/21 trading days. The FPOSNN models consistently outperformed the VPRS rule based model for all tested combinations of calibration and forecast periods for the S&P index. For S&P and DAX time series the $FPOSNN_{VQRS}$ model delivered the best average accuracy of 53.11 % and 52.14 %, respectively, vs. the respective average performance of the VPRS rule based model of 51.54 % and 50.90 %. For the HSI index the $FPOSNN_{VQRS}$ and $FPOSNN_{T_L}$ models delivered an average classification performance of 50.17 % and 50.43, vs. 51.17 % delivered by the VPRS model. The length of calibration and forecast periods had an impact on the classification efficiency for all tested models. The use and increasing length of the calibration sample had a positive impact only on performance of the VPRS model, when predicting movements of the HSI index. For other indices the increasing length of the calibration sample had none to negative impact on classification accuracy of the tested models. This shows that the time distance between the classified sample and the ruleset or the nearest neighbour base sample (i.e. the training sample) was more important than the grid search based tuning of model parameters. The latter required a calibration sample, thus increasing the time distance between the training and test samples. One can conclude that consideration of time distance as a factor in the classification process would most likely improve it [12]. It is also worth noticing that both, DAX and S&P500 had a strong upwards trend in more than a half of the testing period, whereas HSI growth was relatively moderate with a sidewards trend in the same time period (see Fig. 2). Thus, DAX and S&P500 time series supported trend following prediction models, and by extension the nearest neighbour algorithm taking decisions based on a relatively small number of k samples. Prediction of sidewards movements (so called *whipsaws*), like it was the case for the HSI time series, is recognised as challenging in the field of technical analysis. In this case the rule based algorithm was relatively better. The classification accuracy alone is however not sufficient to deliver a profitable predictive model. A decisive factor is the *timing* of the correct prediction, which directly impacts the financial performance.

3.2 Financial Performance

The profit and loss backtesting using the market timing strategy outlined in Sect. 2.2 as well as the *buy and hold* benchmark strategy were executed for all

Fig. 2. Cumulated return time series (calibration sample of 63 trading days) (Color figure online)

combinations of the training, calibration and test sample lengths. The trading simulation used tracking ETFs associated with the underlying indices as the traded equity instrument. The trading costs and slippage were ignored. Figure 2 displays the cumulated return time series of tested models vs. buy and hold strategy for test runs using the calibration period of 63 trading days[5], whereas Table 2 displays related financial performance summary. The S&P500 and DAX indices were in a strong upwards trend for the second half of the testing period, reflected in the total return of 85.39 % and 72.69 % of the respective tracking ETFs (and associated buy and hold strategy). Consequently, beating these indices was expected to be difficult. This was confirmed for the majority of the calibration/test sample length combinations, where the predictive models underperformed the market, although still generating positive returns above 30 % (DAX) and 40 % (S&P500). An exception was the $FPOSNN_{VQRS}$ model, consistently beating the S&P500 performance when using the calibration period od 21 days, and delivering the peak return of 99.88 % using the 21/21 combination of calibration/test sample lengths. A similar exceptional performance in case of the DAX benchmark was delivered by the VPRS model with 90.13 % and 78.68 % return using the 42/3 and 42/5 trading days combination of calibration/test, respectively. In case of the S&P500 index, the decrease of financial return was coupled with the increased length of the calibration period, reflecting the similar observation done with regards to the classification accuracy. For the DAX index, this trend was not present, with the VPRS model performing the best on the calibration sample of 42 days. It should be noted, that DAX movements were more volatile than these of S&P500, although both indices shared the same strong upward trend in the second half of the testing period. This may explain the reason for the lack of the clear dependencies visible in the case of the S&P500 index, i.e. the US VIX volatility index might not be able to fully reflect the volatility of the German DAX index. Thus, it may be necessary to define more localized input variables for the tested models, so the inherent capability of Rough Sets to select reducts is applied for each underlying market. The HSI index reflects the above observations amplified by the fact that it had a relatively weaker upwards trend, with multiple whipsaws, resulting in the total return of the tracking ETF of only 24.36 % over the testing period. Consequently, the correct prediction of the index movement played larger role and can be seen in an excellent performance of $FPOSNN_{VQRS}$ and VPRS models outperforming the Hong Kong market for all combinations of tested lengths of calibration and test samples. Both underlying models, namely VPRS and VQRS are closely related, aiming at reduction of classification noise caused by outliers. On the other hand, the $FPOSNN_{TL}$ model underperformed the market, which would indicate that this model is actually trend following and so unable to work in a nontrending market. Even though the tested models mostly underperform DAX and S&P500 markets due to the above described dependencies, the total return alone gives a biased view of the performance without considering the associated risk.

---

[5] The full data set was not shown due to space limitations but can be obtained from the authors.

Table 2. Financial performance of tested models on a holdout sample

Model	Cum. Return			Information Ratio			Max. Drawdown		
	HSI	DAX	S&P500	HSI	DAX	S&P500	HSI	DAX	S&P500
Calibration period = 63, Test period = 3									
$VPRS$	**57.96**	35.50	58.16	*0.46*	-0.43	-0.37	18.18	**23.01**	*12.17*
$FPOSNN_{VQRS}$	54.92	44.54	**72.68**	0.21	-0.19	*-0.11*	**14.76**	32.03	33.66
$FPOSNN_{TL}$	20.92	**51.62**	63.25	−0.07	**-0.14**	-0.13	15.90	33.27	32.54
Calibration period = 63, Test period = 5									
$VPRS$	**48.94**	49.34	68.83	**0.34**	-0.26	-0.22	18.18	20.60	**16.50**
$FPOSNN_{VQRS}$	45.22	46.60	63.47	0.15	-0.18	**-0.17**	14.76	32.03	35.24
$FPOSNN_{TL}$	22.40	49.19	57.45	−0.04	**-0.16**	-0.18	**14.64**	33.27	34.02
Calibration period = 63, Test period = 21									
$VPRS$	44.14	**58.88**	42.17	**0.28**	-0.15	-0.60	16.90	**20.82**	14.96
$FPOSNN_{VQRS}$	*62.42*	58.25	**57.34**	0.25	*-0.09*	-0.22	*13.21*	32.45	33.47
$FPOSNN_{TL}$	18.92	57.37	56.87	−0.10	-0.10	**-0.18**	14.87	33.68	33.07
Buy&Hold	24.36	*72.69*	*85.39*	-	-	-	32.99	32.65	18.61

Especially, the information ratio and maximal drawdowns [2] showed the level of risk associated with the permanent exposure required by the buy&hold strategy. The FPOSNN based models had risk similar to that of DAX and S&P but performed well vs. the HSI index, with less than 50 % of the drawdown caused by the index. The VPRS rule based model also delivered performance slightly lower than DAX and S&P500 indices but with a much lower level of risk (drawdown). Like the $FPOSNN_{VQRS}$ model, it outperformed the HSI index with the half of the risk, reflected in the 50 % lower drawdown and positive information ratio.

4   Conclusions

Variable Precision Rough Sets and Fuzzy Rough Sets were used to generate binary classifiers applied to real life market data of multiple stock indices. Both, the classification accuracy and ability to support profitable market timing strategies were evaluated. The experimental results showed that Fuzzy Rough Sets based models were able to outperform the VPRS based model in terms of classification accuracy in the majority of experiments. In terms of financial performance, the VPRS and $FPOSNN_{VQRS}$ models were able to outperform the buy&hold strategy applied to the HSI index. Considering the simplicity of the used strategy and the strong upwards trend of S&P500 and DAX, the models were robust when applied to these times series. The application of Rough Sets models to portfolios representing a specific strategy can be simulated by tracking certain indices or exchange traded funds and is planned to be included in the future research. In general, the effectiveness of the delivered trading signals can be further improved by considering the time distance in classification algorithms and wider set of input variables.

References

1. Achelis, S.: Technical Analysis from A to Z, 2nd edn. McGraw-Hill Education, New York (2000)
2. Bacon, C.R.: Practical Risk-Adjusted Performance Measurement. Wiley, Oxford (2012)
3. Cornelis, C., De Cock, M., Radzikowska, A.M.: Vaguely quantified rough sets. In: An, A., Stefanowski, J., Ramanna, S., Butz, C.J., Pedrycz, W., Wang, G. (eds.) RSFDGrC 2007. LNCS (LNAI), vol. 4482, pp. 87–94. Springer, Heidelberg (2007)
4. Dubois, D., Prade, H.: Rough fuzzy sets and fuzzy rough sets. Int. J. Gen. Syst. 17(2–3), 191–209 (1990)
5. Jensen, R., Cornelis, C., Shen, Q.: Hybrid fuzzy-rough rule induction and feature selection. In: 2009 IEEE International Conference on Fuzzy Systems, FUZZ-IEEE 2009, pp. 1151–1156. IEEE (2009)
6. Kaastra, I., Boyd, M.: Designing a neural network for forecasting financial and economic time series. Neurocomputing 10(3), 215–236 (1996)
7. Kinlay, J., Rico, D.: Can machine learning techniques be used to predict market direction?-the 1,000,000 model test (2011)
8. Komorowski, J., Pawlak, Z., Polkowski, L.: Rough sets: a tutorial. In: Pal, S.K., Skowron, A. (eds.) Rough Fuzzy Hybridization: A New Trend In Decision-Making, pp. 3–98. Springer, Singapore (1999)
9. Lee, S.J., et al.: Using rough set to support investment strategies of real-time trading in futures market. Appl. Intell. 32(3), 364–377 (2010)
10. Leung, M.T., Daouk, H., Chen, A.S.: Forecasting stock indices: a comparison of classification and level estimation models. Int. J. Forecast. 16(2), 173–190 (2000)
11. Øhrn, A.: Rosetta technical reference manual, Department of Computer and Information Science, pp. 1–66. Norwegian University of Science and Technology (NTNU), Trondheim (2000)
12. Podsiadło, M., Rybiński, H.: Financial time series forecasting using rough sets with time-weighted rule voting. Rep. of Inst. of Comp. Sci., 1/2015, Warsaw University of Technology, subm. to. Eur. J. Oper. Res
13. Podsiadło, M., Rybiński, H.: Rough sets in economy and finance. In: Peters, J.F., Skowron, A. (eds.) Transactions on Rough Sets XVII. LNCS, vol. 8375, pp. 109–173. Springer, Heidelberg (2014)
14. Radzikowska, A.M., Kerre, E.E.: A comparative study of fuzzy rough sets. Fuzzy sets syst. 126(2), 137–155 (2002)
15. Riza, L.S., et al.: Implementing algorithms of rough set theory and fuzzy rough set theory in the R package RoughSets. Inf. Sci. 287, 68–89 (2014)
16. Verbiest, N., Cornelis, C., Jensen, R.: Fuzzy rough positive region based nearest neighbour classification. In: 2012 IEEE International Conference on Fuzzy Systems (FUZZ-IEEE), pp. 1–7. IEEE (2012)
17. Ziarko, W.: Variable precision rough set model. J. comput. syst. sci. 46(1), 39–59 (1993)

A New Post-processing Method to Detect Brain Tumor Using Rough-Fuzzy Clustering

Shaswati Roy and Pradipta Maji$^{(\boxtimes)}$

Biomedical Imaging and Bioinformatics Lab, Machine Intelligence Unit,
Indian Statistical Institute, Kolkata, India
{shaswatiroy_t,pmaji}@isical.ac.in

Abstract. Automatic and accurate brain tumor segmentation from MR images is one of the important problems in cancer research. However, the lack of shape prior and weak contrast at boundary make unsupervised brain tumor segmentation more challenging. In this background, a new brain tumor segmentation method is being developed, integrating judiciously the merits of multiresolution image analysis technique and rough-fuzzy clustering. One of the major issues of the clustering based segmentation method is how to extract brain tumor accurately, since tumors may not have clearly defined intensity or textural boundaries. In this regard, this paper presents a new post-processing method for clustering based brain tumor detection. It combines the merits of mathematical morphology and the concept of rough set based region growing approach to refine the result obtained after clustering, thereby ensuring the accurateness of brain tumor segmentation application. The performance of the proposed approach, along with a comparison with related methods, is demonstrated on a set of synthetic and real brain MR images.

1 Introduction

Automatic segmentation of healthy and pathologic brain tissues from MRI plays an important role in brain tumor detection application. Early and accurate brain tumor segmentation from MR images is a difficult task in many cancer research applications. Brain tumors may be of any size, may have a variety of shapes, may appear at any location in brain, and may appear in different image intensities. Some tumors also deform other structures and appear together with edema that changes intensity properties of the nearby region.

The challenges associated with automatic brain tumor segmentation have given rise to many different approaches [2,7,8,10–12]. Classification based tumor detection algorithms are widely used in brain tumor detection applications. These methods are constrained to the supervised [7,9] or unsupervised [1]. Menze et al. [7] combined a healthy brain atlas with a tumorous brain atlas to segment tumors using a generative probabilistic model and spatial regularization. Bauer et al. [9] combined support vector machine using multispectral intensities and textures with subsequent hierarchical regularization based on

© Springer International Publishing Switzerland 2015
M. Kryszkiewicz et al. (Eds.): PReMI 2015, LNCS 9124, pp. 407–417, 2015.
DOI: 10.1007/978-3-319-19941-2_39

conditional random fields. This method requires four modalities, namely, T1-weighted, T1-weighted with contrast agent, T2-weighted, and FLAIR to classify the tumor region. Fletcher-Heath et al. [1] combined fuzzy clustering and integrated domain knowledge to improve the tumor segmentation applied on T1-, T2-, and PD-weighted images. A block diagram illustrating the main steps during the clustering based brain tumor segmentation pipeline is shown in Fig. 1. The pre-processing step may include denoising, skull stripping, correction of intensity inhomogeneity, or registration for multi-modal input images. Feature vectors are then generated for clustering algorithm to extract the tumor class.

Fig. 1. Block diagram of the clustering based brain tumor detection method

In this regard, a texture-based brain MR image segmentation method is presented in [6] that can be used for detecting brain tumor from MR images. It judiciously integrates the merits of multiresolution image analysis and rough-fuzzy clustering. The multiresolution wavelet analysis is used to extract scale-space feature vector for each pixel of the given brain MR image. Since the boundary between brain and skull is relatively strong on T1 scan, a skull stripping algorithm is used to extract the brain tissues and remove non-cerebral tissues. However, the use of wavelet decomposition may give rise to some irrelevant and insignificant features. Hence, an unsupervised feature selection method is used to reduce the dimensionality of feature space by maximizing both relevance and significance of the selected features. Finally, the robust rough-fuzzy c-means algorithm [5] is used for segmentation of brain MR image. While the integration of both membership functions of fuzzy sets enables efficient handling of overlapping classes in noisy environment, the concept of lower and upper approximations of rough sets deals with uncertainty, vagueness, and incompleteness in class definition. In effect, it groups similar textured tissue classes contained in the image.

However, the lack of intensity or shape priors and weak contrast at boundary make unsupervised brain tumor segmentation more challenging. Once the tumor class is extracted by clustering algorithm, the result can be further improved by a sequence of post-processing steps as shown in Fig. 1. Hsieh et al. [2] proposed a simple region-growing and knowledge-based post-processing step after the fuzzy classification applied on non-contrasted T1- and T2-weighted MR images, followed by a morphology operator. Iscan et al. [3] developed a filtering and region growing based post-processing approach on segmented tumorous tissue.

In this background, this paper presents a new post-processing method. It integrates the merits of morphological operations and the notion of rough set theory in formulation of region growing method. Since tumors may not have clearly defined intensity or textural boundaries, there may be some ambiguity at boundary region of a tumor class. This uncertainty is handled by incorporating the rough set theory into region growing approach. Hence, this post-processing step plays an important role in order to ensure the correctness of the diagnosis in brain tumor segmentation applications. The performance of the proposed approach, along with a comparison with related methods, is demonstrated on a set of synthetic and real brain MR images both qualitatively and quantitatively.

2 Proposed Post-Processing Method for Tumor Detection

This section briefly presents rough-fuzzy clustering and wavelet based brain MR image segmentation method reported in [6]. It consists of five steps as follows:

1. Generation of mask from brain MR image for identification of brain region;
2. Dyadic wavelet analysis of MR image using Daubechies 6-tap filter;
3. Generation of feature vectors for brain region using the mask;
4. Unsupervised feature selection to select relevant and significant features for clustering; and
5. Rough-fuzzy clustering to generate segmented image.

However, the above algorithm results in oversegmentation or undersegmentation of brain tumor due to the ambiguities at tumor boundaries. Hence, a new post-processing method, based on morphology and rough set based region growing method is used to handle efficiently these ambiguities. Due to partial volume effect, the brain tumor segmentation based on clustering algorithm may produce many residual areas around the tumor region. Again, since the tumor is not homogeneous, clustering based segmentation may produce small holes within the tumor class. Hence, the morphological operations, namely, closing and opening, in the order, followed by finding largest connected component, are applied to eliminate these residual areas, as well as to eliminate any discontinuity within tumor mass. The closing and opening morphology use square shaped structuring element of different dimensions. Two-pass connected component labelling is embedded for finding the largest connected component. Then, a rough set based region growing method is proposed to efficiently handle the ambiguities at anatomical tumor boundaries.

The theory of rough sets begins with the notion of an approximation space, which is a pair $< U, R >$, where U be a non-empty set, the universe of discourse, and R an equivalence relation on U. Given an arbitrary set $X \in 2^U$, in general, it may not be possible to describe X precisely in $< U, R >$. One may characterize X by a pair of lower and upper approximations. The lower approximation $\underline{R}(X)$ is the union of all the elementary sets which are subsets of X, and the upper approximation $\overline{R}(X)$ is the union of all the elementary sets which have a non-empty intersection with X. The interval $< \underline{R}(X), \overline{R}(X) >$ is the representation of X in the approximation space $< U, R >$ or simply called the rough set of X. The accuracy of X, denoted by $\alpha_R(X)$, is the ratio of the number of objects in its lower approximation to that in its upper approximation; namely

$$\alpha_R(X) = |\underline{R}(X)||\overline{R}(X)|^{-1} \tag{1}$$

Note that the higher the accuracy of a subset, the better is its approximation.

The proposed region growing technique is based on the concept of rough set theory. Let T be the tumor class. Application of various morphological operations produces the region that possibly belongs to T, that is, upper approximation of the tumor class, $\overline{A}(T)$. The lack of shape prior and weak contrast at anatomical boundary induce uncertainty of belongingness of boundary region pixels. Hence, it puts some non-tumorous cells into $\overline{A}(T)$ because they are characterized as indiscernible with other actual tumorous pixels, using the available information.

The accuracy of approximation of T as in (1), in fact, provides a measure of how closely the rough set is approximating the target set with available knowledge. Hence, the non-tumorous tissues in $\overline{A}(T)$ are discarded in this methodology in order to increase the accuracy of approximation. In this regard, a region growing approach based on the gradient information of input image is proposed, which concentrates on elimination of non-tumorous healthy brain tissues from pathologic region appropriately, thereby maximizing the rough set accuracy.

The center of gravity (CoG) is computed within $\overline{A}(T)$. The CoG of $\overline{A}(T)$ is assumed to certainly belong to the tumor region, and hence the CoG is assigned initially in the lower approximation of the tumor class $\underline{A}(T)$. Then, a simple region growing approach is followed with CoG as the seed point. An edge map of the input brain MR image is constructed within the region of $\overline{A}(T)$. In this approach, Sobel's gradient operator is used. It starts growing initially from the CoG to spread over the edge map of $\overline{A}(T)$. The region of interest $\underline{A}(T)$ expands until a certain criterion is met which is based on the mean gradient magnitude of input image within the region of $\overline{A}(T)$.

Let $\delta(x)$ be gradient magnitude of pixel x. The threshold Δ is computed as:

$$\Delta = \frac{1}{|\overline{A}(T)|} \sum_{x \in \overline{A}(T)} \delta(x), \tag{2}$$

Based on the value of Δ, the unassigned pixel is included in $\underline{A}(T)$ using the pixel assignment rule for lower approximation described as below:

$$\underline{A}(T) \leftarrow \underline{A}(T) \cup \{x \in \overline{A}(T) | \delta(x) \leq \Delta \wedge (\exists y \in \underline{A}(T) \text{ s.t. } x \in \mathcal{N}(y))\}, \tag{3}$$

where $\mathcal{N}(p)$ represents the neighbors of the pixel p using eight-connectivity. This region growing method is repeated until no more changes occurred. After building $\underline{A}(T)$, the boundary of $\underline{A}(T)$ is refined using the lower approximation refinement rule as presented below:

$$\underline{A}(T) \leftarrow \underline{A}(T) \cup \{x \in \overline{A}(T) | \delta(x) \geq \Delta \wedge (\exists y \in \underline{A}(T) \text{ s.t. } x \in \mathcal{N}(y))\}. \tag{4}$$

This assignment rule includes the pixels that represent the boundary edge region of tumor. Figure 2 shows an example of building the rough approximation $\overline{A}(T)$ and 2D region growing approach applied on CoG of $\overline{A}(T)$ to approximate the corresponding tumor class. As it can be seen in the tumor result of Fig. 2g,

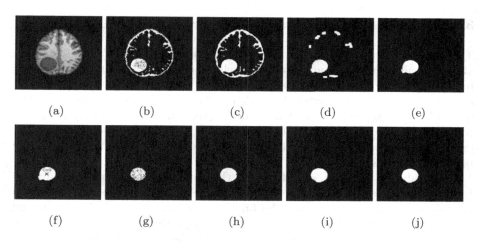

Fig. 2. An example of detection of tumor region T: (a) input, (b) after clustering, (c) closing, (d) opening, (e) largest connected component, i.e., $\overline{A}(T)$, (f) CoG marking, (g) $\underline{A}(T)$, (h) $\underline{A}(T)$ refinement, (i) final closing, and (j) ground truth

there are several holes inside the brain region caused by the higher gradient magnitude values for some pixels within tumor region than the threshold Δ computed, due to its noise factor. Also, the gradient magnitude on tumor edge boundaries may reach above the threshold value Δ. These are refined using the lower approximation refinement rule of (4). Finally, the morphological closing operation is used to smooth the surface of the tumor class. The main steps of the proposed morphology and rough set based post-processing method proceed as follows:

1. Apply closing morphology with a window of dimension 5×5 on the tumor class extracted by clustering algorithm.
2. Apply opening morphology with a window of size 7×7 in order to separate the tumor region from healthy brain tissues.
3. Find the largest connected component using connected component labelling and consider it as rough approximation of tumor $\overline{A}(T)$.
4. Construct an edge map from input brain MR image for region of $\overline{A}(T)$.
5. Compute the threshold value Δ using (2).
6. Compute the CoG of $\overline{A}(T)$ and consider it as seed point for region growing method.
7. Initially, $\underline{A}(T) \leftarrow$ CoG.
8. The region $\underline{A}(T)$ grows by incorporating boundary region pixels by pixel assignment rule of (3).
9. Refine the boundary of lower approximation using the lower approximation refinement rule of (4).
10. Assign the lower approximation of T, $\underline{A}(T)$, to T.
11. Finally, apply closing morphology with a window of dimension 5×5 to smooth the surface of the tumor class.

3   Experimental Results and Discussions

This section presents the performance of the proposed brain tumor detection algorithm, along with a comparison with related methods. The proposed method uses robust rough-fuzzy c-means (rRFCM) [5] for segmentation of brain MR images. The methods compared are $\mathcal{M}_1$, $\mathcal{M}_2$, $\mathcal{M}_3$, and BraTumIA [9]. The methods $\mathcal{M}_1$, $\mathcal{M}_2$, and $\mathcal{M}_3$ use hard c-means (HCM), fuzzy c-means (FCM), and rough-fuzzy c-means (RFCM) [4], respectively, while mask generation, feature extraction and selection, and post-processing steps are same as those of the proposed method.

The value of fuzzifier is 2.00, while that of weight parameter for rough-fuzzy clustering algorithms, which represents relative importance of lower and boundary region is set to 0.51. The final cluster prototypes of HCM are used as the initial centroids of other clustering algorithms. Brain tumor image data used in this work were obtained from the MICCAI 2012 Challenge on Multimodal Brain Tumor Segmentation (http://www.imm.dtu.dk/projects/BRATS2012) organized by B. Menze A. Jakab, S. Bauer, M. Reyes, M. Prastawa, and K. Van Leemput. The challenge database contains fully anonymized images from the following institutions: ETH Zurich, University of Bern, University of Debrecen, and University of Utah. For a quantitative comparison of the performance of the proposed method with other methods, the ground truth of brain tumor for this data set is obtained from its corresponding website.

Based on the region of interest to be extracted in the output image and that in the ground truth, the false positive (FP), false negative (FN), true positive (TP), and true negative (TN) counts can be computed for each segmented image. The quantitative measures, namely, Dice coefficient and Youden index, are described as follows with the help of these counts:

– The Dice coefficient measures the overlap between the ground truth and the result, expressed as:

$$DC = \frac{2.\text{TP}}{(\text{FP} + \text{TP}) + (\text{FN} + \text{TP})}.$$

– The Youden index is defined as:

$$YI = \text{sensitivity} + \text{specificity} - 1,$$

$$\text{where sensitivity} = \frac{\text{TP}}{\text{TP} + \text{FN}} \text{ and specificity} = \frac{\text{TN}}{\text{TN} + \text{FP}}.$$

Higher numbers of these metrics represent better overlapping in segmented image and ground truth image, indicating the significance of underlying algorithm. The metrics are calculated here for brain tumor that includes active tumor and edema. Figures 3 and 4 present the effectiveness of the proposed brain tumor segmentation algorithm for few images, while the heatmaps in Fig. 5 depict brain tumor segmentation results for fifty images quantitatively. From 3rd to 7th columns of Figs. 3 and 4, the TP, FP, and FN regions are represented in

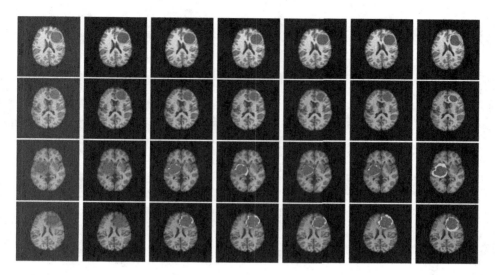

Fig. 3. Segmented images for simulated brain tumor images: input image, ground truth, proposed, $\mathcal{M}_1$, $\mathcal{M}_2$, $\mathcal{M}_3$, and BraTumIA (from left to right) (Color figure online)

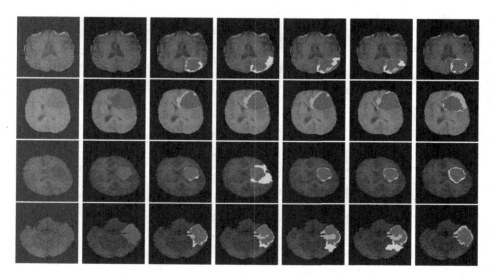

Fig. 4. Segmented images for real brain tumor images: input image, ground truth, proposed, $\mathcal{M}_1$, $\mathcal{M}_2$, $\mathcal{M}_3$, and BraTumIA (from left to right) (Color figure online)

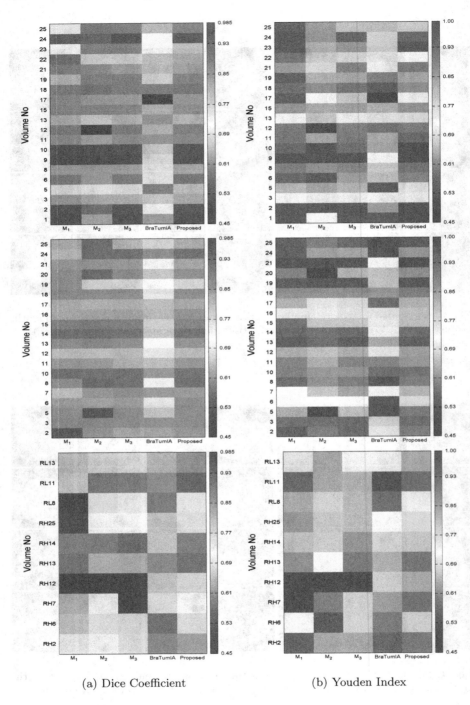

(a) Dice Coefficient   (b) Youden Index

Fig. 5. Simulated high graded, simulated low graded, and real brain tumor

red, orange, and blue color, respectively, while red color in the images at second column represents region of interest to be segmented. In heatmaps for Dice coefficient and Youden index, the results below 0.45 are put into the bin numbered 1, while the results, starting from 0.45, lying within equally spaced intervals, are put into the corresponding bins.

3.1   Importance of Robust Rough-Fuzzy C-Means

The performance of the proposed method is extensively compared with that of the methods $\mathcal{M}_1$, $\mathcal{M}_2$, and $\mathcal{M}_3$. The qualitative results reported in columns 3rd to 6th of Figs. 3 and 4 compare the performance of the proposed method with that of the methods $\mathcal{M}_1$, $\mathcal{M}_2$, and $\mathcal{M}_3$, while Fig. 5 compares it quantitatively with respect to two quantitative indices on several brain tumor images. The proposed method attains better performance than $\mathcal{M}_1$, $\mathcal{M}_2$, and $\mathcal{M}_3$ in 56, 73, 62 cases, respectively, out of 100 cases, irrespective of segmentation validation indices used. From the output images reported in Figs. 3 and 4, it can also be seen that there is a significant improvement in the segmentation results obtained using the proposed method as compared to other clustering methods.

From all these figures, it can also be concluded that the proposed post-processing method works well irrespective of clustering algorithms used. But, the clustering result actually forms the base for post-processing. So, more efficient is the clustering technique, more accurate the tumor segmentation. The best performance of the proposed method using rRFCM clustering algorithm is achieved due to the fact that the probabilistic membership function of the rRFCM handles efficiently overlapping partitions, while the possibilistic membership function of lower approximation of a cluster helps to discover arbitrary shaped cluster. In effect, good segmented regions are obtained using the proposed brain MR image segmentation algorithm.

3.2   Comparative Performance with Existing Method

Finally, the performance of the proposed method is compared with that of Bra-TumIA. The third and seventh columns of Figs. 3 and 4 represent the comparison between proposed segmentation results and the existing BraTumIA software tool qualitatively. It can be seen from Fig. 5 that the proposed method significantly gives better result than existing BraTumIA software tool. The proposed method attains higher values than BraTumIA in 42 and 30 cases with respect to Dice coefficient and Youden index, respectively, out of 50 cases each.

4   Conclusion

The main contribution of this paper is to introduce a post-processing step for brain tumor detection from MRI. The proposed post-processing method judiciously integrates the merits of morphological operations and the notion of rough set theory embedded into region growing technique. It improves the performance of the brain tumor segmentation method. Formulation of this method using rough set enables efficient handling of ambiguities at anatomical pathologic boundaries. The proposed post-processing method works significantly well, irrespective of clustering algorithms as shown by the experimental results. Hence, the proposed morphology and rough set based region growing post-processing method can be applied to any clustering based brain tumor segmentation approach. Several quantitative measures are used to evaluate the performance of the proposed brain tumor segmentation method. Finally, the effectiveness of the proposed method is demonstrated both qualitatively and quantitatively, along with a comparison with other related algorithms, on a set of synthetic and real brain MR images.

Acknowledgments. This work is partially supported by the Indian National Science Academy, New Delhi, India (grant no. SP/YSP/68/2012).

References

1. Fletcher-Heath, L.M., Hall, L.O., Goldgof, D.B., Murtagh, F.R.: Automatic segmentation of non-enhancing brain tumors in magnetic resonance images. Artif. Intell. Med. **21**, 43–63 (2001)
2. Hsieh, T.M., Liu, Y.-M., Liao, C.-C., Xiao, F., Chiang, I.-J., Wong, J.-M.: Automatic segmentation of meningioma from non-contrasted brain MRI integrating fuzzy clustering and region growing. BMC Med. Inform. Decis. Mak. **11**, 54 (2011)
3. Iscan, Z., Dokur, Z., Ölmez, T.: Tumor detection by using zernike moments on segmented magnetic resonance brain images. Expert Syst. Appl. **37**(3), 2540–2549 (2010)
4. Maji, P., Pal, S.K.: Rough set based generalized fuzzy C-Means algorithm and quantitative indices. IEEE Trans. Syst. Man Cybern. Part B: Cybern. **37**(6), 1529–1540 (2007)
5. Maji, P., Paul, S.: Rough-fuzzy clustering for grouping functionally similar genes from microarray data. IEEE/ACM Trans. Comput. Biol. Bioinf. **10**(2), 286–299 (2013)
6. Maji, P., Roy, S.: Rough-Fuzzy Clustering and Unsupervised Feature Selection for Wavelet Based MR Image Segmentation. PLoS ONE **10**(4), e0123677 (2015). doi:10.1371/journal.pone.0123677
7. Menze, B.H., Leemput, K., Lashkari, D., Weber, M.-A., Ayache, N., Golland, P.: A generative model for brain tumor segmentation in multi-modal images. Med. Image Comput. Comput.-Assist. Interv. **6362**, 151–159 (2010)
8. Nie, J., Xue, Z., Liu, T., Young, G.S., Setayesh, K., Guo, L., Wong, S.T.: Automated brain tumor segmentation using spatial accuracy-weighted hidden markov random field. Comput. Med. Imaging Graph. **33**(6), 431–441 (2009)
9. Porz, N., Bauer, S., Pica, A., Schucht, P., Beck, J., Verma, R.K., Slotboom, J., Reyes, M., Wiest, R.: Multi-modal Glioblastoma segmentation: man versus machine. PLoS ONE **9**(5), e96873 (2014). doi:10.1371/journal.pone.0096873
10. Stadlbauer, A., Moser, E., Gruber, S., Buslei, R., Nimsky, C., Fahlbusch, R., Ganslandt, O.: Improved delineation of brain tumors: an automated method for segmentation based on pathologic changes of H-MRSI metabolites in gliomas. NeuroImage **23**(2), 454–461 (2004)

11. Sung, Y.-C., Han, K.-S., Song, C.-J., Noh, S.-M., Park, J.-W.: Threshold estimation for region segmentation on MR image of brain having the partial volume artifact. In: Proceeding of the 5th International Conference on Signal Processing, vol. 2, pp. 1000–1009 (2000)
12. Xie, K., Yang, J., Zhang, Z.G., Zhu, Y.M.: Semi-automated brain tumor and edema segmentation Using MRI. Eur. J. Radiol. **56**(1), 12–19 (2005)

Rough Set Based Modeling and Visualization of the Acoustic Field Around the Human Head

Piotr Szczuko[✉], Bożena Kostek, Józef Kotus,
and Andrzej Czyżewski

Faculty of Electronics, Telecommunications and Informatics,
Multimedia Systems Department and Audio Acoustics Laboratory,
Gdańsk University of Technology, Narutowicza 11/12,
80-233 Gdańsk, Poland
{szczuko,kostek,joseph,andcz}@sound.eti.pg.gda.pl

Abstract. The presented research aims at modeling acoustical wave propagation phenomena by applying rough set theory in a novel manner. In a typical listening environment sound intensity is determined by numerous factors: a distance from a sound source, signal levels and frequencies, obstacles' locations and sizes. Contrarily, a free-field is characterized by direct, unimpeded propagation of the acoustical waves. The proposed approach is focused on processing sound field measurements performed in an anechoic chamber, collected by a dedicated acoustic probe, comprising thousands of datapoints for six signal frequencies, with and without the presence of a dummy head in a free-field. The rough set theory is applied for modeling the influence of an obstacle that a dummy head creates in a free-field and the effects of the head acoustic interferences, shading and diffraction. A data pre-processing method is proposed, involving coordinate system transformation, data discretization, and classification. Four rule sets are acquired, and achieved accuracy and coverage are assessed. Final results allow simplification of the model and new method for visualization.

Keywords: Rough sets · Imprecision · Acoustical field visualization

1 Introduction

It is important to emphasize that, in general, analytic models as well as acoustic field measurements provide useful information about pressure acoustics, pressure fields, but none currently offers a full vector mapping of the acoustic energy flow in front of and behind obstacles. Interference, diffraction, and scattering of waves make the real field very complex and challenging in terms of creating truly faithful theoretical frameworks. Taking these facts into consideration, the authors decided to conduct measurements of the vector acoustic field around a so-called dummy head imitating human head, using the sound intensity technique [4] and to enable resulting data processing employing a soft computing approach originating from the rough set methodology [13].

© Springer International Publishing Switzerland 2015
M. Kryszkiewicz et al. (Eds.): PReMI 2015, LNCS 9124, pp. 418–427, 2015.
DOI: 10.1007/978-3-319-19941-2_40

2   Measurement Methodology

Measurements were carried out in an anechoic chamber which is characterized by the free-field conditions, to avoid the problem of reflections coming from walls and obstacles. The acoustic intensity distribution was obtained using a sound intensity measurement technique, namely a 3D Acoustic Vector Sensor (AVS) [2, 5, 8–10, 14] was employed. AVS measures the acoustic particle velocity directly, instead of a sound pressure, which is captured by conventional microphones [6, 16]. The AVS senses air flow across two resistive strips of platinum that are heated up to approximately 200°C to provide temperature difference being the result of cooling by air flow [3]. The sensor itself is very miniature: typical dimensions of the wires are 5 µm in diameter and 1 to 3 mm in length, thus giving a nearly pin-point measurement, at the same time not causing the acoustic field disturbance. It operates in a flow range of 10 nm/s up to about 1 m/s. Each sensor is sensitive in only one direction, therefore, three orthogonally placed transducers are used. In combination with pressure measurement by a microphone, the sound field in a single point is fully characterized, and the acoustic intensity vector, which is the product of pressure and particle velocity, can be determined. This intensity vector indicates the acoustic energy flow.

Sound intensity is the average rate at which sound energy is transmitted through the unit area perpendicular to the specified direction at the considered point. The intensity in a certain direction is the product of sound pressure (scalar) $p(t)$ and the particle velocity (vector) component in that direction $u(t)$. The time-averaged intensity I in a single direction is given by (1) [4]:

$$I = \frac{1}{T} \int_T p(t)u(t)dt \tag{1}$$

In the applied algorithm of sound intensity calculation, the averaging time T was 4096 samples (with the sampling frequency of 48 kHz). It means that the measured value was updated more than 10 times per second. A single intensity measurement takes 1 s, being an average of values.

Using the AVS, the particular sound intensity components can be obtained based on Eq. (1). The sound intensity vector in three dimensions is composed of the acoustic intensities in three orthogonal directions (x, y, z) (2):

$$\vec{I} = I_x \vec{e}_x + I_y \vec{e}_y + I_z \vec{e}_z \tag{2}$$

Sound intensity calculation can be performed in the time domain or in the frequency domain [8, 9]. Due to the fact that the multi-harmonic signal was employed in the measurements, a method for calculating the sound intensity in the frequency domain was applied for this purpose. The detailed information about algorithm and methodology of the measurement performed can be found in [10].

The full three dimensional sound intensity vector field can be determined within the audible frequency range from 20 Hz up to 20 kHz [2, 14]. As an example of the measurement results, Fig. 1 shows sound intensity distribution characteristics for 1000 Hz.

The signal was radiated from the center loudspeaker in a 5.1 surround sound setup (top-middle of images), without and with a dummy head present in the field, and additionally a difference between these two measurements is visualized. The wave phenomena such as diffraction and interference are clearly visible. It is also easy to observe local compression of sound energy (in front of and behind the head), as well as the increase of the intensity level on both sides of the head.

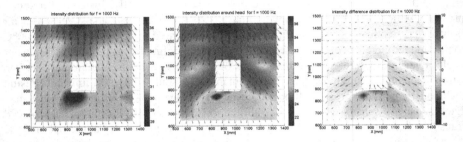

Fig. 1. Sound intensity distribution for 1000 Hz, signal from the center loudspeaker. From left: free-field and no dummy head; dummy head present in the field; the difference expressing influence of the head

The research presented in this paper shows an attempt to model acoustical wave propagation phenomena by applying rough set theory to acoustic field visualization. The approach proposed is focused on processing sound field measurement performed in an anechoic chamber, collected by a dedicated acoustic probe. Datapoints for six signal frequencies with and without the presence of a dummy head in a free-field, resulted from the measurements, are analyzed employing the rough set theory. However, first, a data pre-processing method, involving coordinate transformation is proposed. Also, data discretization for modeling the influence of an obstacle that a dummy head creates in a free-field [7] is performed before the rough set-based analysis. In consequence, a rule set is acquired, allowing for the model simplification and visualization.

3   Acoustic Field Parameterization

The acoustic field was measured by a vector sensor, capable of registering values of particle velocity flow in x, y, z directions. This sensor was positioned precisely in given coordinates by a dedicated Cartesian robot arm [11]. As a result the sound intensity at a certain point in space has been obtained.

Measurements were conducted in two conditions: (1) with the speaker emitting test signals in an anechoic chamber environment; (2) the same speaker and signals emitted but along with a dummy head, positioned in the center of the measured space. Collected measurement results enabled to visualize directional characteristic of the speaker and open-space sound propagation phenomena [12], on one hand; and impact of the head on sound propagation, compression and rarefaction on the other hand [15] (Fig. 2).

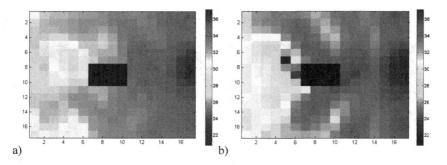

Fig. 2. Results of acoustic field measurements for 1000 Hz test signal: (a) field without the head, (b) field with the head

Measurements covered horizontal grid of 17 × 17 points, spaced evenly in interval of 50 mm, positioned at the level of ears of the dummy head.

3.1 Vector Field Data Preprocessing

The Cartesian coordinate system is convenient for positioning the robot arm and uniformly covering large spaces. Nevertheless, for further processing polar coordinates are to be used. A conversion from Cartesian to polar systems was made, assuming that the center of polar system is the head center, and direction 0 degrees is in front of it, the same as the speaker direction. For polar representation of acoustic field it was assumed that it is sufficient to analyze only 20 different directions (360 deg split into 18 degree wide slices), and 20 grades of distance from the center (Fig. 3), resulting in a discretization of the measured space.

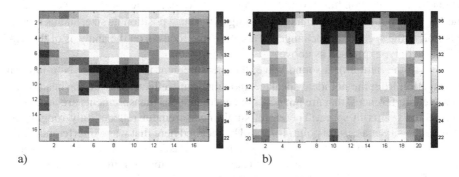

Fig. 3. Conversion of coordinates: (a) acoustic field with head present for 2000 Hz the Cartesian coordinate system, (b) in polar coordinates

For particular frequency of emitted sound, two measurements are available – empty space $I_{env}(x,y)$ and with the dummy head $I_{w\_head}(x,y)$ for every point in the space. A difference between these two reflects the influence of the head presence on sound propagation (see Fig. 4 and Eq. (3)).

$$I_{env} - I_{w\_head} = I_{influence} \tag{3}$$

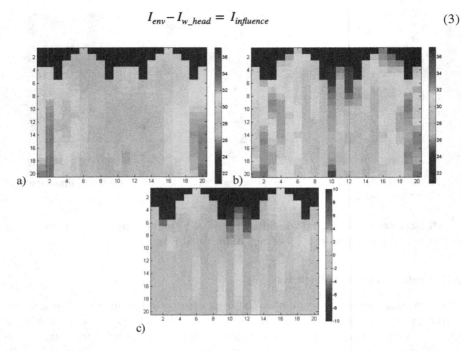

a)   b)   c)

Fig. 4. Influence of head on sound propagation: (a) acoustic field for 2000 Hz in an empty space, (b) field with the head present, (c) difference reflecting the influence

It is assumed, that in different conditions the factor I_{env}, reflecting the propagation in the environment and the speaker characteristics, would be different. The presence of the head in such a new condition would also result in different values of I_{w_head}. Nevertheless the $I_{influence}$ should remain unchanged, as long as the head position its size, and its relative position to the speaker is unchanged. Therefore, the focus is on the analysis of the head influence only, expressed as above.

Due to the nature of a sound wave, the measurement is not precise. Non-uniform propagation from the not ideal speaker[1] and numerous reflections from the head, result in wave interferences, that are not observed accurately, as measurements are taken only in selected points in space. Therefore, single measurement reflects only roughly the sound intensity at a certain point. It should be recalled here that wave interference is the interaction of two sound waves traveling in the same medium. The interference of waves causes the shape of the pressure wave to combine in the medium into a complex waveform. If the *compressions* of two similar sounds meet, they will combine and have twice the amplitude. If two *rarefaction* sections intersect, they will combine in the medium and a *rarefaction* of twice the amplitude will result [17].

---

[1] The ideal sound source is a point emitting in all directions with the same level.

Due to the low precision of measured values, it is advisable to introduce a general and rough representation of the sound intensity. Thus, a discretization of measurements was performed, focused on forming classes reflecting the sound intensity changes, i.e. (1) high *rarefaction*, (2) medium *rarefaction*, (3) unchanged, (4) medium *compression*, (5) high *compression* (Fig. 5).

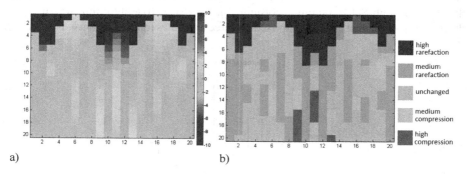

a)   b)

Fig. 5. Sound intensity changes for 2000 Hz: (a) original values, (b) discretized into five classes

The discretization is based on the expert knowledge: e.g. the intensity decrease of -3 dB or more is regarded as high *rarefaction*, a decrease in range of $(-3$ dB, 0 dB) is not significant, therefore regarded as medium *rarefaction*, an increase in intensity level by 1 dB can be either measurement error or the result of a minor interference, therefore is medium *compression*, and finally, an increase larger than $+2$ dB is a major interference, and is high *compression* (Table 1).

Table 1. Discretization of sound intensity values

Range of values	$I_{influence} \leq -3$	$-3 < I_{influence} < 0$	$0 \leq I_{influence} < 1$	$1 \leq I_{influence} < 2$	$I_{influence} \geq 2$
Assigned class	hraref.	mraref.	unch.	mcomp.	hcomp.

Class names:
    hraref. - high rarefaction
    mraref. - medium rarefaction
    unch. - unchanged
    mcompr.- medium compression
    hcompr. - high compression

As a result of the described preprocessing, the input data from sound level measurements are discretized in terms of direction (20 angles), distance (20 grades), and value (5 classes).

3.2 Rough Set Based Reasoning

A decision table containing four discrete attributes from all measurements was created: angle, distance, frequency, $I_{influence}$, the last one treated as the decision attribute. 400 points were provided for six frequencies of: 250 Hz, 500 Hz, 1000 Hz, 2000 Hz, 4000 Hz, and 8000 Hz, resulting in 2400 objects in the table (Fig. 6). The table was processed in the RSES Rough Set Exploration System [1]. An exhaustive algorithm was used for determining the reduct, and it comprised all three attributes (angle, distance, frequency).

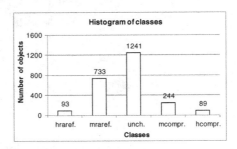

Fig. 6. Histogram of classes in the 2400 objects decision table

Analyzing all objects in the table, a set of rules was derived, resulting in 2034 rules: 97 % of rules included three attributes, and 3 % only two attributes. Mean support was equal to 1.2.

The next step was generalization of rules, with a factor of 0.9. This resulted in the same number of rules, but with greatly increased mean support, equal to 57.3. One of the rules with the highest support of 180 objects, is related to region of unchanged intensity region, present in almost all frequencies (4):

$$\begin{aligned}
&\text{IF}\ (\text{angle} = \{1\ \text{or}\ 2\ \text{or}\ 3\}) \\
&\text{AND}\ (\text{dist} = \{1\text{or}\ 2\ \text{or}\ 3\ \text{or}\ 9\ \text{or}\ 10\ \text{or}\ 11\ \text{or}\ 12\ \text{or}\ 13\ \text{or}\ 19\ \text{or}\ 20\}) \qquad (4) \\
&\text{THEN}\ \left(\text{value} = \{''\text{unch.}''\}\right)
\end{aligned}$$

Over 2000 rules explaining relations observed in a database of 2400 objects is definitely too high number, close to overfitting. Therefore a series of experiments were performed, testing accuracy of classification with reduced rule sets.

4 Experiments and Results

Rules with the lowest support were removed, using three thresholds (minimal support of 3, 10 and 20), and classification tests were performed on the decision table with all 2400 objects. A comparison of performance reveals that satisfactory level of generalization was obtained, as reduction of number of rules to 29 % of initial number still yields high accuracy and coverage (Table 2) (Fig. 7).

Table 2. Comparison of initial and filtered rule sets: support, number of rules and achieved accuracy and coverage

	Rule sets			
	Initial set	Filtered A	Filtered B	Filtered C
Minimal support of rules	1	3	10	20
Maximal support of rules	20	194	194	194
Mean support of rules	1.2	50.1	60.4	69.9
Number of rules (percent of initial set)	2034 (100 %)	588 (29 %)	476 (23 %)	396 (19 %)
Accuracy	0.95	0.937	0.916	0.885
Coverage	1.0	0.999	0.936	0.854

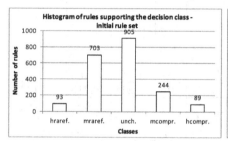 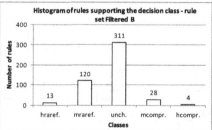

Fig. 7. Number of rules supporting decision classes in initial rule set and Filtered B set

Detailed results of classification of all objects using 588 rules (rule set "Filtered A") reveal confusion between medium rarefaction/compression/unchanged cases, where the observed value of the sound intensity change varies by ± 1 dB, and can be explained by the measurement imprecision. Adequate high accuracy and coverage are achieved (Table 3). Reduction of rules to 476 (rule set "Filtered B") shows a significant loss of coverage, especially for high compression cases, and the total performance is questionable (Table 4).

Summarizing, the resulted 588 "Filtered A" rule set, is a good compromise between accuracy and the ability to describe dependencies of angle, distance, frequency, and result compression/rarefaction in the sound field. Confusion between medium changes of ± 1 dB is negligible, by means of perception of the sound by the human ear.

Table 3. Summary of results for rules with minimal support of 3

		Predicted class							
		un-changed.	med. compr.	med. rare-faction	high compr.	high rare-faction	No. of obj.	Accu-racy	Cov-erage
Actual class	unchanged	**1,224**	1	16	0	0	1,241	0.986	1
	med. compr.	16	**228**	0	0	0	244	0.934	1
	med. rarefaction	108	0	**624**	0	1	733	0.851	1
	high compr.	0	0	0	**87**	0	89	1	0.978
	high rarefaction	1	0	7	0	**85**	93	0.914	1
	True positive rate	0.91	1	0.96	1	0.99			

Table 4. Summary of results for rules with minimal support of 10

		Predicted class							
		un-changed.	med. compr.	med. rare-faction	high compr.	high rare-faction	No. of obj.	Accu-racy	Cov-erage
Actual class	unchanged	**1,224**	1	16	0	0	1,241	0.986	1
	med. compr.	48	**124**	2	0	0	244	0.713	0.713
	med. rarefaction	109	0	**607**	0	1	733	0.847	0.978
	high compr.	2	0	0	**24**	0	89	0.923	0.292
	high rarefaction	1	0	8	0	**80**	93	0.899	0.957
	True positive rate	0.88	0.99	0.96	1	0.99			

Accuracy: TP/(TP + FN), e.g. 24/(24 + 2) = 0.923 for high compression.
Coverage: (TP + FN)/(No. of obj.), e.g. (24 + 2)/89 = 0.292 for high compression.
True positive rate: TP/(TP + FP), e.g. 607/(16 + 2+607 + 8) = 0.96 for med. rarefaction.

5   Conclusions

A novel approach to modeling and visualization of the sound field was proposed in the paper employing the rough set-based mechanism for the delivery of a rule set helping to study acoustical phenomena. The overall image of the acoustic field resulted from measurement comprises the effect of the obstacles appearing in a source radiation path, as well as the influence of scattered reflections together with their phase and amplitude relationship. The aforementioned measurements employed an acoustic intensity probe. The observation of acoustic wave distribution around the human head illustrates that phenomena occurring in the real acoustic field are more complex than those typically shown in acoustic field simulations. The wave phenomena such as diffraction and additional interferences are clearly visible. The experiments conducted helped to verify the method for reducing the complexity of data resulted from acoustic field measurements.

Acknowledgements. The project was funded by the National Science Centre allocated on the basis of the decision DEC-2012/05/B/ST7/02151.

References

1. Bazan, J., Szczuka, M.S., Wróblewski, J.: A new version of rough set exploration system. In: Alpigini, J.J., Peters, J.F., Skowron, A., Zhong, N. (eds.) RSCTC 2002. LNCS (LNAI), vol. 2475, pp. 397–404. Springer, Heidelberg (2002)
2. de Bree, H.-E.: The microflown: an acoustic particle velocity sensor. Acoust. Aust. **31**(3), 91–94 (2003)
3. Cengarle, G., Mateos, T.: Comparison of Anemometric Probe and Tetrahedral Microphones for Sound Intensity Measurements. AES 130th Convention, Paper No. 8363, London, UK, 13–16 May 2011
4. Fahy, F.J.: Sound Intensity. Elsevier Applied Science, London (1989)
5. Jacobsen, F.: Sound intensity and its measurement and applications. Acoustic Technology, Department of Electrical Engineering Technical University of Denmark (2011)
6. Jacobsen, F., de Bree, H.-E.: Measurement of sound intensity: p-u probes versus p-p probes. In: Proceedings of NOVEM (2005)
7. Kahana, Y.: Numerical modelling of the head-related transfer function. Ph.D. Thesis, University of Southampton, UK (2000)
8. Kotus, J.: Application of passive acoustic radar to automatic localization, tracking and classification of sound sources. Inf. Technol. **18**, 111–116 (2010)
9. Kotus, J.: Multiple sound sources localization in free field using acoustic vector sensor. Multimedia Tools Appl., June 2013. doi:10.1007/s11042-013-1549-y
10. Kotus, J., Kostek, B.: Measurements and visualization of sound intensity around the human head in free field using acoustic vector sensor. J. Audio Eng. Soc. **63**(1/2), 99–109 (2015). doi:http://dx.doi.org/10.17743/jaes.2015.0009
11. Kotus, J., Plewa, M., Kostek, B.: Measurements and visualization of sound intensity around the human head using acoustic vector sensor. e-Brief 154, 136th AES Convention, April 26–29 Berlin, Germany (2014)
12. Merimaa, J., Lokki, T., Peltonen, T., Karjalainen, M.: Measurement, Analysis, and Visualization of Directional Room Responses. 111 AES Convention, New York, USA (2001)
13. Pawlak, Z.: Rough sets. Int. J. Comput. Inform. Sci. **11**(5), 341–356 (1982)
14. Raangs, R., Druyvesteyn, W.F., de Bree, H.E.: A low cost Intensity Probe. 110 AES Convention, Paper no. 5292, Amsterdam (2001)
15. Weyna, S.: Identification of reflection and scattering effects in real acoustic flow field. Arch. Acoust. **28**(3), 191–203 (2003)
16. Woszczyk, W., Iwaki, M., Sugimoto, T., Ono, K., de Bree, H.-E.: Anechoic Measurements of Particle-Velocity Probes Compared to Pressure Gradient and Pressure Microphones. 122 AES Convention: May 2007, Paper Number: 7107 (2007)
17. Sound interference explanation. http://www.sound-physics.com/Sound/Interference/

Global Optimization of Exact Association Rules Relative to Coverage

Beata Zielosko[✉]

Institute of Computer Science, University of Silesia,
39, Będzińska St., 41-200 Sosnowiec, Poland
beata.zielosko@us.edu.pl

Abstract. In the paper, an application of dynamic programming app-roach to global optimization of exact association rules relative to coverage is presented.

Keywords: Rough sets · Dynamic programming · Coverage · Decision rules · Association rules

1 Introduction

Association rules are popular form of knowledge representation. They are used to discover patterns, associations, correlations among sets of items in data set. There are many approaches for mining association rules. The most popular, is Apriori algorithm based on frequent itemsets [1].

During years, many new algorithms were designed which are based on, e.g., closed itemset lattices [16], partitioning the data [19], and others [7,15].

In the paper, an application of dynamic programming approach to optimization of association rules relative to coverage is presented. Coverage is one of popular rule's evaluation measure which denotes how many objects are covered by the rule. It allows one to discover major patterns in the data and also reduces the number of generated rules. Unfortunately, the problem of construction of rules with maximum coverage is NP-hard [6]. The most part of approaches, with the exception of brute-force and Apriori algorithm, cannot guarantee the construction of rules with the maximum coverage. The proposed approach allows one to construct optimal rules, i.e., rules with the maximum coverage.

Application of rough sets theory to the construction of rules for knowledge representation or classification tasks are usually connected with the usage of decision table [17] as a form of input data representation. In such a table one attribute is distinguished as a decision attribute and it relates to a rule's conse-quence. However, in the last years, associative mechanism of rule construction, where all attributes can occur as premises or consequences of particular rules, is popular. Association rules can be defined in many ways. In the paper, a special kind of association rules is studied, i.e., they relate to decision rules. Such rules can be interpreted as Horn rules [5,8] where literals are coded using *attribute-value* pairs. This form of rule representation is widely used in data mining. Simi-lar approach was considered in [13,14], where greedy algorithm for minimization of length of association rules was investigated.

© Springer International Publishing Switzerland 2015
M. Kryszkiewicz et al. (Eds.): PReMI 2015, LNCS 9124, pp. 428–437, 2015.
DOI: 10.1007/978-3-319-19941-2_41

This paper is an essential extension of the paper [3] in which decision tables are considered that do not contain equal rows with different decisions (consistent tables). When association rules for information systems are studied and each attribute is sequentially considered as the decision one, inconsistent tables are often obtained, i.e., tables containing equal rows with different decisions. So, the approach considered in [3] is extended to the case of inconsistent decision tables. It required changes in definitions, algorithms (new conditions of stop), proofs of algorithm correctness, and, especially, in the software.

The aim of the paper is to create a research tool which is applicable to medium sized decision tables and allows one to construct exact association rules with maximum coverage that cover given object (row). To this end, a directed acyclic graph is constructed. Based on this graph it is possible to describe all irredundant association rules with given attribute on the right-hand side. This graph is used also in an optimization procedure which changes the initial graph such that it describes, for each row, all association rules with the considered attribute on the right-hand side that cover given row, and have maximum cardinality.

Methods and results considered in this paper have common points with Formal Concept Analysis (FCA) [18]. In particular, for a given decision table T, a directed acyclic graph is constructed. It is used for the description and optimization of association rules. Nodes of this graph are so-called separable subtables of the table T given by systems of equations of the kind "attribute=value". The considered construction is connected with FCA. Let the decision table T contain only conditional attributes with values from the set $\{0, 1\}$ and, for each attribute, contain also its negation. Then each subtable (node) from the directed acyclic graph corresponds to a concept for the context consisting of the set of rows of T and the set of conditional attributes of T. More common points are in results. In particular, exact association rules (called "implications") are studied in FCA [9,10,15,16].

The paper consists of six sections. Section 2 contains main notions. In Sect. 3, algorithm for construction of a directed acyclic graph is presented. Section 4 contains a description of optimization procedure relative to coverage. Section 5 contains experimental results for decision tables from UCI Machine Learning Repository, and Sect. 6 - conclusions.

2   Main Notions

An *information system* I is a rectangular table with $n + 1$ columns labeled with attributes $f_1, \ldots, f_{n+1}$. Rows of this table are filled by nonnegative integers which are interpreted as values of attributes.

An association rule for I is a rule of the kind

$$f_{i_1} = a_1 \wedge \ldots \wedge f_{i_m} = a_m \rightarrow f_j = a,$$

where $f_j \in \{f_1, \ldots, f_{n+1}\}$, $f_{i_1}, \ldots, f_{i_m} \in \{f_1, \ldots, f_{n+1}\} \setminus \{f_j\}$, and $a, a_1, \ldots, a_m$ are nonnegative integers.

The notion of an association rule for I is based on the notion of a decision table and decision rule.

A *decision table* T is a rectangular table with n columns labeled with (conditional) attributes $f_1, \ldots, f_n$. Rows of this table are filled by nonnegative integers which are interpreted as values of conditional attributes. Each row is labeled with a nonnegative integer (decision) which is interpreted as a value of a decision attribute. It is possible that T contains equal rows with the same or different decisions.

For each attribute $f_i \in \{f_1, \ldots, f_{n+1}\}$, the information system I is transformed into a decision table I_{f_i}. The column f_i is removed from I and a table with n columns labeled with attributes $f_1, \ldots, f_{i-1}, f_{i+1}, \ldots, f_{n+1}$ is obtained. Values of the attribute f_i are attached to the rows of the obtained table which will be denoted by I_{f_i}.

The set $\{I_{f_1}, \ldots, I_{f_{n+1}}\}$ of decision tables obtained from the information system I is denoted by Φ. Let $T \in \Phi$. For simplicity, it is assumed that $T = I_{f_{n+1}}$.

The table T is called *degenerate* if it is empty or all rows of T are labeled with the same decision, or all rows of T are equal.

A minimum decision value that is attached to the maximum number of rows in T will be called the *most common decision for* T.

A table obtained from T by the removal of some rows is called a *subtable* of the table T. Let $f_{i_1}, \ldots, f_{i_m} \in \{f_1, \ldots, f_n\}$ and $a_1, \ldots, a_m$ be nonnegative integers. A subtable of the table T, which contains only rows of T that have numbers $a_1, \ldots, a_m$ at the intersection with columns $f_{i_1}, \ldots, f_{i_m}$, is denoted by $T(f_{i_1}, a_1) \ldots (f_{i_m}, a_m)$. Such subtables (including the table T) are called *separable subtables* of T.

The set of attributes from $\{f_1, \ldots, f_n\}$ which are nonconstant in T is denoted by $E(T)$. For any $f_i \in E(T)$, the set of values of the attribute f_i in T, is denoted by $E(T, f_i)$.

The expression

$$f_{i_1} = a_1 \wedge \ldots \wedge f_{i_m} = a_m \rightarrow f_{n+1} = d \tag{1}$$

is called a *decision rule over* T if $f_{i_1}, \ldots, f_{i_m} \in \{f_1, \ldots, f_n\}$, and $a_1, \ldots, a_m, d$ are nonnegative integers. It is possible that $m = 0$. In this case (1) is equal to the rule

$$\rightarrow f_{n+1} = d. \tag{2}$$

Let $r = (b_1, \ldots, b_n)$ be a row of T. Rule (1) will be called *realizable for* r, if $a_1 = b_{i_1}, \ldots, a_m = b_{i_m}$. If $m = 0$ then rule (2) is realizable for any row from T.

Rule (1) will be called *true for* T if the table $T' = T(f_{i_1}, a_1) \ldots (f_{i_m}, a_m)$ is degenerate and d is the most common decision for T'. If $m = 0$ then rule (2) is true for T if T is degenerate and d is the most common decision for T.

If rule (1) is true for T and realizable for r, then (1) will be called a *decision rule for* T *and* r.

Decision rules for T and r will be called (f_{n+1})-association rules for I and r. In general case, the notion of (f_i)-*association rule for* I *and* r coincides with the notion of decision rule for I_{f_i} and r, $i = 1, \ldots, n + 1$. The union of sets of

(f_i)-association rules, $i = 1, \ldots, n+1$, will be considered as the set of association rules for I and r.

Let $T = I_{f_{n+1}}$ and (1) be a decision rule over T. Rule (1) will be called an *irredundant* rule for T and r if (1) is a decision rule for T and r and the following conditions hold if $m > 0$:

(i) $f_{i_1} \in E(T)$, and if $m > 1$ then $f_{i_j} \in E(T(f_{i_1}, a_1) \ldots (f_{i_{j-1}}, a_{j-1}))$ for $j = 2, \ldots, m$;
(ii) if $m = 1$ then the table T is nondegenerate, and if $m > 1$ then the table $T(f_{i_1}, a_1) \ldots (f_{i_{m-1}}, a_{m-1})$ is nondegenerate.

If $m = 0$ then rule (2) is an *irredundant* decision rule for T and r if (2) is a decision rule for T and r, i.e., if T is degenerate and d is the most common decision for T.

Let $T = I_{f_{n+1}}$, τ be a decision rule over T, and τ be equal to (1). The *coverage* of τ is the number of rows in T for which τ is realizable and which are labeled with the decision d. It is denoted by $c(\tau)$. If $m = 0$ then the coverage of the rule τ is equal to the number of rows in T which are labeled with the decision d.

3 Algorithm for Directed Acyclic Graph Construction

In this section, an algorithm for construction of a directed acyclic graph for a given decision table T is presented. Based on this graph it is possible to describe the set of irredundant rules for T and for each row r of T. This algorithm is repeated for each decision table I_{f_i}, $i = 1, \ldots, n+1$, obtained from the information system I.

Let $T = I_{f_{n+1}}$. The constructed graph is denoted by $\Delta(T)$. Nodes of the graph are some separable subtables of the table T. During each step, the algorithm processes one node and marks it with the symbol *. At the first step, the algorithm constructs a graph containing a single node T which is not marked with *. Let the algorithm have already performed p steps. Now the step $(p + 1)$ will be described. If all nodes are marked with the symbol * as processed, the algorithm finishes its work and presents the resulting graph as $\Delta(T)$. Otherwise, choose a node (table) Θ, which has not been processed yet. If Θ is degenerate then mark the considered node with symbol * and proceed to the step $(p + 2)$. Otherwise, for each $f_i \in E(\Theta)$, draw a bundle of edges from the node Θ. Let $E(\Theta, f_i) = \{b_1, \ldots, b_t\}$. Then draw t edges from Θ and label these edges with pairs $(f_i, b_1), \ldots, (f_i, b_t)$ respectively. These edges enter to nodes $\Theta(f_i, b_1), \ldots, \Theta(f_i, b_t)$. If some of nodes $\Theta(f_i, b_1), \ldots, \Theta(f_i, b_t)$ are absent in the graph then add these nodes to the graph. Each row r of Θ is labeled with the set of attributes $E_{\Delta(T)}(\Theta, r) = E(\Theta)$. Mark the node Θ with the symbol * and proceed to the step $(p + 2)$.

The graph $\Delta(T)$ is a directed acyclic graph. A node Θ of this graph will be called *terminal* if there are no edges leaving this node, i.e., Θ is degenerate.

Later, a local optimization of the graph $\Delta(T)$ relative to the coverage will be described. As a result, a graph $G(T)$, with the same sets of nodes and edges as in $\Delta(T)$, is obtained. The only difference is that any row r of each nonterminal node

Θ from $G(T)$ is labeled with a nonempty set of attributes $E_{G(T)}(\Theta, r) \subseteq E(\Theta)$, possibly different from $E(\Theta)$.

Now, for each node Θ of $G(T)$ and for each row r of Θ, a set of rules $Rul_{G(T)}(\Theta, r)$ will be described. It is possible that $G(T) = \Delta(T)$.

Let Θ be a terminal node of $G(T)$. In this case Θ is a degenerate table and

$$Rul_{G(T)}(\Theta, r) = \{\to f_{n+1} = d\}.$$

Let now Θ be a nonterminal node of $G(T)$ such that for each child Θ' of Θ and for each row r' of Θ', the set of rules $Rul_{G(T)}(\Theta', r')$ is already defined. Let $r = (b_1, \ldots, b_n)$ be a row of Θ. For any $f_i \in E_{G(T)}(\Theta, r)$, the set of rules $Rul_{G(T)}(\Theta, r, f_i)$ is defined as follows:

$$Rul_{G(T)}(\Theta, r, f_i) = \{f_i = b_i \wedge \gamma \to f_{n+1} = s : \gamma \to f_{n+1} = s \in Rul_{G(T)}(\Theta(f_i, b_i), r)\}.$$

Then $Rul_{G(T)}(\Theta, r) = \bigcup_{f_i \in E_{G(T)}(\Theta, r)} Rul_{G(T)}(\Theta, r, f_i)$.

One can prove the following statement.

Theorem 1. *For any node Θ of $\Delta(T)$ and for any row r of Θ, $Rul_{\Delta(T)}(\Theta, r)$ is equal to the set of all irredundant rules for Θ and r.*

The algorithm for the directed acyclic graph construction is repeated for each decision table I_{f_i}, $i = 1, \ldots, n+1$, obtained from the information system I. In general, the obtained graph is denoted by $\Delta(I_{f_i})$, $i = 1, \ldots, n+1$. As a result, for $i = 1, \ldots, n+1$, a set $Rul_{\Delta(I_{f_i})}(I_{f_i}, r)$ of irredundant decision rules for I_{f_i} and r is obtained. This set will be called the set of irredundant (f_i)-association rules for I and r, $i = 1, \ldots, n+1$. The union of sets $Rul_{\Delta(I_{f_i})}(I_{f_i}, r)$ forms a set $Rul(I, r)$ of irredundant association rules for I and r:

$$Rul(I, r) = \bigcup_{i=1,\ldots,n+1} Rul_{\Delta(I_{f_i})}(I_{f_i}, r).$$

To illustrate the presented algorithm the information system I depicted in Fig. 1 will be considered. Set $\Phi = \{I_{f_1}, I_{f_2}, I_{f_3}\}$ contains three decision tables obtained from I. Figure 2 presents a directed acyclic graph for decision table I_{f_1}. Based on the graph $\Delta(I_{f_1})$ the sets of rules attached to rows of I_{f_1} are described.

$Rul_{\Delta(I_{f_1})}(I_{f_1}, r_1) = \{f_2 = 2 \wedge f_3 = 0 \to f_1 = 0, f_3 = 0 \to f_1 = 0\}$;

$Rul_{\Delta(I_{f_1})}(I_{f_1}, r_2) = \{f_2 = 2 \wedge f_3 = 1 \to f_1 = 1, f_3 = 1 \to f_1 = 1\}$;

$Rul_{\Delta(I_{f_1})}(I_{f_1}, r_3) = \{f_2 = 1 \to f_1 = 1, f_3 = 1 \to f_1 = 1\}$.

Finally, based on the graphs $\Delta(I_{f_1})$, $\Delta(I_{f_2})$ and $\Delta(I_{f_3})$, it is possible to describe sets $Rul(I, r)$ of irredundant association rules for I and each row r of I.

$Rul(I, r_1) = \{f_2 = 2 \wedge f_3 = 0 \to f_1 = 0, f_3 = 0 \to f_1 = 0, f_1 = 0 \to f_2 = 2, f_3 = 0 \to f_2 = 2, f_1 = 0 \to f_3 = 0, f_2 = 2 \wedge f_1 = 0 \to f_3 = 0\}$;

$$I = \begin{array}{c|ccc} & f_1 & f_2 & f_3 \\ \hline r_1 & 0 & 2 & 0 \\ r_2 & 1 & 2 & 1 \\ r_3 & 1 & 1 & 1 \end{array} \Rightarrow I_{f_1} = \begin{array}{c|cc|c} & f_2 & f_3 & \\ \hline r_1 & 2 & 0 & 0 \\ r_2 & 2 & 1 & 1 \\ r_3 & 1 & 1 & 1 \end{array}$$

Fig. 1. Decision table I_{f_1} obtained from information system I

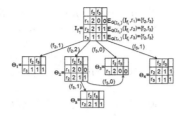

Fig. 2. Directed acyclic graph $\Delta(I_{f_1})$

$Rul(I, r_2) = \{f_2 = 2 \wedge f_3 = 1 \rightarrow f_1 = 1, f_3 = 1 \rightarrow f_1 = 1, f_1 = 1 \rightarrow f_2 = 1, f_3 = 1 \rightarrow f_2 = 1, f_1 = 1 \rightarrow f_3 = 1, f_2 = 2 \wedge f_1 = 1 \rightarrow f_3 = 1\};$
$Rul(I, r_3) = \{f_2 = 1 \rightarrow f_1 = 1, f_3 = 1 \rightarrow f_1 = 1, f_1 = 1 \rightarrow f_2 = 1, f_3 = 1 \rightarrow f_2 = 1, f_1 = 1 \rightarrow f_3 = 1, f_2 = 1 \rightarrow f_3 = 1\}.$

4   Optimization Relative to Coverage

In this section, two kinds of optimization relative to the coverage c are presented: local optimization and global optimization.

Local optimization relative to the coverage is an optimization of the directed acyclic graph $\Delta(I_{f_i})$, $I_{f_i} \in \Phi$, constructed for a given decision table I_{f_i}.

Global optimization relative to the coverage is made for an information system I, i.e., among all graphs $G(I_{f_1}), \ldots, G(I_{f_{n+1}})$ constructed for decision tables from the set Φ and optimized locally relative to the coverage.

Let $T = I_{f_{n+1}}$. Now, a procedure of local optimization of the graph $\Delta(T)$ relative to the coverge will be described. For each node Θ in the graph $\Delta(T)$, this procedure assigns to each row r of Θ the set $Rul^c_{\Delta(T)}(\Theta, r)$ of decision rules with the maximum coverage from $Rul_{\Delta(T)}(\Theta, r)$ and the number $Opt^c_{\Delta(T)}(\Theta, r)$ – the maximum coverage of decision rule from $Rul_{\Delta(T)}(\Theta, r)$.

The algorithm moves from the terminal nodes of the graph $\Delta(T)$, which are degenerate tables, to the node T. It will attach the set $E_{G(T)}(\Theta, r)$ to each row r in Θ if Θ is a nonterminal node of $\Delta(T)$. The obtained graph is denoted by $G(T)$.

Let Θ be a terminal node of $\Delta(T)$. Then each row r of Θ has assigned the number $Opt^c_{\Delta(T)}(\Theta, r)$ that is equal to the number of rows in Θ which are labeled with the most common decision for Θ.

Let Θ be a nonterminal node of $\Delta(T)$ and all children of Θ have already been treated. Let $r = (b_1, \ldots, b_n)$ be a row of Θ. The number

$$Opt^c_{\Delta(T)}(\Theta, r) = \max\{Opt^c_{\Delta(T)}(\Theta(f_i, b_i), r) : f_i \in E(\Theta, r)\}$$

is assigned to the row r in the table Θ and

$$E_{G(T)}(\Theta, r) = \{f_i : f_i \in E_{\Delta(T)}(\Theta, r), Opt^c_{G(T)}(\Theta(f_i, b_i), r) = Opt^c_{G(T)}(\Theta, r)\}.$$

Theorem 2. *For each node Θ of the graph $G(T)$ and for each row r of Θ, the set $Rul_{G(T)}(\Theta, r)$ is equal to the set $Rul^c_{\Delta(T)}(\Theta, r)$ of all rules with the maximum coverage from the set $Rul_{\Delta(T)}(\Theta, r)$.*

Now, a global optimization relative to the coverage is presented. It is made for the information system I.

The set of irredundant association rules for I and r with the maximum coverage from $Rul(I, r)$ is denoted by $Rul^c(I, r)$, and the maximum coverage of an association rule from $Rul(I, r)$ is denoted by $Opt^c(I, r)$.

To make global optimization relative to the coverage, the directed acyclic graph is constructed for each decision table $I_{f_i} \in \Phi$, and local optimization relative to the coverage of the graph $\Delta(I_{f_i})$, $i = 1, \ldots, n+1$, is made. As a result, the graph $G(I_{f_i})$ is obtained and each row r of I_{f_i}, $i = 1, \ldots, n+1$, has assigned the set $Rul_{G(I_{f_i})}(I_{f_i}, r)$ of (f_i)-association rules for I and r with the maximum coverage from $Rul_{\Delta(I_{f_i})}(I_{f_i}, r)$ and the number $Opt^c_{\Delta(I_{f_i})}(I_{f_i}, r)$, which is the maximum coverage of (f_i)-association rule from $Rul_{\Delta(I_{f_i})}(I_{f_i}, r)$.

Then, the value $Opt^c(I, r)$ is obtained, such that,

$$Opt^c(I, r) = \max\{Opt^c_{\Delta(I_{f_i})}(I_{f_i}, r) : i = 1, \ldots, n+1\},$$

and among all numbers $Opt^c_{\Delta(I_{f_i})}(I_{f_i}, r)$, $i = 1, \ldots, n+1$, only these are selected, where

$$Opt^c_{\Delta(I_{f_i})}(I_{f_i}, r) = Opt^c(I, r).$$

The set $Rul^c(I, r)$ is the union of sets $Rul_{G(I_{f_i})}(I_{f_i}, r)$ for which the last equality holds:

$$Rul^c(I,r) = \bigcup_{Opt^c_{\Delta(I_{f_i})}(I_{f_i}, r) = Opt^c(I, r), Opt^c(I,r) = \max\{Opt^c_{\Delta(I_{f_i})}(I_{f_i}, r) : i=1,\ldots,n+1\}} Rul_{G(I_{f_i})}(I_{f_i}, r).$$

As a result of the global optimization relative to the coverage each row r of I has assigned the set $Rul^c(I, r)$ of association rules with the maximum coverage and the number $Opt^c(I, r)$.

Below one can find the sets of association rules, for information system I (depicted in Fig. 1) and row r, after global optimization relative to coverage, and the value $Opt^c(I, r)$.

$Rul^c(I, r_1) = \{f_2 = 2 \wedge f_3 = 0 \rightarrow f_1 = 0, f_3 = 0 \rightarrow f_1 = 0, f_1 = 0 \rightarrow f_2 = 2, f_3 = 0 \rightarrow f_2 = 2, f_1 = 0 \rightarrow f_3 = 0, f_2 = 2 \wedge f_1 = 0 \rightarrow f_3 = 0\}$, $Opt^c(I, r_1) = 1$;

$Rul^c(I, r_2) = \{f_3 = 1 \rightarrow f_1 = 1, f_1 = 1 \rightarrow f_3 = 1\}$, $Opt^c(I, r_2) = 2$;

$Rul^c(I, r_3) = \{f_3 = 1 \rightarrow f_1 = 1, f_1 = 1 \rightarrow f_3 = 1\}$, $Opt^c(I, r_3) = 2$.

The problem of rule coverage maximization is NP-hard [6]. The algorithms considered in this paper have polynomial time complexity depending on the size of decision table and the number of separable subtables in it. In general case, the number of separable subtables grows exponentially with the growth of table size. However, in [11,12] classes of decision tables are described for each of which the number of separable subtables in tables from the class is bounded from above by a polynomial on the size of decision table.

5   Experimental Results

Experiments were made using data sets from UCI Machine Learning Repository [4] and modified software system Dagger [2].

Each data set was considered as information system I and, for each attribute $f_i \in \{f_1, \ldots, f_{n+1}\}$, the system I was transformed into a decision table I_{f_i}. The column f_i was removed from I and a table with n columns labeled with attributes $f_1, \ldots, f_{i-1}, f_{i+1}, \ldots, f_{n+1}$, was obtained. Values of the attribute f_i were attached to the rows of the obtained table I_{f_i}. The set $\{I_{f_1}, \ldots, I_{f_{n+1}}\}$ of decision tables obtained from the information system I is denoted by Φ.

Table 1 presents preliminary results of experiments connected with the maximum coverage of irredundant association rules. For each row r of I, the maximum coverage of an irredundant association rule for I and r was obtained. After that, for rows of I the minimum coverage of an association rule for I and r with the maximum coverage (column "Min"), the maximum coverage of such rule (column "Max") and the average coverage of association rules with the maximum coverage - one for each row (column "Avg") were obtained. Column "Rows" contains the number of rows in I, column "Attr" contains the number of attributes in I.

Table 2 presents the average number of nodes (column "Nodes") and the average number of edges (column "Edges") related to the data set I and the graph $\Delta(I_{f_i})$, $i = 1, \ldots, n + 1$. For each data set I, the set Φ was obtained. For each decision table I_{f_i}, $i = 1, \ldots, n + 1$, the graph $\Delta(I_{f_i})$ was constructed and the number of nodes and edges were calculated. Then, the average number of nodes and edges related to the directed acyclic graphs $\Delta(I_{f_i})$, $i = 1, \ldots, n+1$, and data set I were computed.

The proposed approach of rule induction is based on the analysis of the directed acyclic graph constructed for a given decision table. A structure of the graph depends on data set, i.e., number of attributes, distribution of values of

Table 1. Maximum coverage of rules

Data set	Rows	Attr	Min	Avg	Max
adult-stretch	16	5	4	7.0	8
breast-cancer	266	10	5	41.143	65
cars	1728	7	1	336.319	576
hayes-roth	69	5	2	7.435	12
lenses	24	5	2	7.958	12
monks-1-test	432	7	12	45.0	108
monks-3-test	432	7	36	56.0	108
shuttle-landing	15	7	6	7.8	9
teeth	23	9	8	12.652	15
tic-tac-toe	958	10	12	67.282	90
zoo-data	59	17	31	34.814	35

Table 2. Average number of nodes and edges

Data set	Rows	Attr	Nodes	Edges
adult-stretch	16	5	48.0	104.0
breast-cancer	266	10	6082.0	61063.6
cars	1728	7	4335.3	17697.1
hayes-roth	69	5	190.8	569.0
lenses	24	5	70.8	174.8
monks-1-test	432	7	1734.9	6760.1
monks-3-test	432	7	1584.9	5770.4
shuttle-landing	15	7	73.6	368.6
teeth	23	9	112.3	952.7
tic-tac-toe	958	10	31415.1	264362.9
zoo-data	59	17	3595.4	57868.2

attributes, number of rows. Such graph can be huge for larger data set. Therefore, possibilities of decreasing the size of the graph are studied by the author. In [20], the graph is constructed only for selected values of attributes contained in a decision table.

6   Conclusions

In the paper, an application of dynamic programming to global optimization of exact association rules relative to coverage was presented. It is based on the dynamic programming approach to optimization of decision rules. However, there are differences: (i) the information system is used, (ii) decision table can be inconsistent, (iii) definitions are different, and (iv) global optimization of rules was studied. The presented approach can be considered as research tool which allows one to construct association rules with maximum coverage.

Possible applications of association rules obtained using presented approach are construction of classifiers, inference process in knowledge base system, filling missing values of attributes.

Future works will be connected with the study of approximate association rules and sequential optimization of rules relative to the length and coverage.

Acknowledgements. The author wishes to thank the anonymous reviewers for helpful suggestions.

References

1. Agrawal, R., Imieliński, T., Swami, A.: Mining association rules between sets of items in large databases. In: SIGMOD 1993, pp. 207–216. ACM (1993)

2. Alkhalid, A., Amin, T., Chikalov, I., Hussain, S., Moshkov, M., Zielosko, B.: Dagger: a tool for analysis and optimization of decision trees and rules. In: Ficarra, F.V.C. (ed.) Computational Informatics, Social Factors and New Information Technologies: Hypermedia Perspectives and Avant-Garde Experiences in the Era of Communicability Expansion, pp. 29–39. Blue Herons, Bergamo (2011)
3. Amin, T., Chikalov, I., Moshkov, M., Zielosko, B.: Dynamic programming approach for exact decision rule optimization. In: Skowron, A., Suraj, Z. (eds.) Rough Sets and Intelligent Systems Professor Zdzisław Pawlak in Memoriam Volume 1. Volume 42 of Intelligent Systems Reference Library, pp. 211–228. Springer, Heidelberg (2013)
4. Asuncion, A., Newman, D.J.: UCI Machine Learning Repository (2007). (http:// www.ics.uci.edu/~mlearn/)
5. Balcázar, J.L., Garriga, G.C.: Horn axiomatizations for sequential data. Theor. Comput. Sci. **371**(3), 247–264 (2007)
6. Bonates, T., Hammer, P.L., Kogan, A.: Maximum patterns in datasets. Discret. Appl. Math. **156**(6), 846–861 (2008)
7. Borgelt, C.: Simple algorithms for frequent item set mining. In: Koronacki, J., Raś, Z.W., Wierzchoń, S.T., Kacprzyk, J. (eds.) Advances in Machine Learning II. SCI, vol. 263, pp. 351–369. Springer, Heidelberg (2010)
8. Hammer, P.L., Kogan, A.: Essential and redundant rules in horn knowledge bases. Decis. Support Syst. **16**(2), 119–130 (1996)
9. Kuznetsov, S.O., Obiedkov, S.A.: Some decision and counting problems of the duquenne-guigues basis of implications. Discret. Appl. Math. **156**(11), 1994–2003 (2008)
10. Medina, R., Nourine, L.: Conditional functional dependencies: an FCA point of view. In: Kwuida, L., Sertkaya, B. (eds.) ICFCA 2010. LNCS, vol. 5986, pp. 161–176. Springer, Heidelberg (2010)
11. Moshkov, M., Chikalov, I.: On algorithm for constructing of decision trees with minimal depth. Fundam. Inform. **41**(3), 295–299 (2000)
12. Moshkov, M.J.: On the class of restricted linear information systems. Discret. Math. **307**(22), 2837–2844 (2007)
13. Moshkov, M.J., Piliszczuk, M., Zielosko, B.: Greedy algorithm for construction of partial association rules. Fundam. Inform. **92**(3), 259–277 (2009)
14. Moshkov, M.J., Piliszczuk, M., Zielosko, B.: On construction of partial association rules. In: Wen, P., Li, Y., Polkowski, L., Yao, Y., Tsumoto, S., Wang, G. (eds.) RSKT 2009. LNCS, vol. 5589, pp. 176–183. Springer, Heidelberg (2009)
15. Obiedkov, S.A., Duquenne, V.: Attribute-incremental construction of the canonical implication basis. Ann. Math. Artif. Intell. **49**(1–4), 77–99 (2007)
16. Pasquier, N., Bastide, Y., Taouil, R., Lakhal, L.: Efficient mining of association rules using closed itemset lattices. Inf. Syst. **24**(1), 25–46 (1999)
17. Pawlak, Z., Skowron, A.: Rudiments of rough sets. Inf. Sci. **177**(1), 3–27 (2007)
18. Poelmans, J., Ignatov, D.I., Kuznetsov, S.O., Dedene, G.: Formal concept analysis in knowledge processing: a survey on applications. Expert Syst. Appl. **40**(16), 6538–6560 (2013)
19. Savasere, A., Omiecinski, E., Navathe, S.B.: An efficient algorithm for mining association rules in large databases. In: Dayal, U., Gray, P.M.D., Nishio, S. (eds.) VLDB, pp. 432–444. Morgan Kaufmann (1995)
20. Zielosko, B.: Optimization of decision rules relative to coverage - comparative study. In: Kryszkiewicz, M., Cornelis, C., Ciucci, D., Medina-Moreno, J., Motoda, H., Raś, Z.W. (eds.) RSEISP 2014. LNCS, vol. 8537, pp. 237–247. Springer, Heidelberg (2014)

Bioinformatics

PDP-RF: Protein Domain Boundary Prediction Using Random Forest Classifier

Piyali Chatterjee[1], Subhadip Basu[2(✉)], Julian Zubek[3,4], Mahantapas Kundu[2],
Mita Nasipuri[2], and Dariusz Plewczynski[4,5,6(✉)]

[1] Department of Computer Science and Engineering, Netaji Subhash Engineering College,
Garia, Kolkata 700152, India
[2] Department of Computer Science and Engineering, Jadavpur University, Kolkata 700032, India
subhadip@cse.jdvu.ac.in
[3] Institute of Computer Science, Polish Academy of Sciences, Warsaw, Poland
[4] Centre of New Technologies, University of Warsaw, Warsaw, Poland
d.plewczynski@cent.uw.edu.pl
[5] The Jackson Laboratory for Genomic Medicine, c/o University of Connecticut Health Center,
263 Farmington Avenue, Farmington, CT 06030, USA
[6] Centre for Innovative Research, Medical University of Bialystok, Bialystok, Poland

Abstract. The Domain Boundary Prediction is a crucial task for functional classification of proteins, homology-based protein structure prediction and for high-throughput structural genomics. Each amino acid is represented using a set of physico-chemical properties. Random Forest Classifier is explored for accurate prediction of domain regions by training on the curated dataset obtained from CATH database. The software is tested on proteins of CASP-6, CASP-8, CASP-9 and CASP-10 targets in order to evaluate its prediction accuracy using three fold cross validation experiments. Finally, a consensus approach is used to combine results of the classifiers obtained through the cross-validation experiments. The average recall and precision scores achieved by the developed consensus based Random Forest classifiers (PDP-RF) are 0.98 and 0.88 respectively for prediction of CASP targets. The overall accuracy and F-scores of the PDP-RF are observed as 0.87 and 0.91 respectively.

1 Introduction

A *domain* is a segment of a polypeptide chain that can fold into a three dimensional structure irrespective of the presence of other segments of the chain [1]. Some simple combinations of protein secondary structure elements are referred to as 'super-secondary structure', or 'motifs'. Several motifs pack together to form compact, local, semi-independent units called *domains*. The overall 3D structure of the polypeptide chain is referred to as the protein's tertiary structure, whereas the domain is the fundamental building block of tertiary structure. So, a *domain* is a structural and functional unit of protein. To predict the tertiary structure of a protein, it is useful to segment the protein by identifying domain boundaries in it. A number of methods so far have been developed to identify protein domains starting from their primary sequences which are mainly developed for prediction of multi-domains in protein chains.

© Springer International Publishing Switzerland 2015
M. Kryszkiewicz et al. (Eds.): PReMI 2015, LNCS 9124, pp. 441–450, 2015.
DOI: 10.1007/978-3-319-19941-2_42

Galzitskaya et al. [2] considered conformational entropy for each amino acid and searches for a global minimum on an entropy profile constructed for the whole protein chain from its amino acid sequence. Based on the difference in amino acid compositions between domain and linker regions, a method DOMCUT [3] has been developed to predict linker regions among domains. CHOPnet [4] uses evolutionary information, predicted secondary structure, solvent accessibility, amino acid flexibility and amino acid composition for predicting domains in protein chains. Armadillo [5], the another domain predictor uses any amino acid index named as *Domain Linker propensity Index* (DLI) to convert a protein sequence to a smoothed numeric profile, from which domains and domain boundaries may be predicted. The Position Specific Scoring Matrix (PSSM)of the target protein obtained through PSI-BLAST, has also been used for domain boundary prediction by PPRODO [6] using Artificial neural network as a classifier. A machine learning predictor DOMpro [7] uses a combination of evolutionary information (in the form of profiles), predicted secondary structures, predicted solvent accessibility of the protein chains.

In the work of Sikder and Zomaya [8], the performance of DomainDiscovery of protein domain boundary assignment is improved significantly by including inter domain linker index value along with PSSM, predicted secondary structures, solvent accessibility information. Support Vector Machine (SVM) is used to predict possible domain boundaries for target sequences. Based on the application of secondary structure element alignment (SSEA) and profile-profile alignment (PPA) in combination with InterPro pattern searches, a protein domain prediction approach, called SSEP-Domain, is proposed by Gewehr and Zimmer [9]. Cheng [10] proposed a hybrid domain prediction web service, called DOMAC, by integrating *template-based* and *ab initio* methods. The template-based method is used in DOMAC to predict domains for proteins having homologous template structures in protein Data Bank [11]. If no significant homologous template is found, DOMAC invokes the *ab initio* domain predictor DOMpro to predict domains. To achieve a more accurate and stable predictive performance than the existing state-of–the-art models, a new machine learning based domain predictor, viz., DomNet [12] is trained using a novel compact domain profile, predicted secondary structure, solvent accessibility information and inter-domain linker index. FIFEDom [13] is other type of multi-domain prediction where prediction is done using fuzzy mean operator. This fuzzy operator assigns a membership value for each residue as belonging to a domain boundary thus finding contiguous boundary regions. Eickholt et al. propose a new method DoBo [14] where machine learning approach with evolutionary signals is used. It first extracts putative domain boundary signals from MSA between sequence and its homologs. Then those sites are classified by SVM where sequence profiles, secondary structures or solvent accessibility are used as features. Another SVM predictor DROP [15] empowered with 25 optimal features distinguish linkers from non-linkers effectively. In the first step, a random forest algorithm was used to evaluate 3000 features. In the next step, a selection protocol was used to select optimal features. Based on a creating hinge region strategy, a new approach DomHR [16] predicts domain boundary by means of constructing profiles of domain Hinge-boundary (DHB) features. Besides these, improvement in contact prediction provides a new source of domain boundary prediction. In the work of Sadowski [17], kernel smoothing based method and

methods based on building alpha carbon models onto this contact information. A recent template based method on this field is ThreaDom [18] proposed by Xue et al. in which protein domain boundary information is extracted from multiple threading alignments. The core of the method is use of domain conservation score that combines information from template domain structures and terminal and internal alignment gaps.

It appears from the above discussion that there are still some scopes for improvement in protein domain prediction. The rationale behind the choices of the feature sets and classifiers for prediction of domain boundaries are discussed in the following sections.

2   Materials and Methods

An attempt has been made under the present work to employ Random Forest Classifier as a machine learning algorithm for protein domain boundary prediction on the basis of an effective feature set consisting of hydrophobicity, linker index, polarity, ordered or disordered region of protein sequence and flexibility. Different methods [3, 6, 19] use different sliding window sizes for domain boundary prediction. Studies say that prediction within ± 20 residues from the true boundary position are considered as successful with existing evaluation criteria for boundary prediction methods. These studies motivate us to test the prediction performance of our domain predictor PDP-RF with optimal residue windows, since larger window size is useful to predict multi-domain proteins.

Features Set
Five types of features, viz., *predicted ordered or disordered region, normalized flexibility parameters (B-values), polarity, linker index, modified Kyte-Doolittle hydrophobicity scale* are used for this work. The last four features for the current experiment are chosen from (exactly 544 in the selected version) AAIndex database [20] release 9.0 (http://www.genome.jp/aaindex/). From experimental findings, it is known that large ordered region when they are divided by shorter parts of disordered regions in a protein chain, are likely to be separate domains [21]. For this reason, ordered or disordered region predicted by disprot tool [22] is taken as a feature. On the other hand, the presence of multiple domains in proteins gives rise to a great deal of flexibility and mobility [23]. The Debye-Waller factor (B-value) (ACC No: VINM940101) which measures average flexibility parameters is used as one of the five features. The distribution of polar and non-polar side chains is one of the most important factors governing the folding of a protein into 3D structure [24]. Latest polarity (ACC No: GRAR740102)feature is taken as a feature in this work. To represent the preference for amino acid residues in linker or regions, a parameter called the linker index is defined by Sumaya and Ohara [3]. From the AAINDEX, linker index(Acc No: BAEK050101) is taken as a feature. The more exposed the linker, the more likely it is to contain hydrophilic residues. Greater hydrophobicity is found in more linker connections between two domains. Modified Kyte-Doolittle hydrophobicity scale (Acc No: JURD980101) is taken as a feature in the current work, which is also from the AAIndex dataset.

Experimentation

In this work, we have taken Random Forest (RF) Classifier and a consensus scheme. Random Forest is a popular ensemble algorithm based on decision trees [25]. It is commonly used in bioinformatics, as it is relatively easy to apply and robust against many kinds of noisy and incomplete data characteristic for experimental biological problems [26]. In this work we trained Random Forest with 100 trees with $\sqrt{d}$ attributes considered for each split (d – number of all attributes). The implementation we used came from scikit-learn library [27].

It is conducted in two stages. In the first stage 354 protein chains of the CATH database (version 2.5.1) are used to perform a three-fold cross validation experiment where in each experimental fold, 67% of the positive/negative samples are used for training and the rest of the samples for testing. Each domain region residue is considered a positive sample, and non-domain residues are considered negatives. RF based classifiers are trained to generate three trained classifiers from three cross-validated experiments.

In the second stage of the experiment, we consider a consensus approach on the basis of the trained classifiers to generate test results on 19 protein sequences, taken from the CASP-6 dataset [28], 109 protein sequences from the CASP-8 dataset [29], 100 protein sequences from the CASP-9 dataset [30] and 59 protein sequences from CASP-10 dataset [31]. According to the consensus strategy, for each classifier, $1 - star$, $2 - star$ and $3 - star$ consensus classifiers are designed. At next step, a $n - star$ consensus strategy (here, $n = 3$, as number of classifiers are 3) is applied [32] to three classifiers. Thus we obtained $1 - star$, $2 - star$ and $3 - star$ classifiers. As a result, 3 consensus classifiers are also designed to achieve improved performance. Here we define a 3-star quality consensus scheme as C_n^N, where N is the number of classifiers of a particular type participating in the specific consensus strategy, and n $(1 \leq n \leq N)$ is the quality of prediction [32]. More specifically, $1 - star$ prediction says that any one of possible N classifiers predicts the test sequence to be positive for the domain region under consideration, and $N - star$ represents that all classifiers agreed to the decision. Along this principle, we define the $3 - star$ consensus over 3-variations of training on three fold cross-validation data of a special type classifier. Subsequently, C_n^3 is defined as the consensus among three classifiers. Question arises as to how $n - star$ consensus relates to Random Forest, which is already an ensemble algorithm. In Random Forest decision is made through weighted voting. Our consensus approach is equivalent to standard (equal weights) voting with a variable threshold. This allows choosing a tradeoff between precision and recall of the ensemble.

3   Results and Discussion

The current experiment is conducted in two stages. In the first stage 354 protein chains of the CATH database (version 2.5.1) are used to perform a three-fold cross validation experiment where in each experimental fold, 67% of the positive/negative samples are used for training and the rest of the samples for testing. RF based classifiers are trained to generate three trained classifiers from three cross-validated experiments. In the second stage of the experiment, we consider a consensus approach on the basis of the trained classifiers to generate test results on 19 protein sequences, taken from the CASP-6 dataset [28], 109 protein sequences from the CASP-8 dataset [29],100 protein sequences from the CASP-9 dataset [30] and 59 protein sequences from CASP-10 dataset [31]. According to the consensus strategy, for three classifiers of each classifier, 1 − *star*, 2 − *star* and 3 − *star* consensus classifiers, namely, PDP-RF-1,PDP-RF-2 and PDP-RF-3 are designed. In case of sequence based prediction, the length of sequence fragment whose central amino acid is being predicted as domain or linker region is very crucial. Different methods use different sliding window sizes for domain boundary prediction. Studies say that prediction within ± 20 residues from the true boundary prediction are considered as successful with existing evaluation criteria for domain boundary prediction methods. To determine the length of the sequence fragment or window, prediction results are observed for classifiers only on a single fold among three cross validated datasets. Among 13, 15, 17, 19, 21, 25 and 29 window sizes, performance of classifier at 17 window size is the best. So, this window size is made fixed for this work.

Table 1. Performance of three RF single Classifiers

Performance (single classifiers)	CASP targets	Recall	Precision	Accuracy	F-Scores
RF1	CASP-6	0.996	0.949	0.944	0.971
	CASP-8	0.998	0.913	0.912	0.950
	CASP-9	0.997	0.897	0.894	0.933
	CASP-10	0.993	0.793	0.799	0.849
RF2	CASP-6	0.989	0.948	0.938	0.967
	CASP-8	0.914	0.913	0.911	0.950
	CASP-9	0.993	0.897	0.891	0.932
	CASP-10	0.985	0.793	0.802	0.847
RF3	CASP-6	0.937	0.948	0.890	0.940
	CASP-8	0.977	0.918	0.901	0.943
	CASP-9	0.962	0.902	0.869	0.917
	CASP-10	0.941	0.797	0.800	0.838
Average	CASP-6	0.974	0.948	0.924	0.954
	CASP-8	0.963	0.915	0.908	0.927
	CASP-9	0.984	0.899	0.885	0.904
	CASP-10	0.973	0.794	0.8	0.82

From three cross validated experiments, three classifiers are designed and their performance is observed. Outstanding performance is observed in Random Forest Classifiers in prediction of CASP-6, CASP-8, CASP-9 and CASP-10 targets. For CASP-6, CASP-8 and CASP-9 targets its behavior is found to be consistent whereas prediction results are somewhat less in CASP-10 targets. Table 1 shows the average performance of 3 classifiers.

As three classifiers are taken, so, $1 - star$, $2 - star$, $3 - star$ consensus strategy may be adopted as already defined in the previous section. The performance of consensus classifier must demand the good predictive accuracy in comparison to single classifier. From Table 2, it can be observed that with the introduction of consensus classifier, the performance of each type classifier is increased in a large scale.

Table 2. Average Performance of consensus RF Classifiers

Average performance (consensus classifiers)	CASP targets	Recall	Precision	Accuracy	F-Scores
PDP- RF-1	CASP-6	0.997	0.949	0.997	0.971
	CASP-8	0.999	0.913	0.913	0.950
	CASP-9	0.998	0.900	0.895	0.934
	CASP-10	0.995	0.800	0.800	0.849
PDP- RF-2	CASP-6	0.989	0.948	0.988	0.967
	CASP-8	0.996	0.914	0.912	0.950
	CASP-9	0.993	0.900	0.900	0.932
	CASP-10	0.985	0.794	0.803	0.848
PDP- RF-3	CASP-6	0.937	0.948	0.937	0.940
	CASP-8	0.977	0.918	0.901	0.943
	CASP-9	0.962	0.903	0.869	0.917
	CASP-10	0.941	0.798	0.801	0.838

As performance of single RF classifier is found to be the best whereas consensus classifier uplifts its accuracy up to its highest limit. Table 2 shows overall performance of consensus classifiers of RF. In case of RF classifiers, performance of $1 - star$, $2 - star$ and $3 - star$ consensus schemes are found to be the same which indicate the prediction decisions among three classifiers at higher confidence. In Table 3, it is seen that consensus classifier improves the accuracy of single classifier a little.

Table 3. Improved performance of PDP-RFs over single RF Classifiers

Improved performance (consensus classifiers)	CASP targets	Recall	Precision	Accuracy	F-Scores
PDP- RF-1	CASP-6	0.001	0	0.053	0
	CASP-8	0.001	0	0.001	0
	CASP-9	0.001	0.003	0.001	0.001
	CASP-10	0.002	0.007	0.001	0
PDP- RF-2	CASP-6	0	0	0.05	0
	CASP-8	0.082	0.001	0.001	0
	CASP-9	0	0.003	0.009	0
	CASP-10	0	0.001	0.001	0.001
PDP- RF-3	CASP-6	0	0	0.047	0
	CASP-8	0	0	0	0
	CASP-9	0	0.001	0	0
	CASP-10	0	0.001	0.001	0

As mentioned earlier, we have taken domain as positive class and linker as negative class. Since the proportion of domain and linker in our dataset is not equal i.e., domain residue represents majority class and non-domain or linker residue represents minority class, the prediction results may turn out to be biased towards majority class. For this reason, we reverse the role of domain and linker residue by taking linker residue as positive and domain residue as negative class. The overall performance of PDP-RF is found to be the somewhat less compared to former performance (Accuracy in prediction of CASP targets using majority class training is 0.88 whereas is 0.85 using minority class training).

A Robust Consensus Classifier

In this work, an attempt has been done to choose random Forest, as effective machine learning classifier, to exploit strong multi facet feature sets and by applying a novel consensus approach. Thus objective is to design a strong robust classifier which enables the system to predict targets very efficiently and effectively. In prediction of CASP targets, in most of the cases, RF classifier offers the best predictive ability. Inclusion of the novel 3 − star consensus approaches further improves the classifiers' performances.

We have taken PPRODO [6], DomPro [7], DROP [15], FIFEDom [13], ThreaDom [18] as existing methods for comparison because most of the methods are freely available. PPRODO, DomPro are not recent but they are based on machine learning method. DROP is recent machine learning method as well. On the other hand, Threadom is recent template based method which predicts multi domain proteins of CASP targets very well.

Overall, the successful performance of most of the classifiers in CASP competition is found. Performances of PDP-RF classifiers are analyzed with ThreaDom1, ThreaDom2 [18], FIFEDom [13], Pfam [33], DROP [15], DOMPro [7], PPRODO [6], DoBo [14] in prediction of CASP-9 targets and CASP-10 targets. Finding the appropriate robust machine learning classifier, use of significant feature set, selection of optimal window and finally incorporation of consensus approach into three classifiers of each type of classifier is a very challenging task in prediction of domain boundaries along protein sequences. Learning patterns is a very challenging issue for any classifier in case of binary classification where proportion of positive and negative samples is not equal. Moreover, a novel $3 - star$ consensus approach is applied to further improve the prediction accuracy. We finally conclude that the designed feature set; alongside with Random Forest based classifier based consensus approach effectively predicts the domain regions in multi-domain protein chains. The cross-validated experimental setup with standard CATH database establishes our claims. Prediction decisions from the three experimental folds are combined to design $n - star$ quality consensus strategies. Here, $3 - star$ quality consensus is designed by combining the decisions of the three classifiers from each of the three sets of cross validation experiments. The consensus strategy is found to be superior in comparison with the performances of the best single classifier.

Prediction is done on residue level i.e. whether a residue belongs to domain or linker region but not on domain boundary based. Domain prediction methods vary in the procedure, i.e. either they are template based (e.g., Threadom or FIFEDOM) or ab initio based (e.g. DomPro, DROP etc.). Some Predictors predicts domain boundary (DOMPRo, Threadom) and some of them predicts linkers. The goal of the current state of the art and our proposed method is more or less same but difference lies in the domain boundary definition (e.g. DomPro considers the residues in the range of 20 residues around the center of domain region the domain boundary residues from the CATH assignment). In this work, we take domain regions from CATH by considering domain number starting/end positions of each domain sequentially. As a result, our dataset contains domain residue serving as majority class. So, it cannot be compared with current state of the art in terms of performance metrics. Here, recall scores of PDP-RFs on CASP-9 and CASP-10 targets are 0.98 and 0.97 whereas precision scores of PDP-RFs on the same are 0.89 and 0.79. Template based method Threadom2, Threadom1, FIFEDOM predicts CASP-9 targets at 0.534, 0.397, 0.233 recall scores and 0.764, 0.636, 0.34 precision scores. PFAM, DROP (linker based), and DomPro, PPRODO (ab initio) predict CASP9-targets at recall of 0.548, 0.26, 0.219, 0.397 and precision of 0.5, 0. 679, 0.727 and 0.56.

In prediction of CASP-10 targets, Threadom2 and Threadom1 predict targets well (recall score: 0.625, 0.625 and precision score: 0.796, 0.732). But FIFEDOM predict targets at low recall and precision score (0.188, 0.28). On the other hand, PFAM, DROP (linker based), and DomPro, PPRODO (ab initio) predict CASP10-targets at recall of 0.547, 0.156, 0.109 and 0.406 and precision of 0.466, 0.714, 0.44 and 0.591which is better than that of CASP-9 targets. Recall and precision score of PDP-RF are reported but not compared with these methods as it is not fair to compare a residue based prediction scheme with domain boundary based or linker based prediction method or with template based method where there lies a difference in evaluation criteria.

Methods for building feature importance rankings based on Random Forest can also be used to gain more insights into amino acid properties correlated with domain boundaries. To support validity of our method we also plan to include comparison with other machine learning algorithms in our next work.

Acknowledgments. The paper is co-funded by the European Union from financial resources of the European Social Fund. Project PO KL "Information technologies: Research and their interdisciplinary applications". This work was partially supported by the Polish National Science Centre (Grant number 2014/15/B/ST6/05082 and UMO-2013/09/B/NZ2/00121), and COST BM1405 EU action. Authors are thankful to the "Center for Microprocessor Application for Training Education and Research" for providing infrastructure facility during progress of the work. It is also co-funded by UPE-II, PURSE project, Govt. Of India at Department of Computer Science & Engineering, Jadavpur University

References

1. Mount, D.: Bioinformatics: Sequence and Genome Analysis, p. 416. Cold Spring Harbor Laboratory Press, New York (2004)
2. Melnik, B.S., Galzitskaya, O.V.: Prediction of protein domain boundaries from sequence alone. Protein Sci. **12**, 696–701 (2003)
3. Suyama, M., Ohara, O.: Domcut: prediction of inter-domain linker regions in amino acid sequences. Bioinformatics **19**, 673–674 (2003)
4. Liu, J., Rost, B.: Sequence-based prediction of protein domains. Nucleic Acids Res. **32**, 3522–3530 (2004)
5. Dumontier, M., Yao, R., Feldman, H.J., Hoque, C.W.: Armadillo: domain boundary prediction by amino acid composition. J. Mol. Biol. **350**, 1061–1073 (2005)
6. Sim, J., Kim, S.Y., Lee, J.: PPRODO: prediction of protein domain boundaries using neural networks. Proteins. **59**, 627–632 (2005)
7. Cheng, J., Sweredoski, M.J., Baldi, P.: DOMpro: Protein domain prediction using profiles, secondary structure, relative solvent accessibility, and recursive neural networks. Data Min. Knowl. Discov. **13**, 1–10 (2006)
8. Sikder, A.R., Zomaya, A.Y.: Improving the performance of domaindiscovery of protein domain boundary assignment using inter-domain linker index. BMC Bioinformatics. **7** (Suppl 5), S6 (2006)
9. Gewehr, J.E., Zimmer, R.: SSEP-Domain: Protein domain prediction by alignment of secondary structure elements and profiles. Bioinformatics **22**, 181–187 (2006)
10. Cheng, J.: DOMAC: An accurate, hybrid protein domain prediction server. Nucleic Acids Res. **35**, W354–W356 (2007)
11. Berman, H.M., Westbrook, J., Feng, Z., Gilliland, G., Bhat, T.N., Weissig, H., Shindyalov, I.N., Bourne, P.E.: The Protein Data Bank. Nucleic Acids Res. **28**, 235–242 (2000)
12. Yoo, P.D., Sikder, A.R., Taheri, J., Zhou, B.B., Zomaya, A.Y.: DomNet: protein domain boundary prediction using enhanced general regression network and new profiles. NanoBioSci. IEEE Trans. **7**, 172–181 (2008)
13. Bondugula, R., Lee, M.S., Wallqvist, A.: FIFEDom: a transparent domain boundary recognition system using a fuzzy mean operator. Nucleic Acids Res. **37**, 452–462 (2009)
14. Eickholt, J., Deng, X., Cheng, J.: DoBo: protein domain boundary prediction by integrating evolutionary signals and machine learning. BMC Bioinformatics **12**, 43 (2011)

15. Ebina, T., Toh, H., Kuroda, Y.: DROP: an SVM domain linker predictor trained with optimal features selected by random forest. Bioinformatics **27**, 487–494 (2011)
16. Zhang, X.Y., Lu, L.J., Song, Q., Yang, Q.Q., Li, D.P., Sun, J.M., Li, T.H., Cong, P.S.: DomHR: accurately identifying domain boundaries in proteins using a hinge region strategy. PLoS One **8**, e60559 (2013)
17. Sadowski, M.I.: Prediction of protein domain boundaries from inverse covariances. Proteins **81**, 253–260 (2013)
18. Xue, Z., Xu, D., Wang, Y., Zhang, Y.: ThreaDom : extracting protein domain boundary information from multiple threading alignments. Bioinformatics **29**, 247–256 (2013)
19. Galzitskaya, O.V., Dovidchenko, N.V., Lobanov, M., Garbuzinskii, S.A.: Prediction of protein domain boundaries from statistics of appearance of amino acid residues. Mol. Biol (Mosk). **40**, 96–107 (2006)
20. Kawashima, S., Ogata, H., Kanehisa, M.: AAindex: amino acid index database. Nucleic Acids Res. **27**, 368–369 (1999)
21. Wyrwicz, L.S., Koczyk, G., Rychlewski, L., Plewczynski, D.: ProteinSplit: splitting of multi-domain proteins using prediction of ordered and disordered regions in protein sequences for virtual structural genomics. J. Phys. Condens. Matter **19**, 285222 (2007)
22. Sickmeier, M., Hamilton, J.A., LeGall, T., Vacic, V., Cortese, M.S., Tantos, A., Szabo, B., P, Tompa, Chen, J., Uversky, V.N., Obradovic, Z., Dunker, A.K.: DisProt: The database of disordered proteins. Nucleic Acids Res. **35**, D786–93 (2007)
23. Bu, Z., Callaway, D.J.: Proteins move! protein dynamics and long range allostery in cell signaling. Adv. Protein Chem. Struct. Biol. **83**, 163–221 (2011)
24. Cordes, M.H., Davidson, A.R., Sauer, R.T.: Sequence space, folding and protein design. Curr. Opin. Struct. Biol. **6**, 3–10 (1996)
25. Breiman, L.: Random forests. Mach. Learn. **45**, 5–32 (2001)
26. Yang, P., Yang, Y.H., Zhou, B.B., Zomaya, A.Y.: A review of ensemble methods in bioinformatics. Curr. Bioinform. **5**, 296–308 (2010)
27. Pedregosa, F., Varoquaux, G., Gramfort, A., Michel, V., Thirion, B., Grisel, O., Blondel, M., Prettenhofer, P., Weiss, R., Dubourg, V., Vanderplas, J., Passos, A., Cournapeau, D., Brucher, M., Perrot, M., Duchesnay, E.: Scikit-learn: machine learning in python. J. Mach. Learn. Res. **12**, 2825–2830 (2011)
28. Moult, J., Fidelis, K., Rost, B., Hubbard, T., Tramontano, A.: Critical assessment of methods of protein structure prediction (CASP)–round 6. Proteins. **61**(Suppl 7), 3–7 (2005)
29. Moult, J., Fidelis, K., Kryshtafovych, A., Rost, B., Tramontano, A.: Critical assessment of methods of protein structure prediction (CASP)-round VIII. Proteins **77**, 1–4 (2009)
30. Moult, J., Fidelis, K., Kryshtafovych, A.: Critical assessment of methods of protein structure prediction (CASP)–round IX. Proteins. **79**(Suppl 10), 1–5 (2011)
31. Moult, J., Fidelis, K., Kryshtafovych, A., Schwede, T., Tramontano, A.: Critical assessment of methods of protein structure prediction (CASP)-round X. Proteins. **82**(Suppl 1), 1–6 (2014)
32. Plewczynski, D., Basu, S., Saha, I.: AMS 4.0: consensus prediction of post-translational modifications in protein sequences. Amino Acids **43**, 573–582 (2012)
33. Finn, R.D., Bateman, A., Clements, J., Coggill, P., Eberhardt, R.Y., Eddy, S.R., Heger, A., Hetherington, K., Holm, L., Mistry, J., Sonnhammer, E.L., Tate, J., Punta, M.: Pfam: the protein families database. Nucleic Acids Res. **42**, D222–D230 (2014)

A New Similarity Measure for Identification of Disease Genes

Pradipta Maji[1]([✉]), Ekta Shah[1], and Sushmita Paul[2]

[1] Biomedical Imaging and Bioinformatics Lab, Machine Intelligence Unit,
Indian Statistical Institute, Kolkata, India
pmaji@isical.ac.in, eshah491@gmail.com
[2] Laboratory of Systems Tumor Immunology, Department of Dermatology,
University of Erlangen-Nurnberg, Erlangen, Germany
sushmita.paul@uk-erlangen.de

Abstract. One of the important problems in functional genomics is how
to select the disease genes. In this regard, the paper presents a new similarity measure to compute the functional similarity between two genes.
It is based on the information of protein-protein interaction networks.
A new gene selection algorithm is introduced to identify disease genes,
integrating judiciously the information of gene expression profiles and
protein-protein interaction networks. The proposed algorithm selects a
set of genes from microarray data as disease genes by maximizing the relevance and functional similarity of the selected genes. The performance
of the proposed algorithm, along with a comparison with other related
methods, is demonstrated on colorectal cancer data set.

1 Introduction

Genetic diseases such as Alzheimer's disease, breast cancer, leukemia, colorectal
cancer, down syndrome, and heart disease are caused by abnormalities in genes
or chromosomes. A genetic disease may be heritable disorder or may not be.
While some genetic diseases are passed down from the parent's genes, others are
frequently caused by new mutations or changes to the DNA. In other instances,
the same disease, for example, some forms of cancer, may stem from an inherited
genetic condition in some people, from new mutations in some people, and from
non-genetic causes in other people. As the term genetic disease suggests, these
diseases are caused by the dysfunction of some genes. Therefore, such genes are
better known as disease genes [1].

Recent advancement and wide use of high-throughput biotechnologies have
been producing an explosion in using gene expression phenotype for understanding the function of disease genes [4,8]. Analyzing the difference of gene expression levels in particular cell types may provide an idea about the propensity of
a disease. Specifically, if a set of genes shows a consistent pattern of different
expression levels in sick subjects and a control group, then that gene set is likely
a strong candidate of playing a pathogenic role. Differences in expression levels

© Springer International Publishing Switzerland 2015
M. Kryszkiewicz et al. (Eds.): PReMI 2015, LNCS 9124, pp. 451–461, 2015.
DOI: 10.1007/978-3-319-19941-2_43

can be detected primarily by microarray studies. In this background, microarray gene expression data has been widely used for identification of disease genes using different feature selection algorithms [5,16,18,22].

In [2,6], it has been shown that the genes associated with the same disorder tend to share common functional features, reflecting that their protein products have a tendency to interact with each other. Hence, another indicative trait of a disease gene is that its protein product is strongly linked to other disease-gene proteins. In this background, the protein-protein interaction (PPI) data have been used in various studies to identify disease genes [13,20]. Individually microarray data or the PPI network data can be used to identify potential disease genes, although there is a limited chance of finding novel disease genes from such an analysis. In this regard, data integration methods have been developed to identify pleiotropic genes involved in the physiological cellular processes of many diseases.

The integrated approaches assume that the protein products of disease genes tend to be close to differentially expressed genes in the protein interaction network. Chao et al. [24] developed a method by integrating gene expression data and the PPI network data to prioritize cancer-associated genes. Zhao et al. [25] also proposed an approach by integrating gene expression data and the PPI network data to select disease genes. Jia et al. [10] developed a dense module searching method to identify disease genes for complex diseases by integrating the association signal from genome wide association studies data sets into the human PPI network. Li and Li [17] developed another approach to identify candidate disease genes, where heterogeneous genomic and phenotype data sets are used. In this method, separate gene networks are first developed using different types of data sets. The various genomic networks are then merged into a single graph, and disease genes are identified using random walk. In [16], minimum redundancy-maximum relevance (mRMR) [5] approach has been used to select a set of genes from expression data, while maximum relevance-maximum significance (MRMS) criterion [18] has been used in [22]. The selected gene set is then used for identification of intermediate genes between a pair of selected genes using the PPI network data. However, all the methods reported earlier consider gene expression and PPI data separately while selecting candidate genes.

In this regard, this paper presents a new gene selection algorithm to identify disease genes. It selects a set of disease genes by maximizing the relevance and functional similarity of the selected genes. A new similarity measure is introduced to compute the functional similarity between two genes. The proposed algorithm judiciously integrates the information of gene expression profiles and PPI networks. The mutual information is employed to compute the relevance of the genes with respect to class labels based on gene expression profiles, while the PPI data is used to calculate the functional similarity between two genes. The mutual information is used to select differentially expressed genes as disease genes using gene expression profiles, on the other hand, the functional protein association network is used to study the mechanism of diseases. The performance of the proposed algorithm, along with a comparison with other related methods, is demonstrated on colorectal cancer data set. An important finding is that the proposed algorithm is shown to be effective for selecting relevant and functionally similar genes from microarray data, and the identified genes are significantly linked with colorectal cancer. Extensive experimental study on colorectal cancer establishes the fact that the genes identified by the proposed method have more colorectal cancer genes than those identified by existing methods and using the gene expression profiles alone, irrespective of any gene selection algorithm. All the results indicate that the proposed method is quite promising and may become a useful tool for identifying disease genes.

2   A New Gene-Gene Similarity Measure

In general, the genes, which are associated with the same disorder, tend to share common functional features. The protein products of these genes also have a tendency to interact with each other [2,6]. Hence, an important characteristic of a disease gene is that its protein product is strongly linked to other disease-gene proteins. It has also been observed that proteins with short distances to each other in the network are more likely to involve in common biological functions [3, 14,21], and that interactive neighbors are more likely to have identical biological function than non-interactive ones [11,15]. This is because the query protein and its interactive proteins may form a protein complex to perform a particular function or be involved in a same pathway. Accordingly, a quantitative measure is required that can efficiently compute the similarity between two genes. In this paper, the information of PPI networks is used to calculate the functional similarity.

The PPI networks are commonly represented as graphs (Fig. 1), with nodes corresponding to proteins and edges representing PPIs. The weight of the edge in graph depends on experimental as well as predicted interaction information. Let $\mathcal{N}_i$ be the set of interactive neighbors or successor genes of a candidate gene $\mathcal{A}_i$ and $\omega_{ij} \in [0,1]$ is the weight value of the edge between gene $\mathcal{A}_j \in \mathcal{N}_i$ and candidate gene $\mathcal{A}_i$. The set of successors $\mathcal{N}_i$ of gene $\mathcal{A}_i$ and corresponding weight value ω_{ij} can be obtained from the information of PPI network. Let $\mathcal{N}_{ik}$ be the set of genes, which are successors of both genes $\mathcal{A}_i$ and $\mathcal{A}_k$, that is, $\mathcal{N}_{ik} = \mathcal{N}_i \cap \mathcal{N}_k$. Define $\tilde{\mathcal{N}}_i = \mathcal{N}_i \setminus \mathcal{N}_{ik}$ as the set of genes those are successors of gene $\mathcal{A}_i$ but not of gene $\mathcal{A}_k$. The functional similarity between two genes $\mathcal{A}_i$ and $\mathcal{A}_k$, having sets of successor genes $\mathcal{N}_i$ and $\mathcal{N}_k$, respectively, is as follows:

$$S(\mathcal{A}_i, \mathcal{A}_k) = \frac{\sum\limits_{\mathcal{A}_j \in \mathcal{N}_{ik}} \min\{\omega_{ij}, \omega_{kj}\}}{\sum\limits_{\mathcal{A}_j \in \tilde{\mathcal{N}}_i} \omega_{ij} + \sum\limits_{\mathcal{A}_j \in \mathcal{N}_{ik}} \max\{\omega_{ij}, \omega_{kj}\} + \sum\limits_{\mathcal{A}_j \in \tilde{\mathcal{N}}_k} \omega_{kj}}. \tag{1}$$

Hence, if the interactive neighbors and the corresponding edge weights of two genes are same, then the functional similarity between these two genes is high. On the other hand, two genes are functionally dissimilar if they have no common interactive neighbors. The following properties can be stated about the measure:

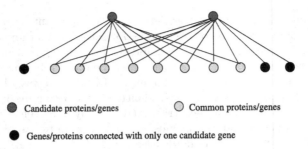

Fig. 1. An example of protein-protein interaction network

1. $0 \leq \mathcal{S}(\mathcal{A}_i, \mathcal{A}_k) \leq 1$.
2. $\mathcal{S}(\mathcal{A}_i, \mathcal{A}_k) = 1$ if and only if two sets $\mathcal{N}_i$ and $\mathcal{N}_k$ contain exactly same set of successor genes, that is, $\mathcal{N}_{ik} = \mathcal{N}_i = \mathcal{N}_k$, and weight value $\omega_{ij} = \omega_{kj}, \forall \mathcal{A}_j \in \mathcal{N}_{ik}$.
3. $\mathcal{S}(\mathcal{A}_i, \mathcal{A}_k) = 0$ if and only if $\mathcal{N}_{ik} = \emptyset$.
4. $\mathcal{S}(\mathcal{A}_i, \mathcal{A}_k) = \mathcal{S}(\mathcal{A}_k, \mathcal{A}_i)$ (symmetric).

In this regard, it should be noted that if the weight value $\omega_{ij} \in \{0, 1\}$, then the proposed similarity measure reduces to

$$\mathcal{S}(\mathcal{A}_i, \mathcal{A}_k) = \frac{|\mathcal{N}_i \cap \mathcal{N}_k|}{|\mathcal{N}_i \cup \mathcal{N}_k|} \tag{2}$$

which is Jaccard index $J(\mathcal{A}_i, \mathcal{A}_k)$.

3   Proposed Disease Gene Selection Algorithm

Recent advancement and wide use of high-throughput biotechnologies have been producing huge amount of gene expression profiles data, which have been widely used in different studies to understand the function of disease genes. If a set of genes shows a consistent pattern of different expression levels in sick subjects and a control group, then that gene set is likely a strong candidate of playing a pathogenic role. The difference of gene expression levels in particular cell types can be studied to get an idea about the propensity of a disease. On the other hand, the genes associated with the same disease tend to share common functional features. Also, the protein products of disease genes have a tendency to interact with other disease-gene proteins.

In this regard, the paper presents a new gene selection algorithm, integrating judiciously the gene expression and PPI data, to identify pleiotropic genes involved in the physiological cellular processes of the disease. The proposed method assumes that the protein products of disease genes tend to be close to differentially expressed genes in the protein interaction network. Hence, the proposed gene selection algorithm selects a set $\mathbb{S}$ of disease genes from the whole gene set $\mathbb{C}$ of the given microarray gene expression data set by maximizing both relevance and functional similarity of genes present in $\mathbb{S}$. Let $\mathbb{C} = \{\mathcal{A}_1, \cdots, \mathcal{A}_i, \cdots, \mathcal{A}_j, \cdots, \mathcal{A}_m\}$ be the set of m genes of a given microarray gene expression data set and $\mathbb{S}$ is the set of selected genes. Define $\gamma_{\mathcal{A}_i}(\mathbb{D})$ as the relevance of the gene $\mathcal{A}_i$ with respect to the class labels $\mathbb{D}$ while $\mathcal{S}(\mathcal{A}_i, \mathcal{A}_j)$ as the functional similarity between two genes $\mathcal{A}_i$ and $\mathcal{A}_j$. Hence, the total relevance of all selected genes is

$$\mathcal{J}_{\text{relevance}} = \sum_{\mathcal{A}_i \in \mathbb{S}} \gamma_{\mathcal{A}_i}(\mathbb{D}), \tag{3}$$

while the total functional similarity among the selected genes is

$$\mathcal{J}_{\text{similarity}} = \sum_{\mathcal{A}_i \neq \mathcal{A}_j \in \mathbb{S}} \mathcal{S}(\mathcal{A}_i, \mathcal{A}_j). \tag{4}$$

Hence, the problem of selecting a set $\mathbb{S}$ of relevant and functionally similar genes from the whole set $\mathbb{C}$ of m genes is equivalent to maximizing both $\mathcal{J}_{\text{relevance}}$ and $\mathcal{J}_{\text{similarity}}$, that is, to maximize the objective function

$$\mathcal{J} = a\mathcal{J}_{\text{relevance}} + (1-a)\mathcal{J}_{\text{similarity}}, \tag{5}$$

where a is a weight parameter. To solve the above problem, following greedy algorithm is used in the current study:

1. Initialize $\mathbb{C} \leftarrow \{\mathcal{A}_1, \cdots, \mathcal{A}_i, \cdots, \mathcal{A}_j, \cdots, \mathcal{A}_m\}, \mathbb{S} \leftarrow \emptyset$.
2. Calculate the relevance $\gamma_{\mathcal{A}_i}(\mathbb{D})$ of each gene $\mathcal{A}_i \in \mathbb{C}$.
3. Select the gene $\mathcal{A}_i$ as the most relevant gene that has the highest relevance value $\gamma_{\mathcal{A}_i}(\mathbb{D})$. In effect, $\mathcal{A}_i \in \mathbb{S}$ and $\mathbb{C} = \mathbb{C} \setminus \mathcal{A}_i$.
4. Repeat the following two steps until the desired number of genes is selected.
5. Calculate the functional similarity between each of the remaining genes of $\mathbb{C}$ with respect to the selected genes of $\mathbb{S}$ and remove it from $\mathbb{C}$ if it has zero functional similarity value with respect to any one of the selected genes.
6. From the remaining genes of $\mathbb{C}$, select gene $\mathcal{A}_j$ that maximizes the following condition:

$$a\gamma_{\mathcal{A}_j}(\mathbb{D}) + \frac{(1-a)}{|\mathbb{S}|} \sum_{\mathcal{A}_i \in \mathbb{S}} \mathcal{S}(\mathcal{A}_i, \mathcal{A}_j). \tag{6}$$

   As a result of that, $\mathcal{A}_j \in \mathbb{S}$ and $\mathbb{C} = \mathbb{C} \setminus \mathcal{A}_j$.
7. Stop.

The mutual information [22] can be used to calculate the relevance of a gene with respect to class labels, while the proposed similarity measure, based on the information of PPI data, can be used for computing functional similarity between two genes. However, in microarray gene expression data sets, the class labels of samples are represented by discrete symbols, while the expression values of genes are continuous. Hence, to measure the gene-class relevance of a gene with respect to class labels using mutual information, the continuous expression values of a gene are divided into several discrete partitions. The *a prior* (marginal) probabilities and their joint probabilities are then calculated to compute the gene-class relevance using the definitions for discrete cases. In this paper, the discretization method reported in [5, 22] is employed to discretize the continuous gene expression values. The expression values of a gene are discretized using mean μ and standard deviation σ computed over n expression values of that gene: any value larger than $(\mu+\sigma/2)$ is transformed to state 1; any value between $(\mu-\sigma/2)$ and $(\mu+\sigma/2)$ is transformed to state 0; any value smaller than $(\mu-\sigma/2)$ is transformed to state -1. These three states correspond to the over-expression, baseline, and under-expression of genes. On the other hand, the STRING (Search Tool for the Retrieval of Interacting Genes) is an online database resource that provides both experimental as well as predicted PPI information, along with a confidence score. In the current work, STRING is used for computing functional similarity between two genes considering confidence score as the weight value.

4   Experimental Results and Discussions

This section presents the performance of the proposed maximum relevance-maximum functional similarity (MRMFS) criterion based proposed gene selection algorithm, along with a comparison with other related methods. The algorithms compared are t-test, MR (maximum relevance), mRMR [5], MRMS [18], MR+PPIN [22], mRMR+PPIN [16], and MRMS+PPIN [22]. The mutual information is used to compute the relevance, redundancy, and significance of the genes. The value of a in (6) is set to 0.5.

In this study, the gene expression data from the colorectal cancer study of Hinoue et al. [7] is used. The gene expression profiling of 26 colorectal tumors and matched histologically normal adjacent colonic tissue samples were retrieve from the NCBI Gene Expression Omnibus (www.ncbi.nlm.nih.gov/geo/) with the accession number of GSE25070. The number of genes and samples in this data set are 24526 and 52, respectively. The data set is pre-processed by standardizing each sample to zero mean and unit variance.

The performance of different methods is compared with respect to the degree of overlapping with three gene lists, namely, LIST-1, LIST-2, and LIST-3. The LIST-1 contains 742 cancer related genes, which are collected from the Cancer Gene Census of the Sanger Centre, Atlas of Genetics and Cytogenetic in Oncology [9], and Human Protein Reference Database [12]. On the other hand, both LIST-2 and LIST-3 consist of colorectal cancer related genes. While the LIST-2 is retrieved from the study of Sabatas-Bellver et al. [23], the LIST-3 is prepared from the work of Nagaraj and Reverter [19]. While LIST-2 contains 438 colorectal cancer genes, LIST-3 consists of 134 colorectal cancer genes.

4.1   Performance of Different Gene Selection Algorithms

This section presents the comparative performance analysis of different gene selection algorithms with respect to the degree of overlapping with the three gene lists. The algorithms compared are t-test, MR, mRMR [5], MRMS [18], and the proposed MRMFS. Results are reported for first twenty genes selected by different algorithms.

Table 1. Twenty top-ranked genes and overlapping with known disease genes

t-Test		MR		mRMR		MRMS		Proposed	
Gene	Y/N	Gene	Y/N	Gene	Y/N	Gene	Y/N	Gene	Y/N
GUCA2B	y	GUCA2B	y	GUCA2B	y	GUCA2B	y	GUCA2B	y
ADH1B	y	BEST2	n	PI16	n	BCHE	y	GUCA2A	y
SCARA5	y	TMIGD1	n	CDH3	y	CLDN8	y	BEST2	n
ESM1	n	CLDN8	y	SPIB	y	PI16	n	CLCA4	y
TSPAN7	n	PI16	n	BEST2	n	BEST2	n	SCNN1B	y
CA7	y	SCNN1B	y	HMGCLL1	n	TMIGD1	n	NR3C2	y
LGI1	n	CLCA4	y	CILP	n	CILP	n	CA4	y
CEMIP	n	ADH1B	y	NR3C2	y	CLCA4	y	CA1	y
GLTP	n	CA1	y	ADH1B	y	ADH1B	y	ELANE	n
CLDN1	y	CA4	y	BOP1	n	SCNN1B	y	AQP8	y
TMIGD1	n	SCARA5	y	ECI2	n	ECI2	n	GCG	y
ACKR2	n	GNG7	n	CXCL8	n	CA1	y	PLCD1	n
NR3C2	y	NR3C2	y	CLCA4	y	CXCL8	n	CFD	n
PLAC9	y	ECI2	n	TEP1	n	TMEM37	n	C7	y
PCOLCE2	n	CXCL8	n	LRP8	n	GNG7	n	BGN	y
MMP7	y	CILP	n	GCG	y	CA4	y	CDK4	y
CLEC3B	y	TMEM37	n	WISP2	n	AFF3	y	PRPH	n
BEST4	n	CLEC3B	y	TMIGD1	n	NR3C2	y	TGFBI	y
AQP8	y	ELANE	n	CFD	n	SCARA5	y	KLF4	n
RUNDC3B	n	HEPACAM2	n	C16ORF62	n	WISP2	n	MMP3	y

Table 1 presents the lists of genes selected by different gene selection algorithms, along with their degree of overlapping with any one of the three cancer gene lists. From the results reported in Table 1, it can be seen that the proposed method provides better results than that of other methods with respect to degree of overlapping with known gene lists. Out of 20 selected genes, 14 genes selected by the proposed algorithm overlap with known disease genes, while t-test, MR, mRMR, and MRMS algorithms can identify 10, 10, 7, and 11 disease genes.

4.2 Performance of Different Disease Gene Identification Methods

Finally, the performance of the proposed algorithm is compared with two algorithms, namely, MR+PPIN [22] and mRMR+PPIN [16], which combine gene expression and PPIN data for selection of disease genes. The results are reported in Table 2 considering 41 genes as both MR+PPIN and mRMR+PPIN methods consider 41 genes for their analysis. Table 2 also presents the statistical significance test of the gene sets selected by the MR+PPIN, mRMR+PPIN, and proposed methods with respect to the genes of LIST-1, LIST-2, and LIST-3. Using the Fisher's exact test, statistical analysis of the overlapped genes is performed.

Table 2. Degree of overlapping and fisher's exact test

Methods/Algorithms	LIST-1		LIST-2		LIST-3		LIST 2-3
	Overlap	P-Value	Overlap	P-Value	Overlap	P-Value	
MR+PPIN	9	2.84E-05	7	2.10E-05	5	5.01E-06	10
mRMR+PPIN	8	1.91E-04	4	1.06E-02	3	2.02E-03	5
Proposed	5	2.33E-02	16	2.20E-16	8	1.29E-10	19

Table 3. KEGG enrichment analysis

KEGG ID	Term	Count	%	P-Value	Benjamin
05216	Thyroid cancer	5	0.42955	3.33E-04	3.37E-02
00910	Nitrogen metabolism	4	0.34364	2.33E-03	1.14E-01
05200	Pathways in cancer	11	0.94502	4.53E-03	1.44E-01
05219	Bladder cancer	4	0.34364	1.29E-02	2.85E-01
05222	Small cell lung cancer	5	0.42955	1.67E-02	2.94E-01
05210	Colorectal cancer	5	0.42955	1.67E-02	2.94E-01
04062	Chemokine signaling pathway	7	0.60137	2.20E-02	3.17E-01
05223	Non-small cell lung cancer	4	0.34364	2.53E-02	3.14E-01
04916	Melanogenesis	5	0.42955	2.87E-02	3.12E-01
04060	Cytokine-cytokine receptor interaction	8	0.68729	3.32E-02	3.21E-01
04115	p53 signaling pathway	4	0.34364	4.56E-02	3.81E-01

Out of total 41 genes selected by the proposed method, 16 and 8 genes are related to colorectal cancer with respect to the LIST-2 and LIST-3, respectively, while only 7 and 5 genes obtained using MR+PPIN are colon cancer related genes. On the other hand, only 4 and 3 genes selected using mRMR+PPIN are related to colon cancer with respect to two lists. Hence, the Fisher's exact test for the proposed method generates lower p-values for both LIST-2 and LIST-3, which are significantly better than the p-values obtained by other two methods. However, the degree of overlapping by the proposed algorithm with cancer related genes of LIST-1 is lower than that by existing methods. The last column of Table 2 depicts the degree of overlapping with respect to the two colorectal cancer gene lists. While the proposed method can identify 19 colorectal cancer related genes, only 10 and 5 disease genes are identified by the MR+PPIN and mRMR+PPIN methods.

4.3   KEGG Enrichment Analysis

The hundred genes selected by the proposed method are further analyzed using the functional annotation tool of David. The enriched p-value was corrected to control family-wide false discovery rate under certain threshold (for example, <0.05) with Benjamin multiple testing correction method. Table 3 represents the KEGG pathway enrichment analysis of the gene set obtained by the proposed algorithm. From the table, it is seen that most of the networks are associated with cancer. Various processes, those are associated with colon cancer like p53 signaling pathway and colorectal cancer, are also observed in the result. Moreover, the gene set is found to be highly associated with colorectal cancer disease according to the OMIM disease database as analyzed by the functional annotation tool of David.

5   Conclusion

The main contribution of the paper is to present a new gene selection algorithm to identify disease genes. The proposed algorithm integrates judiciously the information of gene expression profiles and protein-protein interaction networks. It selects a set of genes from microarray data as disease genes by maximizing the relevance and functional similarity of the selected genes. A new similarity measure is introduced to compute the functional similarity between two genes. It is based on the information of protein-protein interaction networks. The performance of the proposed algorithm, along with a comparison with other related methods, is demonstrated on colorectal cancer data set. Extensive experimental study on colorectal cancer establishes the fact that the genes identified by the proposed method have more colorectal cancer genes than the genes identified by the existing gene selection algorithms. All these results indicate that the proposed method is quite promising and may become a useful tool for identifying disease genes.

References

1. Altshuler, D., Daly, M.J., Lander, E.S.: Genetic mapping in human disease. Science **322**(5903), 881–888 (2008)
2. Barrenas, F., Chavali, S., Holme, P., Mobini, R., Benson, M.: Network properties of complex human disease genes identified through genome-wide association studies. PLoS ONE **4**(11), e8090 (2009)
3. Bogdanov, P., Singh, A.K.: Molecular function prediction using neighborhood features. IEEE/ACM Trans. Comput. Biol. Bioinform. **7**(2), 208–217 (2010)
4. Cai, Yu-Dong, Huang, T., Feng, K.-Y., Hu, L., Xie, L.: A unified 35-gene signature for both subtype classification and survival prediction in diffuse large B-cell lymphomas. PLoS ONE **5**(9), e12726 (2010)
5. Ding, C., Peng, H.: Minimum redundancy feature selection from microarray gene expression data. J. Bioinf. Comput. Biol. **3**(2), 185–205 (2005)
6. Goh, K.-I., Cusick, M.E., Valle, D., Childs, B., Vidal, M., Barabsi, A.-L.: The human disease network. Proc. National Acad. Sci. USA **104**(21), 8685–8690 (2007)
7. Hinoue, T., Weisenberger, D.J., Lange, C.P.E., Shen, H., Byun, H.M., Van Den Berg, D., Malik, S., Pan, F., Noushmehr, H., van Dijk, C.M., Tollenaar, R.A.E.M., Laird, P.W.: Genome-scale analysis of aberrant dna methylation in colorectal cancer. Genome Res. **22**(2), 271–282 (2012)

8. Huang, T., Chen, L., Cai, Y.-D., Chou, K.-C.: Classification and analysis of regulatory pathways using graph property, biochemical and physicochemical property, and functional property. PLoS ONE **6**(9), e25297 (2011)

9. Huret, J.L., Dessen, P., Bernheim, A.: Atlas of genetics and cytogenetics in oncology and haematology. Nucleic Acids Res. **31**(1), 272–274 (2003)

10. Jia, P., Zheng, S., Long, J., Zheng, W., Zhao, Z.: dmGWAS: dense module searching for genome-wide association studies in protein-protein interaction networks. Bioinformatics **27**(1), 95–102 (2011)

11. Karaoz, U., Murali, T.M., Letovsky, S., Zheng, Y., Ding, C., Cantor, C.R., Kasif, S.: Whole-genome annotation by using evidence integration in functional-linkage networks. Proc. National Acad. Sci. USA **101**(9), 2888–2893 (2004)

12. Keshava Prasad, T.S., Goel, R., Kandasamy, K., Keerthikumar, S., Kumar, S., Mathivanan, S., Telikicherla, D., Raju, R., Shafreen, B., Venugopal, A., Balakrishnan, L., Marimuthu, A., Banerjee, S., Somanathan, D.S., Sebastian, A., Rani, S., Ray, S., Harrys Kishore, C.J., Kanth, S., Ahmed, M., Kashyap, M.K., Mohmood, R., Ramachandra, Y.L., Krishna, V., Rahiman, B.A., Mohan, S., Ranganathan, P., Ramabadran, S., Chaerkady, R., Pandey, A.: Human protein reference database-2009 update. Nucleic Acids Res. **37**(suppl 1), D767–D772 (2009)

13. Kohler, S., Bauer, S., Horn, D., Robinson, P.N.: Walking the interactome for prioritization of candidate disease genes. Am. J. Hum. Gen. **82**(4), 949–958 (2008)

14. Kourmpetis, Y.A.I., van Dijk, A.D.J., Bink, M.C.A.M., van Ham, R.C.H.J., ter Braak, C.J.F.: Bayesian markov random field analysis for protein function prediction based on network data. PLoS ONE **5**(2), e9293 (2010)

15. Letovsky, S., Kasif, S.: Predicting protein function from protein/protein interaction data: a probabilistic approach. Bioinformatics **19**(suppl 1), i197–i204 (2003)

16. Li, B.-Q., Huang, T., Liu, L., Cai, Y.-D., Chou, K.-C.: Identification of colorectal cancer related genes with mrmr and shortest path in protein-protein interaction network. PLoS ONE **7**(4), e33393 (2012)

17. Li, Y., Li, J.: Disease gene identification by random walk on multigraphs merging heterogeneous genomic and phenotype data. BMC Genomics **13**(Suppl 7), S27 (2012)

18. Maji, P., Paul, S.: Rough set based maximum relevance-maximum significance criterion and gene selection from microarray data. Int. J. Approximate Reasoning **52**(3), 408–426 (2011)

19. Nagaraj, S., Reverter, A.: A boolean-based systems biology approach to predict novel genes associated with cancer: application to colorectal cancer. BMC Syst. Biol. **5**(1), 35 (2011)

20. Navlakha, S., Kingsford, C.: The power of protein interaction networks for associating genes with diseases. Bioinformatics **26**(8), 1057–1063 (2010)

21. Ng, K.-L., Ciou, J.-S., Huang, C.-H.: Prediction of protein functions based on function-function correlation relations. Comput. Biol. Med. **40**(3), 300–305 (2010)

22. Paul, S., Maji, P.: Gene expression and protein-protein interaction data for identification of colon cancer related genes using f-information measures. Natural Computing (2015). doi:10.1007/s11047-015-9485-6

23. Sabates-Bellver, J., Van der Flier, L.G., de Palo, M., Cattaneo, E., Maake, C., Rehrauer, H., Laczko, E., Kurowski, M.A., Bujnicki, J.M., Menigatti, M., Luz, J., Ranalli, T.V., Gomes, V., Pastorelli, A., Faggiani, R., Anti, M., Jiricny, J., Clevers, H., Marra, G.: Transcriptome profile of human colorectal adenomas. Mol. Cancer Res. **5**(12), 1263–1275 (2007)

24. Chao, W., Zhu, J., Zhang, X.: Integrating gene expression and protein-protein interaction network to prioritize cancer-associated genes. BMC Bioinform. **13**(1), 182 (2012)
25. Zhao, J., Yang, T.-H., Huang, H., Holme, P.: Ranking candidate disease genes from gene expression and protein interaction: a katz-centrality based approach. PLoS ONE **6**(9), e24306 (2011)

MaER: A New Ensemble Based Multiclass Classifier for Binding Activity Prediction of HLA Class II Proteins

Giovanni Mazzocco[1,5], Shib Sankar Bhowmick[2], Indrajit Saha[3,4](✉),
Ujjwal Maulik[2], Debotosh Bhattacharjee[2], and Dariusz Plewczynski[1]

[1] Center of New Technologies, University of Warsaw, Warsaw, Poland
{g.mazzocco,d.plewczynski}@cent.uw.edu.pl
[2] Department of Computer Science and Engineering,
Jadavpur University, Kolkata, India
shibsankar.ece@gmail.com, umaulik@cse.jdvu.ac.in, debotosh@ieee.org
[3] Institute of Informatics and Telematics, National Research Council, Pisa, Italy
indrajit.saha@iit.cnr.it
[4] Institute of Computer Science, University of Wroclaw, Wroclaw, Poland
[5] Institute of Computer Science, Polish Academy of Sciences, Warsaw, Poland

Abstract. Human Leukocyte Antigen class II (HLA II) proteins are crucial for the activation of adaptive immune response. In HLA class II molecules, high rate of polymorphisms has been observed. Hence, the accurate prediction of HLA II-peptide interactions is a challenging task that can both improve the understanding of immunological processes and facilitate decision-making in vaccine design. In this regard, during the last decade various computational tools have been developed, which were mainly focused on the binding activity prediction of different HLA II isotypes (such as DP, DQ and DR) separately. This fact motivated us to make a humble contribution towards the prediction of isotypes binding propensity as a multiclass classification task. In this regard, we have analysed a binding affinity dataset, which contains the interactions of 27 HLA II proteins with 636 variable length peptides, in order to prepare new multiclass datasets for strong and weak binding peptides. Thereafter, a new ensemble based multiclass classifier, called <u>M</u>eta<u>E</u>nsemble<u>R</u> (MaER) is proposed to predict the activity of weak/unknown binding peptides, by integrating the results of various heterogeneous classifiers. It pre-processes the training and testing datasets by making feature subsets, bootstrap samples and creates diverse datasets using principle component analysis, which are then used to train and test the MaER. The performance of MaER with respect to other existing state-of-the-art classifiers, has been estimated using validity measures, ROC curves and gain value analysis. Finally, a statistical test called Friedman test has been conducted to judge the statistical significance of the results produced by MaER.

G. Mazzocco, S. S. Bhowmick, and I. Saha—Joint first authors and contributed equally

M. Kryszkiewicz et al. (Eds.): PReMI 2015, LNCS 9124, pp. 462–471, 2015.
DOI: 10.1007/978-3-319-19941-2_44

Keywords: HLA class II proteins · Machine learning · MHC · Peptide binding · T cell epitopes

1 Introduction

T-cells [1] are specialized immune cells, playing a crucial role in activation of the adaptive immune system. Once the HLA II proteins have established a stable binding with the exogenous peptide antigens (T-cell epitopes), they are transported on the extracellular domain of the antigen-presenting cell (APCs). The T-cell receptors (TCRs) located on the surface of the T-cells, interacts with the HLA II-peptide constituting trimeric complexes responsible for the activation of the T-cell CD4+. HLA II proteins are encoded in three different genetic loci: HLA-DP, HLA-DQ and HLA-DR (also called isotypes) and are constituted of two separate protein chains: α and β [2]. They contain an open-ended binding cleft which allows the peptides to accommodate using multiple binding frames [3]. Thus the complexity in binding prediction problem is significantly increased.

In this regard, different computational techniques have been developed to predict HLA class II binding. Among them sequence-based and structure-based approaches are the most popular. Sequence-based methods include matrix models [4], binding motif recognition [5], artificial neural network [6], quantitative matrices [7], hidden markov models [8], support vector machines [9] and QSAR [10] based methods. Structure-based methods involve threading algorithms [11], peptide docking [12] and molecular dynamics [13]. Sophisticated methods, such as an iterative meta-search algorithm [14] and ant colony search [15] have been developed to resolve the dynamic variable length problem of HLA class II proteins prediction. Apart from this, some of the recent approaches has also significantly outperformed more traditional methods [16–18].

In this article, sequence-based information is used to predict the HLA II isotype binding propensity of peptides. For this purpose, new multiclass training and testing datasets of HLA II proteins are prepared from raw binding affinity interaction dataset of 27 HLA II proteins and 636 peptides. Thereafter, an ensemble based multiclass classifier, called MetaEnsembleR (MaER) is developed for the classification of peptides into different HLA II isotypes. The development of an ensemble based approach is basically motivated by the improving performance of similar other methods in bioinformatics [18–23]. The MaER classifier is build on the top of widely used four heterogeneous multiclass classifiers like Support Vector Machine (SVM) [24], Decision Tree (DT) [25], Naive Bayes (NB) [26] and K-Nearest Neighbor (K-NN) [27]. Before the classification task, MaER splits the training dataset into different number of rotational non-overlapping subsets. Subsequently, bootstrap sampling and principal component analysis are used for each subset. All the major principal components (in terms of coefficient and depending on eigenvalues) for all subsets are retained to create an informative set that preserve the diversity of the original training data. After that, such informative set is multiplied with original training and testing datasets before being classified. In MaER, finally a consensus of ensemble results is produced

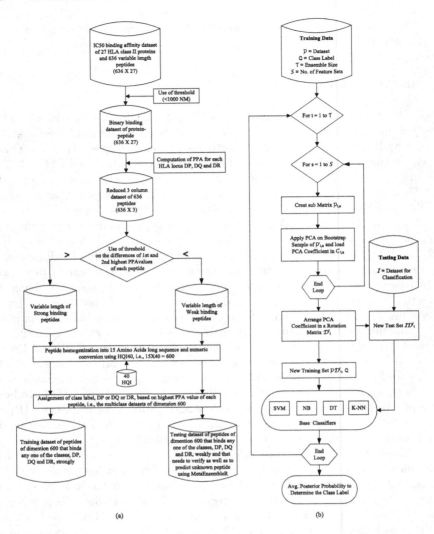

Fig. 1. Block diagrams of (a) Dataset Preparation and (b) MaER Algorithm

and for this purpose, selection of classifier is done randomly for different values of ensemble size. The performance of the MaER is reported in comparison with the individual classifies like SVM, DT, NB and K-NN in terms of average accuracy, precision or positive predictive value (PPV), recall, F-measure, Matthews correlation coefficient (MCC) and area under the ROC curve (AUC) values. Finally, Friedman test [28] has been conducted to judge the statistical significance of the results produced by MaER.

Table 1. Statistics of dataset used for MaER

Threshold (k)	Strong binding peptides				Weak binding peptides				Percentage of Strong binding peptides
	DP	DQ	DR	Total	DP	DQ	DR	Total	
0.02	101	195	270	566	61	9	0	70	88.99%
0.04	90	168	240	498	72	36	30	138	78.30%
0.06	90	168	240	498	72	36	30	138	78.30%
0.08	81	148	222	451	81	56	48	185	70.91%
0.10	79	129	202	410	83	75	68	226	64.47%
0.12	78	128	202	408	84	76	68	228	64.15%
0.15	**65**	**116**	**169**	**350**	**97**	**88**	**101**	**286**	**55.03%**
0.17	49	79	154	282	113	125	116	354	44.34%
0.20	43	72	138	253	119	132	132	383	39.78%

2   Materials and Methods

2.1   Preparation of Dataset

The dataset contains IC50 binding-affinity values originally measured between 27 HLA II proteins and 636 peptides derived from Phleumpratense [29]. The raw affinity dataset is transformed into a binary binding matrix defining binding and non-binding events solely. The IC50 value of 1000 nM is used as a threshold, since it represents the common reference to define HLA II-peptide binding events in literature [20,29]. Thereafter, the percentage of positive activity (PPA) is separately computed for each isotype ϕ (DR, DQ and DP) from the binary binding matrix. Initially, the highest number of HLA proteins capable of binding a single peptide x, is used as a reference, e.g., $\rho(DR)$, within each isotype. Consequently, three (one for each isotype) PPAs values are calculated for each given peptide. The PPA value for a given peptide, and for a given isotype is simply defined as the proportion of positive values that this peptide has for that particular isotype, with respect to the reference value for the same isotype, e.g., $\rho(DR)$.

$$PPA(x_i)_\phi = \frac{\sum_{j=1}^{|\phi|} a_{i,j}}{\rho(\phi)} \qquad \forall\, i \le |X|, \forall \phi \in \Phi \tag{1}$$

where

$$\rho(\phi) = \max_{1 < i < |X|} \sum_{j=1}^{|\phi|} a_{i,j} \qquad \forall\, i \le |X|,\ \ and\ \ \Phi = \{DR, DQ, DP\} \tag{2}$$

Each peptide x, does now have three PPAs: one for each isotype ϕ. The maximum $max_1(x)$ and second highest maximum $max_2(x)$ PPA values are identified for each peptide. The difference value $\Delta x = max_1(x) - max_2(x)$ is used to determine if the given peptide has either strong or weak binding properties, with respect to a given threshold k. If $\Delta x \geq k$ then the corresponding peptide is consider a strong binder for the class $\phi(x) \mapsto max_1(x)$ otherwise it is considered as a weak binder for the classes $(\phi_1(x), \phi_2(x)) \mapsto (max_1(x), max_2(x))$.

The peptide classification into multiple isotype is defined with respect to the threshold value k. As the lower k value increases the number of peptides, defined as strong HLA binders. The statistics about the effect of variation of k both in terms of isotype ratio and percentage of strong binders, are given in Table 1 and has taken into account for the choice of k. Ultimately, the value of k has been set to 0.15, in order to either maintain a comparable ratio among the isotypes, and grant a equal definition of strong and weak binding peptides.

Since the classification technique requires a common number of features for each peptide, a common length of 15 AAs is adopted. In the homogenized dataset, the edging AAs of peptides longer than 15 AA are shorted. The dissection is performed upon an accurate comparative analysis of the less conserved residues within the original peptides. In order to represent the entire pool of 636 peptides in a numerical form, a 40 high-quality AA indices (HQI40) [20,30,31] are used. Therefore, the length of the peptide sequence is 15×40=600. The block diagram representation of the experiment with data generation is given in Fig. 1(a).

2.2 The Proposed MetaEnsembleR

MetaEnsembleR (MaER) is an ensemble based classifier, where four heterogeneous classifiers like support vector machine, naive bayes, decision tree and K-nearest neighbor are used. It creates a diverse set of training points by preparing different non-overlapping sets of features. In order to discuss MaER, some notations are introduced here. Let $\mathcal{P}$ is a matrix of size $n \times M$, which consists of M input attributes or features values for each training instance and $\mathcal{Q}$ be an one dimensional column vector contains the output attribute of each training instance in $\hbar$. Therefore, $\hbar$ can be expressed as after concatenating $\mathcal{P}$ and $\mathcal{Q}$ horizontally, i.e., $\hbar = [\mathcal{P}\mathcal{Q}]$. Also let $\mathcal{F} = \{\mathcal{P}_1, \mathcal{P}_2, \ldots, \mathcal{P}_M\}$ and $\mathcal{T}$ are the set of features ($M \geq 4$) and ensemble size. Therefore, it can be assumed that a training set of n labelled instances $\hbar = \{p_j, q_j\}_{j=1}^n$ in which each instance (p_j, q_j) is described by M input attributes or features and an output attribute, i.e., $p \in \mathbb{R}^n$ and $q \in \mathbb{R}$ where q takes a value from the label space $\{L_1, L_2, \ldots, L_c\}$. In a classification task, the goal is to use the information only from $\hbar$ to construct a classifier which can perform well on the unseen data. Note that in MaER, the feature set, $\mathcal{F} = \{\mathcal{P}_1, \mathcal{P}_2, \ldots, \mathcal{P}_M\}$, splits into $\mathcal{S}$ number of feature subsets, where $\mathcal{S} \in [2, \lfloor \frac{M}{2} \rfloor]$. Also from the pool of classifiers, one classifier is randomly selected for the each value of $\mathcal{T}$. In order to construct the training and testing datasets for a classifier in ensemble, the following necessary steps are performed.

Step1: Randomly split $\mathcal{F}$ into $\mathcal{S}$ number of subsets, i.e.,$\mathcal{S}_{s,t}$ for simplicity, where t counts the ensemble size and s signifies the current attribute or feature subset. As $\mathcal{S} \in [2, \lfloor \frac{M}{2} \rfloor]$, therefore, the minimum number of subsets is 2 with at least 2 features in each subset is considered.

Step2: Repeat the following steps $\mathcal{S}$ times for each subset, i.e., $s = 1, 2, \ldots, \mathcal{S}$.
  (a) A new submatrix $\mathcal{P}_{s,t}$ is constructed which corresponds to the data in matrix $\mathcal{P}$.
  (b) From this new submatrix a bootstrap sample $\mathcal{P}'_{s,t}$ is drawn where the sample size is generally smaller than $\mathcal{P}_{s,t}$.
  (c) Thereafter, $\mathcal{P}'_{s,t}$ is used for PCA and the coefficients of all computed principal components are stored into a new matrix $\mathcal{C}_{t,s}$.
Step3: In order to have a matrix of same size of feature, arrange each $\mathcal{C}_{t,s}$ into a block diagonal sparse matrix $\mathcal{D}_t$. Once the coefficients in $\mathcal{C}_{t,s}$ are placed in to the block diagonal sparse matrix $\mathcal{D}_t$, the rows of $\mathcal{D}_t$ are rearranged so that the order of them corresponds to the original attributes in $\mathcal{F}$. During this rearrangement, columns with all zero values are removed from the sparse matrix.
Step4: The rearranged rotation matrix $\mathcal{D}_t^r$ is then used as $[\mathcal{PD}_t^r; \mathcal{Q}]$ and $[\mathcal{ID}_t^r]$ for training and test sets of classifier, where $\mathcal{I}$ is a given test sample.
Step5: In the testing phase, let $MaER_{v,t}(\mathcal{ID}_t^r)$ be the posterior probability produced by the classifier $MaER_t$ on the hypothesis that $\mathcal{I}$ belongs to class L_v. Then the confidence for a class is calculated by the average posterior probability of combination base classifiers:

$$\mathcal{Q}_v(\mathcal{I}) = \frac{1}{\mathcal{T}} \sum_{t=1}^{\mathcal{T}} MaER_{v,t}(\mathcal{ID}_t^r), \quad where \quad v = 1, 2, \ldots, c \qquad (3)$$

Thereafter, $\mathcal{I}$ is assigned to the class with the largest confidence. Note that all the five steps will repeat for $t = 1, 2, \ldots, \mathcal{T}$.

This is to be noted that due to the process of random feature subdivision, in each iteration, the selected classifier in MaER will have new sets of training and testing data, which will help to diversify the ensemble of classifiers in order to get better classification results. The MaER is applied to predict the multiclass binding activity of HLA class II protein, i.e., DP, DQ and DR at a time. For this purpose, based on threshold value $k=0.15$, multiclass dataset of strong binding peptides is used to train the MaER to predict the classes of weak binding peptides. Note that here the train and test datasets are normalized, where each input data is normalized to the range [0,1]. The flowchart of MaER is shown in Fig. 1(b).

3   Results and Discussions

The MaER contains four basic classifiers like SVM, ND, DT and K-NN. Therefore, setting the parameters of these classifiers is equivalent to set the parameters of MaER. Thus the parameters of SVM such as γ for kernel function and the soft margin $\mathcal{C}$ (cost parameter), are set to be 0.5 and 2.0, respectively. Note that, for SVM and DT, RBF (Radial Basis Function) kernel and C4.5 classifier are used. The value of K for the K-NN classifier, ensemble size $\mathcal{T}$ and number of feature subsets $\mathcal{S}$ are set to 7, 10 and 50, experimentally. The MaER classifier has been executed 20 times and the performance of MaER classifier is evaluated

Table 2. Performance comparison of MaER based HLA II-peptide binding predictor with other classifiers in terms of average Accuracy, Precision, Recall, F-measure, MCC and AUC values

Algorithm	Accuracy (%)	Precision or PPV	Recall or Sensitivity	F-measure	MCC	AUC
MaER	86.84	81.81	78.59	79.70	0.70	0.92
SVM	84.99	77.75	77.53	77.52	0.66	0.89
DT	75.22	65.88	59.31	60.24	0.43	0.78
NB	67.19	49.81	50.29	49.94	0.25	0.58
K-NN	84.47	76.28	75.98	75.80	0.64	0.89

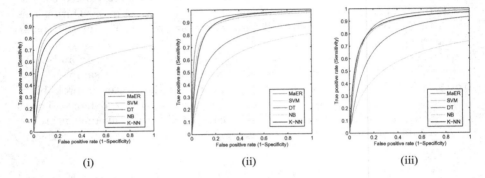

(i)   (ii)   (iii)

Fig. 2. Best ROC plots of HLA class II protein-peptide binding prediction for (i) DP, (ii) DQ and (iii) DR

in term of average accuracy, precision or PPV, recall or sensitivity, F-measure, MCC and area under the ROC curve (AUC) values. To compute these metrics the result of three class problem has been decomposed into three indivisible two class results by setting class-1 as an active and other classes are inactive, similarly, for class-2 and class-3. Thereafter, the average is computed and that also used to computer the average of 20 runs. Moreover, the effectiveness of MaER results has also been justified in terms of gain value. The gain is calculated in percentage as follows.

$$Gain = \frac{(MaER\ Predicted\ Accuracy - Referred\ Classifier\ Accuracy)}{(Referred\ Classifier\ Accuracy)} \times 100$$

(4)

Table 2 reports the performance of all classifiers including MaER on HLA II-peptide binding prediction. From the results, it can be clearly stated that the MaER performing better than the other classifiers. Moreover, the result is also suggesting that it can be used as a potentially computation tool for discovering multiclass HLA class II binding epitopes that has a great importance in vaccinology. The best ROC curves of all the classifiers are shown in Fig. 2. The curves produced by MaER for DP, DQ and DR, are showing the average AUC values like 0.92, 0.93 and 0.91. Moreover, The MaER classifier achieves the average gain values of 2.18 %, 15.45 %, 29.25 % and 2.81 % over SVM, DT, NB and K-NN classifiers, respectively. Finally, Friedman test based has been conducted average accuracy values of all classifiers to judge the statistical significance of the predicted results. The test produced a Chi-square value of 86.81 and p-value of 0.126×10^{-5} at $\alpha = 0.05$ significance level. The results provide a strong evident in order to rejecting the null hypothesis, that means there is a significant difference in the results produced by various classifies, while MaER produced the best results among them.

4   Conclusions

In this article, the binding prediction of HLA II isotypes is considered as a multiclass problem. For this purpose, new multiclass training and testing datasets of HLA II proteins have been prepared after analysing raw binding affinity interaction dataset of 27 HLA II proteins and 636 peptides. Thereafter, an ensemble based multiclass classifier, named as MetaEnsembleR (MaER) has been developed for the same problem. MaER is an ensemble based classifier which avoids the weakness of a single classifier while improving the prediction performance by integrating the outputs of multiple heterogeneous classifiers. It generally preprocesses the original training and testing datasets by making feature subsets, bootstrap samples and creates diverse datasets using principle component analysis. The efficacy of the developed MaER has been demonstrated in comparison with support vector machine, decision tree, naive bayes and K-nearest neighbor on newly generated test data in terms of average accuracy, precision, recall, F-measure, MCC, area under the ROC curve (AUC) and gain values. It is observed that MaER achieves the maximum gain of 29.25 % over Naive Bayes classifier. Finally, the statistical significance of the results produced by MaER has been justified by the Friedman test.

The application of the MaER method could be particularly beneficial wherever the informations about HLA isotype propensity and coverage are crucial. One example is the design of peptide-based vaccines, where the identification of the epitopes able to interact with different HLA isotypes, is a crucial factor for vaccine efficacy and population coverage. Another case is the study of autoimmune diseases, where the detection of self epitopes showing extensive cross reactivity with several HLA isotypes, is indeed a central issue. Apart from this, MaER can be used to find the potential markers from gene expression data [32,33].

Acknowledgments. This work was supported by grants from the Polish National Science Centre (2014/15/B/ST6/05082 and UMO-2013/09/B/NZ2/00121), COST BM1405 EU action and European Union Seventh Framework Program (FP7/2007-2013) under the grant agreement no.: 246016.

References

1. Flower, D.R. (ed.): Bioinformatics for Vaccinology. Wiley-Blackwel, Oxford (2008)

2. Janeway, C.A., Travers, P., Walport, M., Capra, J.D.: Immunobiology: The Immune System in Health and Disease. Garland Publications, New York (1999)
3. Rudensky, A., Janeway, C.A.: Studies on naturally processed peptides associated with MHC class II molecules. Chem. Immunol. **57**, 134–351 (1993)
4. Sturniolo, T., Bono, E., Ding, J., Raddrizzani, L., Tuereci, O., Sahin, U., Braxenthaler, M., Gallazzi, F., Protti, M.P., Sinigaglia, F., Hammer, J.: Generation of tissue-specific and promiscuous HLA ligand databases using DNA microarrays and virtual HLA class II matrices. Nat. Biotech. **17**(6), 555–561 (1999)
5. Sette, A., Buus, S., Appella, E., Smith, J.A., Chesnut, R., Miles, C., Colon, S.M., Grey, H.M.: Prediction of major histocompatibility complex binding regions of protein antigens by sequence pattern analysis. Proc. National Acad. Sci. **86**, 3296–3300 (1989)
6. Brusic, V., Rudy, G., Honeyman, M., Hammer, J., Harrison, L.: Prediction of MHC class II-binding peptides using an evolutionary algorithm and artificial neural network. Bioinformatics **14**(2), 121–130 (1998)
7. Hammer, J., Bono, E., Gallazzi, F., Belunis, C., Nagy, Z., Sinigaglia, F.: Precise prediction of major histocompatibility complex class II-peptide interaction based on peptide side chain scanning. J. Exp. Med. **180**, 2353–2358 (1994)
8. Noguchi, H., Kato, R., Hanai, T., Matsubara, Y., Honda, H., Brusic, V., Kobayashi, T., Biosci, J.: Hidden markov model-based prediction of antigenic peptides that interact with MHC class II molecules. J. Biosci. Bioeng. **94**(3), 264–270 (2002)
9. Wan, J., Liu, W., Xu, Q., Ren, Y., Flower, D.R., Li, T.: SVRMHC prediction server for MHC-binding peptides. BMC Bioinform. **7**, 463 (2006)
10. Dimitrov, I., Garnev, P., Flower, D.R., Doytchinova, I.: Peptide binding to the HLA-DRB1 supertype: a proteochemometric analysis. J. Med. Chem. **45**(1), 236–243 (2010)
11. Adrian, P.E., Rajaseger, G., Mathura, V.S., Sakharkar, M., Kangueane, P.: Types of inter-atomic interactions at the MHC-peptide interface: Identifying commonality from accumulated data. BMC Struct. Biol. **2**, 2 (2002)
12. Atanasova, M., Dimitrov, I., Flower, D.R., Doytchinova, I.: MHC Class II binding prediction by molecular docking. Mol. Inf. **30**, 368–375 (2011)
13. Oytchinova, I.D., Petkov, P., Dimitrov, I., Atanasova, M., Flower, D.R.: HLA-DP2 binding prediction by molecular dynamics simulations. Protein Sci. **20**, 1918–1928 (2011)
14. Mallios, R.R.: Predicting class II MHC peptide multi-level binding with an iterative stepwise discriminant analysis meta-algorithm. Bioinformatics **17**(10), 942–948 (2001)
15. Karpenko, O., Shi, J., Dai, Y.: Prediction of MHC class II binders using the ant colony search strategy. Artif. Intell. Med. **35**, 147–156 (2005)
16. Salomon, J., Flower, D.R.: Predicting class II MHC-peptide binding: a kernel based approach using similarity scores. BMC Bioinform. **7**, 501 (2006)
17. Zhang, W., Liu, J., Niu, Y.: Quantitative prediction of MHC-II binding affinity using particle swarm optimization. Artif. Intel. Med. **50**(2), 127–132 (2010)
18. Bhowmick, S.S., Saha, I., Mazzocco, G., Maulik, U., Rato, L., Bhattacharjee, D., Plewczynski, D.: Application of RotaSVM for HLA class II protein-peptide interaction prediction. In: Proceedings of the Fifth International Conference on Bioinformatics Models, Methods and Algorithms (BIOINFORMATICS 2014), pp. 178–185 (2014)
19. Bhowmick, S.S., Saha, I., Rato, L., Bhattacharjee, D.: RotaSVM: a new ensemble classifier. Adv. Intel. Syst. Comput. **227**, 47–57 (2013)

20. Saha, I., Mazzocco, G., Plewczynski, D.: Consensus classification of human leukocyte antigen class II proteins. Immunogenetics **65**, 97–105 (2013)
21. Pio, G., Malerba, D., D'Elia, D., Ceci, M.: Integrating microRNA target predictions for the discovery of gene regulatory networks: a semi-supervised ensemble learning approach. BMC Bioinform. **15**(Suppl 1), S4 (2014)
22. Marbach, D., Costello, J.C., Kuffner, R., et al.: Wisdom of crowds for robust gene network inference. Nat. Methods **9**, 796–804 (2012)
23. Saha, I., Zubek, J., Klingstrom, T., Forsberg, S., Wikander, J., Kierczak, M., Maulik, U., Plewczynski, D.: Ensemble learning prediction of protein-protein interactions using proteins functional annotations. Mol. BioSyst. **10**, 820–830 (2014)
24. Vapnik, V.N.: The Nature of Statistical Learning Theory. Springer, New York (1995)
25. Quinlan, J.R.: C4.5: Programs for Machine Learning. Morgan Kaufma, California (1993)
26. George, H., Langley, J.P.: Estimating continuous distributions in bayesian classifiers. In: Proceedings of the Eleventh Conference on Uncertainty in Artificial Intelligence, pp. 338–345 (1995)
27. Cover, T., Hart, P.: Nearest neighbor pattern classification. IEEE Trans. Inf. Theor. **13**(1), 21–27 (1967)
28. Friedman, M.: A comparison of alternative tests of significance for the problem of m rankings. Ann. Math. Stat. **11**, 86–92 (1940)
29. Greenbaum, J., Sidney, J., Chung, J., Brander, C., Peters, B., Sette, A.: Functional classification of class II human leukocyte antigen (HLA) molecules reveals seven different supertypes and a surprising degree of repertoire sharing across supertypes. Immunogenetics **63**(6), 325–335 (2011)
30. Saha, I., Maulik, U., Bandyopadhyay, S., Plewczynski, D.: Fuzzy clustering of physicochemical and biochemical properties of amino acids. Amino Acid **43**(2), 583–594 (2011)
31. Plewczynski, D., Basu, S., Saha, I.: AMS 4.0: consensus prediction of post-translational modifications in protein sequences. Amino Acid **43**(2), 573–582 (2012)
32. Saha, I., Maulik, U., Bandyopadhyay, S., Plewczynski, D.: Improvement of new automatic differential fuzzy clustering using SVM classifier for microarray analysis. Expert Syst. Appl. **38**(12), 15122–15133 (2011)
33. Saha, I., Plewczynski, D., Maulik, U., Bandyopadhyay, S.: Improved differential evolution for microarray analysis. Int. J. Data Min. Bioinform. **6**(1), 86–103 (2012)

Selection of a Consensus Area Size for Multithreaded Wavefront-Based Alignment Procedure for Compressed Sequences of Protein Secondary Structures

Dariusz Mrozek[✉], Bożena Małysiak-Mrozek, Bartek Socha, and Stanisław Kozielski

Institute of Informatics, Silesian University of Technology, Akademicka 16,
44-100 Gliwice, Poland
Dariusz.Mrozek@polsl.pl

Abstract. Multithreaded wavefront-based alignment procedure is used in the PSS-SQL language that allows for flexible scanning of databases of protein secondary structures and finding similarities among protein molecules. Efficiency of the process depends on several factors, including the way how the similarity matrix, calculated during the process, is divided into areas, the number of CPU cores possessed by the computer hosting the database with PSS-SQL extension, and structural patterns submitted by users in PSS-SQL queries. In this paper, we show how we achieved consensus values of area sizes for the multithreaded wavefront-based alignment procedure by a series of experimental trials. Availability: PSS-SQL extension for Microsoft SQL Server database management system can be downloaded from PSS-SQL project home page available at: http://www.zti.aei.polsl.pl/w3/dmrozek/science/pss-sql.htm.

Keywords: Proteins · Secondary structure · Structural patterns · Structural bioinformatics · Similarity searching · Structural alignment · Databases · SQL · Relational databases · Query language

1 Introduction

Proteins, molecules of life, have usually complex construction that is described on four representation levels, from primary to quaternary structure [5]. Protein secondary structure as a representation level describes protein construction in terms of regularly occurring shapes, including α-helices, β-strands, loop, turns, or coils (Fig. 1a), that protein amino acid chain can adopt in some of its regions. This representation level is very important for studying potential functions of protein molecules in living cells, determination of domain boundaries, topology recognition, and family assignment. These processes are usually realized through scanning databases of protein structures and finding similarities between the specified structural pattern and candidate database structures. We have developed a dedicated query language called PSS-SQL, which allows to perform such similarity searches, when protein secondary structures are collected in relational database [3].

© Springer International Publishing Switzerland 2015
M. Kryszkiewicz et al. (Eds.): PReMI 2015, LNCS 9124, pp. 472–481, 2015.
DOI: 10.1007/978-3-319-19941-2_45

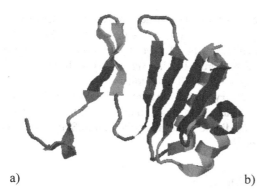

UEEEEEEEECUUHHHHHHHHHHCEEEE
CCEEEEEEUHHHHHHHHHHHHHHUCCC
EEEEEEEUCCCCCEEEEEEECCUUEEE
EECCEEEEEEUUUHHHCCUUUU

U1 E8 C1 U2 H10 C1 E4 C2 E6
U1 H14 U1 C3 E7 U1 C5 E7 C2
U2 E5 C2 E6 U3 H3 C2 U4

a)   b)

Fig. 1. Crystal structure of the CAS1-CAS2 complex from *Escherichia Coli* [12], PDB ID: 4P6I, chain A: (a) tertiary (3D) structure with secondary structures (spiral α-helices, wavy β-strands, and loops) exposed in RasMol [14], (b) two representations of secondary structures in relational databases used by PSS-SQL: (*top*) full, showing types of secondary structures for corresponding residues, and (*bottom*) compressed, describing types and lengths of secondary structures identified; types of secondary structures: H – α-helix, E – β-strand, C – loop, turn or coil, U – undetermined structure.

1.1 PSS-SQL for Scanning Protein Secondary Structures in RDBMS

PSS-SQL (Protein Secondary Structure - Structured Query Language) [10, 11] is a query language that allows execution of similarity searches against sequences of protein secondary structures stored in relational database management systems (RDBMS). PSS-SQL was initially developed in Institute of Informatics at the Silesian University of Technology in Gliwice, Poland in 2009 and later enriched by several improvements. Technically, PSS-SQL is an extension to the Transact-SQL language implemented in Microsoft SQL Server RDBMS. The great power of PSS-SQL lies in the fact that it is a declarative language. While writing PSS-SQL queries, programmers and bioinformaticians, equipped with a library of functions and procedures extending standard searching capabilities of the database engine, specify the structural pattern of a protein for which they want to find similar proteins, provide what they want to display, where the data are stored, i.e. in which tables of the database, and how to filter them. For example, a simple SELECT statement, which is used to retrieve and display proteins that are similar to the given structural pattern, may have the following form:

```
SELECT p.protAC AS AC, p.name, s.matchingSeq, p.secondary
FROM ProteinTbl AS p CROSS APPLY dbo.sequenceMatch(p.id, 'secondary',
   'e(1;10),c(0;5),h(5;6),c(0;5),e(1;10),c(5)') AS s
WHERE p.name LIKE '%Staphylococcus aureus%' AND p.length > 150
ORDER BY AC
```

This sample query returns Accession Numbers (*AC*) and names of proteins from *Staphylococcus aureus* having the length greater than 150 residues and structural region containing β-strand of the length from 1–10 elements, optional loop up to 5 elements, an α-helix of the length 5 to 6 elements, optional loop up to 5 elements, a β-strand of the length 1 to 10 elements and a 5 element loop. The structural region is defined by structural pattern `e(1;10),c(0;5),h(5;6),c(0;5),e(1;10),c(5)` passed to the *sequenceMatch* table function invoked in the FROM clause (*dbo* is a database schema name). Secondary structures are stored in the *secondary* column of the *ProteinTbl* in the form presented in Fig. 1b (*top*). Two additional columns, namely *matchingSeq* and already mentioned *secondary*, which are listed in the SELECT clause, show the exact pattern sequence that was matched to a candidate protein from the database and full secondary structure of the candidate protein. The matchingSeq column is returned by the *sequenceMatch* function.

1.2 Multithreaded Wavefront-Based Alignment

Due to the approximate nature of the similarity searching appropriate dynamic programming alignment procedure underlies all similarity searches and matches realized in PSS-SQL queries. More efficient scanning is achieved when the alignment procedure is preceded by multiple scanning of the segment index (MSSI) created for the table column storing secondary structures [10]. The index uses the representation of protein secondary structures shown in Fig. 1b (*bottom*). The alignment procedure makes use of results of the multiple scanning of the segment index. Therefore, this variant of the alignment procedure is marked as +MSSI.

In the +MSSI variant of the alignment the candidate database sequence of secondary structure S^C is represented as follows:

$$S^C = s_1^C l_1, s_2^C l_2, s_3^C l_3, \ldots, s_n^C l_n, \tag{1}$$

where single element $s_j^C \in \{H, E, C, U\}$ corresponds to single secondary structure (segment) identified in the protein, l_j is the length of the jth segment measured in residues, and n is the number of segments. Example of such a sequence is shown in Fig. 1b (*bottom*).

Query protein structure S^Q is represented as follows:

$$S^Q = s_1^Q(l_1;u_1), s_2^Q(l_2;u_2), s_3^Q(l_3;u_3), \ldots, s_m^Q(l_m;u_m), \tag{2}$$

where single element $s_i^Q \in \{H, E, C, U\}$ corresponds to single secondary structure (segment) of the structural pattern provided by a user, l_i, u_i are lower and upper limits for the length of the ith segment, both measured in residues, and m is the number of segments in the query pattern. Example of such defined pattern is shown in the sample PSS-SQL query presented in Sect. 1.1.

In the +MSSI variant of the alignment procedure, we calculate the similarity matrix D according to the following formulas.

$$D_{i,0} = 0 \text{ for } i \in [0;m] \text{ and } D_{0,j} = 0 \text{ for } j \in [0;n] \tag{3}$$

$$D_{i,j} = max \begin{cases} 0 \\ D_{i-1,j-1} + d_{i,j} \\ E_{i,j} \\ F_{i,j} \end{cases} , \text{ for } i \in [1;m], j \in [1;n] \tag{4}$$

where $d_{i,j}$ is the matching degree between elements of both sequences:

$$d_{i,j} = \begin{cases} \omega_+ \text{ if } s_i^Q = s_j^C \wedge l_j \geq l_i \wedge l_j \leq u_i \\ \omega_- \text{ otherwise} \end{cases}, \tag{5}$$

where ω_+ is a matching award, and ω_- is a mismatch penalty. Additional matrices E and F are used for fast calculation of horizontal and vertical gaps:

$$E_{i,j} = max \begin{cases} E_{i-1,j} - \delta \\ D_{i-1,j} - \sigma \end{cases} \text{ and } F_{i,j} = max \begin{cases} F_{i,j-1} - \delta \\ D_{i,j-1} - \sigma \end{cases} \tag{6}$$

for:

$$E_{i,0} = 0 \, dla \, i \in [0;m], F_{i,0} = 0 \, dla \, i \in [0;m]$$
$$E_{0,j} = 0 \, dla \, j \in [0;n], F_{0,j} = 0 \, dla \, j \in [0;n]$$

where: σ is the penalty for opening a gap in the alignment, and δ is the penalty for extending the gap.

Since the alignment procedure is computationally costly, we have developed the multithreaded (+MT) implementation of the procedure. This implementation improves efficiency of the computational procedure, while the computational complexity of the alignment algorithm remains unchanged.

While performing the multithreaded, pairwise alignments the search engine of the PSS-SQL language calculates the similarity matrix D. Particular cells of the matrix depend on each other (each cell $D_{i,j}$ can be calculated only, if cells $D_{i-1,j-1}$, $D_{i,j-1}$, $D_{i-1,j}$ have already been calculated), so calculations must occur in a particular order. To avoid costly synchronizations of threads, the similarity matrix is divided into areas, which are assigned to working threads. Calculations are performed diagonally for areas, according to the wavefront approach [1, 6], as it is presented in Fig. 2.

While details of the alignment algorithm and general conclusions on performance of PSS-SQL queries were reported in [7, 10], the question remains how the size of the area influences the efficiency of PSS-SQL queries. Moreover, there are additional three factors that influence efficiency and should be taken into considerations: (1) the number of CPU cores possessed by the computer hosting the DBMS with the PSS-SQL extension, (2) structural patterns submitted by users in PSS-SQL queries, and (3) characteristics of the data stored in the database. Since each user may work on its own repository and we are unable to predict the characteristics of data collected in it, we decided to investigate the first and the second of the mentioned factors. In this paper, we show how we achieved consensus values of area sizes for both variants of the multithreaded wavefront-based alignment procedure by a series of experimental trials.

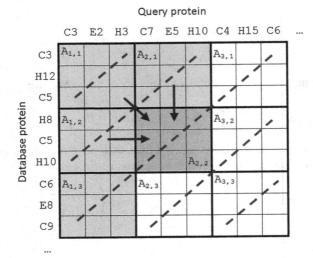

Fig. 2. Wavefront-based calculation of similarity matrix divided into areas ($A_{u,v}$). Compressed sequences of secondary structures (described by type and length of the structure) are located along vertical and horizontal edges of the matrix. We assumed that query protein structural pattern is defined precisely. Arrows show dependences between calculated areas. Areas already calculated are marked in grey, the area currently being calculated is marked in orange ($A_{2,2}$). Steps of the wavefront approach are marked by dashed lines.

2   Consensus on the Area Size

Area sizes for the +MT+MSSI variant of the alignment procedure were chosen experimentally taking into account various patterns and varying number of CPU cores. Tests were performed on the Microsoft SQL Server 2012 EE working on nodes of the virtualized cluster controlled by the HyperV hypervisor hosted on Microsoft Windows 2008 R2 Datacenter Edition 64-bit. The host server had the following parameters: 2x Intel Xeon CPU E5620 2.40 GHz, RAM 92 GB, 3x HDD 1 TB 7200 RPM. Cluster nodes were configured to use from 1 up to 4 CPU cores and 4 GB RAM per node, and worked under the Microsoft Windows 2008 R2 Enterprise Edition 64-bit operating system. Tests were performed on the database storing 6 360 protein structures and for PSS-SQL queries containing patterns representing various classes:

- *class 1:* short patterns and patterns frequently occurring in the database, e.g., *c(10;20)*, *h(2;5)*, *c(2;40)*;
- *class 2:* patterns with precisely defined regions, e.g., region *e(15)* in the sample pattern *e(4;20)*, *c(3;10)*, *e(4;20)*, *c(3;10)*, *e(15)*, *c(3;10)*, *e(1;10)*, *c(3;10)*, *e(5;12)*;
- *class 3:* patterns with unique regions, e.g., region *h(243)* in the sample pattern *h(10;20)*, *c(1;10)*, *h(243)*, *c(1;10)*, *h(5;10)*, *c(1;10)*, *h(10;15)*;
- *class 4:* patterns with undefined type of secondary structure (wildcard symbol ?) and with unlimited length of one of its regions (wildcard symbol *), e.g., region *?(1;30)* and *e(5;*)*, in the sample pattern *c(10;20)*, *h(2;5)*, *c(2;40)*, *?(1;30)*, *e(5;*)*.

Query patterns chosen for our tests had different characteristics and were representatives of possible patterns that can be entered by users. During experiments we have not observed any dependency between types of the secondary structures specified in patters and the execution time. The aim of the series of tests was to determine the possible best size of the area, which was assigned to every single thread. We have checked different area sizes for the following heights and widths: 1, 2, 3, ..., 9. We tested popular configurations of 1, 2, 3, and 4 threads working in parallel, which corresponded to 1, 2, 3, and 4 CPU cores available for the database management system. This gave $(9^2) \cdot 4 = 324$ combinations that were examined for each query pattern.

In Table 1 we can see execution times for the sample PSS-SQL query containing sample pattern from *class 2*. The class represents complex patterns that consist of many segments. As we can notice in Table 1 smaller area sizes (especially the size 1×1) result in higher execution times. Increasing the area size above 2×2 reduces the execution time. However, changes of the area size (above 3×3) do not affect the execution time significantly. The same tendency was observed in all tested cases, i.e. while testing various numbers of threads for different query patterns.

Table 2 shows relative execution times for the same pattern class. The table is presented as a heat map, where particular cells are colored using the red color (worst result), through yellow, and green color (best results). Results are expressed as a percentage - 0 % denotes the longest execution time, 100 % denotes the shortest execution time. Values are calculated according to the following expression:

$$t_{i,j}^{\%} = \frac{max\,(t) - t_{i,j}}{max\,(t) - min(t)} \cdot 100\% \tag{7}$$

where: $t_{i,j}$ is measured execution time of the sample query taken from Table 1 for the corresponding area size, $max(t)$, $min(t)$ are maximal and minimal execution times of the query (in Table 1).

Table 1. Execution times (s) for PSS-SQL query with sample pattern from *class 2* parallelized on 4 threads for various sizes (W × H, H for query sequence, W for database sequence) of area

H\W	1	2	3	4	5	6	7	8	9
1	7.074	5.580	5.209	4.712	4.622	5.293	4.437	4.490	4.770
2	5.825	4.955	4.806	4.514	4.422	4.365	4.521	4.358	4.399
3	5.672	4.945	4.599	4.519	4.491	4.428	4.483	4.377	4.372
4	5.569	4.891	4.643	4.566	4.602	4.397	**4.295**	4.444	4.398
5	5.384	5.026	5.336	4.565	4.571	5.870	4.379	4.487	4.684
6	5.296	4.712	4.872	5.620	4.522	4.535	4.490	4.402	4.462
7	5.325	5.052	4.618	4.558	4.568	4.512	4.382	4.534	4.484
8	5.289	4.733	4.866	4.567	6.289	4.524	4.536	4.456	4.375
9	5.276	4.581	4.791	4.538	4.543	6.133	4.376	4.423	4.510

Table 2. Relative execution times $t_{i,j}^{\%}$ for PSS-SQL query with sample pattern from *class 2* parallelized on 4 threads for various sizes (W × H) of area (%)

H\W	1	2	3	4	5	6	7	8	9
1	0.00	53.76	67.11	84.99	88.23	64.09	94.89	92.98	82.91
2	44.94	76.25	81.61	92.12	95.43	97.48	91.87	97.73	96.26
3	50.45	76.61	89.06	91.94	92.95	95.21	93.23	97.05	97.23
4	54.16	78.55	87.48	90.25	88.95	96.33	100.00	94.64	96.29
5	60.81	73.70	62.54	90.28	90.07	43.32	96.98	93.09	86.00
6	63.98	84.99	79.24	52.32	91.83	91.36	92.98	96.15	93.99
7	62.94	72.76	88.38	90.54	90.18	92.19	96.87	91.40	93.20
8	64.23	84.24	79.45	90.21	28.25	91.76	91.33	94.21	97.12
9	64.70	89.71	82.15	91.26	91.08	33.86	97.09	95.39	92.26

The heat map (Table 2) reveals preferred and recommended area sizes (green) and those that should be avoided (red and orange). We have to remember that various query patterns and the number of possessed CPU cores may move the best point in any direction. Therefore, we have made these types of statistics, as presented in Tables 1 and 2, for all tested patterns in all tested classes for all tested n-core CPU configurations and area sizes. Then, in order to determine a universal area size for common patterns we have calculated the value of weighted arithmetic mean taking into account the participation of popular n-core processors in the market (n = 1, 2, 3, 4) and possible assignment of logical CPU cores in virtualized environments. We have arbitrary chosen the following values of weights: 15 % for 1-core CPUs (1 core = 1 thread execution), 40 % for 2-core CPUs, 5 % for 3-core CPUs, and 40 % for 4-core CPUs. Results are presented in Table 3. Best execution times (relative to the worst case) were obtained for the area width greater than 6. Height of the area has no significant impact, since compressed query sequences of secondary structures are usually short, while database sequences are relatively longer. In Fig. 3 we can see the histogram of the number of secondary structures identified in proteins stored in tested database. It shows that most of the proteins have less than 100 secondary structures (segments). Therefore, widths of whole similarity matrices after compression of sequences of secondary structures should be lower than 100 elements in most cases. Heights of whole similarity matrices depend on the query pattern. For example, the sample query pattern from pattern *class 2* contains 9 segments, and consequently, the height of the similarity matrix is 9. Therefore, on the basis of our experiments, we have chosen 3 × 7 (H × W) for the area size, i.e., 3 for the query pattern and 7 for database sequence of secondary structures.

Table 3. Weighted arithmetic mean for relative execution times of all PSS-SQL queries parallelized on various number of threads (1, 2, 3, 4) depending on the sizes (W × H) of area (%)

H\W	1	2	3	4	5	6	7	8	9
1	36,49	45,79	63,88	70,29	73,26	75,49	77,51	79,44	79,46
2	44,13	52,33	67,74	72,89	75,43	77,18	79,25	80,29	80,31
3	56,68	62,91	73,89	76,78	79,04	80,28	**82,61**	82,45	82,58
4	61,38	65,6	74,35	77,04	77,54	77,87	82,57	80,83	79,67
5	64,27	67,12	72,63	75,1	75,81	77,35	78,28	79,01	79,45
6	65,67	68,07	72,64	75,42	75,95	77,27	77,65	78,54	78,93
7	66,72	69,47	73,91	73,48	75,49	77,66	79,75	80,44	80,96
8	66,97	69,76	76,11	75,51	74,9	75,98	79,46	79,57	80,47
9	68	70,54	76,36	75,28	74,54	75,95	79,84	80,07	80,39

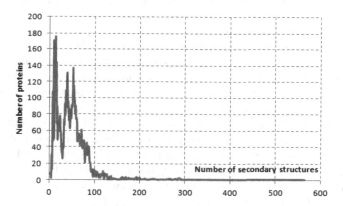

Fig. 3. Histogram of the number of secondary structures (segments) identified in proteins deposited in tested database.

3 Related Works

Advantages of a declarative processing of biological data with the use of the SQL query language were noticed in the last decade, which resulted in the development of various SQL extensions. One of the first extensions was ODM BLAST [15], which was developed for Oracle RDBMS. ODM BLAST allows to align and match amino acid sequences of proteins (primary structures) and nucleotide sequences of DNA and RNA acids. Similarly, BioSQL [2], which makes use of particular modules of BioJava [13], provides a generic relational model for persistent storage of bio-molecular sequences, features, sequence and feature annotation, a reference taxonomy, and ontologies (or controlled vocabularies). Various bio-oriented projects may utilize the information by means of object-relational mapping (ORM). Recently reported P3D-SQL [9] extends Oracle

PL/SQL capabilities by providing a set of functions and procedures that allow to perform structural alignments and superposition of 3D (tertiary) protein structures. Declarative processing of protein secondary structures is available by means of PSS-SQL [10, 11] and the query languages developed by Hammel and Patel [4] and Tata et al. [16]. These languages allow searching on the secondary structure of protein sequences. All mentioned projects confirm that for bio-database developers, highly skilled users, also those working in the domain of structural bioinformatics, the SQL language became an important communication interface.

4   Discussion and Concluding Remarks

By providing the PSS-SQL libraries to Microsoft SQL Server (version 2012 or higher) we have extended standard capabilities of the Transact-SQL language towards processing biological data. PSS-SQL extensions allow to store, index, process, compare, align and match protein structures based on their secondary structures in the relational database management system. These operations can be especially beneficial for database developers, data analysts, data scientists and programmers working in the domain of structural bioinformatics.

Efficiency of these operations, especially PSS-SQL queries, depend on several factors, including the internal configuration of the multithreaded wavefront-based alignment procedure, which is one of the two main phases of the query execution process. In this paper, we tried to show how we achieved the consensus area size for the alignment procedure by a series of experimental trials. We took into account two important factors that may influence the execution time, i.e., various query patterns (their lengths, uniqueness, vagueness) and the number of available CPU cores. These investigations allowed to develop efficient, domain-specific query language. Results of our experiments made PSS-SQL competitive to similar solutions implemented by Hammel and Patel [4] and Tata et al. [16], although, both mentioned solutions do not utilize alignment at all. Detailed performance evaluation and comparisons between mentioned query languages for protein secondary structures reported in Mrozek et al. [10] show that even with the computationally costly alignment the PSS-SQL is able to find protein similarities efficiently becoming a successful example of a DBMS-side processing. In such a way, PSS-SQL joins a narrow group of bio-oriented SQL extensions, such as BioSQL, ODM BLAST, and P3D-SQL, complementing the group with the important capability of finding similarities among proteins on the basis of their secondary structures. Thereby, PSS-SQL becomes a declarative, domain-specific query language for protein similarity searching and protein function identification on commonly available workstations without specialized equipment.

There is still a possibility to improve the efficiency of the PSS-SQL, e.g., by a parallelization of the alignment procedure on GPU devices. We have successfully developed such an alignment procedure for secondary structures in the GPU-CASSERT method [7, 8]. In the GPU-CASSERT the alignment of secondary structures alone (phase 1) takes just a fraction of seconds. However, the data must be prepared appropriately before the alignment begins, and data transfers between the host workstation and global memory of the GPU device also take some time.

Availability. PSS-SQL is free for scientific and testing purposes. It is available from PSS-SQL project home page at: http://zti.polsl.pl/w3/dmrozek/science/pss-sql.htm.

References

1. Anvik, J., MacDonald, S., Szafron, D., Schaeffer, J., Bromling, S., Tan, K.: Generating parallel programs from the wavefront design pattern. In: Proceedings of the 7th International Workshop on High-Level Parallel Programming Models and Supportive Environments (HIPS 2002), Fort Lauderdale, Florida, April 2002, pp. 1–8 (2002)
2. BioSQL. http://biosql.org/. Accessed 13 January 2015
3. Date, C.: An Introduction to Database Systems, 8th edn. Addison-Wesley, New York (2003)
4. Hammel, L., Patel, J.M.: Searching on the secondary structure of protein sequences. In: 28th International Conference on Very Large Data Bases VLDB 2002, Hong Kong, China, pp. 634–645 (2002)
5. Lesk, A.M.: Introduction to Protein Science: Architecture, Function, and Genomics, 2nd edn. Oxford University Press, New York (2010)
6. Liu, W., Schmidt, B.: Parallel design pattern for computational biology and scientific computting applications. In: Proceedings of the 2003 IEEE International Conference on Cluster Computing, Hong Kong, China, pp. 456–459 (2003)
7. Mrozek, D.: High-Performance Computational Solutions in Protein Bioinformatics. Springer Briefs in Computer Science. Springer International Publishing, Heidelberg (2014). http://dx.doi.org/10.1007/978-3-319-06971-5
8. Mrozek, D., Brożek, M., Małysiak-Mrozek, B.: Parallel implementation of 3D protein structure similarity searches using a GPU and the CUDA. J. Mol. Model. **20**(2), 2067 (2014)
9. Mrozek, D., Małysiak-Mrozek, B., Adamek, R.: P3D-SQL: extending oracle PL/SQL capabilities towards 3d protein structure similarity searching. In: Ortuño, F., Rojas, I. (eds.) IWBBIO 2015, Part I. LNCS, vol. 9043, pp. 548–556. Springer, Heidelberg (2015)
10. Mrozek, D., Socha, B., Kozielski, S., Małysiak-Mrozek, B.: An efficient and flexible scanning of databases of protein secondary structures with the segment index and multithreaded alignment. J. Intell. Inf. Syst. (in press). http://dx.doi.org/10.1007/s10844-014-0353-0
11. Mrozek, D., Wieczorek, D., Małysiak-Mrozek, B., Kozielski, S.: PSS-SQL: protein secondary structure - structured query language. In: 32th Annual International Conference of the IEEE Engineering in Medicine and Biology Society, EMBS 2010, Buenos Aires, Argentina, pp. 1073–1076 (2010)
12. Nunez, J.K., Kranzusch, P.J., Noeske, J., Wright, A.V., Davies, C.W., Doudna, J.A.: Cas1-Cas2 complex formation mediates spacer acquisition during CRISPR-Cas adaptive immunity. Nat. Struct. Mol. Biol. **21**, 528–534 (2014)
13. Prlić, A., et al.: BioJava: an open-source framework for bioinformatics in 2012. Bioinformatics **28**, 2693–2695 (2012)
14. Sayle, R.: RasMol, Molecular Graphics Visualization Tool. Biomolecular Structures Group, Glaxo Welcome Research & Development, Stevenage, Hartfordshire, 5 February 2013, 1998. http://www.umass.edu/microbio/rasmol/
15. Stephens, S.M., Chen, J.Y., Davidson, M.G., Thomas, S., Trute, B.M.: Oracle database 10 g: a platform for BLAST search and regular expression pattern matching in life sciences. Nucl. Acids Res. **33**(Suppl 1), D675–D679 (2005). doi:10.1093/nar/gki114
16. Tata, S., Friedman, J.S., Swaroop, A.: Declarative querying for biological sequences. In: 22nd International Conference on Data Engineering, pp. 87–98. IEEE Computer Society, Atlanta (2006)

Supervised Cluster Analysis of miRNA Expression Data Using Rough Hypercuboid Partition Matrix

Sushmita Paul[(✉)] and Julio Vera

Laboratory of Systems Tumor Immunology, Department of Dermatology,
University of Erlangen-Nürnberg, Hartmannstr. 14, 91052 Erlangen, Germany
{sushmita.paul,julio.vera-gonzalez}@uk-erlangen.de

Abstract. The microRNAs are small, endogenous non-coding RNAs found in plants and animals, which suppresses the expression of genes post-transcriptionally. It is suggested by various genome-wide studies that a substantial fraction of miRNA genes is likely to form clusters. The coherent expression of the miRNA clusters can then be used to classify samples according to the clinical outcome. In this background, a new rough hypercuboid based supervised similarity measure is proposed that is integrated with the supervised attribute clustering to find groups of miRNAs whose coherent expression can classify samples. The proposed method directly incorporates the information of sample categories into the miRNA clustering process, generating a supervised clustering algorithm for miRNAs. The effectiveness of the rough hypercuboid based algorithm, along with a comparison with other related algorithms, is demonstrated on three miRNA microarray expression data sets using the $B.632+$ bootstrap error rate of support vector machine. The association of the miRNA clusters to various biological pathways are also shown by doing pathway enrichment analysis.

Keywords: MicroRNA · Co-expressed miRNAs · Clustering · Rough sets

1 Introduction

Micro RNAs/miRNAs are a class of short approximately 22-nucleotide non-coding RNAs found in many plants and animals. They inhibit the expression of mRNA expression post-transcriptionally. It has been shown by [1] that the miRNAs on a genome tend to present in a cluster. Large scale surveys [2] have established the fact that miRNAs have tendency to present in clusters. Existence of co-expressed miRNAs is also demonstrated using expression profiling analysis in [3]. These findings suggest that members of a miRNA cluster, which are at a close proximity on a chromosome, are highly likely to be processed as co-transcribed units. In [4,15], different approaches are introduced to discover

© Springer International Publishing Switzerland 2015
M. Kryszkiewicz et al. (Eds.): PReMI 2015, LNCS 9124, pp. 482–494, 2015.
DOI: 10.1007/978-3-319-19941-2_46

miRNA cluster patterns. Expression data of miRNAs can be used to detect clusters of miRNAs as it is suggested that co-expressed miRNAs are co-transcribed, so they should have similar expression pattern.

Several unsupervised clustering techniques like hierarchical clustering algorithms [8] and self organizing maps [2] are used to cluster a miRNA expression data. However, the groups of miRNAs discovered by these unsupervised clustering algorithms are not potential enough to do tissue classification [5], as the miRNAs are grouped based on their similarity without incorporating the class label information. In this regard, several supervised clustering algorithms are proposed to cluster gene expression data [5,10,11]. In [5], genes are clustered by incorporating the knowledge of tissue. On the other hand, hierarchical clustering is employed on the gene expression data and the average of resultant clustering solutions are further used to do sample classification. Only in the later part, information of the class label is incorporated [10]. In [11], a fuzzy-rough supervised gene clustering algorithm is described. The algorithm uses fuzzy equivalence classes to compute relevance of the clusters, that makes the algorithm sensitive to the fuzzy parameter. However, none of the works has addressed the problem of supervised clustering of miRNAs.

However, one of the main problems in expression data analysis is uncertainty. Some of the sources of this uncertainty include imprecision in computations and vagueness in class definition. In this background, the rough set [16] provides a mathematical framework to capture uncertainties associated with human cognition process. In [11,13,14], rough sets have been successfully used to analyze a microarray expression data.

In this regard, this paper presents a new rough hypercuboid based supervised clustering algorithm. It is developed by integrating the concepts of rough hypercuboid equivalence partition matrix [12,14] and supervised attribute clustering algorithm [11]. It finds coregulated clusters of miRNAs whose collective expression is strongly associated with the sample categories. Using the concept of rough hypercuboid equivalence partition matrix, the degree of dependency is calculated for miRNAs, which is used to compute both relevance and significance of the miRNAs. Hence, the only information required in the proposed method is in the form of equivalence classes for each miRNA, which can be automatically derived from the data set. A new measure is developed for calculating similarity between two miRNAs. Based upon the similarity values, the miRNAs are grouped into cluster. The new supervised clustering algorithm divides the miRNA expression data in distinct clusters. In each cluster, the first selected miRNA has high relevance value with respect to the class label and it is the representative of the cluster. The representative is modified in such a way that the averaged expression value has high relevance value with the class label. Finally, the proposed method generates a set of clusters, whose coherent average expression levels allow perfect discrimination of tissue types. The concept of B.632+ error rate [7] is used to minimize the variability and biasedness of the derived results. The support vector machine is used to compute the B.632+ error rate as well as several other types of error rates as it maximizes the margin between data samples in different classes. The effectiveness of the proposed approach, along with a comparison with other related approaches, is demonstrated on several miRNA expression data sets.

2   Rough Hypercuboid Based Supervised Attribute Clustering

In this paper, a new algorithm is developed based on rough hypercuboid equivalence partition matrix. Every clustering algorithm need a distance or similarity measure to group objects. Accordingly, a new rough hypercuboid based similarity measure is proposed. The concept of rough hypercuboid was presented in [20], while that of rough hypercuboid equivalence partition matrix was proposed in [12,14]. It has also been successfully applied for feature/gene/miRNA selection in [12,14]. The relevance of a cluster is calculated using rough hypercuboid equivalence partition matrix based dependency measure. The proposed rough hypercuboid based supervised similarity measure is integrated into the supervised attribute clustering algorithm developed by Maji [11]. Prior to describe about the new supervised attribute clustering algorithm, next the concept of rough hypercuboid equivalence partition matrix is described.

2.1   Rough Hypercuboid Equivalence Partition Matrix

Let $\mathbb{U} = \{s_1, \cdots, s_i, \cdots, s_n\}$ be the set of n objects or samples and $\mathbb{C} = \{\mathscr{M}_1, \cdots, \cdots, \mathscr{M}_m\}$ denotes the set of m attributes or miRNAs of a given microarray data set. Let $\mathbb{D}$ be the set of class labels or sample categories of n samples.

If $\mathbb{U}/\mathbb{D} = \{\beta_1, \cdots, \beta_i, \cdots, \beta_c\}$ denotes c equivalence classes or information granules of $\mathbb{U}$ generated by the equivalence relation induced from the decision attribute set $\mathbb{D}$, then c equivalence classes of $\mathbb{U}$ can also be generated by the equivalence relation induced from each condition attribute or miRNA $\mathscr{M}_k \in \mathbb{C}$. If $\mathbb{U}/\mathscr{M}_k = \{\mu_1, \cdots, \mu_i, \cdots, \mu_c\}$ denotes c equivalence classes or information granules of $\mathbb{U}$ induced by the condition attribute or miRNA $\mathscr{M}_k$ and n is the number of objects in $\mathbb{U}$, then c-partitions of $\mathbb{U}$ are the sets of (cn) values $\{h_{ij}(\mathscr{M}_k)\}$ that can be conveniently arrayed as a $(c \times n)$ matrix $\mathbb{H}(\mathscr{M}_k) = [h_{ij}(\mathscr{M}_k)]$. The matrix $\mathbb{H}(\mathscr{M}_k)$ is denoted by

$$\mathbb{H}(\mathscr{M}_k) = \begin{pmatrix} h_{11}(\mathscr{M}_k) & h_{12}(\mathscr{M}_k) & \cdots & h_{1n}(\mathscr{M}_k) \\ h_{21}(\mathscr{M}_k) & h_{22}(\mathscr{M}_k) & \cdots & h_{2n}(\mathscr{M}_k) \\ \cdots & \cdots & \cdots\cdots \\ h_{c1}(\mathscr{M}_k) & h_{c2}(\mathscr{M}_k) & \cdots & h_{cn}(\mathscr{M}_k) \end{pmatrix} \tag{1}$$

$$\text{where}\quad h_{ij}(\mathscr{M}_k) = \begin{cases} 1 \text{ if } L_i \leq x_j(\mathscr{M}_k) \leq U_i \\ 0 \text{ otherwise.} \end{cases} \tag{2}$$

The tuple $[L_i, U_i]$ represents the interval of ith class β_i according to the decision attribute set $\mathbb{D}$. The interval $[L_i, U_i]$ is the value range of condition attribute or miRNA $\mathscr{M}_k$ with respect to class β_i. It is spanned by the objects with same class label β_i. That is, the value of each object s_j with class label β_i falls within interval $[L_i, U_i]$. This can be viewed as a supervised granulation process, which utilizes class information.

On employing a condition attribute or miRNA $\mathscr{M}_k$ a $c \times n$ matrix $\mathbb{H}(\mathscr{M}_k)$ termed as hypercuboid equivalence partition matrix is generated. The $c \times n$ matrix $\mathbb{H}(\mathscr{M}_k)$ is termed as hypercuboid equivalence partition matrix of the condition attribute or miRNA $\mathscr{M}_k$. Each row of the matrix $\mathbb{H}(\mathscr{M}_k)$ is a hypercuboid equivalence partition or class. Here $\mathrm{h}_{ij}(\mathscr{M}_k) \in \{0, 1\}$ represents the membership of object s_j in the class β_i satisfying following two conditions:

$$1 \leq \sum_{j=1}^{n} \mathrm{h}_{ij}(\mathscr{M}_k) \leq n, \forall i; \quad 1 \leq \sum_{i=1}^{c} \mathrm{h}_{ij}(\mathscr{M}_k) \leq c, \forall j. \tag{3}$$

The above axioms should hold for every equivalence partition, which correspond to the requirement that an equivalence class is non-empty. However, in real data analysis, uncertainty arises due to overlapping class boundaries. Hence, such a granulation process does not necessarily result in a compatible granulation in the sense that every two class hypercuboids or intervals may intersect with each other. The intersection of two hypercuboids also forms a hypercuboid, which is referred to as implicit hypercuboid. The implicit hypercuboids encompass the misclassified samples or objects those belong to more than one classes. The degree of dependency of the decision attribute set or class label on the condition attribute set depends on the cardinality of the implicit hypercuboids. The degree of dependency increases with the decrease in cardinality.

Using the concept of hypercuboid equivalence partition matrix, the misclassified objects of boundary region present in the implicit hypercuboids can be identified based on the confusion vector defined next

$$\mathbb{V}(\mathscr{M}_k) = [\mathrm{v}_1(\mathscr{M}_k), \cdots, \cdots, \mathrm{v}_n(\mathscr{M}_k)]; \text{ where } \mathrm{v}_j(\mathscr{M}_k) = \min\{1, \sum_{i=1}^{c} \mathrm{h}_{ij}(\mathscr{M}_k) - 1\}. \tag{4}$$

In rough sets if an object s_j belongs to the lower approximation of any class β_i, then it does not belong to the lower or upper approximations of any other classes and $\mathrm{v}_j(\mathscr{M}_k) = 0$. On the other hand, if the object s_j belongs to the boundary region of more than one classes, then it should be encompassed by the implicit hypercuboid and $\mathrm{v}_j(\mathscr{M}_k) = 1$. Hence, the hypercuboid equivalence partition matrix and corresponding confusion vector of the condition attribute $\mathscr{M}_k$ can be used to define the lower and upper approximations of the ith class β_i of the decision attribute set $\mathbb{D}$. Let $\beta_i \subseteq \mathbb{U}$. β_i can be approximated using only the information contained within $\mathscr{M}_k$ by constructing the M-lower and M-upper approximations of β_i:

$$\underline{M}(\beta_i) = \{s_j| \ \mathrm{h}_{ij}(\mathscr{M}_k) = 1 \text{ and } \mathrm{v}_j(\mathscr{M}_k) = 0\}; \quad \overline{M}(\beta_i) = \{s_j| \ \mathrm{h}_{ij}(\mathscr{M}_k) = 1\}; \tag{5}$$

where equivalence relation M is induced from attribute $\mathscr{M}_k$. The boundary region of β_i is then defined as

$$BN_M(\beta_i) = \{s_j| \ \mathrm{h}_{ij}(\mathscr{M}_k) = 1 \text{ and } \mathrm{v}_j(\mathscr{M}_k) = 1\}. \tag{6}$$

Dependency. The dependency between condition attribute $\mathscr{M}_k$ and decision attribute $\mathbb{D}$ can be defined as follows:

$$\gamma_{\mathscr{M}_k}(\mathbb{D}) = \frac{1}{n}\sum_{i=1}^{c}\sum_{j=1}^{n} h_{ij}(\mathscr{M}_k) \cap [1 - v_j(\mathscr{M}_k)]; \text{ that is, } \gamma_{\mathscr{M}_k}(\mathbb{D}) = 1 - \frac{1}{n}\sum_{j=1}^{n} v_j(\mathscr{M}_k), \quad (7)$$

where $0 \leq \gamma_{\mathscr{M}_k}(\mathbb{D}) \leq 1$. If $\gamma_{\mathscr{M}_k}(\mathbb{D}) = 1$, $\mathbb{D}$ depends totally on $\mathscr{M}_k$, if $0 < \gamma_{\mathscr{M}_k}(\mathbb{D}) < 1$, $\mathbb{D}$ depends partially on $\mathscr{M}_k$, and if $\gamma_{\mathscr{M}_k}(\mathbb{D}) = 0$, then $\mathbb{D}$ does not depend on $\mathscr{M}_k$. The $\gamma_{\mathscr{M}_k}(\mathbb{D})$ is also termed as the relevance of attribute $\mathscr{M}_k$ with respect to class $\mathbb{D}$.

Significance. The resultant hypercuboid equivalence partition matrix $\mathbb{H}(\{\mathscr{M}_k, \mathscr{M}_l\})$ of size $c \times n$ can be computed from $\mathbb{H}(\mathscr{M}_k)$ and $\mathbb{H}(\mathscr{M}_l)$ as follows:

$$\mathbb{H}(\{\mathscr{M}_k, \mathscr{M}_l\}) = \mathbb{H}(\mathscr{M}_k) \cap \mathbb{H}(\mathscr{M}_l); \text{ where } h_{ij}(\{\mathscr{M}_k, \mathscr{M}_l\}) = h_{ij}(\mathscr{M}_k) \cap h_{ij}(\mathscr{M}_l). \quad (8)$$

The significance of the attribute $\mathscr{M}_k$ with respect to the condition attribute set $\{\mathscr{M}_k, \mathscr{M}_l\}$ is given by

$$\sigma_{\mathrm{M}}(\mathbb{D}, \mathscr{M}_k) = \frac{1}{n}\sum_{j=1}^{n} [v_j(\mathrm{M} - \{\mathscr{M}_k\}) - v_j(\mathrm{M})]; \quad (9)$$

where $0 \leq \sigma_{\{\mathscr{M}_k, \mathscr{M}_l\}}(\mathbb{D}, \mathscr{M}_k) \leq 1$. Hence, the higher the change in dependency, the more significant the attribute $\mathscr{M}_k$ is. If significance is 0, then the attribute is dispensable.

2.2 Rough Hypercuboid Based Supervised Similarity Measure

The simple concepts of rough hypercuboid based dependency and significance is used to calculate distance between two miRNAs and then the non-linear transformation of the distance is used to calculate similarity between two miRNAs. This subsection presents the proposed rough hypercuboid based supervised similarity measure.

Let $\mathbb{C} = \{\mathscr{M}_1, \cdots, \mathscr{M}_i, \cdots, \mathscr{M}_j, \cdots, \mathscr{M}_{\mathcal{D}}\}$ denotes the set of $\mathcal{D}$ condition attributes or miRNAs of a given data set. Define $R_{\mathscr{M}_i}(\mathbb{D})$ as the relevance of the condition attribute $\mathscr{M}_i$ with respect to the class label or decision attribute $\mathbb{D}$. The dependency function of rough hypercuboid can be used to calculate the relevance of condition attributes or miRNAs. Hence, the relevance $R_{\mathscr{M}_i}(\mathbb{D})$ of the condition attribute $\mathscr{M}_i$ with respect to the decision attribute $\mathbb{D}$ using rough hypercuboid can be calculated as follows:

$$R_{\mathscr{M}_i}(\mathbb{D}) = \gamma_{\mathscr{M}_i}(\mathbb{D}) \quad (10)$$

where $\gamma_{\mathscr{M}_i}(\mathbb{D})$ represents the degree of dependency between condition attribute or miRNA $\mathscr{M}_i$ and decision attribute or class label $\mathbb{D}$ that is given by (7).

At first, the distance between two miRNAs $\mathscr{M}_i$ and $\mathscr{M}_j$ is calculated using rough hypercuboid based approach. Then the non-linear transformation of the distance is done for getting the similarity between these two miRNAs. The non-linear transformation is done to detect nonlinear interdependencies between the underlying two miRNAs. The rough hypercuboid based significance (9) is used to compute similarity between two miRNAs and it is defined next.

Definition 1. *The rough hypercuboid based similarity measure between two attributes or miRNAs $\mathscr{M}_i$ and $\mathscr{M}_j$ is defined as follows:*

$$\psi(\mathscr{M}_i, \mathscr{M}_j) = \frac{1}{\sqrt{\kappa^2 + 1}}; \quad \text{where} \quad \kappa = \left\{ \frac{\sigma_{\mathscr{M}_i}(\mathbb{D}, \mathscr{M}_j) + \sigma_{\mathscr{M}_j}(\mathbb{D}, \mathscr{M}_i)}{2} \right\} \quad (11)$$

Hence, the supervised similarity measure $\psi(\mathscr{M}_i, \mathscr{A}_j)$ directly takes into account the information of sample categories or class labels $\mathbb{D}$ while computing the similarity between two attributes or miRNAs $\mathscr{M}_i$ and $\mathscr{M}_j$. If attributes $\mathscr{M}_i$ and $\mathscr{M}_j$ are completely correlated with respect to class labels $\mathbb{D}$, then $\kappa = 0$ and so $\psi(\mathscr{M}_i, \mathscr{M}_j)$ is 1. If $\mathscr{M}_i$ and $\mathscr{M}_j$ are totally uncorrelated, $\psi(\mathscr{M}_i, \mathscr{M}_j) = \frac{1}{\sqrt{2}}$. Hence, $\psi(\mathscr{M}_i, \mathscr{M}_j)$ can be used as a measure of supervised similarity between two miRNAs $\mathscr{M}_i$ and $\mathscr{M}_j$.

2.3 Supervised miRNA Clustering Algorithm

In this work the proposed rough hypercuboid based similarity measure is incorporated into the Fuzzy-Rough Supervised Attribute Clustering Algorithm [11]. In the proposed method a new rough hypercuboid based similarity measure is developed to calculate similarity between two miRNAs. Whereas, in [11] a fuzzy-rough supervised similarity measure is proposed. However, the fuzzy-rough supervised similarity measure is sensitive to the fuzzy parameter that is used to calculate the similarity between two objects.

Let $\mathbb{C}$ represents the set of miRNAs of the original data set, while $\mathbb{S}$ and $\bar{\mathbb{S}}$ are the set of actual and augmented attributes, respectively, selected by the miRNA clustering algorithm. Let $\mathbb{V}_i$ is the coarse cluster associated with the miRNA $\mathscr{M}_i$ and $\bar{\mathbb{V}}_i$, the finer cluster of $\mathscr{M}_i$, represents the set of miRNAs of $\mathbb{V}_i$ those are merged and averaged with the attribute $\mathscr{M}_i$ to generate the augmented cluster representative $\bar{\mathscr{M}}_i$. The main steps of the integrated miRNA clustering algorithm are reported next.

1. Initialize $\mathbb{C} \leftarrow \{\mathscr{M}_1, \cdots, \mathscr{M}_i, \cdots, \mathscr{M}_j, \cdots, \mathscr{M}_{\mathcal{D}}\}$, $\mathbb{S} \leftarrow \emptyset$, and $\bar{\mathbb{S}} \leftarrow \emptyset$.
2. Calculate the rough hypercuboid based relevance value $\mathrm{R}_{\mathscr{M}_i}(\mathbb{D})$ of each miRNA $\mathscr{M}_i \in \mathbb{C}$.
3. Repeat the following nine steps (steps 4 to 12) until $\mathbb{C} = \emptyset$ or the desired number of attributes are selected.
4. Select miRNA $\mathscr{M}_i$ from $\mathbb{C}$ as the representative of cluster $\mathbb{V}_i$ that has highest rough hypercuboid based relevance value. In effect, $\mathscr{M}_i \in \mathbb{S}$, $\mathscr{M}_i \in \mathbb{V}_i$, $\mathscr{M}_i \in \bar{\mathbb{V}}_i$, and $\mathbb{C} = \mathbb{C} \setminus \mathscr{M}_i$.

5. Generate coarse cluster $\mathbb{V}_i$ from the set of existing attributes/miRNAs of $\mathbb{C}$ satisfying the following condition:

$$\mathbb{V}_i = \{\mathscr{M}_j | \psi(\mathscr{M}_i, \mathscr{M}_j) \geq \delta; \mathscr{M}_j \neq \mathscr{M}_i \in \mathbb{C}\}. \tag{12}$$

6. Initialize $\bar{\mathscr{M}}_i \leftarrow \mathscr{M}_i$.
7. Repeat following four steps (steps 8–11) for each miRNA $\mathscr{M}_j \in \mathbb{V}_i$.
8. Compute two augmented cluster representatives by averaging $\mathscr{M}_j$ and its complement with the attributes of $\bar{\mathbb{V}}_i$ as follows:

$$\bar{\mathscr{M}}_{i+j}^+ = \frac{1}{|\bar{\mathbb{V}}_i| + 1} \left\{ \sum_{\mathscr{M}_k \in \bar{\mathbb{V}}_i} \mathscr{M}_k + \mathscr{M}_j \right\}; \bar{\mathscr{M}}_{i+j}^- = \frac{1}{|\bar{\mathbb{V}}_i| + 1} \left\{ \sum_{\mathscr{M}_k \in \bar{\mathbb{V}}_i} \mathscr{M}_k - \mathscr{M}_j \right\} \tag{13}$$

9. The augmented cluster representative $\bar{\mathscr{M}}_{i+j}$ after averaging $\mathscr{M}_j$ or its complement with $\bar{\mathbb{V}}_i$ is as follows:

$$\bar{\mathscr{M}}_{i+j} = \begin{cases} \bar{\mathscr{M}}_{i+j}^+ \text{ if } R_{\bar{\mathscr{M}}_{i+j}^+}(\mathbb{D}) \geq R_{\bar{\mathscr{M}}_{i+j}^-}(\mathbb{D}) \\ \bar{\mathscr{M}}_{i+j}^- \text{ otherwise.} \end{cases} \tag{14}$$

10. The augmented cluster representative $\bar{\mathscr{M}}_i$ of cluster $\mathbb{V}_i$ is $\bar{\mathscr{M}}_{i+j}$ if $R_{\bar{\mathscr{M}}_{i+j}}(\mathbb{D}) \geq R_{\bar{\mathscr{M}}_i}(\mathbb{D})$, otherwise $\bar{\mathscr{M}}_i$ remains unchanged.
11. Select attribute $\mathscr{M}_j$ or its complement as a member of the finer cluster $\bar{\mathbb{V}}_i$ of attribute $\mathscr{M}_i$ if $R_{\bar{\mathscr{M}}_{i+j}}(\mathbb{D}) \geq R_{\mathscr{M}_i}(\mathbb{D})$.
12. In effect, $\bar{\mathscr{M}}_i \in \bar{\mathbb{S}}$ and $\mathbb{C} = \mathbb{C} \setminus \bar{\mathbb{V}}_i$.
13. Stop.

3   Experimental Results

The performance of the proposed rough hypercuboid equivalence partition matrix based supervised miRNA clustering (RH-SAC) method is extensively studied and compared with that of some existing feature selection and clustering algorithms on three miRNA expression data sets GSE17846, GSE21036, and GSE28700. The algorithms compared are mutual information based Info-Gain [17] and minimum redundancy-maximum relevance (mRMR) algorithm [6], method proposed by Golub et al. [9], rough set based maximum relevance-maximum significance (RSMRMS) algorithm [13], μHEM [14], fuzzy-rough supervised attribute clustering algorithm (FR-SAC) [11]. The error rate of support vector machine (SVM) [18] is used to evaluate the performance of different algorithms. To compute the error rate of SVM, bootstrap approach ($B.632+$ error rate) [7] is performed on each miRNA expression data set. For each training set, a set of differential miRNA groups is first generated, and then SVM is trained with the selected coherent miRNAs. After the training, the information of miRNAs those were selected for the training set is used to generate test set and then the class label of the test sample is predicted using the classifier. The maximum number of features selected by the new integrated supervised miRNA clustering algorithm are 50.

3.1   Optimal Value of δ Parameter

The threshold δ in (12) plays an important role in the performance of the proposed supervised miRNA clustering algorithm. It controls the size of a cluster. Hence, it has direct influence in the performance of the proposed algorithm. Higher the value of δ sparse the cluster becomes. To find the optimal value of δ parameter the proposed algorithm is implemented on three data sets. The value for which the $B.632+$ error rate is minimum is considered to be the optimum δ value for the corresponding data set.The value of δ is varied from 0.90 to 1.00. Hence, the optimum value of δ for three miRNA data sets are calculated using the following relation:

$$\delta^{\star} = \arg\min_{\delta}\{B.632 + \text{error}\}. \tag{15}$$

The optimum values of δ^{*} obtained using (15) are 0.99, 1.00, 0.95 for GSE17846, GSE21036, and GSE28700 data sets, respectively. The number of miRNAs at which optimal δ^{*} value is obtained for miRNA data sets are 31, 49, and 43 for GSE17846, GSE21036, and GSE28700 data sets, respectively.

3.2   Different Types of Errors

This section describes about the different types of errors generated by the SVM classifier. The importance of $B.632+$ error over apparent error (AE), gamma error (γ), and bootstrap $(B1)$ error is also established. All the errors are calculated using the SVM for the proposed method. The results are presented for the optimum values of δ. Figure 1 represents different types of errors obtained for three different data sets. From the figure it is seen that the γ error rate is higher than any other type of errors for each data set, while $B1$ error is lower than the γ error rate but higher than the $B.632+$ error and AE. The average of $B1$ error and AE leads to $B.632+$ error rate lower than the $B1$ error but higher than AE. Table 1 represents minimum values of different types of errors and corresponding number of miRNAs at which the error is obtained for each miRNA data sets. From the table it is seen that the $B.632+$ estimator rectifies the upward bias of $B1$ error and downward bias of AE.

Table 1. Comparative analysis of different types of errors for proposed method

Microarray data sets	AE		B1 Error		γ Error		B.632+ Error	
	Error	miRNAs	Error	miRNAs	Error	miRNAs	Error	miRNAs
GSE17846	0.000	5	0.087	31	0.458	2	0.059	31
GSE21036	0.000	41	0.062	49	0.397	7	0.041	49
GSE28700	0.000	2	0.250	43	0.466	27	0.197	43

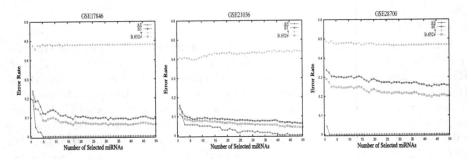

Fig. 1. Different error rates of the proposed algorithm on different data sets obtained using the SVM averaged over 50 random splits

3.3   Comparative Performance Analysis

In this section comparative performance analysis of the proposed supervised miRNA clustering algorithm has been shown. The proposed algorithm has been compared with some popular feature selection and supervised attribute clustering algorithms.

Table 2 represents the different types of error obtained by different methods at their optimal parameters. It also contains the number of miRNAs at which the corresponding lowest error rate is obtained by each method. From the table it is seen that the almost all the algorithms generate AE equal to zero. However, the RSMRMS generates non zero AE in 2 cases. From the table it is seen that the proposed supervised miRNA clustering algorithm generates $B.632+$ error rate lower than any other method except in one case. Only in one case the μ-HEM miRNA selection algorithm generates better result than the proposed method.

3.4   Pathway Enrichment Analysis of Obtained miRNAs

In this section biological importance of the obtained miRNAs using proposed supervised miRNA clustering algorithm is described. Those miRNAs which are selected by the proposed method in all the 50 bootstrap samples were used for further analysis. The association of those miRNAs with different biological pathways were determined. The DIANA-miRPath v2.0 [19] interface has been used to identify the miRNA-pathway relationship. The server performs an enrichment analysis of miRNA gene targets in KEGG pathways. The tool first identifies the target genes of the uploaded miRNAs.

The DIANA-miRPath v2.0 has been applied on the selected miRNAs of miRNA data sets. Those pathways are selected whose P-value is lower than 0.05. The miRNA-pathway relation is represented by a heatmap. Figure 2 represents the heatmap of the miRNA-pathways which are found to be statistically significant. The darker colors represent that the miRNA is associated with the pathway more significantly. In data set GSE17846 the miRNA profiling of total blood of multiple sclerosis and control samples is performed. From the figure it is

Table 2. Comparative performance analysis of different algorithms

Microarray data sets	Algorithms/ Methods	Apparent Error		B1 Error		γ Error		B.632+ Error	
		Error	miRNAs	Error	miRNAs	Error	miRNAs	Error	miRNAs
GSE17846	Golub	0.0000	6	0.1165	48	0.4795	48	0.0809	48
	InfoGain	0.0000	7	0.0930	37	0.4799	37	0.0630	37
	mRMR	0.0000	3	0.1010	48	0.4798	48	0.0690	48
	RSMRMS	0.0000	2	0.0930	39	0.4792	39	0.0640	39
	μ-HEM	0.0000	2	0.0870	49	0.4790	49	0.0590	49
	FR-SAC	0.0000	2	0.2340	47	0.4659	18	0.1803	47
	RH-SAC	0.0000	5	0.0870	31	0.4580	2	**0.0588**	31
GSE21036	Golub	0.0000	35	0.0694	48	0.4370	39	0.0466	48
	InfoGain	0.0000	39	0.0730	50	0.4452	44	0.0490	50
	mRMR	0.0000	19	0.0640	49	0.4400	50	0.0430	49
	RSMRMS	0.0500	5	0.0890	5	0.4173	5	0.0750	5
	μ-HEM	0.0000	42	0.0580	47	0.4440	47	**0.0390**	47
	FR-SAC	0.0000	41	0.0785	50	0.4020	1	0.0530	50
	RH-SAC	0.0000	41	0.0620	49	0.3970	7	0.0410	49
GSE28700	Golub	0.0000	27	0.3004	27	0.4736	3	0.2482	27
	InfoGain	0.0000	35	0.3090	8	0.4678	8	0.2710	21
	mRMR	0.0000	21	0.3330	49	0.4728	7	0.2850	49
	RSMRMS	0.0230	34	0.3310	19	0.4715	15	0.2850	19
	μ-HEM	0.0000	25	0.3060	4	0.5000	4	0.2570	4
	FR-SAC	0.0000	24	0.3362	50	0.4650	43	0.2888	50
	RH-SAC	0.0000	2	0.2500	43	0.4660	27	**0.1969**	43

seen the miRNAs selected by the proposed method are statistically related with 29 pathways. Multiple Sclerosis is a autoimmune disorder and from the Fig. 2 it is seen that around 7 pathways are significant and they are related to autoimmune disorder. They are Cell adhesion molecules (CAMs), TGF-beta signaling pathway, PI3K-Akt signaling pathway, Leukocyte transendothelial migration, MAPK signaling pathway, Fc gamma R- mediated phagocytosis, and Calcium signaling pathway. On the other hand around 48 pathways-miRNAs relationship are found to be statistically significant for GSE21036 data set. This data set is generated using metastatic prostate cancer samples and normal adjacent benign prostate. From Fig. 2 it is seen that the proposed method is able to select those miRNAs that are associated with prostate cancer. In addition to that it is also able to identify other significant pathways like Progestrone-mediated oocyte maturation, Inositol phosphate metabolism, mTOR signaling pathway, and so forth. Similarly, several significant miRNA-pathway relations are obtained using the DIANA-miRPath tool for the data set GSE28700. In this data set, expression profiles of microRNAs in gastric cancer are stored. From Fig. 2 it is clear several cancer related pathways are found to be significant using the proposed method. From the figure it is seen that total 22 pathways are found to be significant and few of them are Colorectal cancer, Pancreatic cancer, Non-small cell lung cancer, Chronic myeloid leukemia, Hepatitis B, Small cell lung cancer, HIF-1 signaling pathway, Focal adhesion, Prostate cencer, Pathways in cancer.

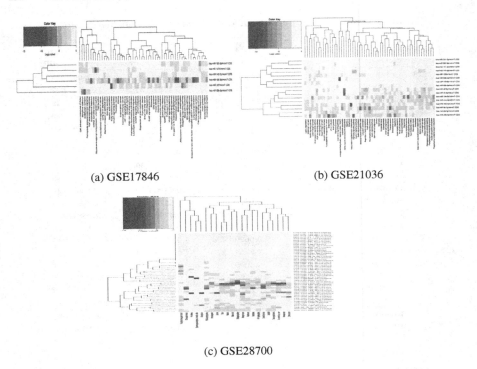

(a) GSE17846   (b) GSE21036

(c) GSE28700

Fig. 2. miRNAs versus pathways heat map for different miRNA data sets

4   Conclusion

The paper presents a new rough hypercuboid based supervised similarity measure that is incorporated into the supervised miRNA clustering algorithm. It uses the concept of rough hypercuboid for calculating similarity between two miRNAs and thus improves the performance of the method. The rough hypercuboid based similarity measure uses the information of class label for calculating similarity between two miRNAs and hence, makes it a supervised measure. The proposed method fetches cluster of miRNAs whose collective expressions are strongly associated with the class label. The effectiveness of the proposed rough hypercuboid based supervised miRNA clustering algorithm is shown and compared with other existing methods on three miRNA expression data sets. The selected miRNAs are also found to be significantly associated with different important pathways that are related to the data set.

Acknowledgements. The authors want to acknowledge Dr. Pradipta Maji of Indian Statistical Institute, Kolkata, India for his valuable suggestions. This work was supported by the German Federal Ministry of Education and Research as part of the projects eBio:miRSys [0316175A to JV]. Julio Vera is funded by the Erlangen University Hospital (ELAN funds, 14-07-22-1-Vera-Gonzlez) and the German Research

Foundation through the project SPP 1757/1 (VE 642/1-1 to JV). Sushmita Paul is funded by the Erlangen University Hospital.

References

1. Altuvia, Y., Landgraf, P., Lithwick, G., Elefant, N., Pfeffer, S., Aravin, A., Brownstein, M.J., Tuschl, T., Margalit, H.: Clustering and conservation patterns of human microRNAs. Nucleic Acids Res. **33**, 2697–2706 (2005)
2. Bargaje, R., Hariharan, M., Scaria, V., Pillai, B.: Consensus miRNA expression profiles derived from interplatform normalization of microarray data. RNA **16**, 16–25 (2010)
3. Baskerville, S., Bartel, D.P.: Microarray profiling of microRNAs reveals frequent coexpression with neighboring miRNAs and host genes. RNA **11**, 241–247 (2005)
4. Chan, W.C., Ho, M.R., Li, S.C., Tsai, K.W., Lai, C.H., Hsu, C.N., Lin, W.C.: MetaMirClust: discovery of miRNA cluster patterns using a data-mining approach. Genomics **100**(3), 141–148 (2012)
5. Dettling, M., Buhlmann, P.: Supervised clustering of genes. Genome Biol. **3**(12), 1–15 (2002)
6. Ding, C., Peng, H.: Minimum redundancy feature selection from microarray gene expression data. J. Bioinform. Comput. Biol. **3**(2), 185–205 (2005)
7. Efron, B., Tibshirani, R.: Improvements on cross-validation: the.632+ bootstrap method. J. Am. Stat. Assoc. **92**(438), 548–560 (1997)
8. Enerly, E., Steinfeld, I., Kleivi, K., Leivonen, S.K., Aure, M.R., Russnes, H.G., Ronneberg, J.A., Johnsen, H., Navon, R., Rodland, E., Makela, R., Naume, B., Perala, M., Kallioniemi, O., Kristensen, V.N., Yakhini, Z., Dale, A.L.B.: miRNA-mRNA integrated analysis reveals roles for miRNAs in primary breast tumors. PLoS ONE **6**(2), e16915 (2011)
9. Golub, T.R., Slonim, D.K., Tamayo, P., Huard, C., Gaasenbeek, M., Mesirov, J.P., Coller, H., Loh, M.L., Downing, J.R., Caligiuri, M.A., Bloomfield, C.D., Lander, E.S.: Molecular classification of cancer: class discovery and class prediction by gene expression monitoring. Science **286**(5439), 531–537 (1999)
10. Hastie, T., Tibshirani, R., Botstein, D., Brown, P.: Supervised harvesting of expression trees. Genome Biol. **1**, 1–12 (2001)
11. Maji, P.: Fuzzy-rough supervised attribute clustering algorithm and classification of microarray data. IEEE Trans. Syst. Man Cybern. B Cybern. **41**(1), 222–233 (2011)
12. Maji, P.: A rough hypercuboid approach for feature selection in approximation spaces. IEEE Trans. Knowl. Data Eng. **26**(1), 16–29 (2014)
13. Maji, P., Paul, S.: Rough set based maximum relevance-maximum significance criterion and gene selection from microarray data. Int. J. Approximate Reasoning **52**(3), 408–426 (2011)
14. Paul, S., Maji, P.: μHEM for identification of differentially expressed miRNAs using hypercuboid equivalence partition matrix. BMC Bioinform. **14**(1), 266 (2013)
15. Paul, S., Maji, P.: City block distance and rough-fuzzy clustering for identification of co-expressed MicroRNAs. Mol. BioSyst. **10**(6), 1509–1523 (2014)
16. Pawlak, Z.: Rough Sets: Theoretical Aspects of Resoning About Data. Kluwer, Dordrecht (1991)
17. Quinlan, J.R.: C4.5: Programs for Machine Learning. Morgan Kaufmann, San Francisco (1993)

18. Vapnik, V.: The Nature of Statistical Learning Theory. Information Science and Statistics. Springer, New York (1995)
19. Vlachos, I.S., Kostoulas, N., Vergoulis, T., Georgakilas, G., Reczko, M., Maragkakis, M., Paraskevopoulou, M.D., Prionidis, K., Dalamagas, T., Hatzigeorgiou, A.G.: DIANA miRPath v. 2.0: investigating the combinatorial effect of microRNAs in pathways. Nucleic Acids Res. **40**(W1), W498–W504 (2012)
20. Wei, J.-M., Wang, S.-Q., Yuan, X.-J.: Ensemble rough hypercuboid approach for classifying cancers. IEEE Trans. Knowl. Data Eng. **22**(3), 381–391 (2010)

Analysis of AmpliSeq RNA-Sequencing Enrichment Panels

Marek S. Wiewiorka[1], Alicja Szabelska[2], and Michal J. Okoniewski[3]([⊠])

[1] Institute of Computer Science, Warsaw University of Technology, Warsaw, Poland
marek.wiewiorka@gmail.com
[2] Department of Mathematical and Statistical Methods,
Poznan University of Life Sciences, Poznan, Poland
alicja.szabelska@gmail.com
[3] Scientific IT Services, Swiss Federal Institute of Technology (ETH Zurich),
Zurich, Switzerland
michal.okoniewski@id.ethz.ch

Abstract. This study presents a proof of concept of encoding genomic signatures in the AmpliSeq technology. The samples of patients with a disease and healthy ones have been processed using an AmpliSeq RNA sequencing kit of a custom design, that include 290 amplicons, sequenced using an IonTorrent machine. The read count data show the sufficient coverage in most of the chosen amplicons, which results in a good separability between the disease patients and healthy donors. In addition, several amplicons allow for checking useful genomics variants (SNPs), whenever the coverage level permits. The paper presents a machine-learning classifier evaluation of the answer to the question of difference between the patients and healthy donors, based upon the AmpliSeq panel data. The outcome confirms the potential utility of similar RNA amplicon kits in the research and clinical practice to encode gene expression signatures of diseases and their phenotypes.

Keywords: Genomics · Transcriptomics · Amplicon sequencing · Classification · Genomic signatures

1 Introduction

Next generation sequencing techniques, which have already become the driving force in molecular biology, are recently being introduced into various areas of applications in medicine. To achieve a focused insight maintaining economically affordable costs, it is essential to apply specialized kits for enrichment of particular sequences, e.g. exome kits [1,2]. Such enrichment kits exist now for both DNA and RNA sequencing, and can be ordered in a fully customized way, running the design process via the specialized web interfaces of the technical solution providers. The other very popular technique of precise measurements in genomics and transcriptomics is RT-PCR. The primers can be freely designed or ordered in pre-defined panels, specific for a given application, e.g. TaqMan [3].

© Springer International Publishing Switzerland 2015
M. Kryszkiewicz et al. (Eds.): PReMI 2015, LNCS 9124, pp. 495–500, 2015.
DOI: 10.1007/978-3-319-19941-2_47

The statistical analysis of such data described in [4]. The amplicon enrichment kits for RNA sequencing are a solution that combines the advantages of both: enriched sequencing and RT-PCR approaches. Amplicon sequencing has been done recently on the sequencing platforms of all three generations [5], still combining many amplicons in one PCR run and preparing an RNA sequencing library from such an amplified product is a novel technique. The example of such technique is AmpliSeq, introduced by LifeTech in early 2013. As with other products of modern nanotechnology, the biological hardware often precedes the methodologies for in-depth analysis of data, simply because by the amount and variety of data that it produces. This paper presents a study on the technical applicability of AmpliSeq kits in the area of autoimmune disease, in particular to verify the gene expression signature [6,7] that differentiates patients of various diseases and the healthy donors. It can also be a technical proof that will encourage researchers from other areas of medical research to encode their gene expression signatures into the amplicon sequencing panels, especially if speed, precision and cost of analysis prove to be competitive.

2   Materials and Methods

2.1   Panel Design

The amplicon panel has been designed for 289 amplicons of 284 genes known from the medical literature to be specific for the disease. 12 amplicons included a SNP in the coding region.

2.2   RNA Samples

The blood samples have been isolated from the blood of 8 patients and 8 healthy donors, matching the patients by age and gender. RNA extraction was done using RNeasy Mini Kit (Quiagen, cat. no. 74104) with subsequent purification by precipitation and ethanol washes. Concentration and purity was measured with NanoDrop 1000. Integrity of RNA was measured with Agilent 2200 TapeStation resulting RIN* values between 8.6 and 9.3. The sequencing libraries were prepared with Ion AmpliSeq RNA Library kit (Life Technologies, cat. no. 4482335) using custom primers (Life Technologies), designed as above, and Ion Xpress Barcode Adapters (Life Technologies, cat. no. 4474518), according to the manufacturers protocol. Barcoded libraries were pooled in equimolar amounts, diluted to the concentration of 20 pM and used for subsequent template preparation with Ion PGM Template OT2 200 Kit (Life Technologies, cat. no. 4480974), according to the manufacturer's protocol.

2.3   Sequencing and Data Acquisition

The sequencing reactions were performed using a Personal Genome Machine (PGM) System with Ion PGM Sequencing 200 Kit v2 (Life Technologies, cat. no.

4482006) and Ion 318 Chip Kit (Life Technologies, cat. no. 4466617), according to the manufacturer's instructions. The sequenced data were primarily processed using TorrentServer software (TorrentSuite v 3.6, LifeTech). Reads data were extracted from the chip using FastqCreator plugin v3.6.0-r57238.

2.4 Mapping of Reads and Variant Calling

The mapping to the canonic transcripts list (LifeTech, RefSeq based) was done using ion-alignment v 3.6,3-1, and counts of the reads in amplicons using the coverageAnalysis plugin v3.6.58977, with the BED file describing the amplicons. In addition, variant calling was done using variantCaller plugin v3.6.59049. An alternative analysis path was done by mapping of the reads by tophat mapper (v. 2.0.8b) [9] to the current human genome reference (hg19) instead, then the counts table has been generated using R, in particular the RSamtools and rnaSeqMap [10] libraries from Bioconductor. Checking of the encoded SNPs and variants was done using the variantCaller plugin (v3.6.59049) of Torrent Server, and alternatively by the samtools mpileup (v0.1.18) [11]. As the distribution of counts does not allow to apply a classic t-test, and the number of measured amplicons is too small to properly estimate a negative binomial distribution for RNA-sequencing tests [12] , the differential expression has been measured by a log2 fold change between the two groups and the non-parametric test from SAMseq [13]. The classification of the count data for the patients and healthy donors was done with an algorithms available in the MLInterfaces library of Bioconductor. The classifiers was trained on the count data, also in variants with the gender attribute and information about the presence of SNPs. Leave-one-out cross-validation based on five different classifiers was applied to count the correctness of classification.

3   Results and Discussion

The sequencing of 16 samples with the 318 chip resulted in 2853777 total reads, out of which 2.851M (97 % aligned bases) could be mapped to the canonic transcripts list. Detection levels. Out of 289 amplicons, 235 had the detection level of 10 in at least one of the samples. The coverage presenting the fraction of amplicons with a useful range is presented as the Fig. 1A.

3.1 Fold Change and Differential Expression

The Fig. 1B presents the distribution of fold changes and the plot of SAMseq results.

The combined results of detection level check and the differential expression analysis prove that the majority of amplicons was designed in such a way that it is useful for differentiating between patients and healthy donors. Detection of SNPs is possible only when the depth of coverage of the reads is sufficient. In the case of RNA it depends on the gene expression, thus is never guaranteed.

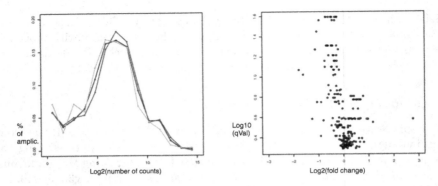

Fig. 1. The normalized distribution of log2 of counts in the amplicons (A/left). The blue line is a distribution for all the amplicons and all samples as mapped with tophat to hg19, green - in the least abundant sample, red - the sample with the biggest total number of reads. If the cutoff level for detection is 10, then ca 10 % of values is below the detection level. Globally, 235 amplicons in the panel has the expression over this detection level in at least one of the 16 samples. The volcano plot (B/right), shows the relationship between the log2 fold change and the q-value from the samSeq test for the amplicons in the panel, while comparing the patients and healthy donors (Color figure online).

Nonetheless, some of the SNPs was detected and reported in VFC files in a systematic way. The utility of the gene expression signature is confirmed by the machine learning approach. The initial clustering (Fig. 2A and B - subset of 77 amplicons with the highest absolute log fold change and 75 with highest variance) shows that most of the patients and healthy donors cluster together. The gender seems to have some influence on the clustering. There is an outlier – a healthy donor turned out to be a person from different geographical zone than the others, thus could have the immune system trained in a completely different way than the other patients and donors, coming all from Europe. The results of classification summarized in the Fig. 3, show that most of the samples are correctly classified. Adding the gender information increases the predictive power of classification. In a similar way, adding the attribute describing the SNPs found, increases the number of the samples correctly classified. All the results of the analyses described above support the claim that the gene signatures can be efficiently encoded in the AmpliSeq panel. The counts of reads representing the expression levels of genes may be used in a combined way as predictors and also in combination with clinical parameters (e.g. gender, age) and the genotyping results, that can be obtained in case of some genes from the same RNA panel.

The results described above show that such approach is feasible and may render medically useful results. There is still a room for improvement, especially in the area of custom design of the panel. In particular, tuning the selection of amplicons, can be used to distinguish between disease phenotypes in the cases that can be diagnosed from peripheral blood samples as we have proven in the case of the disease.

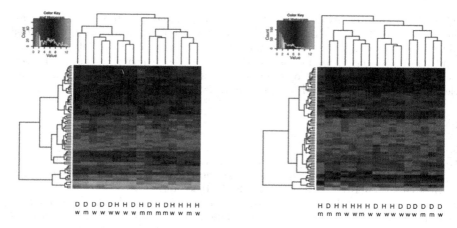

| D D D D DH H D H D H DH H H H | H D H H HH H D H H D DD D D D |
| w m w w w w w w mm mm w w m w | m m m w wm w w w ww www m m w |

Fig. 2. The heatmaps of the amplicon counts most differentiating patients and healthy donors by log2 fold change (A/left) and of the amplicon counts most variable across all the samples (B/right). The settings of the heatmap.2 functions are default, color scale set to red-green, with marginal 15 % of values in the full color saturation. The heatmaps give a proof that even without a particular tuning of the classifier, the samples are grouped mainly by the disease status (Healthy/Disease) and gender (man/woman) (Color figure online).

The approach turned out to be sufficiently precise and customizable, that it soon may become competitive to already classic RT-PCR panels measuring RNA expression. In addition, the chance of variant calling can give some extra genotyping hints, that can be used for classification of samples, and in consequence, to support the medical diagnostics.

4   Software Availability

The software is available as the Bioconductor R package ampliQueso:
http://www.bioconductor.org/packages/release/bioc/html/ampliQueso.html

	Correct classification - number of samples,leave-one-out cross-validation				
	naiveBayes	knn*	svm+	nnet*	randomFor
100 top var amplicons	10	8	7	10	9.1
100 top var amplicons with gender	10	8	7	10	9.4
100 top var amplicons with SNPs	9	8	8	10	8.9
100 top var amplicons with SNPs and gender	9	8	8	10.1	8.9
all amplicons expression	12	10	12	11.2	11.2
all amplicons expression, with gender	12	10	11	11.5	11.6
all amplicons expression, with SNPs	13	10	12	11.5	11.4
all amplicons expression, with SNPs and gender	13	10	12	11.8	11.6

* - averaged 10 iterations
+ excluding the low coverage amplicons

Fig. 3. 1 The number of correctly classified samples by various classifier function from MLInterfaces. The classification can be based upon the gene expression signature limited to 100 amplicons, with additionally gender, SNP attributes or both

Acknowledgements. We are grateful to Kelli Bramlett, Jeoffrey Schageman, and Daniel Williams from LifeTech for discussion on AmpliSeq technology, to Andreas Tobler for coordinating the collaboration and to Marzanna Künzli-Gontarczyk, Daria Bochenek and Josias Brito Frazao for the help in the sequencing library prep and discussion on the lab aspects of the study. This work was supported by the grants Sciex.ch (nr. 11.182 to AS and MO, and nr 12.289 to MW and MO).

References

1. Clark, M.J., Chen, R., Lam, H.Y.K., Karczewski, K.J., Chen, R., Euskirchen, G., et al.: Performance comparison of exome DNA sequencing technologies. Nat. Biotechnol. **29**(10), 908–914 (2011). doi:10.1038/nbt.1975
2. Sulonen, A.-M., Ellonen, P., Almusa, H., Lepistoe, M., Eldfors, S., Hannula, S., et al.: Comparison of solution-based exome capture methods for next generation sequencing. Genome Biol. **12**(9), R94 (2011). doi:10.1186/gb-2011-12-9-r94
3. Rachwal, P.A., Rose, H.L., Cox, V., Lukaszewski, R.A., Murch, A.L., Weller, S.A.: The potential of TaqMan array cards for detection of multiple biological agents by real-time PCR. PloS One **7**(4), e35971 (2012). doi:10.1371/journal.pone.0035971
4. Yuan, J., Reed, A., Chen, F., Stewart, C.N.: Statistical analysis of real-time PCR data. BMC Bioinform. **7**(1), 85–85 (2005). doi:10.1186/1471-2105-7-85
5. Okoniewski, M.J., Meienberg, J., Patrignani, A., Szabelska, A., Matyas, G., Schlapbach, R.: Precise breakpoint localization of large genomic deletions using PacBio and Illumina next-generation sequencers. BioTechniques **54**(2), 98–100 (2013). doi:10.2144/000113992
6. Veer, L.V., Dai, H., Van De Vijver, M.J., He, Y.D.: Gene expression profiling predicts clinical outcome of breast cancer. Nature **415**(6871), 530–536 (2002)
7. Van De Vijver, M.J., He, Y.D., Veer, L.V., et al.: A gene-expression signature as a predictor of survival in breast cancer. New Engl. J. Med. N **347**, 1999–2009 (2002). doi:10.1056/NEJMoa021967
8. Zhou, T., Zhang, W., Sweiss, N.J., Chen, E.S., Moller, D.R., Knox, K.S., et al.: Peripheral blood gene expression as a novel genomic biomarker in complicated sarcoidosis. PloS One **7**(9), e44818 (2012). doi:10.1371/journal.pone.0044818
9. Trapnell, C., Pachter, L., Salzberg, S.L.: TopHat: discovering splice junctions with RNA-Seq. Bioinformatics (Oxford, England) **25**(9), 1105–1111 (2009). doi:10.1093/bioinformatics/btp120
10. Leniewska, A., Okoniewski, M.J.: rnaSeqMap: a bioconductor package for RNA sequencing data exploration. BMC Bioinform. **12**, 200 (2011). doi:10.1186/1471-2105-12-200
11. Li, H., Handsaker, B., Wysoker, A., Fennell, T., Ruan, J., Homer, N., et al.: The sequence Alignment/Map format and SAMtools. Bioinformatics (Oxford, England) **25**(16), 2078–2079 (2009). doi:10.1093/bioinformatics/btp352
12. Anders, S., McCarthy, D.J., Chen, Y., Okoniewski, M., Smyth, G.K., Huber, W., Robinson, M.D.: Count-based differential expression analysis of RNA sequencing data using R and bioconductor. Nat. Protoc. **8**, 1765–1786 (2013). http://arxiv.org/abs/1302.3685
13. Li, J., Tibshirani, R.: Finding consistent patterns: a nonparametric approach for identifying differential expression in RNA-seq data. Stat. Methods Med. Res. **22**(5), 519–536 (2011)

Consensus-Based Prediction of RNA and DNA Binding Residues from Protein Sequences

Jing Yan and Lukasz Kurgan[(⊠)]

Electrical and Computer Engineering,
University of Alberta, Edmonton T6G 2V4, Canada
lkurgan@ece.ualberta.ca

Abstract. Computational prediction of RNA- and DNA-binding residues from protein sequences offers a high-throughput and accurate solution to functionally annotate the avalanche of the protein sequence data. Although many predictors exist, the efforts to improve predictive performance with the use of consensus methods are so far limited. We explore and empirically compare a comprehensive set of different designs of consensuses including simple approaches that combine binary predictions and more sophisticated machine learning models. We consider both DNA- and RNA-binding motivated by similarities in these interactions, which should lead to similar conclusions. We observe that the simple consensuses do not provide improved predictive performance when applied to sequences that share low similarity with the datasets used to build their input predictors. However, use of machine learning models, such as linear regression, Support Vector Machine and Naïve Bayes, results in improved predictive performance when compared with the best individual predictors for the prediction of DNA- and RNA-binding residues.

Keywords: RNA-binding proteins · DNA-binding proteins · Prediction · Consensus · Machine learning

1 Introduction

Interactions between proteins and DNA/RNA are crucial for many cellular functions including regulation of gene expression, genome maintenance, recombination, replication and transcription, to name a few [1, 2]. The DNA-binding and RNA-binding proteins occupy a relatively large fraction of eukaryotic genomes, in the order of 3 to 5 % [3] and 2 to 8 % [1], respectively. However, only a small fraction of these interactions was annotated so far, primarily since the experimental methods that are used to determine the protein-DNA and protein-RNA interactions are technically challenging and relatively expensive. These methods are unable to keep pace with the rapid accumulation of the protein, DNA and RNA sequences; the current NCBI's RefSeq database includes over 10 million DNA and RNA transcripts and about 52 million non-redundant proteins from over 51 thousand organisms. As a solution, the currently available experimental data are used to develop time- and cost-efficient computational tools that predict these interactions for the millions of the uncharacterized proteins.

© Springer International Publishing Switzerland 2015
M. Kryszkiewicz et al. (Eds.): PReMI 2015, LNCS 9124, pp. 501–511, 2015.
DOI: 10.1007/978-3-319-19941-2_48

Many computational predictors of the protein-DNA and protein-RNA interactions from the protein sequence and structure have been published and reviewed in the literature over the past several years [1, 4–11]. We focus on the prediction from protein chains since these methods can find the binding proteins and residues in the vast and rapidly growing sequence databases. Differences in the design and outcomes generated by various predictors can be exploited to build consensus-based predictors that take outputs generated by several individual predictors as the inputs. Research in related fields, such as sequence-based prediction of secondary structure and intrinsic disorder, shows that consensuses offer improved predictive performance when compared to the use of individual methods [12–17]. The differences in the design are also characteristic to the sequence-based prediction of DNA- and RNA-binding residues. The inputs to these methods, which represent information about each residue in the input protein sequence, differ in the scope and type of information used. The scope is defined based on the size of sequence segments centered on the predicted residues that are used to generate inputs, which varies widely between 3 and 41 residues [18, 19]. The considered types include various combinations of information about amino acid composition, physiochemical properties of the input amino acids, evolutionary profiles, sequence conservation, and structural characteristics that are predicted from the sequence, such as secondary structure and solvent accessibility. Past methods also utilized different types of predictive models, primarily generated by machine learning algorithms including neural network [18, 19], Support Vector Machine (SVM) [11, 20–22], Naïve Bayes [23], regression [24], decision tree [25], and random forest [26–28].

Consequently, a couple of studies investigated development of consensuses. Si et al. [29] developed MetaDBSite consensus that combines six DNA-binding predictors: DBS-pred [18], BindN [30], DP-Bind [24], DISIS [31], DNABindR [32], and BindN-RF [28] using SVM model. This consensus was shown to outperform each of the six predictors [29]. Similarly, Puton et al. [10] proposed Meta2 consensus that combines three RNA-binding predictors: PiRaNhA [33], Pprint [34], and BindN+ [20]. Although this approach merges the input predictions based on a simple weighted average, it still outperforms each of the three input predictors [10]. However, these two studies have drawbacks. First, some of the methods that they combine are no longer maintained and thus cannot be used. For instance, the current version of MetaDBSite combines only BindN and DP-Bind. Second, they did not compare and explore different ways to generate the consensuses but simply demonstrated that a given design is successful.

To this end, we explore and empirically compare different ways to generate consensuses and we apply only the currently available and well-maintained input predictors. We investigate the use of simple consensuses and more sophisticated machine learning models. We consider the prediction of both the DNA-binding and the RNA-binding motivated by similarities in the main characteristics of these interactions, e.g., these binding residues in the protein are positively charged and have strong propensity to interact with the negatively charged phosphate backbone of DNA or RNA [35, 36]. In other words, we expect similar conclusions for both types of binding.

2   Materials and Methods

2.1   Selection of Methods Included in the Consensus

We selected eight out of 30 methods for the prediction of DNA- and RNA-binding residues. These methods were available as reliably working (i.e., able to predict large protein set) webservers as of Dec 2013 (when we collected the data) characterized by relatively low runtime (i.e., they predict a protein with 200 residues in under 10 min). We applied the most recent versions of predictors that have multiple versions. The eight methods include five predictors of DNA-binding residues: DBS-PSSM [37], two versions of DP-Bind [24, 35], ProteDNA [22], and BindN+ [20]; and three predictors of the RNA-binding residues: Pprint [34], BindN+ [20], and RNABindR [11, 21, 36]. For the DP-Bind, we use two "default" versions based on the kernel logistic regression (KLR), DP-Bind(klr), and an ensemble of three classifiers, DP-Bind(maj). For Prote-DNA that has two modes, we use the balanced version, ProteDNA(B), that provides a better balance between sensitivity and specificity [22].

2.2   Datasets and Evaluation Protocols

Datasets were collected from the protein-DNA and protein-RNA complexes deposited in the Protein Data Bank (PDB) [38] as of Sept 2013. We annotated binding residues utilizing the most prevalent approach based on a cut-off distance at 3.5 Å, i.e., a given residue is defined as binding if at least one of its atoms is closer than 3.5 Å from an atom of the RNA/DNA [18]. We collected all 1935 DNA-binding and 981 RNA-binding chains which have high-quality X-ray structures, i.e., resolution better than 2.5 Å. Next, we improved the annotations of the binding residues by transferring these annotations between homologous proteins using procedure introduced in ref. [39]. Consequently, the number of annotated DNA- and RNA-binding residues was enlarged by 13.7 % and 9.7 %, respectively. The original redundant datasets were reduced to the non-redundant set 531 DNA- and RNA-binding chains. We divided this dataset into two subsets, the TRAINING and TEST datasets. The former is used to design our consensuses and includes 445 chains that were deposited into PDB before Sept 2010, the date when the most recent dataset used to build the considered eight predictors was collected. The latter dataset includes newer depositions to assure that we test on independent data that were not used to design the considered predictors. The dataset was clustered at 30 % similarity using CD-HIT [40] and we removed from the TEST dataset all proteins that end up in clusters that include any of the proteins from the TRAINING set. This way the final version of the TEST dataset includes 65 chains that share low, <30 %, similarity with the chains that are used to design our consensuses and that were used to design the input methods. The datasets are available at http://biomine.ece.ualberta.ca/ConsRNADNA/ConsRNADNA.htm.

The predictors of DNA- and RNA-binding residues output either only the binary prediction (binding vs. non-binding) or binary prediction together with a real-valued propensity for binding. We evaluate both outputs and exclude residues with missing atomic coordinates in the source structure files (i.e., disordered residues) since we could not complete their annotation of binding. The binary predictions are assessed using accuracy = (TP + TN)/(TP + TN + FP + FN), sensitivity = TP/(TP + FN), specificity = TN/(FP + TN), and MCC = (TP × TN-FN × FP)/√[(TP + FN) × (TP + FP) × (TN + FP) × (TN + FN)], where TP is the number of true positives (correctly predicted binding residues), FN is the number of false negatives (incorrectly predicted binding residues), FP is the number of false positives (incorrectly predicted non-binding residues), and TN is the number of true negatives (correctly predicted non-binding residues). We primarily rely on the MCC given the unbalanced nature of our datasets, i.e., the number of binding residues is lower than the number of non-binding residues. The propensities are evaluated using Receiver Operating Curve (ROC), which is a plot of false positive rate (FPR = 1 − specificity), against the true positive rate (TPR = sensitivity). These two rates are computed by binarizing the propensities using thresholds and we report the area under the ROC curve (AUC).

2.3   Considered Consensus Designs

We consider a comprehensive set of simple consensuses designed as the best performing (highest MCC on the TRAINING dataset) combinations of k methods, $k = 2$, ..., N where N is the number of considered predictors of RNA- or DNA-binding residues. The binary predictions of the k methods are combined using logical OR and logical AND. The latter design assumes that a given residues is predicted as binding only if all k methods predict it as binding; otherwise this residue is predicted as non-binding. The former design predicts a given residue as binding if any of the k methods predicts it as binding. We used these two operators individually and mixed them together. For instance, given $N = 3$ for the prediction of the RNA-binding residues, we explore designs that include "1 AND 2", "1 AND 2 AND 3", "1 OR 3", "1 OR 2 OR 3", "(1 AND 3) OR 2", "1 AND (2 OR 3)", etc. In total, we considered 10 and 116 designs for the prediction of RNA-binding residues ($N = 3$) and DNA-binding residues ($N = 5$), respectively. We select one, best-performing consensus (i.e., consensus that provides the highest value of MCC on the TRAINING dataset) for the prediction of DNA-binding residues and for the prediction of RNA-binding residues.

We also utilize more sophisticated designs where the predictions for a given residue, including both binary values and propensities, from the N methods are combined using predictive models generated by five different popular types of machine learning algorithms. We include the linear logistic regression (LLR), C4.5 decision tree (C4.5), k-nearest neighbor (kNN), SVM, and Naïve Bayes (NB) using the implementations from the WEKA platform [41]. Each of these classifiers was parameterized based on five-fold cross validation on the TRAINING dataset. We use grid search to select parameters that provide the maximal value of MCC. For LLR, we adjust the number of boosting iterations $n = \{0, 1, ..., 10\}$; for C4.5 we parameterize confidence factor $c = \{0.05, 0.1, ..., 0.5\}$ and minimal number of instances per leaf node $m = \{1, 2, ..., 5\}$ that are used for pruning; for kNN we optimize number of neighbors $k = \{1, 2, ..., 30\}$; for SVM we use the Gaussian kernel and find the best values of complexity parameter

Table 1. Results of empirical assessment of predictors of the DNA- or RNA-binding residues on the TEST dataset. +(=) in the Sig column denotes that the difference was (was not) significant at p-value <0.05. The highest MCC and AUC values for each type of binding are given in **bold** font. Individual predictors are denoted with *italics*.

	Method	Accuracy	Sensitivity	Specificity	MCC	Sig	AUC	Sig
DNA-binding	ML consensus LLR	0.857	0.594	0.873	**0.304**		**0.839**	
	ML consensus C4.5	0.889	0.485	0.915	0.301	=	0.789	+
	ML consensus kNN	0.810	0.682	0.818	0.287	+	0.826	+
	ML consensus SVM	0.823	0.648	0.834	0.286	+	0.742	+
	ML consensus NB	0.805	0.664	0.814	0.273	+	0.829	+
	Simple consensus	0.890	0.424	0.919	0.267	+		
	DBS-PSSM	*0.771*	*0.721*	*0.774*	*0.266*	+	*0.810*	+
	BindN+	*0.865*	*0.482*	*0.888*	*0.256*	+	*0.806*	+
	DP-Bind(maj)	*0.810*	*0.598*	*0.823*	*0.247*	+		
	DP-Bind(klr)	*0.814*	*0.590*	*0.828*	*0.246*	+	*0.794*	+
	MetaDBSite consensus	0.898	0.325	0.933	0.221	+		
	ProteDNA(B)	*0.937*	*0.093*	*0.990*	*0.158*	+		
RNA-binding	ML consensus LLR	0.920	0.257	0.939	**0.128**		**0.731**	
	ML consensus SVM	0.919	0.249	0.938	0.123	+	0.618	+
	ML consensus NB	0.931	0.215	0.952	0.121	=	0.727	+
	Meta2 consensus	0.768	0.526	0.774	0.116	+		
	ML consensus kNN	0.927	0.218	0.947	0.115	+	0.711	+
	BindN+	*0.841*	*0.399*	*0.854*	*0.114*	+	*0.706*	+
	Simple consensus	0.915	0.244	0.933	0.113	+		
	RNABindR	*0.714*	*0.575*	*0.718*	*0.105*	+	*0.712*	+
	ML consensus C4.5	0.942	0.154	0.965	0.100	+	0.610	+
	Pprint	*0.773*	*0.433*	*0.782*	*0.084*	+	*0.667*	+

$C = \{2^{-3}, 2^{-1}, ..., 2^{3}\}$ and width of the kernel gamma $= \{2^{-2}, 2^{0}, ..., 2^{8}\}$. Since all these consensuses generate real-values propensity as the output, we binarize it to obtain the binary prediction (binding vs. non-binding) by selecting a threshold that gives maximal value of MCC on the TRAINING dataset.

3   Results and Discussion

The predictive performance of the considered individual methods, the best performing simple consensus and the considered five machine learning consensuses on the TEST dataset for the prediction of the DNA-binding and the RNA-binding residues is summarized in Table 1. The methods are sorted by their MCC values. We include results for the two published consensuses: MetaDBSite [29] and Meta2 [10]; their predictions were collected using the corresponding webservers. Significance of the difference in MCC and AUC values between the best performing method and other methods for a given binding type was assessed based on 10 tests that utilize 70 % of randomly chosen proteins; if the measurements are normal, based on the Anderson–Darling test at 0.05 significance, we use the paired t-test; otherwise we use the Wilcoxon rank sum test. AUC values could not be computed for DP-Bind(maj), MetaDBSite, ProteDNA(B), Meta2, and the two simple consensuses since these methods provide only the binary predictions.

The selected simple consensuses (with the best predictive performance on the TRAINING dataset) include the AND-based combinations: "BindN+ AND DBS-PSSM" for the prediction of DNA-binding residues, and "BindN+ AND RNABindR AND Pprint" for the RNA-binding residues. The simple consensus for the DNA-binding residues includes two methods that have the highest MCC on the TRAINING dataset (0.3 and 0.26) and excludes the other three predictors, which suggests that these three methods do not provide further value for the consensus. The simple consensus for the RNA-binding combines all three considered predictors.

Although these simple consensuses provide improvements in predictive quality when compared with the individual predictors on the TRAINING dataset (MCC higher by 0.01 and 0.04 for the DNA- and RNA-binding, respectively), Table 1 reveals that this does not translate into the TEST dataset. The simple consensuses obtain the same predictive performance as the best individual method, MCC of 0.267 vs. 0.266 of the best individual method DBS-PSSM for the DNA-binding and 0.113 vs. 0.114 of the best BindN+ for the RNA-binding. The reason is that TEST shares low sequence similarity with TRAINING set. This results in differences in predictions of individual methods between the two datasets that negatively affect accuracy of the simple consensus designs. In fact, the simple consensuses that obtain the best results on the TEST dataset for the DNA-binding "BindN+ OR DBS-PSSM AND DP-Bind(klr) OR ProteDNA(B)" and for the RNA-binding "BindN+ AND RNABindR" secure higher MCCs that equal 0.291 and 0.118, respectively, on that dataset. We conclude that the consensuses that rely on the simple designs are unlikely to provide improved predictive performance when applied to sequences that share low similarity with the datasets used to build their input predictors.

Table 1 demonstrates that consensuses based on certain machine learning models offer improved predictive performance when compared with the best individual predictors. In particular, the linear regression (LLR model) secures the highest MCC and AUC values for prediction of both RNA- and DNA-binding residues, and these values are significantly higher than the values offered by the individual predictors. The ROCs of the LLR consensus and the corresponding individual predictors that generate real-values propensities are compared in Fig. 1. These curves reveal that this consensus outperforms the other methods for virtually entire range of the FPR values, except for the low FPR <0.04 for the RNA-binding where Pprint offers slightly higher TPR values. Two other machine learning models, SVM and NB, also offer improvements for the prediction of RNA- and DNA-binding residues. The other two models, C4.5 and kNN, provide improvements for the prediction of DNA-binding residues but not for the prediction of the RNA-binding residues. To sum up, we observe that consensuses that rely on certain more sophisticated models provide improved predictive performance, even when tested using chains that share low sequence similarity with proteins that were used to build their input predictors.

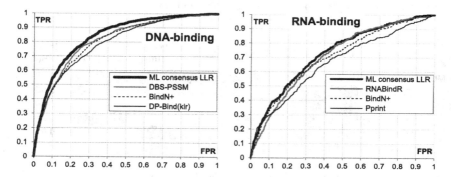

Fig. 1. ROC generated on the TEST dataset for the best performing ML consensus and the considered individual predictors that generate real-values propensity scores for the prediction of DNA-binding and RNA-binding residues

Figure 2 provides insights that may explain why consensuses are successful. It gives values of the Phi correlation coefficient (PhiCC), which is an equivalent of the Pearson correlation coefficient for a pair of binary variables, between the binary predictions of all pairs of the individual methods and between the binary predictions of our LLR consensus and each individual predictor. Except for the pair of DP-Bind(maj) and DP-Bind(klr) methods that share very similar design [24, 35] and consequently secure high correlation close to 0.9, the predictions of the other individual methods are only modestly correlated with the PhiCC values < 0.5 for the DNA-binding and < 0.4 for the RNA-binding. This could be explained by substantial differences in the design of these methods. For instance, BindN+ uses information concerning physiochemical properties of the input amino acids, sequence alignment, evolutionary profiles, and the SVM model. DP-Bind uses regression model and inputs that solely rely on the evolutionary profiles. DBS-PSSM also uses the evolutionary profiles but with the neural network model. RNABindR applies SVM model and the evolutionary profiles. We also note the low correlations for any pair of methods for the prediction of DNA-binding residues that includes ProteDNA(B). This method predicts a subset of DNA-binding residues that bind transcription factors, which is why it secures low sensitivity (Table 1) and has low correlations. The modest levels of correlations between individual predictors are exploited by the consensus. In other words, since all individual predictors offer relatively good predictive performance and their predictions are substantially different (modestly correlated), these predictions likely complement each other. A similar observation was made in the context of the sequence-based prediction of intrinsic disorder [42]. Figure 2 reveals that the LLR-based consensus has higher correlations with the individual methods compared to the correlations between these methods (except for the DP-Bind); values in the first row or column in the heat maps are higher than the remaining values in the same column or row, respectively. More specifically, the correlations with the consensus predictions are >0.57 for the prediction of DNA-binding residues and >0.39 for the prediction of RNA-binding residues, except for ProteDNA(B) that under-predicts the binding residues. This combined with the fact that our consensus obtains higher predictive performance means that it effectively takes advantage of this complementarity between the input predictors.

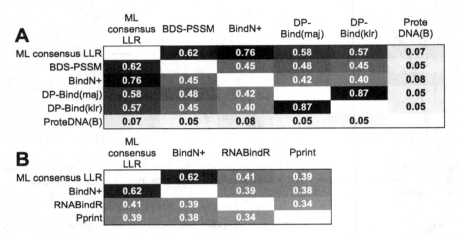

Fig. 2. Correlations between pairs of individual predictors and between the best performing ML consensus and each individual predictor for the prediction of DNA-binding residues (panel A) and RNA-binding residues (panel B). We use heat maps where darker colors correspond to higher Phi correlation coefficients, values of correlations are given for each pair of methods.

Finally, we analyze predictive performance of the two existing consensuses. The MCC of MetaDBSite is relatively low and lower than MCC of some of the considered individual predictors (Table 1). The reason is that this approach is currently implemented a simple consensus "BindN AND DP-Bind" since the other four predictors that it was originally designed to include are no longer available. The Meta2 consensus for the prediction of RNA-binding residues outperforms its input predictors Pprint and BindN+ (Table 1). This consensus is based on a weighted average, which is more complex than our simple consensus designs, but is less sophisticated than our machine learning designs. Correspondingly, Meta2 provides lower predictive performance than our consensuses based on LLR, SVM and NB models.

4   Conclusions and Future Work

To conclude, our empirical study suggests that sequence-based prediction of RNA- and DNA-binding residues would benefit from the use of machine learning consensuses. Such consensuses exploit complementarity between individual predictors to generate predictions with significantly higher predictive quality when compared with the individual predictors, even for the chains characterized by low sequence similarity with the proteins used to develop these predictors.

As a potential future work, a majority vote based consensus and other classifiers, including SVMs with other types of kernels, could be considered. Moreover, potential overlap between predictions of RNA-binding and DNA-binding residues (i.e., whether and how many RNA-binding residues are predicted by predictors of DNA-binding residues and vice versa) should be investigated.

References

1. Re, A., et al.: RNA-protein interactions: an overview. Meth. Mol. Biol. **1097**, 491–521 (2014)
2. Dey, B., et al.: DNA-protein interactions: methods for detection and analysis. Mol. Cell. Biochem. **365**(1–2), 279–299 (2012)
3. Charoensawan, V., Wilson, D., Teichmann, S.A.: Genomic repertoires of DNA-binding transcription factors across the tree of life. Nucleic Acids Res. **38**(21), 7364–7377 (2010)
4. Zhao, H., Yang, Y., Zhou, Y.: Prediction of RNA binding proteins comes of age from low resolution to high resolution. Mol. BioSyst. **9**(10), 2417–2425 (2013)
5. Fornes, O., et al.: On the use of knowledge-based potentials for the evaluation of models of protein-protein, protein-DNA, and protein-RNA interactions. Adv. Protein. Chem. Struct. Biol. **94**, 77–120 (2014)
6. Kauffman, C., Karypis, G.: Computational tools for protein-DNA interactions. Data Min. Knowl. Disc. **2**(1), 14–28 (2012)
7. Liu, L.A., Bradley, P.: Atomistic modeling of protein-DNA interaction specificity: progress and applications. Curr. Opin. Struct. Biol. **22**(4), 397–405 (2012)
8. Gromiha, M.M., Nagarajan, R.: Computational approaches for predicting the binding sites and understanding the recognition mechanism of protein-DNA complexes. Adv. Protein. Chem. Struct. Biol. **91**, 65–99 (2013)
9. Ding, X.M., et al.: Computational prediction of DNA-protein interactions: a review. Curr. Comput. Aided Drug Des. **6**(3), 197–206 (2010)
10. Puton, T., et al.: Computational methods for prediction of protein-RNA interactions. J. Struct. Biol. **179**(3), 261–268 (2012)
11. Walia, R.R., et al.: Protein-RNA interface residue prediction using machine learning: an assessment of the state of the art. BMC Bioinform. **13**, 89 (2012)
12. Yan, J., Marcus, M., Kurgan, L.: Comprehensively designed consensus of standalone secondary structure predictors improves Q_3 by over 3 %. J. Biomol. Struct. Dyn. **32**(1), 36–51 (2014)
13. Zhang, H., et al.: Critical assessment of high-throughput standalone methods for secondary structure prediction. Brief Bioinform. **12**(6), 672–688 (2011)
14. Fan, X., Kurgan, L.: Accurate prediction of disorder in protein chains with a comprehensive and empirically designed consensus. J. Biomol. Struct. Dyn. **32**(3), 448–464 (2014)
15. Kozlowski, L.P., Bujnicki, J.M.: MetaDisorder: a meta-server for the prediction of intrinsic disorder in proteins. BMC Bioinform. **13**, 111 (2012)
16. Walsh, I., et al.: Comprehensive large-scale assessment of intrinsic protein disorder. Bioinformatics **31**(2), 201–208 (2015)
17. Albrecht, M., et al.: Simple consensus procedures are effective and sufficient in secondary structure prediction. Protein Eng. **16**(7), 459–462 (2003)

18. Ahmad, S., Gromiha, M.M., Sarai, A.: Analysis and prediction of DNA-binding proteins and their binding residues based on composition, sequence and structural information. Bioinformatics **20**(4), 477–486 (2004)
19. Jeong, E., Chung, I.F., Miyano, S.: A neural network method for identification of RNA-interacting residues in protein. Genome Inform. **15**(1), 105–116 (2004)
20. Wang, L., et al.: BindN+ for accurate prediction of DNA and RNA-binding residues from protein sequence features. BMC Syst. Biol. **4**(Suppl. 1), S3 (2010)
21. Terribilini, M., et al.: RNABindR: a server for analyzing and predicting RNA-binding sites in proteins. Nucleic Acids Res. **35**, W578–W584 (2007)
22. Chu, W.Y., et al.: ProteDNA: a sequence-based predictor of sequence-specific DNA-binding residues in transcription factors. Nucleic Acids Res. **37**, W396–W401 (2009)
23. Lee, J.H., et al.: Striking similarities in diverse telomerase proteins revealed by combining structure prediction and machine learning approaches. Pac. Symp. Biocomput. **13**, 501–512 (2008)
24. Hwang, S., Gou, Z.K., Kuznetsov, I.B.: DP-Bind: a web server for sequence-based prediction of DNA-binding residues in DNA-binding proteins. Bioinformatics **23**(5), 634–636 (2007)
25. Carson, M.B., Langlois, R., Lu, H.: NAPS: a residue-level nucleic acid-binding prediction server. Nucleic Acids Res. **38**, W431–W435 (2010)
26. Ma, X., et al.: Sequence-based prediction of DNA-binding residues in proteins with conservation and correlation information. IEEE-ACM Trans. Comput. Biol. Bioinform. **9**(6), 1766–1775 (2012)
27. Ma, X., et al.: Prediction of RNA-binding residues in proteins from primary sequence using an enriched random forest model with a novel hybrid feature. Proteins **79**(4), 1230–1239 (2011)
28. Wang, L.J., Yang, M.Q., Yang, J.Y.: Prediction of DNA-binding residues from protein sequence information using random forests. BMC Genom. **10**(Suppl. 1), S1 (2009)
29. Si, J., et al.: MetaDBSite: a meta approach to improve protein DNA-binding sites prediction. BMC Syst. Biol. **5**(Suppl. 1), S7 (2011)
30. Wang, L.J., Brown, S.J.: BindN: a Web-based tool for efficient prediction of DNA and RNA binding sites in amino acid sequences. Nucleic Acids Res. **34**, W243–W248 (2006)
31. Ofran, Y., Mysore, V., Rost, B.: Prediction of DNA-binding residues from sequence. Bioinformatics **23**(13), I347–I353 (2007)
32. Yan, C.H., et al.: Predicting DNA-binding sites of proteins from amino acid sequence. BMC Bioinform. **7**, 262 (2006)
33. Murakami, Y., et al.: PiRaNhA: a server for the computational prediction of RNA-binding residues in protein sequences. Nucleic Acids Res. **38**, W412–W416 (2010)
34. Kumar, M., Gromiha, A.M., Raghava, G.P.S.: Prediction of RNA binding sites in a protein using SVM and PSSM profile. Proteins **71**(1), 189–194 (2008)
35. Kuznetsov, I.B., et al.: Using evolutionary and structural information to predict DNA-binding sites on DNA-binding proteins. Proteins **64**(1), 19–27 (2006)
36. Terribilini, M., et al.: Prediction of RNA binding sites in proteins from amino acid sequence. RNA **12**(8), 1450–1462 (2006)
37. Ahmad, S., Sarai, A.: PSSM-based prediction of DNA binding sites in proteins. BMC Bioinform. **6**, 33 (2005)
38. Berman, H.M., et al.: The protein data bank. Nucleic Acids Res. **28**(1), 235–242 (2000)
39. Chen, K., et al.: A critical comparative assessment of predictions of protein-binding sites for biologically relevant organic compounds. Structure **19**(5), 613–621 (2011)
40. Huang, Y., et al.: CD-HIT suite: a web server for clustering and comparing biological sequences. Bioinformatics **26**(5), 680–682 (2010)

41. Frank, E., et al.: Weka-a machine learning workbench for data mining. In: Maimon, O., Rokach, L. (eds.) Data Mining and Knowledge Discovery Handbook, 2nd edn., pp. 1269–1277. Springer, Heidelberg (2010)
42. Peng, Z., Kurgan, L.: On the complementarity of the consensus-based disorder prediction. Pac. Symp. Biocomput. **8**, 176–187 (2012)

Applications of Artificial Intelligence

Fusion of Static and Dynamic Parameters at Decision Level in Human Gait Recognition

Marcin Derlatka[✉] and Mariusz Bogdan

Faculty of Mechanical Engineering, Bialystok University of Technology,
Wiejska Street 45C, 15-351 Bialystok, Poland
{m.derlatka,m.bogdan}@pb.edu.pl

Abstract. This paper presents the bimodal biometric system based on human gait data of different type: dynamic - ground reaction forces and static - some anthropometric data of human body derived by means of Kinect. The innovation of this work is the use of unprecedented hitherto in the literature set of signals. The study was conducted on a group of 31 people (606 gait cycles). Kistlers force plates and Kinect device as well as the authors software were used to measure and process data. The following anthropometric parameters were used here: torso, hip width, length of left thigh, length of right thigh and body height. These signals have been combined at decision level of the biometric system. Our biometric system in gait recognition process involves both k-nearest neighbour classifier as well as majority voting system. In case of users the False Rejected Rate (FRR) reaches the level of 4.55 % and False Accepted Rate (FAR) is equal to 0.85 %. In the case of impostors it has been possible to reject 26 cases previously classified by 5NN. The presented biometric system fills the gaps in the existing studies and confirms the superiority of systems based fusion over typical methods of human gait recognition.

Keywords: Gait recognition · Data fusion · Ground reaction forces · Kinect

1 Introduction

The accuracy of biometric systems is extremely important. Therefore, there are currently many approaches, in which the biggest emphasis is put on reducing the classification error as much as possible. As the result, more and more popular are becoming the methods which use fusion of two or more sensors, traits or classifiers. A number of studies show that in this manner better results are obtained than with conventional biometrics [5,12,14,19,21]. Among a range of biometric methods human gait deserves a special attention [2,3]. Walking is a natural human activity, which is the result of consistent interaction between nervous, skeletal and muscular systems. It is a unique phenomenon that allows identification of people. Compared to most other biometric methods, depending on the used measurement method, human gait can be measured at a distance and it does not require the cooperation of the person undergoing this type of

© Springer International Publishing Switzerland 2015
M. Kryszkiewicz et al. (Eds.): PReMI 2015, LNCS 9124, pp. 515–524, 2015.
DOI: 10.1007/978-3-319-19941-2_49

procedure. In work [4] all attempts to recognize human gait have been divided dependently on the origin of the data. Gafurov has indicated the methods based on the data obtained from: video [8], floor sensor [3] and portable sensors [4]. In the first approach, the measurement data derive from the equipment recording image sequences. There are also many works [1,10,11,13,17] in which, on the basis of human movement and its anthropometric characteristics obtained by means of Kinect sensor, a person is identified. At least a few approaches and problems associated with the correct recognition can be distinguished here with:

- the choice of the type of neural network (classifier) and with the reduction of features in the input vector [11,13],
- specifying the number of static (anthropometric features) and dynamic (changing the angles between selected pairs of kinematic) recognition parameters [10,11,13,17],
- identifying the quality of recognition and the parameters describing this quality (factors of evaluation) [10,17],
- type of movement (walking, running, etc.) [10,11,16].

Recognition of persons based on the data derived from the floor sensor assumes that during walking, the person tested will walk through the measurement path, in which a hidden device registers such parameters as ground reaction forces or the distribution of pressure while placing foot on the ground. Obviously, these methods record signals generated in the gait supporting phase only. Systems of this type can be used everywhere where examined people are forced to pass through hidden (e.g. in floor) force plates, for example through different kinds of gates at the airport. The last group of methods involves the measurement of parameters by devices (most often accelerometers) placed directly on the examined person. Studies of this type require full cooperation of the person that is going to be recognized. In the literature we can find works that, to identify a person by the their gait, use methods of join fusion based methods. In [15] people were recognized based on data derived from eleven cameras that were positioned at a different angle ($0°$, $18°$, $36°$, ... , $180°$) in relation to the person examined. In [18] dynamic and static characteristics derived from the images of people in motion were combined to improve the quality of the recognition. Whereas in [7], after dividing the gait cycle into a number of parts, submits each of them to identification using the nearest neighbor classifiers (NN). The global decision is the result of majority voting based on the classification results of the NN.

2   Selected Signals Describing Human Movement

Human gait is a multidimensional and highly nonlinear phenomenon described by multiple data collection containing kinematic, dynamic, anthropomorphic, electromyographic and metabolic variables. Thus, the need to use a variety of signals for its description and analysis is something desirable and natural.

2.1 Ground Reaction Forces (GRF)

In human walking, during the support phase of gait, forces are generated between the foot and the ground. These forces are called ground reaction forces. There are three components of the GRF (Fig. 1): vertical, anterior-posterior and lateral. Ground reaction forces depend on the subject's weight, but also on the cadence, the person velocity and the footwear. Ground reaction forces in the framework of this study were measured by the use of two KISTLER platforms operating at the frequency of 1000 Hz.

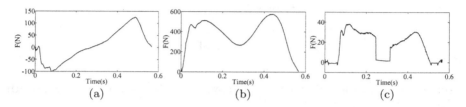

Fig. 1. Components GRF in: (a) anterior/posterior, (b) vertical, (c) medial/lateral direction, during the support phase of the left lower limb

2.2 Anthropometric Measures

Human gait recognition based on anthropometric data began to experience a renaissance with the spread of video cameras and the ability to process a sequence of images on-line. Therefore, a number of works have been published, such as [1], in which the so-called static data have been used for the purpose of identifying a person. It is important to note that, according to the results presented in [3], taking account of the height of a person's body in human gait recognition based on GRF should reduce the number of false recognitions by half. In our work we chose anthropometric parameters which have the greatest values. Thanks to this, possible measurement error associated with the non-perpendicular setting of Kinect system will have a relatively smaller influence on the results obtained. Hip width parameter has been added to ensure an odd number of parameters, and so, the clear result of the operation of majority voting method, and to facilitate the distinction between the sexes of the subjects. Here we have used the following parameters averaged over a gait cycle: torso, hip width, length of left thigh, length of right thigh and body height (Fig. 2).

3 Movement Tracking by Means of Kinect Device

The SDK Software (version 1.8) contains a library of NUI Skeleton, which allows to obtain information about the locations of 20 body parts (joints) towards the sensor - determining their coordinates (Fig. 2). In order to examine the factors influencing the quality and sensitivity of both detection and parametric description of the captured figure movements (based on the SDK), the GUI (Fig. 3) application has been built. This application allows for:

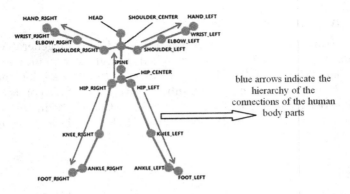

Fig. 2. Body joints in SDK Kinect

- simultaneous capture of image data stream from the RGB camera and depth camera of the Kinect controller,
- skeletal tracking,
- the choice of image resolution from RGB camera and a depth camera,
- a description of the figure movement - calculating and displaying the static parameters of the registered figure (the length of body segments, body height), the angles between the parts of the body (kinematic pairs), the coordinates of points describing the parts of the skeleton,
- displaying graphs from earlier collected data - modification of the static parameters in time and the angles between the parts of the body,
- to record specific (significant) parameters to an Excel file or unformatted text file.

First, the value of the filtration parameters of raw image data were established. In order to stabilize the joint position - minimize occurring vibrations, the information about tracking skeletal connections can be adjusted in each frame. Kinect for Windows SDK guarantees a mechanism that smoothens the joint position in the frame. What is more, the implemented algorithm can adjust the position of the joints in each frame depending on the fixed action - a filtering method. The smoothing filter used is based on the Holt Double Exponential Smoothing method [6]. The advantage of this filter is operating with low delay while ensuring proper smoothing. The filter can be controlled by five smoothing parameters (Fig. 3):

- smoothing (range of changes 0.0–1.0, selected value 0.5) - along with the increased value of the parameter the smoother frame is obtained,
- correction (range of changes 0.0–1.0, selected value 0.5) - lower values are slower to correct towards the raw data and appear smoother, while higher values will correct toward the raw data more quickly,
- prediction (range of changes ≥ 0, selected value 0.5) - the values should be chosen wisely because exceeding the value of 0.5 can cause overshooting effect - too rapid changes. Small values of Max Deviation Radius reductively affect large values of this parameter,

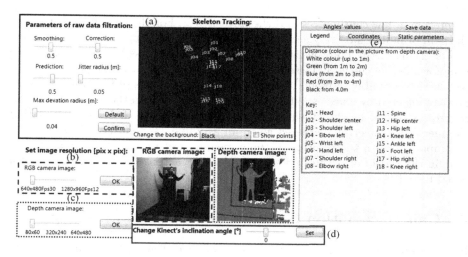

Fig. 3. Main window Kinect application: (a) preview of the tracked skeleton and sliders to change the filter parameters, (b) displaying and change of the image resolution from RGB camera, (c) displaying and change of the image resolution from depth camera, (d) change of the controllers inclination angle, (e) panel with tabs

- jitter radius (range of changes ≥0, selected value 0.05) - this parameter is responsible for the reduction of jitter. It is expressed in the unit of meters,
- max deviation radius (range of changes ≥0, selected value 0.05) - This parameter is expressed in meters. The designated position after filtration may differ from original position.

For the purposes of this article, it was decided to choose only the static parameters (selected anthropometric characteristics). It should be noted that static data are fixed i.e. it is not dependent on the type of human gait (it is often of non-constant speed and non-constant frequency), on its characteristics (speed of locomotion, stride length, etc.). The test of the correctness of the calculation of static parameters depended on the completion of measurements of the selected parts of the body, e.g. body height (BH) and comparing them with the results obtained from the application. The test was performed repeatedly, once with the controller located at the height equal to half of the measured persons height, and the second time at the height equal to 75 % of BH; the examined person moved in accordance with the optical axis of the controller. The measurements commenced at the time when the user was within 2 m from the controller. Then the person had to slowly regress until the sensor lost the sight of its skeleton, and then again returned to the starting location (2 m from the sensor). On the basis of data collected during the test, it can be concluded that the values obtained during the measurement differ little from the real ones. Measurement errors of individual limbs are as follows: left arm - 0.35 %, right forearm - 1.6 %, the left thigh - 2 %, right calf - 2.5 %. The results obtained are very good and the errors are small. In the case of body height, in most frames, the obtained results were lowered by a few centimetres and the error is 3 %. For the calculation of height the sum of individual body sections (among others the length of feet) was used. The sections of the body were connected by points tracked by Kinect. During the movement of the user the sensor could incorrectly identify the position of the feet, which probably influenced the outcome. Calculations of the static parameters are affected by a very small error, in most cases it is in the range of 0.5–2 %, regardless of the height at which the controller is located. It was noted that the measurement error increased when an examined person moved in direction incompatible with the position of the optical axis of the camera sensor. Therefore, in the rest of our study we adopted possible acceptable measurement error equal to ± 5 %.

4   Gait Recognition Based on GRF Data and Kinect

Natural differences in the dynamics of gait influence the support phase duration. As a result, the time series describing the ground reaction forces have different lengths. Comparing these time series is possible thanks to DTW (dynamic time warping). DTW is a well-known algorithm that determines the minimum cost of the 'fit' of one time series to the other [20]. This cost is the smaller, the more similar to each other the time series are. In this study, DTW was used to determine the similarity of the individual components of the GRF. The investigation takes into consideration all (L = 6) components of ground reaction forces. The total distance was calculated as:

$$\rho = \sum_{l=1}^{L} DTW_l \tag{1}$$

where: DTW_l is the DTW distance between the same components of GRF of the same lower limb. From some data patterns were created - biometric signatures characteristic to people who were to be recognized (so called users). In the case of GRF these were time series describing the individual components, which at the time of identification were used to calculate the distance according to the formula (1). In case of anthropometric measurements the average value over gait cycle of measured parameters was taken into account. Thanks to this it was possible to eliminate accidental measurement errors.

Identification of persons was held in two stages (Fig. 4). In the first phase ρ distances between the measured components of GRF and models included in the database of GRFs were calculated. Next, based on the 5NN classifier, the candidate User ID was selected. Of course, there is a number of more sophisticated methods than the one used here, in terms of both classification [9] and resistance to falsification [3]. However, the main aim was to assess the discriminatory abilities of selected features and the impact of phase II on the obtained results. Next step was followed by the verification of the choice made by the 5NN classifier, based on anthropometric measurements provided by Kinect. It should be emphasized that in the second phase it was not possible to change the decision,

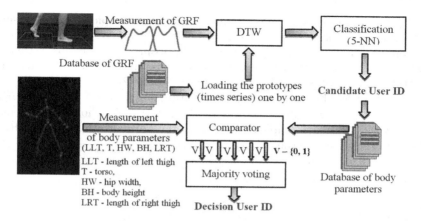

Fig. 4. The scheme of the human gait recognition by means of the typical 5NN classifier based on GRF and body parameters derived from Kinect device

only to accept or deny it. The measured values were compared with the models of those parameters that were stored in the database for the candidate user. If the value of the parameter measured was within ± 5 % of the value stored in the database, it was treated as a vote for acceptance of the selection. Otherwise, it was treated as a voice for the negation of the 5NN choice. The decision was taken in accordance with the choice of the majority.

5   Result and Discussion

The measurements were made in the Bialystok University of Technology on a group of 31 subjects (14 men and 17 women) with the use of the two Kistler force plates and a Kinect device at the same time. The persons taking part in the research were at ages 21.93 ± 0.92, BW: 69.54 ± 15.10 kg and BH 170.95 ± 9.56 cm. It should be noted that the subjects were a rather homogeneous group, characterized by similar physical form and the lack of serious illnesses or injuries which could permanently affect the change in the motion pattern. During the research subjects walked freely in their own sports shoes along the measurement path in which two force plates were hidden. In the distance of 2 m from the measurement path Kinect device was hidden which captured and registered the movement of the subject in the sagittal plane. The gait sequences were repeated several times so a little more than 600 gait cycles were recorded. 224 strides from 28 users (8 strides per each user) were stored in the database and used as so called prototype points. Three subjects (30 strides) were treated as impostors who wanted to falsify the biometric system and obtain an access to protected resources. Their data were used for testing jointly with the rest of users data. The whole set for testing contained of 382 strides (gait cycles). As a result of the operation of 5NN classifier in 5 cases (1.20 %) incorrect recognition (False Acceptance Rate - FAR) was obtained in users group. In the case of these mistakes the classifier most frequently confuses people of different sexes, however, these are often people of similar body weight (BW). As many as three cases concern the confusion between the same people. Of course 5NN classifier is 100 % wrong (FAR) in the case of persons whose patterns are not present in the database (impostors). Here as well, most people confused with each other have a similar BW. This observation is not surprising because, as stated earlier, the mass of the subject affects the values of the GRF. We should also mention here the work [22] which demonstrated, that the normalization of the GRF to BW significantly reduces the quality of recognition.

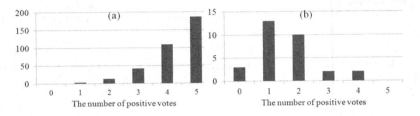

Fig. 5. Histogram with voting results for: (a) users, (b) impostors

Voting results at the second stage of identification are shown in Fig. 5. It is easy to notice that, with the case of users, in general, the biometric system does not have any problems with the confirmation of the selection it has made earlier (at the first stage). In more than half cases, all measured anthropometric parameters fit the set limit of 5 % (Fig. 5a). In the case of impostors (Fig. 5b), the presented biometric system captures differences in selected anthropometric parameters, without any major problems. However, only in a few cases (3) all the parameters differ by more than 5 % of the data stored in the database. There was not such a case, in which the number of votes was 5. Most often the above mentioned differences appear here for data 3 or 4. To sum up, this reflects a well-chosen value of tolerance with respect to the accuracy of the Kinect system. In relation to the group of users, the second stage reduces the FAR rejecting 2 of incorrectly diagnosed cases (FAR =0.85 %). Now, all incorrectly diagnosed gait cycles belong to the same person. The mistake occurring here refers to the persons of similar weight (64.4 kg and 63.2 kg) and body height (167.1 cm and 164.7 cm). Unfortunately, as a result of the operation of comparator a few earlier correctly diagnosed cases have been rejected (exactly 14). Essentially, the False Rejected Rate (FRR) reaches the level of 4.55 % (Table 1). In the case of impostors it has been possible to reject 26 cases previously classified by 5NN. The remaining falsely accepted strides belong to one person (108 kg and 181.6 cm) and are mistaken for another person with a mass of 90.1 kg and body height 188.2 cm. A deeper analysis of these studies clearly show that the misclassification of these people at the first stage is not a result of a close resemblance between them, but rather the fact that people have the highest body weights amongst all of the examined. Hence, it has been difficult to find more similar person for 5NN classifier (when using 5NN classifier). Unfortunately, also in the case of anthropometric parameters it is possible to notice a certain resemblance between these people. This resemblance along with the measuring error of the Kinect system causes the incorrect acceptance of a few strides.

Table 1. FAR, FRR for users and impostors

	Users	Impostors
FAR	0.85%	13.33%
FRR	4.55%	

6   Conclusions

The proposed bimodal biometric system works on the basis of the two kinds of signals: dynamic GRF and static the length parameters of the subjects bodies. The performed experiment shows the potential of human gait as biometrics and on the other hand the superiority of systems based on fusion over standard methods of human gait recognition. In work [11], the proposed method based on full-body motion and anthropometric biometrics has an average ROC EER (Receiver Operating Characteristic Equal Error Rate) of 13 % and an average Cumulative Match Curve Rank-1 identification rate of 90 % ROC. In this work we have achieved comparable results (with Impostors). The authors of presented article are aware of some deficiencies of the presented research. It is necessary to extend the number of investigated persons which could increase the potential application capabilities of the system. In addition, more sophisticated classification algorithms should be used to reduce the level of recognition errors at the first stage of the operation. The use of the second Kinect device set in the frontal plane of a walking subject would be worth considering, as it would improve the accuracy of the used anthropometric parameters.

References

1. Bobick, A.F., Johnson, A.Y.: Gait recognition using static, activity-specific parameters. In: Proceedings of the 2001 IEEE Computer Society Conference on Computer Vision and Pattern Recognition, CVPR 2001, vol. 1, p. I-423 (2001)
2. Boulgouris, N.V., Hatzinakos, D., Plataniotis, K.N.: Gait recognition: a challenging signal processing technology for biometric identification. IEEE Signal Process. Mag. **22**(6), 78–90 (2005)
3. Derlatka, M.: Modified kNN algorithm for improved recognition accuracy of biometrics system based on gait. In: Saeed, K., Chaki, R., Cortesi, A., Wierzchoń, S. (eds.) CISIM 2013. LNCS, vol. 8104, pp. 59–66. Springer, Heidelberg (2013)
4. Gafurov, D., Bours, P., Snekkenes, E.: User authentication based on foot motion. SIViP **5**(4), 457–467 (2011)
5. Gudavalli, M., Babu, A.V, Raju, S.V, Kumar, D.S.: Multimodal biometrics-sources, architecture and fusion techniques: an overview. In: IEEE International Symposium on Biometrics and Security Technologies (ISBAST), pp. 27–34 (2012)

6. Goodwin, P.: The holt-winters approach to exponential smoothing: 50 years old and going strong. In: FORESIGHT Fall, p. 3034 (2010)

7. Lee, B., Hong, S., Lee, H., Kim, E.: Gait recognition system using decision-level fusion. In: 2010 5th IEEE Conference on Industrial Electronics and Applications (ICIEA), pp. 313–316 (2010)

8. Lee, L., Grimson, W.E.L.: Gait analysis for recognition and classification. In: Proceedings of the Fifth IEEE International Conference on Automatic Face and Gesture Recognition, pp. 148–155 (2002)

9. Lu, J., Zhang, E.: Gait recognition for human identification based on ICA and fuzzy SVM through multiple views fusion. Pattern Recognit. Lett. **28**(16), 2401–2411 (2007)

10. Milovanovic, M., Minovic, M., Starcevic, D.: Walking in colors: human gait recognition using Kinect and CBIR. IEEE Multimedia **20**(4), 28–36 (2013)

11. Munsell, B.C., Temlyakov, A., Qu, C., Wang, S.: Person identification using full-body motion and anthropometric biometrics from kinect videos. In: Fusiello, A., Murino, V., Cucchiara, R. (eds.) ECCV 2012 Ws/Demos, Part III. LNCS, vol. 7585, pp. 91–100. Springer, Heidelberg (2012)

12. Poh, N., Ross, A., Lee, W., Kittler, J.: A user-specific and selective multimodal biometric fusion strategy by ranking subjects. Pattern Recognit. **46**(12), 3341–3357 (2013)

13. Preis, J., Kessel, M., Werner, M., Linnhoff-Popien, C.: Gait recognition with kinect. In: 1st International Workshop on Kinect in Pervasive Computing (2012)

14. Proenca, H., Alexandre, L.A.: Toward covert iris biometric recognition: experimental results from the nice contests. IEEE Trans. Inf. Forensics Secur. **7**(2), 798–808 (2012)

15. Sharma, S., Shukla, A., Tiwari, R., Singh, V.: View variations effect in gait recognition and performance improvement using fusion. In: 1st IEEE International Conference on Recent Advances in Information Technology (RAIT), pp. 892–896 (2012)

16. Shum, H.P.H., Ho, E.S.L., Jiang, Y., Takagi, S.: Real-time posture reconstruction for microsoft kinect. IEEE Trans. Cybern. **43**(5), 1357–1369 (2013)

17. Sinha, A., Chakravarty, K., Bhowmick, B.: Person identification using skeleton information from Kinect. In: The Sixth International Conference on Advances in Computer-Human Interactions ACHI 2013, Nice, France, pp. 101–108 (2013)

18. Veres, G.V., Nixon, M.S., Middleton, L., Carter, J.N.: Fusion of dynamic and static features for gait recognition over time. In: 2005 8th IEEE International Conference on Information Fusion, vol. 2, p. 7 (2005)

19. Wang, L., Ning, H., Tan, T., Hu, W.: Fusion of static and dynamic body biometrics for gait recognition. IEEE Trans. Circuits Syst. Video Technol. **14**(2), 149–158 (2004)

20. Wang, X., Mueen, A., Ding, H., Trajcevski, G., Scheuermann, P., Keogh, E.: Experimental comparison of representation methods and distance measures for time series data. Data Min. Knowl. Disc. **26**(2), 275–309 (2013)

21. Yao, Y.F., Jing, X.Y., Wong, H.S.: Face and palmprint feature level fusion for single sample biometrics recognition. Neurocomputing **70**(7), 1582–1586 (2007)

22. Yao, Z.M, Zhou, X., Lin, E.D, Xu, S., Sun, Y.N.: A novel biometric recognition system based on ground reaction force measurements of continuous gait. In: 2010 3rd IEEE Conference on Human System Interactions (HSI), pp. 452–458 (2010)

Web Search Results Clustering Using Frequent Termset Mining

Marek Kozlowski[(⊠)]

Warsaw University of Technology,
Warsaw, Poland
m.kozlowski@ii.pw.edu.pl
http://www.ii.pw.edu.pl

Abstract. We present a novel method for clustering web search results based on frequent termsets mining. First, we acquire the senses of a query by means of a word sense induction method that identify meanings as trees of closed frequent termsets. Then we cluster the search results based on their lexical and semantic intersection with induced senses. We show that our approach is better or comparable with state-of-the-art classical search result clustering methods in terms of both clustering quality and degree of diversification.

1 Introduction

The exponential growth of the Internet community brought to the production of a vast amount of unstructured data, such as web pages, blogs, social media etc. Such mass of information is unlikely to be analyzed by humans, so there is a strong drive to develop automatic methods capable to retrieve knowledge from the Web. Traditional techniques for text mining show theirs limit when they are applied to such huge collections of data. Most of currently used approaches are based on lexico-syntactic analysis of text, and they are mainly focused on words occurrences. Two main flaws of the approach are: inability to identify documents using different wordings, lack of context-awareness, which leads to retrieval documents which are not pertinent to the user needs. Knowledge of an actual meaning of a polysemous word can significantly improve the quality of the information retrieval process by means of retrieving more relevant documents or extracting relevant information from texts. However, the current generation of search engines still lacks an effective way to address the issue of lexical ambiguity. In a previous study [1] conducted using WordNet and Wikipedia as sources of ambiguous words it was reported that around 3 % of Web queries and 23 % of the most frequent queries are ambiguous. Many search engines are resolving the ambiguity problem by diversifying their results, so as to return Web pages that are not too similar to each other.

In the previous years, Web clustering engines [10] have been proposed as a solution to the issue of lexical ambiguity in Information Retrieval. These systems group search results, by providing a cluster for each specific topic of the input query. Users navigate through the clusters in order to retrieve the pertinent

M. Kryszkiewicz et al. (Eds.): PReMI 2015, LNCS 9124, pp. 525–534, 2015.
DOI: 10.1007/978-3-319-19941-2_50

results. Most of clustering engines group search results on the basis of their lexical similarity, and therefore suffer from semantic lackness e.g. polysemy (different user needs expressed with the same words).

In this paper we present a novel web search results clustering method, which exploits word sense induction method in order to create sense-oriented clusters of results. The core of algorithm is a frequent termsets mining method identifying senses as trees of closed frequent termsets. Then results are labeled by the most similar senses. Our approach is evaluated on two datasets: AMBIENT and MORESQUE, and is better or comparable with the state-of-the-art search result clustering methods.

2   Related Work

The goal of text clustering in information retrieval is to discover groups of semantically related documents. Contextual descriptions (snippets) of documents returned by a search engine are short, often incomplete, and highly biased toward the query, so establishing a notion of proximity between documents is a challenging task that is called Search Result Clustering (SRC). Given a query, a flat list of text snippets returned from one or more commonly-available search engines is clustered using some notion of textual similarity. At the root of the clustering approach lies van Rijsbergens [12] cluster hypothesis: closely associated documents tend to be relevant to the same requests, whereas documents concerning different meanings of the input query are expected to belong to different clusters.

Approaches to search result clustering can be classified as data-centric or description-centric [10].

The data-centric approach focuses more on the problem of data clustering, rather than presenting the results to the user. Scatter/Gather [13] is an example, which divides the dataset into a small number of clusters and, after the selection of a group, performs clustering again and proceeds iteratively using the Buckshot-fractionation algorithm. Other data-centric methods use hierarchical agglomerative clustering [14] that replaces single terms with lexical affinities (2-grams of words) as features, or exploit link information [15].

Description-centric approaches are more focused on the description that is produced for each cluster of search results. This problem is also called descriptive clustering: discovery of diverse groups of semantically related documents associated with meaningful, comprehensible and compact text labels. Accurate and concise cluster descriptions (labels) let the user search through the collection's content faster and are essential for various browsing interfaces. The task of creating descriptive, sensible cluster labels is difficult - typical text clustering algorithms rely on samples of keywords for describing discovered clusters. Among the most popular and successful approaches are phrase-based, which form clusters based on recurring phrases instead of numerical frequencies of isolated terms. STC algorithm employs frequently recurring phrases as both document similarity feature and final cluster description [8,9]. Clustering in STC is treated as finding groups of documents sharing a high ratio of frequent phrases. KeySRC improved STC approach by adding part-of-speech pruning and dynamic selection of the cut-off level of the clustering dendrogram [5]. A different idea of label-driven clustering appeared in clustering with committees algorithm [16], where strongly associated terms related to unambiguous concepts were evaluated using semantic relationships from WordNet. Description-Comes-First (DCF) approach reverses the traditional order of cluster discovery. Description-Comes-First is a special case of description-centric approach. Instead of calculating proximity between documents and then labeling the discovered groups, DCF first attempts to find good, conceptually varied cluster labels and then assign documents to the labels to form groups. DCF approach was introduced in the work [6,7] as an algorithm called Lingo. The Lingo algorithm combines common phrase discovery and latent semantic indexing techniques to separate search results into meaningful groups. Lingo uses singular value decomposition of the term-document matrix to select good cluster labels among candidates extracted from the text (frequent phrases). The algorithm was designed to cluster results from Web search engines (short snippets and fragmented descriptions of original documents) and proved to provide diverse meaningful cluster labels. Lingo bridges existing phrase-based methods with numerical cluster analysis to form readable and diverse cluster descriptions.

Phrase-based methods usually provide good results. They report some problems, when one topic is dominating, or the search results contain different words referring to one meaning. Navigli and Crisafulli, Di Marco [2–4] present a novel approach to Web search result clustering, based on the automatic discovery of word senses from raw text. The proposed method clusters the web search results based on their semantic similarity to the induced query senses. Those kind of methods, called semantic search result clustering systems, need huge external corpora in order to build relevant sense representations.

Diversification of results is another research topic dealing with the query ambiguity and retrieving semantically different documents in the top results. The first diversification algorithms use the similarity functions to estimate the diversity among documents and between query and document [20]. Modern search engines return relatively unsatisfactory results, as they consider the query coverage by each page individually, not a set of pages as a whole. This problem is resolved by Essential Pages [11], which selects fundamental pages, which maximize the amount of information covered for a given query.

In our work we perform word sense induction to dynamically acquire an inventory of senses of the input query. We do not use any external corpora, the sense induction is performed only on the search results. The search results are distributed among matching senses. Finally we also use some diversification techniques in order to rerank clusters and theirs content.

3 Approach

In order to perform comparisons we have built a dedicated WSI-based web search result clustering method called SnSRC. SnSRC consists of the following four steps: (1) preprocessing of results (transforming into a bag of words), (2) word sense induction method, (3) clustering of web search results, and (4) cluster sorting.

In the first step for a given query the interesting documents (snippets) are retrieved, and then iteratively processed into bag-of-words representation. In order to obtain a set of all pertinent documents SnSRC uses full-text search index. Then the documents are processed as follows: sentence segmentation, word tokenization, stop-words cleaning, PoS tagging tokens, filtering tokens by wikipedia proper names and PoS rules (e.g.: noun-phrases), finally tokens are lemmatized. Steps 2, 3 and 4 are described in details in the next sections, respectively.

3.1   SnS - Word Sense Induction Method

This section is devoted to word sense discovery problem, namely word sense induction (WSI). WSI is an approach proposed for discovering senses of words automatically from unannotated text. It could be perceived as a clustering problem, where words are grouped into non-disjoint clusters according to their meanings. Each cluster can be considered as a separate sense of the word [17].

In the papers [18,19] there was proposed the method called SenseSearcher (SnS). It is a word sense induction algorithm based on closed frequent sets and multi-level sense representation. SnS is a knowledge-poor approach, which means it does not need any kind of structured knowledge base about senses as well as the algorithms that have embedded deep language knowledge. Senses induced by SnS characterize better readability (are more intuitive), mainly because SnS discovers a hierarchy of senses showing important relationships between them. In other words the proposed method creates structure of senses, where coarse-grained senses contain related sub-senses (fine-grained senses), rather than flat list of concepts.

The SnS algorithm consists of five phases, which we present below. The pseudocode of the whole algorithm is presented in [18].

In Phase I, we build the index (e.g.: full-text search index) for the corpus.

In Phase II, with a given term we run a query on the index, and find paragraphs related to the term. Then we convert them into context representations (bag of words).

In Phase III contextual patterns are discovered from contexts generated in Phase II. The patterns are closed frequent termsets in the context space. The contexts are treated as transactions (itemsets are replaced by termsets) and the process of mining closed frequent termsets is performed with the use of the CHARM algorithm [21].

Phase IV is devoted to forming contextual patterns into sense frames, building a hierarchical structure of senses. In some exceptional states few sense frames may refer to one sense, it may result from the corpus limitations (lack of representativeness and high synonymity against descriptive terms).

In Phase V, sense frames are clustered. The clusters of sense frames are called senses. Optionally senses can be labelled with some descriptive terms.

To illustrate the SnS work, let us start with the term *apple*, and show how its meanings can be retrieved from the randomly chosen five hundred English Wikipedia paragraphs containing word *apple*. The main senses discovered by SnS are S = (computer(99), music(35), country(31), fruit(30), golden apple(18), film(17), book(13)), where the numbers in parentheses indicate the support of the corresponding senses (number of paragraphs with the given meaning).

3.2   Clustering of Web Search Results

The clustering is performed in two phases: first, simultaneously during sense induction, and then after sense discovery for those results that remained not grouped. The first phase is based on the process of frequent termset mining. Discovered closed frequent termsets have support, and list of results, in which they appear. Senses are grouped sense frames. Each sense frame has the main contextual pattern, so according to sense frames the snippets containing the main pattern are grouped in the corresponding result cluster. Summarizing, for each sense (having sense frames) a corresponding cluster of snippets is constructed. Let us note that after this phase is completed, there may remain snippets which are non-grouped.

In the second phase, non-grouped snippets are tested iteratively against each of the induced sense. Now, the clustering of remaining snippets consists in using the bag-of-words representation and a similarity measure between the snippets representations $b_i \in B$ and sense clusters $\{S_1, ..., S_m\}$. Given a snippet r_i, the sense cluster closest to its bag-of-words representation b_i will be selected as the most likely meaning:

$$Sense(r_i) = \begin{cases} \arg \max_{j=1,...,m} sim(b_i, S_j) & \text{if } sim(b_i, S_j) > 0 \\ m + 1 \, (Group \ Others) & \text{otherwise} \end{cases}$$

where $sim(b_i, S_j)$ is a generic similarity value between b_i and S_j (the cluster *Others* denotes that none of the induced senses is assigned to the result snippet). As a result, we obtain a clustering of results R into $\mathcal{C}$, $\mathcal{C} = (C_1, ..., C_m, C_{m+1})$ such that:

$$C_j = \{r_i \in R : Sense(r_i) = j\}$$

i.e. C_j contains the search results classified with the j-th sense of the query q. In our system two different similarity measures were implemented and tested:

1. Bag-of-Words similarity - it calculates the size of intersection between the bag-of-words of r_i and $bow(S_j)$, where $bow(S_j)$ denotes the bag of words, obtained as the union of all content words of contextual patterns building sense frames clustered in S_j:

$$bowSim(b_i, S_j) = |b_i \cap bow(S_j)|$$

2. Wikipedia similarity - it calculates number of words within bag-of-words of r_i that are none-discriminants with some of the words within $bow(S_j)$. The discrimination is evaluated by the formula:

$$discMes(q, w, w') = \frac{c(q, w, w')}{min(c(q, w), c(q, w'))}$$

where q is a query and $c(q, w, w')$ is the number of co-occurrences of the words q and w, w' in Wikipedia paragraphs. Finally, having δ as a threshold, wikipedia similarity is denoted as:

$$wikiSim(b_i, S_j) = |\{w \in b_i \mid \exists w' \in bow(S_j)$$

$$discMes(q, w, w') > \delta\}|$$

In the experiments above measures are used together, it means that $sim(b_i, S_j)$ is a sum of $bowSim$ and $wikiSim$.

3.3   Cluster Sorting

Not all snippets will have the same degree of relevance for the cluster, and additionally the constructed clusters will show a different quality of sense coherence. Diversification evaluation demands from the SnSRC method to sort results (snippets) within the clusters, and sort the clusters themselves. Within each cluster the snippets are sorted using similarity measured in step 3 ($sim(b_i, S_j)$). Clusters are sorted by the supports of sense frame seeds. Each sense is a group of sense frames, but one sense frame is a seed (the sense frame with the highest support among the frames within the same sense). The senses are ranked according to seed's support (the support of the main contextual pattern of the sense frame), which leads to sorted clusters of results. The order of clusters reflects quality depending on the relevance of the search results therein. The order of results within a cluster reflects degree of relevance for the cluster.

4   Experiments

4.1   Experimental Setup

Test Sets. We conducted our experiments on the data sets: AMBIENT, and MORESQUE. Below some details about those datasets are provided:

1. AMBIENT (AMBIguous ENTries[1]) consists of 44 topics, each with a set of subtopics and a list of 100 ranked documents [23].
2. MORESQUE (MORE Sense-tagged QUEries[2]) consists of 114 topics, each with a set of subtopics and a list of 100 top-ranking documents. The dataset has been developed [4] as a complement for AMBIENT.

Reference Systems. Nonsemantic systems are web clustering engines, which are not WSI-based and working only on search results. Those methods do not

---

[1] http://credo.fub.it/ambient/.
[2] http://lcl.uniroma1.it/moresque/.

exploit external corpus in order to overcome lack of data. Semantic SRC algorithms in order to calculate the co-occurrence graph between words need a large corpus to extract co-occurrence counts and calculate the Dice values. We have to remember that semantic methods induce senses from the co-occurrence graph of terms, which is constructed from words acquired from the external textual resources (as GoogleWeb1T [3]). It may lead to discover dominating senses, which have nothing in common with senses covered by search results (snippets). SnSRC is WSI-based, but is working only on search results (not using any external corpora), which makes it similar to the nonsemantic systems, and the fair comparison can be done only against them.

SnSRC was compared with such nonsemantic methods: (1) Lingo [6,7], (2) Suffix Tree Clustering [8,9] and (3) KeySRC (a state-of-the-art search results clustering algorithm) [5].

Baselines. We compared SnSRC and nonsemantic SRC systems against five baselines: (1) Singletons (each snippet is clustered as a separate singleton), (2) All-in-one (all the snippets are clustered into a single cluster), (3) Wikipedia (snippet is added to the cluster corresponding to the best-matching Wikipedia page representing a meaning), and two diversification baselines, namely (4) Essential Pages (EP) [11] and (5) Yahoo! (the original search results returned by the Yahoo! search engine).

4.2 Scoring

Following [3,22], the systems were evaluated in terms of the clustering quality and the diversification quality. Clustering evaluation is difficult issue. Many evaluation measures have been proposed in the literature so, in order to get exhaustive results, we calculated four distinct measures, namely: Rand Index (RI), Adjusted Rand Index (ARI), Jaccard Index (JI) and F1 measure. Diversification is a technique aimed at reranking search results on the basis of criteria that maximize their diversity. Quantifying the impact of web search result clustering methods on flat-list search engines is measured by *S-recall@K* (sense recall at rank K) and *S-precision@r* (Sense precision at recall r). The above mentioned measures are described in detail in [3,22].

4.3 Results

We conducted experiments with the SnSRC on the union of two tests sets AMBIENT and MORESQUE. We compared SnSRC against nonsemantic systems (i.e., Lingo, STC, and KeySRC) and three baselines (i.e., all-in-one, singleton, and Wikipedia).

In Tables 1 and 2 there were reported all quality measures acquired during evaluations on AMBIENT+MORESQUE data set according to results from [3,4]. An initial finding here is that the SnSRC and KeySRC report very good clustering quality measures. We note that although KeySRC outperforms the other SRC descriptive-centric algorithms in terms of RI, ARI and F1, it attains very low JI results. The same characteristics is connected with the SnSRC

Table 1. A comparison between different search result clustering approaches (percentages) on AMBIENT+MORESQUE data set

Algorithm	ARI	JI	F1	cl
Lingo	−0.53	36.36	16.73	2.0
STC	−7.90	38.23	14.96	2.0
KeySRC	14.34	27.77	63.11	18.5
SnSRC	14.04	22.74	60.16	14.7
All-in-one	0.00	47.12	42.40	1.0
Singleton	0.00	0.00	68.17	100.0
Wikipedia	13.83	56.02	14.33	5.7

Table 2. The comparison of Rand index results (percentages) on AMBIENT+MORESQUE data set

Algorithm	RI
Lingo	55.49
STC	54.29
KeySRC	58.78
SnSRC	58.53

method. SnSRC reports similar measures to KeySRC, especially comparing them with results of Lingo and STC. According to baselines, the singleton baseline produces trivial, meaningless clusterings, as measured by ARI and JI. The all-in-one baseline obtains non-zero JI, its F1 is lower than singleton, because of its lower recall. The Wikipedia baseline fares well compared with the other baselines in terms of ARI and JI, but achieves lower F1, again because of low recall. Finally, KeySRC and SnSRC consistently outperform the other SRC systems in terms of RI, ARI and F1.

Table 3. The results for S-recall@K and S-precision@r on AMBIENT+MORESQUE data set (in %)

Algorithm	S-recall				S-precision			
	K = 5	K = 10	K = 15	K = 20	r = 60	r = 70	r = 80	r = 90
Lingo	-	-	-	-	-	-	-	-
STC	-	-	-	-	-	-	-	-
KeySRC	48.7	61.7	68.2	72.5	22.3	17.7	15.4	12.0
SnSRC	46.7	60.8	69.3	74.2	25.1	18.5	14.1	11.9
EP	44.6	59.2	67.9	73.3	24.9	18.9	16.1	13.2
Yahoo!	51.4	63.4	69.1	73.3	25.7	18.7	15.5	12.6

We performed a second experiment aimed at quantifying the impact of our Web search result clustering systems on flat-list search engines (clustering diversity evaluation). Lingo and STC do not return ranked lists of results and, therefore, they can not be evaluated. The results in terms of S-recall@K and S-precision@r are shown in Table 3. KeySRC and EP are surpassed by the Yahoo! method in the whole range of parameter K. SnSRC report better results than KeySRC, Yahoo! when $K \geq 15$. Regarding S-precision@r, baselines (Yahoo, EP) report better results than KeySRC in the whole range of r. SnSRC scores better than KeySRC when $r < 80$.

5   Conclusions

We have presented a new approach to Web search result clustering. Key to our approach is the idea of inducing senses for the target query automatically by means of a simple, yet effective algorithm based on the frequent termset mining. In this paper we exploit a novel WSI knowledge-poor algorithm SnS, based on text mining approaches, namely closed frequent termsets. It converts simple contexts (based on bag-of-words paragraph representation) into relevant contextual patterns (much more concise, and representative). Using significant patterns SnS build hierarchical structures called sense frames. Discovered sense frames usually are independent senses, but sometimes (e.g. because of too small corpus) can point the same sense. Finally using clustering methods sense frames are grouped in order to find similar ones referring to the same main sense. The results of a Web search engine are then mapped to the query senses and clustered accordingly. The last step is to rerank results within clusters, and sort clusters.

An extensive set of experiments, according to AMBIENT and MORESQUE datasets, confirms that SnSRC provides significant improvements over existing methods like STC and Lingo, and is comparable with the state-of-the-art KeySRC approach. Using a broad scope of measures we have shown experimentally that SnS can be efficiently used as a fundamental step in building clustering engine for end-users.

References

1. Sanderson, M.: Ambiguous queries: test collections need more sense. In: Proceedings of SIGIR, pp. 499–506. ACM, New York (2008)
2. Di Marco, A., Navigli, R.: Clustering web search results with maximum spanning trees. In: Proceedings of the 12th Congress of the Italian Association for Artificial Intelligence, Palermo, pp. 201–212 (2011)
3. Di Marco, A., Navigli, R.: Clustering and diversifying web search results with graph-based word sense induction. Comput. Linguist. **39**(3), 709–754 (2013). MIT Press
4. Navigli, R., Crisafulli, G.: Inducing word senses to improve web search result clustering. In: Proceedings of the 2010 Conference on Empirical Methods in Natural Language Processing, Boston, pp. 116–126 (2010)

5. Bernardini, A., Carpineto, C., DAmico, M.: Full-subtopic retrieval with keyphrase-based search results clustering. In: Proceedings of 2009 IEEE/WIC/ACM International Conference on Web Intelligence, Milan, pp. 206–213 (2009)

6. Osinski, S., Weiss, D.: A concept-driven algorithm for clustering search results. IEEE Intell. Syst. **20**(3), 48–54 (2005). IEEE Press

7. Osinski, S., Stefanowski, J., Weiss, D.: Lingo: search results clustering algorithm based on singular value decomposition. In: Proceedings of the International IIS: IIPWM 2004 Conference held in Zakopane, Zakopane, pp. 359–368 (2004)

8. Zamir, O., Etzioni, O.: Web document clustering: a feasibility demonstration. In: Proceedings of the 21st Annual International ACM SIGIR Conference on Research and Development in Information Retrieval, New York, pp. 46–54 (1998)

9. Zamir, O., Etzioni, O.: Grouper: a dynamic clustering interface to web search results. Comput. Netw. **31**(11–12), 1361–1374 (1999). Elsevier

10. Carpineto, C., Osinski, S., Romano, G., Weiss, D.: A survey of web clustering engines. ACM Comput. Surv. **41**(3), 1–38 (2009). ACM, New York

11. Swaminathan, A., Cherian, M., Kirovski, D.: Essential pages. In: Proceedings of the 2009 IEEE/WIC/ACM International Joint Conference on Web Intelligence and Intelligent Agent Technology, Milan, pp. 173–182 (2009)

12. Van Rijsbergen, C.: Information Retrieval. Butterworths, London (1979)

13. Cutting, D., Karger, D., Pedersen, J., Tukey, J.: Scatter/gather: a cluster based approach to browsing large document collections. In: Proceedings of SIGIR, Copenhagen, pp. 318–329 (1992)

14. Maarek, I., Fagin, R., Pelleg, D.: Ephemeral document clustering for web applications. IBM Research Report RJ 10186 (2000)

15. Zhang, X., Hu, X., Zhou, X.: A comparative evaluation of different link types on enhancing document clustering. In: Proceedings of SIGIR, Singapore, pp. 555–562 (2008)

16. Pantel, P., Lin, D.: Discovering word senses from text. In: Proceedings of the 8th KDD, Edmonton, pp. 613–619 (2002)

17. Denkowski, M.: A survey of techniques for unsupervised word sense induction. Technical report, Language and Statistics II Literature Review (2009)

18. Kozłowski, M., Rybiński, H.: SnS: a novel word sense induction method. In: Kryszkiewicz, M., Cornelis, C., Ciucci, D., Medina-Moreno, J., Motoda, H., Raś, Z.W. (eds.) RSEISP 2014. LNCS, vol. 8537, pp. 258–268. Springer, Heidelberg (2014)

19. Kozlowski, M.: Word sense discovery using frequent termsets. PhD in Warsaw University of Technology (2014)

20. Carbonell, J., Goldstein, J.: The use of MMR, diversity-based reranking for reordering documents and producing summaries. In: Proceedings of SIGIR, Melbourne, pp. 335–336 (1998)

21. Zaki, M., Hsiao, C.: CHARM: an efficient algorithm for closed itemset mining. In: Proceedings 2002 SIAM International Conference on Data Mining, Arlington, pp. 457–472 (2002)

22. Navigli, R., Vannella, D.: SemEval-2013 task 11: word sense induction and disambiguation within an end-user applications. In: Proceedings of 7th International Workshop on Semantic Evaluation (SemEval), in the Second Joint Conference on Lexical and Computational Semantics, Atlanta, pp. 193–201 (2013)

23. Carpineto, C., Romano, G.: AMBIENT dataset. http://credo.fub.it/ambient

Effective Imbalanced Classification of Breast Thermogram Features

Bartosz Krawczyk[1] and Gerald Schaefer[2(⊠)]

[1] Department of Systems and Computer Networks,
Wroclaw University of Technology, Wrocław, Poland
[2] Department of Computer Science, Loughborough University,
Loughborough, UK
gerald.schaefer@ieee.org

Abstract. Breast cancer is the most commonly occurring form of cancer in women, and can be diagnosed using various imaging modalities including thermography. In this paper, we present an approach to analysing breast thermograms based on statistical image features and an effective ensemble method for imbalanced classification problems. We extract a series of features from the images to arrive at indications of asymmetry between left and right breast regions. These then form the input to a classification stage for which we develop a dedicated multiple classifier system that employs neural networks or support vector machines as base classifiers, trains base classifiers on balanced subsets of the training data to address the class imbalance that is typically inherent in medical decision making problems, and fuses the decisions using a neural network combined with a fuzzy diversity measure to remove individual classifiers from the ensemble and to enhance prediction performance. Experimental results, on a large dataset of about 150 breast thermograms, confirm our approach to provide excellent classification performance and to outperform other classifier ensembles designed for imbalanced datasets.

Keywords: Breast cancer · Thermography · Pattern classification · Imbalanced classification · Multiple classifier system

1 Introduction

Thermography uses a camera with sensitivities in the thermal infrared to capture the temperature distribution of the human body or parts thereof. In contrast to other modalities such as mammography, it is a non-invasive, non-contact, passive and radiation-free technique. It is well known that the radiance from human skin is an exponential function of the surface temperature which in turn is influenced by the level of blood perfusion in the skin. Thermal imaging is hence well suited to pick up changes in blood perfusion which might occur due to inflammation, angiogenesis or other causes [14]. Thermography has also been shown to be well suited for the task of detecting breast cancer [3,13]. Here, thermography has advantages in particular when the tumor is in its early stages or in dense tissue. Early detection is crucial as it provides significantly higher chances of

© Springer International Publishing Switzerland 2015
M. Kryszkiewicz et al. (Eds.): PReMI 2015, LNCS 9124, pp. 535–544, 2015.
DOI: 10.1007/978-3-319-19941-2_51

survival [12] and in this respect infrared imaging can outperform the standard method of mammography. While mammography can detect tumors only once they exceed a certain size, even small tumors can be identified using thermal infrared imaging due to the high metabolic activity of cancer cells which leads to an increase in local temperature that can be picked up in the infrared [16].

In our approach, which is a continuation of the work presented in [18], we therefore derive a set of image features that describe possible asymmetries between the bilateral breast regions to capture this effect. These features are then used in a pattern classification stage for which we develop a multiple classifier system (MCS). In particular, we employ neural networks or support vector machines as base classifiers, and, importantly, address the problem of class imbalance, that often occurs in medical data analysis, by training the individual classifiers on balanced data subsets, thus eliminating any unfavourable class distribution. The base classifiers are then combined using an fuser implemented as a one-layer perceptron neural network. Finally, we remove redundant classifiers through an ensemble diversity measure based on fuzziness using an energy approach. Experimental results, on a dataset of about 150 breast thermograms, confirm that our proposed approach works well and gives excellent classification performance. We furthermore show it to statistically outperform not only canonical classifiers but also recent classifier ensembles that are also dedicated to imbalanced classification.

2   Background

Several computer aided diagnostic (CAD) approaches to analysing breast thermograms have been presented in the literature. In [23], an attempt based on asymmetry analysis is presented where, following segmentation based on edge detection and the Hough transform, Bezier histograms are generated and compared to identify cancer cases. In [25], some basic statistical features are extracted and passed to a complementary learning fuzzy neural network (CLFNN) for diagnosis. Reference [26] proposes morphological analysis of "localised temperature increase" amplitudes in thermograms to detect tumors. A series of image features from the breast regions (the same features that we employ in this paper) are extracted in [24] and subsequently analysed by a fuzzy classification method, while [31] uses the same feature set in conjunction with a neural network classifier. The approach in [6] is based on transforming the thermogram into a representation derived from independent component analysis, thresholding and correlating the obtained channels to locate tumor areas. In [1], texture features and support vector machine classifiers are employed, while in [22] wavelet and texture descriptors are used in combination with several classification algorithms.

3   Image (A)symmetry Features

As has been shown, an effective approach to detect breast cancer based on thermograms is to study the symmetry between the left and right breast regions [23].

In the case of cancer presence, the tumor will recruit blood vessels resulting in hot spots and a change in vascular pattern, and hence an asymmetry between the temperature distributions of the two breasts. On the other hand, symmetry typically identifies healthy subjects.

We follow this approach and extract image features that describe bilateral differences between the areas of the left and right breasts extracted from frontal view thermograms. We employ the same image features that were used in [24] (for a more extensive discussion of them, see there), namely:

- Basic statistical features: mean, standard deviation, median, 90-percentile;
- Moment features: centre of gravity, distance between moment centre and geometrical centre;
- Histogram features: cross-correlation between histograms; maximum, number of non-empty bins, number of zero-crossings, energy and difference of positive and negative parts of difference histogram;
- Cross co-occurrence matrix [31] features: homogeneity, energy, contrast, symmetry and the first 4 moments of the matrix;
- Mutual information between the two temperature distributions;
- Fourier spectrum features: the difference maximum and distance of this maximum from the centre.

Each breast thermogram is thus described by 4 basic statistical features, 4 moment features, 8 histogram features, 8 cross co-occurrence features, mutual information and 2 Fourier descriptors. We further apply a Laplacian filter to enhance the contrast and calculate another subset of features (the 8 cross co-occurrence features together with mutual information and the 2 Fourier descriptors) from the resulting images, and consequently end up with a total of 38 features which describe the asymmetry between the two sides and which form the basis for the following pattern classification stage.

4   Imbalanced Pattern Classification Ensemble

In our approach, we employ an ensemble classifier, i.e. perform classification not based on a single algorithm but based on a joint decision of a committee of classifiers [20]. This way, we are able to exploit the strengths of different base classifiers while eliminating their weaknesses, thus leading to more robust and typically better classification performance.

Given a pool of N classifiers $\Psi^{(1)}, \Psi^{(2)}, \ldots, \Psi^{(N)}$, for a given feature vector x, each of the individual classifiers makes a decision with respect to class $i \in \mathbf{M} = \{1, \ldots, M\}$. The classifier ensemble $\bar{\Psi}$ then makes a combined decision based on

$$\bar{\Psi}(x) = i \quad \text{if} \quad \hat{F}(i, x) = \max_{k \in M} \hat{F}(k, x), \tag{1}$$

where

$$\hat{F}(i, x) = \sum_{l=1}^{N} w^{(l)} F^{(l)}(i, x) \quad \text{and} \quad \sum_{l=1}^{N} w^{(l)} = 1, \tag{2}$$

and $F^{(l)}(i, x)$ is a discriminant function for the i-th class and object x used by the l-th classifier. The weights $w^{(l)}$ here play a crucial role for the performance of the ensemble and can be assigned either statically and through training.

In the following, we detail the components of our classifier ensemble.

4.1 Base Classifiers

While in principle any classification approach can serve as base classifier, in our approach we build ensembles of neural network (NN) or support vector machine (SVM) classifiers. For the NN classifier [4], we use the Quickprop algorithm for training, and set the number of hidden neurons to half the sum of input and output neurons. For the SVM classifier [27], we employ a Gaussian RBF kernel, and perform classifier tuning [15] to obtain optimal parameters.

4.2 Imbalanced Classification

In medical diagnosis, there are typically far fewer malignant cases than there are benign ones, and consequently conventional classification approaches often suffer from low sensitivity due to the skewed class distribution. Class imbalance can be addressed in several ways, including oversampling of the minority class [9] and cost-sensitive classification [19].

In our approach, we revert to neither of these, but employ our earlier method from [18] which is based on the principle of object space partitioning to train individual classifiers on balanced subsets of the training data.

In particular, we create a number of subspaces using a random undersampling method. Each of the subspaces contains a smaller number of objects, randomly drawn from the dataset, so that the number of objects from each of the classes are equal. Objects of the majority class are randomly sampled and removed from the training set. Subspaces are then created as long as there are objects in the majority set. Each subspace forms the basis of one of the classifiers; that is, each base classifier is trained on a different (balanced) training subset, hence leading to a heterogeneous ensemble that addresses class imbalance.

To boost recognition performance, a feature selection step is performed. For this purpose, we utilise the fast correlation-based feature filter (FCBF) [30]. In FCBF, the relations between features-classes and between pairs of features are considered. The algorithm proceeds at two levels. First, a ranking algorithm using the symmetric uncertainty coefficient index is employed to estimate class-feature relevance, and a threshold established to select predominant features. In the second part, features redundant to the predominant features are removed. Since feature selection is applied separately for each subspace, and hence each classifier, this step also enhances the heterogeneity of the ensemble.

4.3 Ensemble Diversity

Different base classifiers will have different areas of competence and hence may provide different contributions to the committee. Careful classifier selection should be hence conducted in order to choose the most valuable individual models. Therefore, in this paper and in contrast to [18], we employ a classifier ensemble diversity measure for this purpose. For this purpose, we extend the energy-based fuzzy diversity measure introduced in [17] for one-class classification problems to multi-class classification.

The proposed energy approach provides an effective measure of fuzziness. It uses a threshold $\lambda \in [0,1]$ whose role is to filter insignificant degrees of membership, that may otherwise lead to lowering the stability of the measure. Given N base classifiers in the pool, out of which S correctly classify a given training object x_j, one can define a fuzzy membership function $\mu_{x_j} = \frac{S}{N}$ for the given object, with $0 \leq \mu_{x_j} \leq 1$.

Based on this, the employed energy measure is calculated as

$$DIV = \int_X \sum_{i=1}^{N} f_\lambda(x) dx, \tag{3}$$

where

$$f_\lambda(x) = f(x) \Leftrightarrow \frac{\sum_{k=1}^{N} \delta(\Psi_{i_k}^M(x), \Psi^*(x))}{N} > \lambda, \tag{4}$$

and $\Psi^*(x)$ denotes a classifier correctly classifying object x, and $f(x) : [0,1] \to R_+$ is an increasing function in interval $[0,1]$ for $f(0) = 0$.

The derived measure gives an indication of the diversity of the entire classifier committee in the range $[0,1]$, where 0 corresponds to an ensemble of identical classifiers and 1 to the highest possible diversity respectively.

We perform diversity-based classifier selection through an exhaustive search over all possible combinations of committee members, and selecting the ensemble that yields maximal diversity.

4.4 Classifier Fusion

Classifier fusion is an important aspect of classifier ensembles, and the choice of fusion method, which is responsible for the collective decision making process by determining the weights in Eq. (2), is hence crucial. Instead of traditional approaches such as majority voting or static weight assignment, in this paper we utilise a dynamic approach to combine the outputs of base classifiers created on different object subspaces. In particular, we employ a trained fuser which, although taking longer to achieve its final performance, leads to an increase of the overall classification accuracy [11].

In a training process, the fuser needs to identify $W = \{W_1, W_2, \ldots, W_N\}$ where $W_l = \left[w^{(l)}(1), w^{(l)}(2), \ldots, w^{(l)}(M)\right]^T$ comprises the weights assigned to each classifier and each of the M classes.

The aim is to find a fuser which assures the lowest misclassification rate of $\bar{\Psi}$ for which we employ a neural network with a canonical learning approach [29] as illustrated in Fig. 1. One perceptron fuser is constructed for each of the classes under consideration, and may be trained with any standard procedure used in neural network learning. The input weights established during the learning process are then the weights assigned to each of the base classifiers.

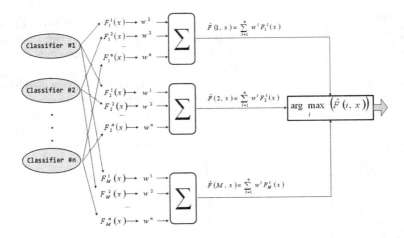

Fig. 1. Classifier fuser implemented as a one-layer neural network

5   Experimental Results

In our experiments, we use a dataset of 146 thermograms of which 29 cases have been confirmed as malignant whereas the other 117 cases were benign. The employed dataset is the same that was used in earlier work [18,24,31]. For all thermograms, the 38 features from Sect. 3 are extracted and serve as input for classification.

For our proposed classification approach, each subspace is designed so as to contain all objects from the minority class and an equal number of samples from the majority class, leading to a pool of 7–9 base classifiers (depending on the fold of CV). To observe the influence of the removal of redundant classifiers, we run our experiments with and without employing the diversity-based classifier selection. The pruned ensembles consist of 4–6 individual classifiers (again, depending on the fold of CV).

In order to put the obtained results into context, we also perform classification using several state-of-the-art ensembles dedicated to imbalanced classification, namely SMOTEBagging [28], SMOTEBoost [9], IIvotes [5] and EasyEnsemble [21], all with support vector machines (with a Gaussian RBF kernel and classifier tuning, as in our approach) as base classifiers. Furthermore, we run the experiments using several canonical classifiers, namely a single SVM [27], bagged SVM [7], boosted SVM [10], and Random Forest [8].

Classification results, based on 5×2 cross validation, are presented in Table 1 where, for each classifier and classifier ensemble, we report sensitivity (i.e. probability that a case identified as malignant is indeed malignant), specificity (i.e. probability that a case identified as benign is indeed benign) and overall classification accuracy (i.e. percentage of correctly classified patterns). We also perform a combined 5×2 CV F test of statistical significance (on sensitivity) [2], and report its results in Table 2.

Table 1. Classification results for all classifiers

Classifier	Sensitivity	Specificity	Accuracy
Single SVM [27]	8.34	86.32	71.23
Bagged SVM [7]	12.68	94.01	79.45
Boosted SVM [10]	19.58	98.00	85.61
Random Forest [8]	22.58	98.29	84.24
SMOTEBagging [28]	77.35	90.50	87.89
SMOTEBoost [9]	79.03	91.00	88.62
IIvotes [5]	79.56	91.89	89.44
EasyEnsemble [21]	80.02	91.00	88.22
Hybrid Ensemble (NN)	78.85	90.82	88.43
Hybrid Ensemble (SVM)	79.85	91.08	88.78
Hybrid Ensemble (NN) + DIV	80.74	90.52	88.56
Hybrid Ensemble (SVM) + DIV	81.96	90.80	89.03

From Table 1, we can see that canonical classification approaches are not able to cope well with the dataset due to the inherent class imbalance, and consequently provide rather poor sensitivity.

The implemented ensembles SMOTEBagging, SMOTEBoost, IIvotes and EasyEnsemble are all specifically designed to address class imbalance in the context of a multiple classifier system. SMOTEBagging and SMOTEBoost do this through oversampling approaches, while IIvotes is also based on the bagging combination idea, but integrates the SPIDER data preprocessing technique with the Ivotes approach. EasyEnsemble carries out a double ensemble learning procedure, combining bagging and boosting. From the results in Table 1, it is apparent that these methods lead to a significant boost in terms of sensitivity. Table 2 shows that all ensembles statistically outperform all single classifiers, while EasyEnsemble gives the best sensitivity.

Now looking at the results of our proposed hybrid ensemble approach, we can notice that it leads to a clear further improvement still. Using the employed subspacing method means that we do not need to create artificial objects or define cost matrices. In the former new artificial objects may be introduced on the basis of already created artificial samples, while for the latter a cost matrix needs to be defined which is often difficult and requires detailed domain knowledge.

Inspecting the differences for employing different base classifiers, it is clear that the SVM-based ensemble, which closely resembles our earlier approach introduced in [18], leads to (statistically) better classification performance. This may be caused by the fact that SVMs tend to work well on small datasets. Each subspace consists of a relatively small number of objects (all minority samples available in the fold and an equal number of majority ones), and consequently

Table 2. Results of statistical significance. A + signifies that the algorithm listed in this row statistically outperforms the algorithm listed in this column (based on sensitivity), a − indicates a statistically inferior performance.

	single SVM	bagged SVM	boosted SVM	Random Forest	SMOTEBagging	SMOTEBoost	IIvotes	EasyEnsemble	Hybrid Ensemble (NN)	Hybrid Ensemble (SVM)	Hybrid Ensemble (NN) + DIV	Hybrid Ensemble (SVM) + DIV
single SVM	▨	−	−	−	−	−	−	−	−	−	−	−
bagged SVM	+	▨	−	−	−	−	−	−	−	−	−	−
boosted SVM	+	+	▨		−	−	−	−	−	−	−	−
Random Forest	+	+		▨	−	−	−	−	−	−	−	−
SMOTEBagging	+	+	+	+	▨	−	−		−		−	−
SMOTEBoost	+	+	+	+	+	▨		−		−	−	−
IIvotes	+	+	+	+	+		▨				−	−
EasyEnsemble	+	+	+	+		+		▨	+	−	−	−
Hybrid Ensemble (NN)	+	+	+	+	+			−	▨	−	−	−
Hybrid Ensemble (SVM)	+	+	+	+		+		+	+	▨	−	−
Hybrid Ensemble (NN) + DIV	+	+	+	+	+	+	+	+	+	+	▨	−
Hybrid Ensemble (SVM) + DIV	+	+	+	+	+	+	+	+	+	+	+	▨

NNs tend to be prone to overfitting in these subspaces, while SVMs are able to handle the dichotomisation process more effectively.

The proposed hybridisation with the fuzzy diversity measure introduced in this paper is shown to give our method a further edge. Comparing the ensembles with and without the classifier selection stage, it is obvious that the former achieve higher classification and sensitivity performance. The overall best results are achieved by an ensemble of support vector machines that gives the maximum diversity using our energy-based measure. This approach yields a sensitivity of 81.96 % which is shown to be statistically better than those of all other methods, while resulting in only a slight drop in terms of specificity, and confirms that our hybrid ensemble algorithm provides an excellent classification method.

The presented approach also clearly outperforms earlier approaches in the literature of breast thermogram analysis. In [24], the same features and dataset were employed together with a cost-sensitive fuzzy if-then rule based classifier optimised by a genetic algorithm and giving a sensitivity of 79.86 % with a specificity of 79.49 %. [31] also used the same data and features and reported a sensitivity and specificity of 79 %.

6   Conclusions

In this paper, we have presented an effective approach to analysing breast thermograms for cancer diagnosis. We extract a set of image features describing bilateral (a)symmetry between the two breast regions from the images, and use these as input to a pattern classification stage. Based on our earlier work, we create a classifier ensemble for classification and address class imbalance by training its base classifiers on balanced data subsets. Using support vector machines and neural networks as individual classifiers and a trained perceptron as classifier fuser, this is shown to provide a powerful decision making system as experimental results on a dataset of about 150 thermograms demonstrate. Crucially though, we additionally perform a classifier selection stage based on a fuzzy diversity measure to eliminate redundant classifiers and identify the best models, and confirm this to lead to even (statistically significant) better classification performance and to yield an ensemble that outperforms not only various canonical classifiers but also several ensemble classifiers designed to address class imbalance.

References

1. Acharya, U.R., Ng, E.Y.K., Tan, J.H., Sree, S.V.: Thermography based breast cancer detection using texture features and support vector machine. J. Med. Syst. **36**(3), 1503–1510 (2012)
2. Alpaydin, E.: Combined 5 × 2 CV F test for comparing supervised classification learning algorithms. Neural Comput. **11**(8), 1885–1892 (1999)
3. Anbar, N., Milescu, L., Naumov, A., Brown, C., Button, T., Carly, C., AlDulaimi, K.: Detection of cancerous breasts by dynamic area telethermometry. IEEE Eng. Med. Biol. Mag. **20**(5), 80–91 (2001)
4. Bishop, C.M.: Neural Networks for Pattern Recognition. Oxford University Press, New York (1995)
5. Błaszczyński, J., Deckert, M., Stefanowski, J., Wilk, S.: Integrating selective preprocessing of imbalanced data with Ivotes ensemble. In: Szczuka, M., Kryszkiewicz, M., Ramanna, S., Jensen, R., Hu, Q. (eds.) RSCTC 2010. LNCS, vol. 6086, pp. 148–157. Springer, Heidelberg (2010)
6. Boquete, L., Ortega, S., Miguel-Jimnez, J.M., Rodrguez-Ascariz, J.M., Blanco, R.: Automated detection of breast cancer in thermal infrared images, based on independent component analysis. J. Med. Syst. **36**(1), 103–111 (2012)
7. Breiman, L.: Bagging predictors. Mach. Learn. **24**(2), 123–140 (1996)
8. Breiman, L.: Random forests. Mach. Learn. **45**(1), 5–32 (2001)
9. Chawla, N.V., Lazarevic, A., Hall, L.O., Bowyer, K.W.: SMOTEBoost: improving prediction of the minority class in boosting. In: 7th European Conference on Principles and Practice of Knowledge Discovery in Database, pp. 107–119 (2003)
10. Dong, Y., Han, K.: Boosting SVM classifiers by ensemble. In: 14th International World Wide Web Conference, pp. 1072–1073 (2005)
11. Duin, R.P.W.: The combining classifier: to train or not to train? In: 16th International Conference on Pattern Recognition, vol. 2, pp. 765–770 (2002)
12. Gautherie, M.: Thermobiological assessment of benign and maligant breast diseases. Am. J. Obstet. Gynecol. **147**(8), 861–869 (1983)

13. Head, J.F., Wang, F., Lipari, C.A., Elliott, R.L.: The important role of infrared imaging in breast cancer. IEEE Eng. Med. Biol. Mag. **19**, 52–57 (2000)

14. Jones, B.F.: A reappraisal of infrared thermal image analysis for medicine. IEEE Trans. Med. Imag. **17**(6), 1019–1027 (1998)

15. Karatzoglou, A., Smola, A., Hornik, K., Zeileis, A.: Kernlab an S4 package for kernel methods in R. J. Stat. Softw. **11**(9), 1–20 (2004)

16. Keyserlingk, J.R., Ahlgren, P.D., Yu, E., Belliveau, N., Yassa, M.: Functional infrared imaging of the breast. IEEE Eng. Med. Biol. Mag. **19**(3), 30–41 (2000)

17. Krawczyk, B.: Diversity in ensembles for one-class classification. In: Pechenizkiy, M., Wojciechowski, M. (eds.) New Trends in Databases and Information Systems. Advances in Intelligent Systems and Computing, vol. 185, pp. 119–129. Springer, Heidelberg (2012)

18. Krawczyk, B., Schaefer, G.: Evolutionary multiple classifier system based on space partitioning for breast thermogram analysis. In: 16th Online World Conference on Soft Computing in Industrial Applications (2011)

19. Krawczyk, B., Wozniak, M., Schaefer, G.: Improving minority class prediction using cost-sensitive ensembles. In: 16th Online World Conference on Soft Computing in Industrial Applications (2011)

20. Kuncheva, L.I.: Combining Pattern Classifiers: Methods and Algorithms. Wiley, Hoboken (2004)

21. Liu, X., Wu, J., Zhou, Z.: Exploratory undersampling for class-imbalance learning. IEEE Trans. Syst. Man Cybern. B Cybern. **39**(2), 539–550 (2009)

22. Mookiaha, M.R.K., Acharyaa, U.R., Ng, E.Y.K.: Data mining technique for breast cancer detection in thermograms using hybrid feature extraction strategy. Quant. InfraRed Thermography J. **9**(2), 151–165 (2013)

23. Qi, H., Snyder, W.E., Head, J.F., Elliott, R.L.: Detecting breast cancer from infrared images by asymmetry analysis. In: 22nd IEEE International Conference on Engineering in Medicine and Biology (2000)

24. Schaefer, G., Zavisek, M., Nakashima, T.: Thermography based breast cancer analysis using statistical features and fuzzy classification. Pattern Recogn. **42**(6), 1133–1137 (2009)

25. Tan, T.Z., Quek, C., Ng, G.S., Ng, E.Y.K.: A novel cognitive interpretation of breast cancer thermography with complementary learning fuzzy neural memory structure. Expert Syst. Appl. **33**(3), 652–666 (2007)

26. Tang, X., Ding, H., Yuan, Y., Wang, Q.: Morphological measurement of localized temperature increase amplitudes in breast infrared thermograms and its clinical application. Biomed. Sig. Process. Control **3**, 312–318 (2008)

27. Vapnik, V.N.: Statistical Learning Theory. Wiley, New York (1998)

28. Wang, S., Yao, X.: Diversity analysis on imbalanced data sets by using ensemble models. In: IEEE Symposium on Computational Intelligence and Data Mining, pp. 324–331 (2009)

29. Wozniak, M., Zmyslony, M.: Designing combining classifier with trained fuser - analytical and experimental evaluation. Neural Netw. World **20**(7), 925–934 (2010)

30. Yu, L., Liu, H.: Feature selection for high-dimensional data: a fast correlation-based filter solution. In: 20th International Conference on Machine Learning, pp. 856–863 (2003)

31. Zavisek, M., Drastich, A.: Thermogram classification in breast cancer detection. In: 3rd European Medical and Biological Engineering Conference, pp. 1727–1983 (2005)

Rician Noise Removal Approach for Brain MR Images Using Kernel Principal Component Analysis

Ashish Phophalia[✉] and Suman K. Mitra

Dhirubhai Ambani Institute of Information and Communication Technology,
382007 Gandhinagar, Gujarat, India
{ashish_phophalia,suman_mitra}@daiict.ac.in

Abstract. It has been observed that the noise accumulated in medical images due to various reasons during acquisition process is Rician in nature. A Rician noise removal method of Brain Magnetic Resonance (MR) Images using Kernel Principal Component Analysis (KPCA) is proposed in this paper. The proposed approach is non-parametric in nature. It explores the image space for *non-local* similar patch search and clusters them accordingly. The basis vectors are then learned using KPCA for each cluster which makes the proposed method data adaptive in nature. The approach has been applied to 2D phantom Brain MR images and experimental results are comparable to the other state-of-the-art methods in terms of various quantitative measures.

Keywords: Kernel Principal Component Analysis (KPCA) · Magnetic Resonance Image (MRI) · Rician noise removal

1 Introduction

Image Restoration is considered as one of the crucial ingredient of Medical Image Analysis systems. The possible sources for addition of noise are various parameters of the acquisition process such as flip angle, scan time, coil resistance, dielectric and inductive losses in sample, patient movement etc. [12]. MRI, being a non-invasive technique, offers many advantages in clinical analysis but the disturbances or *noise* induced in acquisition process degrade the quality of the signal. In Medical Image Denoising problem, the noise model is found to be Rician in nature which is different from commonly used distributions such as Gaussian, Poisson, etc. [8].

It has been shown that the intensities of MR images represent magnitude of underlying complex data which follows Rice distribution [7]. The real and imaginary parts are modeled as independently distributed Gaussian with means a_r and a_i respectively, with same variance σ^2. The probability density function (pdf) of Rician random variable y is defined as follows:

$$f_Y(y|a,\sigma) = \frac{y}{\sigma^2} e^{\left(-\frac{y^2+a^2}{2\sigma^2}\right)} I_0\left(\frac{ya}{\sigma^2}\right), y > 0 \tag{1}$$

© Springer International Publishing Switzerland 2015
M. Kryszkiewicz et al. (Eds.): PReMI 2015, LNCS 9124, pp. 545–553, 2015.
DOI: 10.1007/978-3-319-19941-2_52

where $a = \sqrt{a_r^2 + a_i^2}$ is underlying noise free signal amplitude and $I_n(z)$ is n^{th} order modified bessel function of first kind. When Signal to Noise Ratio (SNR, here it is a/σ) is high, the Rician distribution approaches a Gaussian; when SNR approaches to zero (that is only noise is present, $a \to 0$) the Rician distribution becomes Rayleigh distribution and the pdf turns out to be

$$f_Y(y|a \to 0, \sigma) = \frac{y}{\sigma^2} e^{\left(-\frac{y^2}{2\sigma^2}\right)} \tag{2}$$

Hence, the conventional methods for Rician noise removal first try to find the background portion in the medical images where no signal is assumed. Hence, one can use Rayleigh distribution in background portion and Gaussian distribution in the rest (where SNR is assumed to be high enough) [9,16]. However under the noisy condition, it is difficult to find proper background in the image.

Recent methods use the principle of non-local self similarity for image restoration task, where the first step involves finding out the similar patches (in terms of some predefined criteria such as Euclidean distance) that are similar to a given reference patch from the image [1]. Thereafter, an orthonormal basis is inferred for each patch and shrinkage is performed on the coefficients when the patch is projected on that basis, coefficients are sparse in nature as described in [4,6,14].

Out of recently proposed techniques, BM3D [4] is most popular. BM3D technique creates a 3D stack of similar patches, projects it onto a 3D basis (tensor product of 2D-DCT and 1D-Haar), and performs hard thresholding of these coefficients followed by basis inversion, thereby allowing a coupled update of the coefficients [4]. Another class of methods such as [5,13], first to cluster similar patches and then learn basis for each cluster instead of searching the similar patches for each underlying reference patch. However, due to nature of noise, straight forward implication of natural image denoising methods has not been advocated for medical images. The NLM method has been extended for Medical Image denoising problem in [11] where bias correction needs to be considered. BM3D has been extended using a suitable invertible transformation of the medical data into another domain where data behaves like Gaussian distributed in resultant domain. The most commonly known such kind of transformation for this purpose is Anacombe's Transformation, also known as Variance Stabilization Technique (VST). Recently, VST has been proposed in [7] for Rician distributed data and BM3D method is referred as BM3D+VST method. The BM3D+VST method can be summarized mathematically as follows:

$$\hat{y} = VST^{-1}(BM3D(VST(z,\sigma), \sigma_{VST}), \sigma) \tag{3}$$

where VST^{-1} denotes the inverse VST, σ_{VST} is the stabilized standard deviation induced by VST and z denotes the additive white Gaussian noise whose true intensity is represented by y. However, BM3D+VST is extended to 3D medical data as BM4D method in [10]. This manuscript focuses on 2D data denoising methods only.

The aim of this article is to explore a direct technique that can handle Rician noise suitably giving rise to noise removal as good as BM3D+VST, if not better.

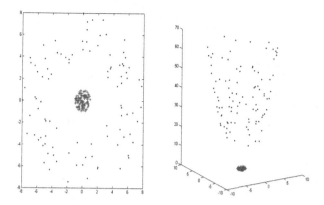

Fig. 1. Transformation of two circular data sets into higher dimension space using kernel method where separation between them is more prominent and can be classified using linear hyper-surface.

We have extended PCA based method using Rough Set based clustering proposed in [13] to Rician noise model and bias term correction is also made, referred as ER-PCA in the paper. We have proposed a new Kernel based PCA (KPCA) method for Rician noise. However, we have adopted the clustering strategy used in [13], which is non-local approach in *true-sense*. As per our knowledge, KPCA has not been applied for Rician noise removal in medical image yet. The kernel based methods can find non-linearity of data in Feature Space. Recently, kernel based methods have been used in Medical imaging in [2,15,19]. However, choice of appropriate kernel for given data is undecidable. In the current proposal, Gaussian kernel is used and the performance of noise removal technique is at par with the state-of-the-art methods.

The paper has been arranged in following manner: Sect. 2 presents proposed method using KPCA. Section 3 compares proposed method with other state-of-the-art methods. The manuscript is concluded in Sect. 4.

2   Proposed Method Using KPCA

A non-parametric variant of PCA, known as Kernal Principal Component Analysis (KPCA) has been explored for Rician noise removal. The KPCA tries to explore structure in the data in Feature Space instead of Image Space itself and tries to capture higher-order dependencies in the data. In Fig. 1, two class data is shown in circular form and transformed to higher dimension for classification purpose, where transformation is $\phi(x) : (x_1, x_2) \rightarrow (x_1, x_2, x_1^2 + x_2^2)$. Hence, one can find a discriminating plane (linear surface) in higher dimensions which is not possible in two dimensions for given data points.

In KPCA, this nonlinearity is introduced by first mapping the data into another space F using a nonlinear map $\phi : R^N \rightarrow F$, before standard linear PCA is carried out in F using the mapped samples $\phi(x_k)$. The map ϕ and the

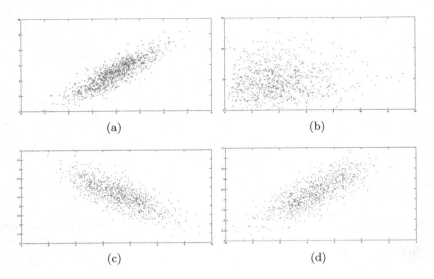

Fig. 2. Reconstruction using PCA and KPCA over synthetic data with Rician noise. (a) Synthetic Data, (b) Rician Noisy Data, (c) Reconstruction using PCA and (d) Reconstruction using KPCA.

space F are determined implicitly by the choice of a kernel function k, which acts as a similarity measure. This mapping computes the dot product between two input samples x and y mapped into F:

$$k(x; y) = \phi(x).\phi(y) \tag{4}$$

One can show that if k is a positive definite kernel, then there exists a map ϕ into a dot product space F such that Eq. 4 holds. The space F then has the structure of a so-called Reproducing Kernel Hilbert Space (RKHS) [2].

The identity Eq. 4 is important for KPCA since PCA in F can be formulated entirely in terms of inner products of the mapped samples. Thus, we can replace all inner products by evaluations of the kernel function. This has two important consequences: first, inner products in F can be evaluated without computing $\phi(x)$ explicitly. This allows to work with a very high-dimensional, possibly infinite-dimensional RKHS F. Second, if a positive definite kernel function is specified, we need to know neither ϕ nor F explicitly to perform KPCA since only inner products are used in the computations. Commonly used positive definite kernel functions are *polynomial kernel* of degree $d \in N, k(\mathbf{x}, \mathbf{y}) = (\mathbf{x}.\mathbf{y})^d$ or $k(\mathbf{x}, \mathbf{y}) = (\mathbf{x}.\mathbf{y} + 1)^d$ or Gaussian kernel of width $\sigma > 0$, $k(\mathbf{x}, \mathbf{y}) = exp\left(-\left\|\mathbf{x} - \mathbf{y}\right\|^2 / 2\sigma^2\right)$. In all the experiments, Gaussian kernel has been used which is isotropic stationary in nature and also satisfies Mercer's Theorem [19].

A synthetic experiment has been performed as shown in Fig. 2 where Rician noise added in the synthetic data. However, KPCA (with Gaussian kernel) is able to preserve orientation of the data in a better way as compared to PCA based reconstruction.

The outline of present work can be described as follows:

1. Get the clusters of patches from the given noisy image using Rough set based method (as described in [13]).
2. For each cluster, get the basis vectors using KPCA method along pixel positions. For patches of size $p \times p$, kernel matrix would be of size $p^2 \times p^2$. Hence, the method is data adaptive in nature.
3. Project the noisy image patches on the obtained basis vectors in the KPCA domain.
4. Apply coefficient shrinkage method on these projected patches to get the denoised patches. Transform them back to image space.
5. Remove the bias term from each pixel of the denoised image.

$$I_{unbiased} = \sqrt{max(\hat{I}(i,j)^2 - 2h^2, 0)} \tag{5}$$

where h is the standard deviation of noise and $\hat{I}$ is the image obtained by step (4).

3 Experimental Results

This Section encompasses the qualitative and quantitative evaluations of the proposed method along with some of the state-of-the-art methods. The experiments

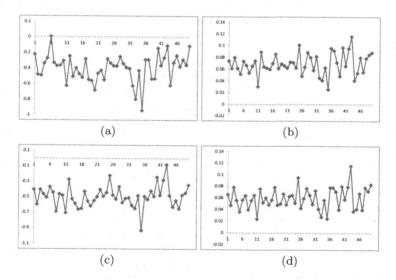

(a)   (b)

(c)   (d)

Fig. 3. Difference comparison of KPCA with reference to BM3D+VST method (at zero level vertically) for 50 slices for noise standard deviation equal to 15 (a) T1 images with PSNR difference values, (b) T1 images with MSSIM difference values, (c) T2 images with PSNR difference values and (d) T2 images with MSSIM difference values.

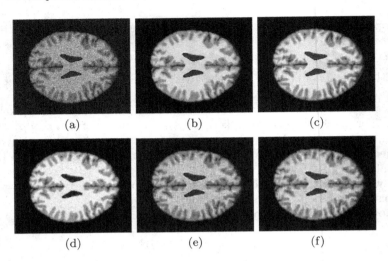

(a)   (b)   (c)

(d)   (e)   (f)

Fig. 4. (a) Synthetic Noisy T1 Image with Rician noise standard deviation=15 and PSNR =22.7220 dB, Denoised image using (b) UNLM method, PSNR = 34.4622 dB, (c) BM3D+VST method, PSNR = 34.2393 dB, (d) RS-NLM method, PSNR = 32.5856 dB, (e) ER-PCA method, PSNR = 33.8155 dB, (f) KPCA method, PSNR = 34.0241 dB.

Table 1. Performance comparison of proposed denoising strategy with different approaches on various quantitative measures under *Rician Noise assumption* in Brain Web database (**slice = 70** & **100**, Modality = T1, image size = 181 × 217 and patch size = 5 × 5). Best figures are shown in Bold.

Noise SD	Methods	Slice 70				Slice 100			
		PSNR	RMSE	MSSIM	FSIM	PSNR	RMSE	MSSIM	FSIM
5	Noisy	32.4293	37.1660	0.6134	0.9296	32.2588	38.6549	0.5564	0.8922
	UNLM [11]	39.0519	8.0889	0.9832	0.9845	40.1551	6.2744	0.9882	0.9887
	BM3D+VST [7]	**40.9727**	**5.1937**	0.9602	0.9843	41.4921	4.6118	0.9602	0.9857
	RS-NLM [13]	39.8595	6.7163	**0.9851**	**0.9853**	41.5829	**4.5164**	**0.9914**	**0.9913**
	ER-PCA	40.4514	5.8606	0.9791	0.9764	39.9719	6.5447	0.9689	0.9563
	KPCA	40.2107	6.1946	0.9197	0.9797	41.2223	4.9073	0.9866	0.9850
10	Noisy	26.4115	148.5702	0.4717	0.8149	26.2398	154.5629	0.4183	0.7567
	UNLM [11]	35.9894	16.3733	0.9608	0.9643	36.9916	12.9993	0.9707	0.9724
	BM3D+VST [7]	**36.3738**	**14.9866**	0.9040	0.9607	36.8590	13.4025	0.9132	0.9653
	RS-NLM [13]	35.8260	17.0011	**0.9631**	**0.9645**	37.2231	**12.3246**	**0.9762**	**0.9770**
	ER-PCA	35.7387	17.3464	0.9389	0.9439	36.3168	15.1846	0.9597	0.9484
	KPCA	36.1061	15.9395	0.9586	0.9522	36.6642	14.0172	0.9682	0.9628
15	Noisy	22.8950	333.8752	0.3744	0.7177	22.7220	347.4434	0.3331	0.6495
	UNLM [11]	33.5147	28.9475	0.9299	**0.9391**	34.4622	**23.2732**	0.9453	**0.9498**
	BM3D+VST [7]	**33.7666**	**27.3162**	0.8583	0.9368	34.2393	24.4992	0.8684	0.9447
	RS-NLM [13]	32.1179	39.9292	0.9273	0.9244	32.5856	35.8523	**0.9472**	0.9448
	ER-PCA	33.2440	30.8093	0.9133	0.9178	33.8155	27.0103	0.9377	0.9287
	KPCA	33.4097	29.6557	**0.9323**	0.9262	34.0241	25.7438	0.9469	0.9404

Table 2. Performance comparison of proposed denoising strategy with different approaches on various quantitative measures under *Rician Noise assumption* in Brain Web database (**slice** = **70** & **100**, Modality = T2, image size = 181 × 217 and patch size = 5 × 5). Best figures are shown in Bold.

Noise SD	Methods	Slice 70				Slice 100			
		PSNR	RMSE	MSSIM	FSIM	PSNR	RMSE	MSSIM	FSIM
5	Noisy	32.4349	37.1185	0.6257	0.9365	32.2639	38.6095	0.5691	0.9052
	UNLM [11]	34.4831	23.1617	0.9822	0.9813	35.2666	19.3385	0.9869	0.9858
	BM3D+VST [7]	**40.4738**	**5.8305**	0.9648	**0.9861**	**41.0752**	**5.0764**	0.9663	0.9885
	RS-NLM [13]	36.9814	13.0300	**0.9856**	0.9835	37.6322	11.2166	**0.9915**	**0.9900**
	ER-PCA	39.8618	6.7127	0.9783	0.9727	39.1934	7.8297	0.9610	0.9473
	KPCA	37.8578	10.6487	0.8002	0.9782	38.1996	9.8429	0.7610	0.9797
·10	Noisy	26.4322	147.8642	0.4956	0.8356	26.2550	154.0201	0.4408	0.7757
	UNLM [11]	32.9818	32.7262	0.9618	0.9623	33.8132	27.0246	0.9710	0.9687
	BM3D+VST [7]	**35.7377**	**17.3504**	0.9181	**0.9681**	35.8044	17.0860	0.9683	0.9637
	RS-NLM [13]	34.5041	23.0502	**0.9691**	0.9676	35.2411	19.4522	**0.9799**	**0.9766**
	ER-PCA	34.8288	21.3894	0.9432	0.9323	34.5457	22.8303	0.9262	0.9008
	KPCA	35.0519	20.3182	0.8527	0.9567	**36.1329**	**15.8413**	0.9184	0.9727
15	Noisy	22.9275	331.3825	0.4131	0.7519	22.7460	345.5293	0.3676	0.6776
	UNLM [11]	31.4832	46.2121	0.9346	0.9408	32.1181	39.9271	0.9472	0.9456
	BM3D+VST [7]	**32.8504**	**33.7321**	0.8769	**0.9496**	**33.1694**	**31.3427**	0.8855	**0.9567**
	RS-NLM [13]	31.9206	41.7849	**0.9446**	0.9452	32.6973	34.9423	**0.9601**	0.9543
	ER-PCA	31.7529	43.4297	0.9034	0.8973	31.7770	43.1894	0.8989	0.8718
	KPCA	32.3606	37.7592	0.9363	0.9346	32.8539	33.7049	0.9516	0.9439
20	Noisy	20.4499	518.2594	0.3540	0.6871	20.2642	611.8738	0.3162	0.6059
	UNLM [11]	30.0502	64.2771	0.9063	0.9205	30.5757	56.9519	0.9199	0.9216
	BM3D+VST [7]	**30.7168**	**55.1319**	0.8426	**0.9303**	**30.9691**	**52.0201**	0.8508	**0.9398**
	RS-NLM [13]	29.4113	74.4654	0.9109	0.9104	30.1086	63.4192	0.9293	0.9137
	ER-PCA	29.6008	71.2860	0.8723	0.8785	29.6002	71.2947	0.8647	0.8527
	KPCA	30.1448	62.8934	**0.9129**	0.9144	30.6031	56.5941	**0.9316**	0.9245
25	Noisy	18.5384	910.4194	0.3095	0.6362	18.3487	951.0736	0.2774	0.5520
	UNLM [11]	28.6394	88.9483	0.8777	0.9012	29.1108	79.7989	0.8914	0.8987
	BM3D+VST [7]	**29.0589**	**80.7598**	0.8109	**0.9114**	**29.2912**	**76.5527**	0.8269	**0.9227**
	RS-NLM [13]	26.4734	146.4670	0.8599	0.8492	26.6486	140.6762	0.8696	0.8372
	ER-PCA	28.0567	101.7219	0.8576	0.8824	28.1251	100.1314	0.8529	0.8685
	KPCA	28.1995	98.4298	**0.8792**	0.8814	28.3402	95.2919	**0.8932**	0.8785

have been carried out on 2D monochrome phantom human brain MRI images obtained from Brain Web Database [3]. The parameters are as follows: RF = 20, protocol = ICBM, slice thickness = 1 mm, volume size = 181 × 217 × 181. The experimental set up considers Rician noise model at different noise levels along with two modalities, namely T1 and T2. The simulated database provides the ground truth image for evaluating denoising performance which most of the time is unavailable with real database. The Rician noise addition and bias correction are done as suggested in [10] and [11] respectively. The evaluation measures used are Peak-Signal-to-Noise Ratio (PSNR), Root Mean Square

Error (RMSE), Mean Structural Similarity Index (MSSIM) [17] and Feature Similarity Index (FSIM) [18].

For comparison purpose, several state-of-the-art methods are considered: Unbiased Non Local Means (UNLM method) presented in [11], BM3D+VST method proposed in [4], Rough Set based Non Local Means (RS-NLM) method proposed in [13] and PCA based method proposed in the [13] has been extended in this work for Rician noise, referred as Extended Rough set based PCA method (ER-PCA). The parameters of all methods are kept default as suggested by respective authors. In all the experiments, patch size is kept as 5×5. The proposed KPCA method does not use VST method. Tables 1 and 2 represent quantitative results for two slices 70 and 100 of T1 MR and T2 MR images respectively. The ER-PCA performance is comparable to UNLM and BM3D+VST methods. The proposed KPCA method outperforms ER-PCA and preserves structure better than other state-of-the-art method. Figure 3 shows difference of PSNR and MSSIM measure for KPCA method with reference to BM3D+PCA (zero level on vertical axis) of 50 slices (from 61st to 110th slice of database mentioned above) with noise standard deviation equal to 15 for both T1 and T2 modalities. Negative value indicates BM3D+VST performs better and, in reverse, positive value is indicator of better performance of KPCA method. From Fig. 3, PSNR of KPCA fall below BM3D+VST method whereas it better preserves structure of the image in terms of MSSIM measure. This is also visually evident in Fig. 4 for the slice 100 of T1 modality at noise level 15.

4   Conclusion

In this paper, an approach for removal of Rician noise from brain MR images using Kernel PCA has been proposed. Being a manifold learning method, KPCA explores a suitable transformation for image representation through sparse bases. This method learns basis vectors from data itself unlike BM3D+VST method where basis vectors are kept fixed. The limitation of KPCA method is the selection of suitable kernel which is yet unanswered. If the nature of data is not known a-prior than one can try various kernels to find a suitable one. However, commonly used Gaussian kernel in KPCA, found to perform comparable with other state-of-the-art methods. The PCA based method proposed in [13] has also been implemented to remove Rician noise, but it fails to attain superior performance over KPCA. The proposed method is implemented on synthetic data for quantitative evaluation since ground truth data is available for the same.

References

1. Buades, A., Coll, B., Morel, J.M.: A non local algorithm for image denoising. In: IEEE Computer Vision and Pattern Recognition, pp. 60–65 (2005)
2. Charpiat, G., Hofmann, M., Schölkopf, B., et al.: Kernel methods in medical imaging. Handbook of Biomedical Imaging (2010)

3. Collins, D.L., Zijdenbos, A., Kollokian, V., Sled, J., Kabani, N., Holmes, C., Evans, A.: Design and construction of a realistic digital brain phantom. IEEE Trans. Med. Imaging **17**(3), 463–468 (1998)

4. Dabov, K., Foi, A., Katkovnik, V., Egiazarian, K.: Image denoising by sparse 3d transform domain collaborative filtering. IEEE Trans. Image Process. **16**(8), 2080–2095 (2007)

5. Dong, W., Li, X., Zhang, D., Shi, G.: Sparsity-based image denoising via dictionary learning and structural clustering. In: 2011 IEEE Conference on Computer Vision and Pattern Recognition (CVPR), pp. 457–464. IEEE (2011)

6. Elad, M., Aharon, M.: Image denoising via sparse and redundant representations over learned dictionaries. IEEE Trans. Image Process. **15**(12), 3736–3745 (2006)

7. Foi, A., Noise estimation and removal in mr imaging: the variance-stabilization approach. In: ISBI, pp. 1809–1814 (2011)

8. Gudbjartsson, H., Patz, S.: The rician distribution of noisy mri data. Magn. Reson. Med. **34**(6), 910–914 (1995)

9. He, L., Greenshields, I.R.: A nonlocal maximum likelihood estimation method for rician noise reduction in mr images. IEEE Trans. Med. Imaging **28**(2), 165–172 (2009)

10. Maggioni, M., Katkovnik, V., Egiazarian, K., Foi, A.: Nonlocal transform domain filter for volumetric data denoising and reconstruction. IEEE Trans. Image Process. **22**(1), 119–133 (2013)

11. Manjon, J.V., Caballero, J.C., Marti, G.G., Marti-Baonmati, L., Robles, M.: Mri denoising using non local means. Med. Image Anal. **12**, 514–523 (2008)

12. McVeigh, E., Henkelman, R., Bronskill, M.: Noise and filtration in magnetic resonance imaging. Med. Phys. **12**, 586 (1985)

13. Phophalia, A., Rajwade, A., Mitra, S.K.: Rough set based image denoising for brain mr images. Sig. Process. **103**, 24–35 (2014)

14. Rajwade, A., Rangarajan, A., Banerjee, A.: Image denoising using the higher singular value decomposition. IEEE Trans Pattern Ana. Mach. Intell. **35**(4), 849–862 (2013)

15. Rubio, E.L., Nunez, M.N.F.: Kernel regression based feature extraction for 3d mr image denoising. Med. Image Anal. **15**, 498–513 (2011)

16. Sijbers, J., den Dekker, A.J., Scheunders, P., Dyck, D.V.: Maximum likelihood estimation of rician distribution parameters. IEEE Trans. Med. Imaging **17**(3), 357–361 (1998)

17. Wang, Z., Bovik, A.C., Sheikh, H.R., Simoncelli, E.P.: Image quality assessment: from error visibility to structural similarity. IEEE Trans. Image Process. **13**(4), 600–612 (2004)

18. Zhang, L., Zhang, L., Mou, X., Zhang, D.: Fsim: A feature similarity index for image quality assessment. IEEE Trans. Image Process. **20**(9), 2378–2386 (2011)

19. Zimmer, V.A., Lekadir, K., Hoogendoorn, C., Frangi, A.F., Piella, G.: A framework for optimal kernel-based manifold embedding of medical image data. Comput. Med. Imaging Graph. **41**, 93–107 (2015)

Climate Network Based Index Discovery for Prediction of Indian Monsoon

Moumita Saha[✉] and Pabitra Mitra

Department of Computer Science and Engineering,
Indian Institute of Technology Kharagpur,
Kharagpur, India
{moumitasaha.2012,pabitra}@gmail.com

Abstract. Identification of climatic indices are vital in essence of their ability to characterize different climatic events. We focus on discovery of climatic indices important for Indian summer monsoon from climatic parameters surface pressure and zonal wind velocity. We use climatic network based community detection approach for discovery of climatic indices. New indices depict better correlation with monsoon than existing indices. Regression and non-linear models are designed using newly discovered climatic indices for prediction of Indian summer monsoon. Models show superior accuracy to existing state of art models.

Keywords: Climatic network · Community detection · Climatic indices · Indian monsoon prediction

1 Introduction

Mechanism behind climatic process is complex. Identification and analysis of different patterns in global climatic system is vital in understanding its intricate nature. The state and dynamics of climatic process are explained by different climatic indices. Climatic indices are based on climatic parameters like sea surface temperature (SST), sea level pressure (SLP), wind velocity, surface pressure (SP), that elucidate specific climatic change. Climatic indices are important for their ability to predict different climatic events. Prediction of Indian summer monsoon rainfall ($ISMR$) is challenging due to its dynamic nature. It is important for economic development of agricultural land like India.

Building and analysis of climatic networks in Earth Sciences is one of the emerging topic with immense future scopes. Complex networks have been widely used in building climatic networks and finding out interesting patterns and interconnections present in the climatic system [1]. Steinhaeuser et al. [2] have proposed use of complex networks in descriptive analysis and predictive modelling of climatic events. Donges et al. [3] have revealed the important internal structure present in the climatic network build upon surface air temperature data and uncover a pattern related to global surface ocean currents. Steinhaeuser et al. [4] have detected community in climatic system, given a climatological

© Springer International Publishing Switzerland 2015
M. Kryszkiewicz et al. (Eds.): PReMI 2015, LNCS 9124, pp. 554–564, 2015.
DOI: 10.1007/978-3-319-19941-2_53

interpretation of the communities and applied the model for discovery of new climatic indices.

Climatic index discovery assists in visualizing different aspects of climatic system. Clustering approaches are used in discovery of climatic indices. Sap and Awan [5] have used kernel k-means algorithm with spatial constraint to identify the spatio-temporal patterns in the system. Similar nearest neighbours-based clustering approach is used for detection of novel climatic indices, which are validated against known climatic indices and are shown to overcome limitations of PCA and SVD approaches [6].

The purpose of our work is two folds– (i) discovery of new climatic indices using climatic network based community detection approach from climatic parameters surface pressure and zonal wind velocity, (ii) utilization of discovered climatic indices as predictors for forecasting Indian summer monsoon rainfall, which acts as validation of our proposed index discovery approach. In our work, climatic networks are formed considering each spatial grid point as a node in the network with time series of climatic parameter in the grid. We use normalized euclidean distance to create weighted edges between the nodes. Three important community detection algorithms are applied for invention of different climatic regions that are significant. Community detection performs better than the traditional clustering method as unlike clustering approach, it also focusses on the structure of the network along with the node attributes. Correlation value between time series of node and Indian monsoon is also included as a node attribute to assists in detecting communities important for prediction of monsoon. The communities found after proper thresholding are shown to be good predictors of Indian monsoon. The discovered climatic indices are compared with established climatic indices of Indian monsoon for validation and they are shown to be more correlated to Indian monsoon than the present climatic indices. Finally, different linear and non-linear models are designed with the newly invented climatic indices as input parameters to predict monsoon. The discovered climatic indices show their imprint and ascertain their superiority in prediction of Indian summer monsoon rainfall.

2 Climatic Network Formation

Climatic networks are built based on two different climatic parameters, namely, surface pressure and zonal wind velocity. Each spatial grid points over the world is considered as a node in the network. Our network consist of $10,512$ nodes. Each node is characterized by its corresponding latitude, longitude, climatic parameter time series values over the temporal scale, and scalar correlation value between the climatic parameter time series and Indian monsoon time series at best lead month. Weighted edges are added studying the strength of bonding between each pair of nodes in the network with normalized euclidean distance measure. Top one percent and five percent edges are considered for networks built for climatic parameters surface pressure (NET_SP), and zonal wind velocity (NET_ZW), respectively. Finally, isolated nodes are removed from the networks to obtain connected networks. *NET_SP* has *1,999* nodes and *23,326* edges, and *NET_ZW* has *4,922* nodes and *6,851* edges.

3   Community Detection and Index Discovery

Three important community detection algorithms, namely, *infomap (Info)*, *walk-trap (Wlktrp)*, and *fastgreedy (Fstgrdy)* are applied on the climatic network to detect communities over the world which will correspond to discovery of novel climatic indices important for prediction of *ISMR*. We have chosen these algorithms guided by requirements as following– (i) ability to utilize edge weights, (ii) suitability for dense networks, (iii) overall computational efficiency, and (iv) inclusion of node weights (in case of *info-map* community detection method).

Info-map Community Detection (Info): The algorithm is based on an information theoretic approach, which use the probability flow of random walks on a network and decompose the network into modules by compressing a description of the probability flow [7]. It discovers community structure in weighted and directed networks, taking into account the node values, weighted edges, and network structure.

Walk-trap Community Detection (Wlktrp): The algorithm employs the concept of random walks through the network for community detection. A node similarity measure based on short walks is used for community detection via hierarchical agglomeration, considering the edge weight and structure of the network. It is efficient in terms of time and space complexity [8].

Fast-greedy Community Detection (Fstgrdy): It is a hierarchical agglomeration algorithm for detecting community structure based on modularity optimization method [9]. It follows greedy optimization in which, starting with each vertex being the sole member of a community of one, two communities are repeatedly join together, whose amalgamation produces the largest increase in modularity value.

The communities found by the above three approaches are evaluated by measure of modularity defined in Sect. 4.2. These communities are utilized for discovery of new climatic indices. We select top few communities by thresholding based on number of nodes present in the community, density of community, correlation of time-series of community with Indian monsoon. Communities filtered out are the representative for new climatic index. We average the time series values over all the nodes present in a specific community and the resulting time-series represents the new climatic index. The correlation of discovered indices with Indian monsoon is studied and compared with correlation of present Indian Meteorological Department's (*IMD*) predictors with Indian monsoon.

Discovered indices show higher correlation than present predictor indices of monsoon. A study of correlation between discovered climatic indices and existing climatic predictors of monsoon is performed as a validation of our discovered

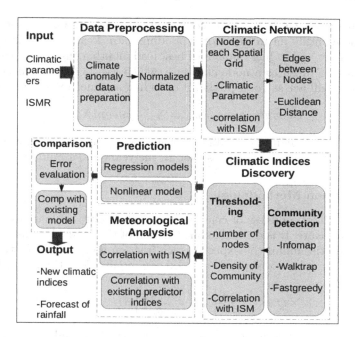

Fig. 1. Block diagram of proposed approach discovery of climatic indices important for Indian summer monsoon rainfall

indices. Finally, discovered climatic indices having high correlation with Indian monsoon are used for prediction of monsoon. Regression and non-linear models are designed with discovered climatic indices as predictors for forecasting annual Indian summer monsoon rainfall. The block diagram of proposed approach of discovery of climatic indices important for Indian monsoon and utilization of indices in forecasting monsoon are shown in Fig. 1.

4   Experimental Evaluation

4.1   Data Sets

Surface pressure and zonal wind velocity are collected from *NCEP* reanalysis data provided by the NOAA/OAR/ESRL (www.esrl.noaa.gov/psd/) [10] at spatial resolution of $2.5° \times 2.5°$ with coverage of $90°N$–$90°S$ and $0°E$–$358°E$. There are *73* latitude and *144* longitude grids, which give *10,512* nodes (73×144) in the network. Annual Indian summer monsoon rainfall (*ISMR*), occurring in months of June, July, August, and September is acquired from Indian Institute of Tropical Meteorology (www.imdpune.gov.in/research/ncc/longrange/data/data.html) [11]. *ISMR* is expressed as percentage of long period average (*LPA*) value of rainfall, which is *878.1 mm* for our period of study *1948–2013*.

As a preprocessing step, data is converted to monthly anomaly data by subtracting the monthly mean from corresponding data. Pearson correlation of climatic parameter and Indian rainfall for best lead month, considering lead of zero to six months is taken as a node attribute, which assists our search of climatic indices which will act as good predictor of Indian summer monsoon rainfall.

$$Climatic\ anomaly_m = X_m - mean(X_m),$$

where, X_m denotes climatic parameter value for month m and mean (X_m) is the average of the parameter values over all the years under study for month m.

4.2   Evaluation Methodology

Modularity. The goodness of communities detected are evaluated in terms of modularity measure. It is defined as the fraction of the edges that fall within the given communities minus the expected such fraction if edges were distributed at random. Higher value corresponds to good community detection. It is shown by Eq. 1.

$$Q = \frac{1}{2e} \sum_{vw} \left[A_{vw} - \frac{k_v k_w}{2e} \right] \delta\left(c_v, c_w\right), \tag{1}$$

where, e represents the number of edges in the graph, v and w are the nodes, $A_{vw} = 1$, if edges present between nodes v and w, 0 otherwise, k_v, k_w are the degree of nodes v and w, $\delta(c_v, c_w) = 1$, if both nodes belong to same community, otherwise 0.

Modularity and number of communities formed by three community detection algorithms for *NET_SP* and *NET_ZW* are shown in Tables 1 and 2, respectively. Communities detected have high modularity measure of *0.93* for surface pressure, and *0.97* for zonal wind velocity by *Fstgrdy* community detection method.

Table 1. Modularity and number of communities detected for network built for surface pressure (*NET_SP*)

Algorithm	Modularity	Number of communities
Info	0.890	512
Wlktrp	0.925	197
Fstgrdy	**0.930**	400

Table 2. Modularity and number of communities detected for network built for zonal wind velocity (*NET_ZW*)

Algorithm	Modularity	Number of communities
Info	0.913	680
Wlktrp	0.977	351
Fstgrdy	**0.978**	358

Selecting Top Communities. Few predictive communities are selected from the obtained communities by thresholding. Three measures are taken as baseline, namely, (i) number of nodes, (ii) density of communities, (iii) communities having correlation with Indian monsoon greater than threshold correlation.

The threshold correlation is ascertained by plotting a histogram of correlation of random *1000* climatic parameter series and Indian monsoon. The result for climatic parameter surface pressure is shown in Fig. 2. It is observed that most of the correlation lies below *0.1*, so we have taken our threshold as *0.13* for surface pressure and similarly *0.15* for zonal wind velocity. The selected predictive communities of both surface pressure and zonal wind velocity are considered as the new discovered climatic indices important for prediction of Indian monsoon.

4.3   Correlation Studies

Discovered climatic indices are evaluated by estimating their correlation with Indian monsoon. Number of selected discovered climatic indices and best correlation of indices with monsoon for all three community detection algorithms are elaborated in Tables 3 and 4 for surface pressure and zonal wind velocity, respectively. Correlation of *0.34* is observed for discovered climatic indices from surface pressure parameter and *0.35* is obtained for zonal wind velocity parameter. Pearson correlation of discovered climatic indices for *NET_SP* by *info-map* community detection method is shown in Fig. 3.

Table 3. Number of discovered climatic indices and their best correlation with Indian monsoon for surface pressure (*NET_SP*)

Algorithm	Number of selected communities	Best correlation
Info	11	0.32
Wlktrp	11	0.32
Fstgrdy	12	**0.34**

Table 4. Number of discovered climatic indices and their best correlation with Indian monsoon for wind velocity (*NET_ZW*)

Algorithm	Number of selected communities	Best correlation
Info	12	**0.35**
Wlktrp	12	0.28
Fstgrdy	14	0.28

4.4   Prediction Performance

Discovered climatic indices are evaluated in terms of their predictability of Indian summer monsoon rainfall. We use climatic indices which have high correlation with Indian monsoon as predictors. The predictor climatic indices obtained from networks built for surface pressure and zonal wind velocity are listed in Tables 5 and 6, respectively. Regression models, namely linear regression, ridge regression model with cross validation, bayesian regression and non-linear model, namely generalized regression neural network (*GRNN*) are built with discovered climatic indices as predictors for forecasting annual Indian summer monsoon rainfall. Test period of twenty years from *1994* to *2013* is considered for evaluation. Mean absolute errors in terms of percentage of long period average value (*LPA*)

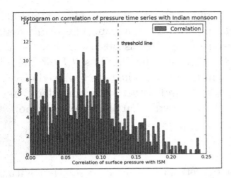

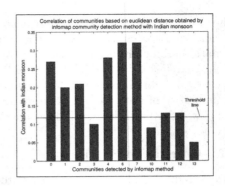

Fig. 2. Histogram for finding baseline *threshold* correlation with Indian monsoon for *NET_SP*

Fig. 3. Correlation of communities with Indian monsoon detected by *Infomap* method for *NET_SP*

of rainfall is presented for regression and non-linear models in Tables 7 and 8 for *NET_SP* and *NET_ZW*, respectively. Climatic indices discovered by *info-map* method give best performance with mean absolute errors of *5.5 %* and *5.4 %* for *NET_SP* and *NET_ZW*, respectively. This verifies the inclusion of correlation of parameter with Indian monsoon as node weight, which is considered by *info-map* technique for discovery of climatic indices.

Table 5. Number of predictors and discovered climatic indices with community id for surface pressure (*NET_SP*)

Algorithm	Number of predictors	Community ids
Info	4	0,4,6,7
Wlktrp	6	1,15,93,103,109,136
Fstgrdy	4	182,186,217,237

Table 6. Number of predictors and discovered climatic indices with community id for wind velocity (*NET_ZW*)

Algorithm	Number of predictors	Community ids
Info	4	1,4,5,8
Wlktrp	4	56,66,67,224
Fstgrdy	6	34,35,56,66,78,184

4.5   Comparisons with Existing Models

The predictability of climatic indices in forecasting Indian monsoon are compared with present Indian Meteorological Department's (*IMD*) models. Models built with indices discovered from network based on surface pressure by all the three community detection methods give better performance than existing *16*-parameter power regression model [12] and *8* and *10*-parameter *IMD* models [13]. Proposed models built with discovered predictor climatic indices by *Info*, *Wlktrp*, and *Fstgrdy* methods give root mean square errors of *4.8 %*, *5.6 %*, and *6.2 %*, respectively, outperforming all three *IMD* models giving *10.8 %*, *6.4 %*, and *7.6 %* errors for period *1996–2002*. Models built from predictor climatic

Table 7. Mean absolute errors (%) for prediction of Indian monsoon by discovered climatic indices from *NET_SP* for test period *1994-2013*

Models	Info	Wlktrp	Fstgrdy
Linear	**5.5**	6.5	5.8
RidgeCV	6.0	**5.7**	**5.7**
Bayesian ridge	6.0	**5.7**	6.0
GRNN	**6.0**	6.3	6.3

Table 8. Mean absolute errors (%) for prediction of Indian monsoon by discovered climatic indices from *NET_ZW* for test period *1994-2013*

Models	Info	Wlktrp	Fstgrdy
Linear	**5.8**	6.5	6.1
RidgeCV	**5.8**	6.6	6.2
Bayesian ridge	**5.4**	6.5	6.2
GRNN	**6.4**	**6.4**	**6.4**

indices discovered from network based on zonal wind velocity by *Info*, *Wlktrp*, and *Fstgrdy* methods give root mean square errors of *7.3 %*, *7.0 %*, and *7.5 %*, respectively, which outrun *IMD*'s *16* and *8*-parameter model, but is greater than *IMD*'s *10*-parameter model having *6.4 %* error. Discovered climatic indices for network based on surface pressure serve as better predictor of Indian monsoon. Therefore, it can be ascertained that surface pressure has more important role than wind velocity for climatic event of monsoon. Comparisons of predictability of models built with discovered climatic indices from *NET_SP* and *IMD* models are shown in Fig. 4.

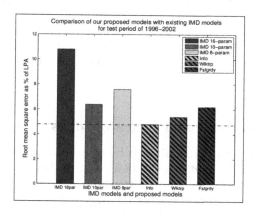

Fig. 4. Comparison of root mean square errors in prediction of Indian monsoon by proposed models based on climatic indices discovered from *NET_SP* and *IMD*'s *16* [12], *10*, *8*-parameter [13] models for period *1996-2002*

5   Meteorological Significance

5.1   Analysis Based on Correlation with ISMR

The Pearson correlation (μ) of discovered climatic indices with Indian monsoon are compared to correlation of existing predictor climatic indices with Indian monsoon [14]. Important predictor of monsoon, as considered by *IMD*, namely, North Atlantic SST (*NA_SST*), Equatorial South Eastern Indian Ocean SST (*ESE_IO_SST*), East Asia surface pressure (*EA_SP*), North Atlantic surface pressure (*NA_SP*), North Central Pacific Ocean zonal wind anomaly (*NC_PO_zonal_wnd*), and North West Europe surface pressure (*NW_Eu_SP*) are considered for validation of the discovered climatic indices. Newly discovered climatic indices are shown to be having higher correlation than *IMD*'s predictor indices. The result for climatic indices discovered for *NET_SP* and *NET_ZW* are shown in Figs. 5 and 6, respectively. High correlation of *0.34* and *0.35* are observed for indices discovered for climatic parameters surface pressure and zonal wind velocity, respectively, which show superior behaviour.

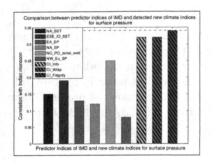 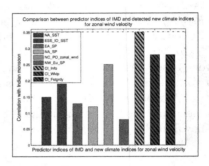

Fig. 5. Comparison of correlation with *ISMR* for *IMD* predictors and discovered climatic indices for *NET_SP*

Fig. 6. Comparison of correlation with *ISMR* for *IMD* predictors and discovered climatic indices for *NET_ZW*

5.2   Validation of Discovered Climatic Indices

New climatic indices (*CI*) are validated by correlation study of the newly discovered indices and *IMD* predictors. Tables 9 and 10 show the best correlation of climatic indices discovered by *Info*, *Wlktrp*, and *Fstgreedy* methods with existing *IMD* predictors as discussed earlier for *NET_SP* and *NET_ZV*, respectively. High correlation value (≥ 0.5) validates the proposed approach of climatic index discovery by inventing the existing indices (highlighted in bold). Medium correlation value ($0.2 \leq \mu < 0.5$) represents invention of new indices, which are related to existing indices, but may act as good predictor than the existing ones (normal font). Low correlation value (< 0.2) represents newly discovered indices different from known indices (highlighted in italics). Discovered climatic index for *NET_SP* shows high correlation with *EA_SP* and *NA_SP*, validate our approach by re-invention of existing predictor indices.

Table 9. Correlation of discovered climatic indices (*CI*) for *NET_SP* with *IMD* predictors for Indian monsoon

Existing Predictor CI	Info CI	Wlktrp CI	Fstgrdy CI
NA_SST	0.09	0.21	0.16
ESE_IO_SST	0.37	0.37	0.39
EA_SP	0.29	0.32	**0.52**
NA_SP	0.32	**0.56**	0.32
NC_PO_ zonal_wnd	0.23	0.32	0.23
NW_Eu_SP	0.24	0.30	0.50

Table 10. Correlation of discovered climatic indices (*CI*) for *NET_ZW* with *IMD* predictors for Indian monsoon

Existing Predictor CI	Info CI	Wlktrp CI	Fstgrdy CI
NA_SST	0.27	0.24	0.24
ESE_IO_SST	0.26	0.16	0.21
EA_SP	0.19	0.21	0.21
NA_SP	0.19	0.21	0.21
NC_PO_ zonal_wnd	0.27	0.18	0.20
NW_Eu_SP	0.22	0.29	0.19

6 Conclusions

New climatic indices important for Indian summer monsoon rainfall are discovered using algorithms of community detection for climatic parameters surface pressure and zonal wind velocity. Indices discovered are shown to have high correlation with Indian monsoon. Their correlation are even better than that of the known predictor indices used by *IMD* for predicting monsoon. Different regression and non-linear models are designed with discovered climatic indices as predictors. Mean absolute error of *5.4%* is achieved, which is appreciable for forecasting complex phenomenon of Indian monsoon. Prediction of monsoon by discovered indices of surface pressure is superior to *IMD*'s existing models. Finally, a study of correlation between discovered indices and predictor indices of Indian monsoon is performed as meteorological validation of our approach.

In future, other climatic parameters can be explored and new climatic indices can be discovered from combination of different climatic parameters which may be highly correlated and act as a better estimator of Indian monsoon.

References

1. Donges, J.F., Zou, Y., Marwan, N., Kurths, J.: Complex networks in climate dynamics. Eur. Phys. J.-Special Topics **174**(1), 157–179 (2009)
2. Steinhaeuser, K., Chawla, N.V., Ganguly, A.R.: Complex networks as a unified framework for descriptive analysis and predictive modeling in climate science. Stat. Anal. Data Min. **4**(5), 497–511 (2011)
3. Donges, J.F., Zou, Y., Marwan, N., Kurths, J.: The backbone of the climate network. Europhys. Lett. (EPL) **87**(4), 48007 (2009)
4. Steinhaeuser, K., Chawla, N.V., Ganguly, A.R.: An exploration of climate data using complex networks. ACM SIGKDD Explor. Newsl. **12**(1), 25–32 (2010)
5. Noor Md Sap, M., Awan, A.M.: Finding spatio-temporal patterns in climate data using clustering. In: Proceeding 2005 International Conference on Cyberworlds, pp. 8–15. IEEE (2005)
6. Steinbach, M., Tan, P.N., Kumar, V., Klooster, S., Potter, C.: Discovery of climate indices using clustering. In: Proceeding of the 9th ACM SIGKDD International Conference on Knowledge Discovery and Data Mining, pp. 446–455. ACM (2003)

7. Rosvall, M., Bergstrom, C.T.: Maps of random walks on complex networks reveal community structure. Proc. Natl. Acad. Sci. **105**(4), 1118–1123 (2008)

8. Pons, P., Latapy, M.: Computing communities in large networks using random walks. In: Yolum, I., Güngör, T., Gürgen, F., Özturan, C. (eds.) ISCIS 2005. LNCS, vol. 3733, pp. 284–293. Springer, Heidelberg (2005)

9. Clauset, A., Newman, M.E., Moore, C.: Finding community structure in very large networks. Phys. Rev. E **70**(6), 066111 (2004)

10. Kalnay, E., Kanamitsu, M., Kistler, R., Collins, W., Deaven, D., Gandin, L., Iredell, M., Saha, S., White, G., Woollen, J., Zhu, Y., Leetmaa, A., Reynolds, R., Chelliah, M., Ebisuzaki, W., Higgins, W., Janowiak, J., Mo, K.C., Ropelewski, C., Wang, J., Jenne, R., Joseph, D.: The NCEP/NCAR 40-year reanalysis project. Bull. Am. Meteorol. Soc. **77**(3), 437–471 (1996)

11. Parthasarathy, B., Munot, A.A., Kothawale, D.R.: Monthly and seasonal rainfall series for All-India homogeneous regions and meteorological subdivisions, 1871–1994. Research Report No. RR-065, Indian Institute of Tropical Meteorology (1995)

12. Gowariker, V., Thapliyal, V., Kulshrestha, S.M., Mandal, G.S., Sen Roy, N., Sikka, D.R.: A power regression model for long range forecast of southwest monsoon rainfall over India. Mausam **42**(2), 125–130 (1991)

13. Rajeevan, M., Pai, D.S., Dikshit, S.K., Kelkar, R.R.: IMD's new operational models for long-range forecast of southwest monsoon rainfall over India and their verification for 2003. Curr. Sci. **86**(3), 422–431 (2004)

14. Rajeevan, M., Pai, D.S., Kumar, R.A., Lal, B.: New statistical models for long-range forecasting of southwest monsoon rainfall over India. Clim. Dyn. **28**(7–8), 813–828 (2007)

Using Patterns in Computer Go

Leszek Stanisław Śliwa[✉]

The Institute of Computer Science, Faculty of Electronics and Information
Technology, Warsaw University of Technology, Warsaw, Poland
l.s.sliwa@stud.elka.pw.edu.pl
http://www.ii.pw.edu.pl

Abstract. Building a good heuristics for a computer program for Go
is difficult. Game tree is highly branched and there is a threat that
the heuristics would eliminate strong moves. Human players often use
patterns to decide where to put stones. Therefore, one of the ideas is to
develop the heuristics based on the database of "good" moves denoted by
patterns. A pattern is a small segment of the board. Each pattern's point
can be vacant, occupied by black or white stone or can be an off-board
point. A potential move is executed in the center of the pattern. Patterns
can be acquired from a human expert or through machine learning. This
paper presents a technique for: (1) retrieving patterns from a collection
of records of games played between human expert players, (2) storing
patterns, (3) implementing patterns in a computer program for Go.

Keywords: Artificial intelligence · Computer Go · MCTS · Heuristics ·
Patterns

1 Introduction and Overview

Construction of computer Go programs has been considered a grand challenge for
Artificial Intelligence. Unlike other two-players games, computer Go programs
does not play at human-master level. Ongoing research aims to develop new
algorithms to replace traditional methods of game tree search. Mini-max and
alpha-beta algorithms, which proved to be effective in many two-players games,
does not work successfully in computer Go programs because these algorithms
require a good evaluation function in order to give satisfactory results. However,
a good quality positional evaluation function, to predict the game state value, has
not been developed and therefore these algorithms are not able to deal with an
enormous size of game tree (which is a natural consequence of the combinatorial
complexity of Go game)[9].

This paper describes the automatic extraction of 3×3 patterns from professional games and is organized as follows. Section 1.1 presents the rules of Go
game. Next, Subsect. 1.2 introduces the algorithm MCTS. Then, in Sect. 1.3, a
short explanation of UCT selection strategy is given. Section 2 is dedicated to
related works. Section 3 presents the method of automatic patterns acquisition.

© Springer International Publishing Switzerland 2015
M. Kryszkiewicz et al. (Eds.): PReMI 2015, LNCS 9124, pp. 565–574, 2015.
DOI: 10.1007/978-3-319-19941-2_54

Section 4 is focused on some development issues. The results of experiments are presented and discussed in Sect. 5. The recommendations for future investigations are given in Sect. 6. The research conclusions are gathered in Sect. 7.

1.1 The Game of Go

Go is one of the oldest games in the world, originated in ancient China more than 2,500 years ago. Rules of Go are quite simple. The two players alternately place black and white stones, on the vacant intersections of a board with a 19×19 grid of lines. (Computer Go is often played on smaller 9×9 and 13×13 boards). Black makes the first move, alternating with White. A player may pass his turn at any time. A stone or orthogonally connected set of stones of one color are captured and removed from the board when all points directly adjacent to it are occupied by the opponent's stones. A player may pass his turn at any time. Stones cannot be placed to repeat a prior board position. The game ends when both players consecutively pass a turn. A player who controls more territory – wins. Generally, a player's territory consists of all the board points occupied or surrounded. Several rule sets exist for the game of Go but the Japanese and the Chinese are most popular. The primary difference between rule sets is the scoring method. Chinese rules are more often implemented in computer Go due to less implementation effort.

1.2 Monte-Carlo Tree Search

In recent years, a new algorithm MCTS (Monte-Carlo Tree Search) was developed [4, 10]. MCTS uses Monte-Carlo simulations as an alternative for positional evaluation function. This approach allows to achieve good results, even in the absence of expert domain knowledge. MCTS was developed initially for computer Go game, but it is a general algorithm and can be applied to solve other problems.

MCTS algorithm consists of two strategies: (1) selection applied recursively until the leaf with the highest scoring move is reached and (2) simulation that selects moves in self-play. In the first strategy, selection of the node corresponding to the most urgent move is the most important decision. The selection attempts to balance between exploitation and exploration because on the one hand, the most promising moves should be favoured and on the other, less promising moves still should be evaluated because their low scores could be a result of unfortunate coincidence (unlucky simulations). Both strategies can be facilitated by applying knowledge [6]. The simulation strategy can be enhanced by transformation of pure random simulations into partly deterministic playouts based on the knowledge. The knowledge can be designed by human experts or learned automatically.

1.3 Upper Confidence Bounds Applied to Trees

The Upper Confidence bounds applied to Trees (UCT) is a variant of the MCTS, introduced by Kocsis and Szepesvári (2006), based on the Upper Confidence Bounds (UCB) algorithm [15]. This strategy is implemented in many programs and it is used in mine as well[1]. The UCT controls the balance between exploitation and exploration. The strategy selects moves that leads to the best results (exploitation), as well as the less promising moves, due to the uncertainty of evaluation (exploration).

2   Related Works

The idea of applying patterns in computer Go is not new. There has been quite a big effort to implement patterns in a Go engine from the early beginnings – Zobrist A.L. (1970)[18]. At the end of the eighties the combination of search, heuristics, expert systems and pattern recognition was the winning methodology. Abramson B. (1990) described pattern acquisition from random generated game records [1]. Boon M. (1990) implemented pattern matching in the computer program GOLIATH [2]. Cazenave's work (2001) consisted in automatic acquisition of patterns generated in a special tactical context – connecting or making eyes and including liberties by using explanation-based learning [5]. Erik van der Werf (2002, 2003) described a neural network approach using professional recorded games to generate local moves [17]. Another approach using Bayesian generation and the K-nearest-neighbor representation can be found in Bouzy's and Chaslot's (2005) work [3]. Gelly et al. (2006) introduced the sequence-like simulation [11]. Wang and Gelly (2007) implemented sequence-like 3×3 patterns in their Go program MOGO [16]. Coulom (2007) searched for useful patterns in Go by computing Elo ratings for patterns, improving program CRAZY STONE [8]. Chaslot et al. (2008), described different methods of learning patterns [7]. Aduard et al. (2009) showed that opening books make a big improvement in play level. Hoock and Teytaud (2010) investigated the use of Bandit-based Genetic Programming to automatically find "better" patterns that should be more often simulated for their computer program MOGO [14]. Chaslot et al. (2010) provided a comprehensive description of common patterns, tactical and strategic rules, progressive widening or progressive unpruning [6].

3   Automatic Pattern Acquisition

In this section the method of automatic pattern acquisition is described and it is structured in the following way. First the training set is introduced. Next, the coding of patterns is described. Subsequently, the collection of raw patterns is characterized. Then, clusters of patterns are introduced. Finally elimination of "poor" patterns is explained.

3.1   The Training Set

A set of Go game records has been copied from the web page [13]. It has contained records of the games played on the K amateur Go Server (KGS, formerly known

---

[1] Development of computer Go program has been one of goals of author's Master's thesis.

Table 1. Number of games, moves and cumulative amount of patterns identified in the training set

Tournaments	Number of games	Number of moves	Cumulated amount of patterns
Jan 2013	1,287	251,812	945
Feb 2013	842	164,641	956
Mar 2013	1,085	211,299	963
Apr 2013	1,146	227,220	970
May 2013	1,053	199,673	974
Jun 2013	1,065	204,844	976
Jul 2013	975	237,782	978
Aug 2013	1,217	267,051	984
Sep 2013	1,384	249,183	985
Oct 2013	1,252	249,183	988
Nov 2013	1,230	248,235	990
Total	12,536	2,447,567	990

as Kiseido Go Server). Participated in the tournaments players represented a master level of playing strength (6–9 dan). For the analysis, tournaments played in 2013 have been chosen – 12,536 games and 2,447,567 moves. All games have been played on 19×19 boards. Table 1 summarizes the number of games, moves and the cumulative amount of discovered patterns.

3.2 Coding of Patterns

"Go Game" computer program has been used to extract patterns from the data set. Game records have been read from SGF[2] files. For every move, a 3×3 pattern has been extracted and coded. The move has always been made in a central point of the pattern. To encode 4 possible states of every point, 2 bits have been used. Bits 00 means a point is a black stone, 01 is a white stone, 10 is an empty point, 11 is a point located outside of the board (off-board). To encode all points Formula 1 has been used, where P_{id} is a pattern's identifier and C_p is the colour code in point p.

There are 3 reasons why 3×3 patterns have been chosen: (1) cardinality of patterns set is small (2) matching is fast and (3) patterns represent enough information about connections and cutting of adjacent chains of stones.

$$P_{id} = (((((((C_1 * 4) + C_2) * 4 + C_3) * 4 + C_4) * 4 + C_5) * 4 + C_6) * 4 + C_7) * 4 + C_8 \tag{1}$$

---

[2] Smart Game Format – tree-based, portable, file format used for storing records of board games.

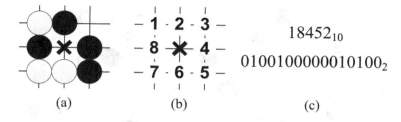

Fig. 1. Coding of 3×3 patterns

Figure 1 shows how the feature patterns have been represented: (a) 3×3 pattern, (b) the sequence of coded points, (c) a pattern's identifier in the decimal and binary code.

3.3   Collection of Raw Patterns

Recognized patterns have been saved in a disk file. Each pattern's record had 2 attributes: the pattern's identifier and the player's colour. Additionally, for every game, additional information has been registered: (1) size of the board, (2) the initials of players, (3) playing strength of opponents, (4) the date of the game, (5) *Komi* points, (6) game's rules, (7) the reason the game has been ended, and (8) the number of handicap stones[3].

3.4   Clusters of Patterns

Collected samples have been loaded from a disk file into the program memory. Then, patterns have been sequentially checked whether each item has already been registered in the database. If the pattern has not been found, it has been added to the database and its frequency counter has been set to 1. Otherwise, its frequency counter has been incremented. During the clustering, to avoid duplication of patterns, operations of: (1) rotation, (2) mirroring and (3) colour exchange have been applied. If the I is the identity, R_x is the rotation of the x degrees, H reflection horizontal and X swap colors, then the same pattern can be represented in 16 identical forms: $I, R_{90}, R_{180}, R_{270}, H, H*R_{90}, H*R_{180}, H*R_{270}, I*X, R_{90}*X, R_{180}*X, R_{270}*X, H*X, H*R_{90}*X, H*R_{180}*X, H*R_{270}*X$.

It has been assumed that the database would contain patterns only for moves made by white. If the registered pattern corresponded to a move made by black, then colour of stones has been inverted (black to white and opposite).

The majority of the patterns (945 out of a total number of 990) have been recognized during the analysis of the first tournament (Jan 2013). During further analysis the number of patterns has been growing logarithmically. On this basis, it has been concluded that: (1) training set does not have to be very large and (2) cardinality of the training set has been sufficient to build the complete database of patterns.

---

[3] Handicap stones – stones placed on a board to balance a game between players of significantly different playing strength.

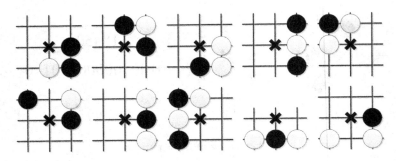

Fig. 2. Top ten high-frequency patterns for White

3.5   Refinement of the Pattern Database

There have been several reasons why collected patterns had to be corrected.

Analysed games have been played by different rules: Japanese, American, Chinese etc. According to Chinese rules a move is illegal, if one or more stones of that player's colour could be captured (prohibition of so called *suicide* move). But in analysed games such moves have been often played. Due to the fact that only Chinese rules have been implemented in "Go Game" computer program, the patterns including *suicide* would be considered as incompatible with the game rules and therefore had to be removed from the pattern database.

Some of the patterns have been found less valuable, eg. the pattern with all empty points. (Interestingly, this pattern has been played most often.) It has been assumed that such moves represented change in "the theatre of war" and attempted to "open another front". In MCTS algorithm, such a strategy is already implemented by random selection from unoccupied points. It has also been assumed that the pattern is interesting if contains at least 3 stones or off-board points.

The first trials of using patterns have shown that the developed set has contained many weak, rarely played moves that couldn't be recommended in any way as "strong". Therefore, the patterns have been sorted in descending order by the frequency counter. Patterns having the highest frequency that cover 80 % of all use cases have been selected. Other patterns have been deleted. After the last operation 231 patterns have left (from the initial set of 990). The result is close to the Pareto rule (80/20). Figure 2 shows the top ten patterns for White with highest frequency of occurrence from the final pattern database.

4   Use of Patterns in Monte-Carlo Simulations

In this section, some development issues are discussed: the algorithm, data structures, and crucial technical details having impact on the program performance.

4.1   The Algorithm and Data Structures

The set of 231 patterns has been stored in CSV file and loaded into the program's memory before start of tournaments. Patterns have been kept in memory in associative container with key-value elements. During the game, after an opponent's move, the list of empty points adjacent to two last moves played on the board has been created. For every adjacent empty point, the corresponding pattern's id (Formula 1) has been calculated and searched in the set of "good" patterns. If several patterns have been found, the pattern having the highest frequency has been selected. The point in the center of the chosen pattern has been a recommended move.

The implemented strategy looks for an interesting move in the proximity of the last two played moves. It is based on insight that in Go game, moves are often played next to each other creating sequence of adjacent moves (mostly at a Manhattan distance of 1).

4.2 Development Details

During thinking time, the program executes tens of thousands of Monte-Carlo simulation. During tests, program with a thinking time of 3 s per move, has been performing from 50,000 simulations at the beginning of a game, up to 100,000 (and more) in the final phase (for games played on 9×9 board). Therefore time of code execution in playouts is critical because reducing the number of simulations may have a negative impact on the playing strength.

Measure of execution time of component operations executed in playouts has showed that the rotation and mirroring have been a bottleneck. Interestingly, searching patterns in the associative container has been the one of fastest operations. The problem has been solved in the following way. The database has contained only 231 patterns, each pattern could appear in maximum 8 variations (during experiments, the computer program has always been playing white), therefore the total number of patterns variations (including rotation and mirroring) has expanded to 1848. To eliminate the bottleneck operations, for fast matching, all possible combinations has been placed in the container. The tests have shown that increase in the number of items in a container from 231 to 1848 didn't affect the number of simulations.

5 Experiments

This section is focused on experiments. In Subsect. 5.1 tournaments without patterns are shortly explained. In Subsect. 5.2, results of games where proposed patterns have been used are discussed in details.

5.1 Reference Tournaments

In order to obtain reference data, tournaments on boards of size 9×9, 13×13, 15×15 and 19×19 have been played against *GnuGo* – another computer program chosen as the opponent [12]. In every tournament 100 games have been played. The final move played in the actual game has been defined by the leaf (in the search tree) with the highest visit count ("robust" leaf). UCT (Upper Confidence Bound applied to trees) strategy has been used to select leaves worth to be estimated by Monte-Carlo simulations. Thinking time has been limited to 3 seconds. The pattern database has not been used.

Table 2. Scores of tournaments verifying effectiveness of "good" patterns heuristics

Parametr		9×9		13×13		15×15		19×19	
		B[a]	W[b]	B	W	B	W	B	W
Number of won games	R[c]	2	98	21	79	35	65	62	38
	H[d]	5	95	8	92	21	79	55	45
Cumulated amount of ahead points	R	16	819	267	1121	456	1282	1700	520
	H	32	1078	150	1363	371	1355	1906	665

[a]B–black (opponent–*GnuGo* program).
[b]W–white (player–*Go Game* program).
[c]R–reference games.
[d]H–games, where "good" pattern heuristics has been used.

5.2   The Assessment of the Pattern Database Heuristics

For tournaments verifying effectiveness of patterns heuristics, computer Go program parameters have been set as for the reference tournaments (number of contents and games, board's size, final move selection, UCT strategy, thinking time), but this time "good" patterns heuristics has been used.

During simulations, the heuristics of "good" patterns could not have the highest priority. Generally, 3×3 patterns are too small to contain information about the wide context of the move. Therefore, capture and *atari* defence moves being more urgent and valuable have been considered as the first option and then followed by patterns adjacent to the last two moves played on the board. Finally, if no move has been selected so far, a point from the list of available and legal moves has been randomly selected. The scores in the tournament games have been shown in Table 2.

The final score for games played on 9×9 board has been slightly worse than the result in the reference tournament. The program has lost 5 games (in the reference tournament only 2), but the difference in the number of ahead points has increased from 819 to 1078. Thus, if the program has won, it has scored more points.

The score of games on 13×13 board has been significantly better. The program has won 13 more games than in the reference tournament. The difference in ahead points has been increased from 1121 to 1363. In games on the 15×15 board the player advantage over the opponent has been maintained. The program has won 14 more games and the difference in the number of ahead points has increased from 1282 to 1355.

Computer programs for Go based on MCTS "do not scale well". They are the most effective at small and medium boards. On large boards they give way to "traditional" programs (based on expert knowledge). In the reference tournament on 19×19 board "Go Game" has won 38 games and has lost 62 games. After the application of "good" pattern heuristics database the playing strength of both programs have been aligned. "Go Game" program has won 45 games and has lost 55.

6 Future Research

I would consider patterns of bigger size – 5 × 5. Some work has already been done, but first attempts based on the method described in this paper and applied to 5 × 5 patterns were not satisfactory. One of ideas is to use n-gram statistical pattern acquisition, similar to NLP methods.

7 Summary

This paper presents the technique for: (1) machine retrieving patterns from a collection of game records played between human expert players, (2) storing patterns and (3) implementing patterns in a computer Go program. Each pattern has been represented by a small 3 × 3 points piece of board with an empty point in its center, where a potential move has been considered. Every pattern identified by an integer number (pattern ID) has been given a weight depending on its frequency. Patterns have been extracted from a training data set containing 2,447,567 moves in 12,536 games (all played on 19 × 19 board). During the clustering, to avoid duplication caused by patterns variations, operations of: (1) rotation, (2) mirroring and (3) color exchange have been applied. For final set, patterns with highest weight, due to the Pareto rule (80/20), have been selected.

To verify how the patterns are effective and improves the playing strength of the computer program, the heuristics based on patterns has been tested in 4 contests and compared to reference tournaments. The results of experiments shows that the heuristics of "good" patterns used in MCTS, in simulations strategy, improves the playing strength on medium and large boards.

References

1. Abramson, B.: Expected-outcome: a general model of static evaluation. IEEE Trans. PAMI **12**, 182–193 (1990)
2. Boon, M.: A pattern matcher for Goliath. Comput. Go **13**, 13–23 (1990)
3. Bouzy, B. and Chaslot, G.: Bayesian generation and integration of K-nearest-neighbor patterns for 19x19 Go. In: Kendall, G., Lucas, S.(eds.) IEEE 2005 Symposium on Computational Intelligence in Games, pp. 176–181, Colchester (2005)
4. Browne, C., et al.: A survey of Monte Carlo Tree Search methods. IEEE Trans. Comput. Intell. AI Games **4**(1), 1–43 (2009)
5. Cazenave, T.: Generation of Patterns with External Conditions for the Game of Go. In: van den Herik, H.J., Monien, B.(eds.) Advances in Computer Games 9, pp. 275–293. Universiteit Maastricht, Maastricht (2001)
6. Chaslot, G., Fiter, C., Hoock, J.-B., Rimmel, A., Teytaud, O.: Adding expert knowledge and exploration in Monte-Carlo Tree Search. In: van den Herik, H.J., Spronck, P. (eds.) ACG 2009. LNCS, vol. 6048, pp. 1–13. Springer, Heidelberg (2010)
7. Chaslot, G., et al.: Progressive strategies for Monte-Carlo Tree Search. New Math. Nat. Comput. **4**(3), 343–357 (2008)

8. Coulom, R.: Computing Elo ratings of move patterns in the game of go. Int. Comp. Games Assoc. J. **30**(4), 198–208 (2007)
9. Drake, P., and Uurtamo, S.: Heuristics in Monte Carlo Go. In: International Conference on Artificial Intelligence, pp. 171–175, Las Vegas (2007)
10. Gelly, S., et al.: The grand challenge of computer go: Monte Carlo Tree Search and extensions. Commun. ACM **55**(3), 106–113 (2012)
11. Gelly, S., et al.: Modifications of UCT with Patterns in Monte-Carlo Go. Technical report 6062, INRIA (2006)
12. GNU Go 3.8. http://www.gnu.org/software/gnugo
13. Görtz, U., Game records in SGF format. http://www.u-go.net/gamerecords
14. Hoock, J.-B., Teytaud, O.: Bandit-based genetic programming. In: Esparcia-Alcázar, A.I., Ekárt, A., Silva, S., Dignum, S., Uyar, A.Ş. (eds.) EuroGP 2010. LNCS, vol. 6021, pp. 268–277. Springer, Heidelberg (2010)
15. Kocsis, L., Szepesvári, C.: Bandit based Monte-Carlo planning. In: Fürnkranz, J., Scheffer, T., Spiliopoulou, M. (eds.) ECML 2006. LNCS (LNAI), vol. 4212, pp. 282–293. Springer, Heidelberg (2006)
16. Wang, Y., Gelly, S.: Modifications of UCT and sequence-like simulations for Monte-Carlo Go. In Proceedings of IEEE Symposium on Computational Intelligence and Games, pp. 175–182, Honolulu (2007)
17. van der Werf, E., Uiterwijk, J.W.H.M., Postma, E.O., van den Herik, H.J.: Local move prediction in go. In: Schaeffer, J., Müller, M., Björnsson, Y. (eds.) CG 2002. LNCS, vol. 2883, pp. 393–412. Springer, Heidelberg (2003)
18. Zobrist, A.L.: Feature Extraction and Representation for Pattern Recognition and the Game of Go. Ph.D. thesis, University of Wisconsin, Madison (1970)

Event Detection from Business News

Ishan Verma[1(✉)], Lipika Dey[1], Ramakrishnan S. Srinivasan[2],
and Lokendra Singh[1]

[1] TCS Innovation Labs, Tata Consultancy Services Ltd., Delhi, India
{ishan.verma,lipika.dey,singh.lokendra1}@tcs.com
[2] TCS Innovation Labs, Tata Consultancy Services Ltd., Cincinnati, USA
ramakrishnan.srinivasan@tcs.com

Abstract. An event is usually defined as a specific happening associated with a particular location and time. Though there has been a lot of focus on detecting events from political and other general News articles, there has not been much work on detecting Business-critical events from Business News. The major difference of business events from other events is that business events are often announcements that may refer to future happenings rather than happenings that have already occurred. In this paper, we propose a method to identify business-critical events within News text and classify them into pre-defined categories using a k-NN method. We also present an event-based retrieval mechanism for business News collections.

Keywords: Event detection · Classification · Event-based news retrieval

1 Introduction

Today's business intelligence is heavily event-driven. In this context, events of interest are those that provide information necessary to strategically exploit and improve an enterprise's processes as well as to take tactical advantage of events as they occur. With a multitude of News sources on the web, it is possible to gather a lot of information about competitors, product failures, and major global economic and political events almost as soon as they occur. Analysis of business or financial news has gained popularity due to the immense potential these articles have for contributing towards effective predictive analytics for companies. Traditionally, the most exploited technology for analyzing business news is sentiment classification that assigns positive or negative polarity to an article and thereby to the entities contained within it. There is a large body of research that studies the correlation between News sentiments and market indices. However, there are many other specific events that can be extracted from business News articles that can provide useful business insights or competitive intelligence for future planning. For example an announcement like *"Hyundai has just launched the third generation of the Santa Fe which has sleeker lines and a very executive look"* may prove to be useful for Hyundai's competitors to plan demand scenarios in similar segments.

An event is usually defined as a specific happening associated with a particular location and time. Though there has been a lot of focus on detecting events from

M. Kryszkiewicz et al. (Eds.): PReMI 2015, LNCS 9124, pp. 575–585, 2015.
DOI: 10.1007/978-3-319-19941-2_55

political and other general News articles, there has not been much work on detecting Business-critical events from Business News. The major difference of business events from other events is that one is only interested in specific types of events that have impact on business intelligence, competitive intelligence and planning. Business events are also of interest to investors and financial planners who make their decisions based on factors like brand image, disclosures on governance and environmental issues etc. Typically, business events may be announcements rather than reporting.

In this paper, we present a system that can be trained to detect and classify business-critical events from business News. We propose a k nearest neighbors (k-NN) based method for classification of business-critical events. The uniqueness of the current work lies in the use of word vectors while computing the k nearest neighbors, thereby exploiting both surface-level syntactic and semantic similarities of sentences for classification. We also present an event-based News retrieval system that helps analyst retrieve News articles of specific interest.

Section 2 provides an overview of earlier work. Sections 3, 4 and 5 present the core contributions of this paper in terms of event classification and event-based news retrieval. A specific application scenario is presented in Sect. 6 for the supply chain process. Section 7 presents results from experiments with a large News collection.

2   Review of Related Work

Event extraction from large volumes of unstructured text like News collections has emerged as one of the popular sub-tasks of information extraction (IE). An overview paper on event extraction from text documents, categorized three predominant approaches to event extraction. Reference [1] Data-driven approaches, as adopted in [2–5] rely solely on quantitative methods to discover relations and thereby events from text. They require large text corpora to develop models that approximate linguistic phenomena. Knowledge-driven event-extraction on the other hand uses patterns derived from expert inputs based on linguistic and lexicographic knowledge. The patterns are either lexico-syntactic patterns [6, 7] or lexico-semantic patterns that are customized for domains like stock-market as proposed in [8]. Reference [9] reported a large-scale event extraction system named REES that extracted 61 pre-specified types of events under different categories like crime-events, business-events, financial-events, political-events etc. REES uses a declarative, lexicon-driven approach where each lexicon entry is defined for a specific type of event. Reference [10] proposed an ontology-based event extraction mechanism to extract violent and natural disaster events from online news, which were first clustered into similar groups.

Many functional approaches use a combination of the above methods to make best use of all the worlds. References [11–13] propose methods for extracting events and event templates or event schemas from large-scale text collections. While events represent a time-stamped single instance of an incident, an event schema is defined as a set of actors that play different roles in an event, such as the perpetrator, victim, and instrument in a bombing event. Reference [14] reported dynamic event discovery mechanisms based on discovery of relationships among co-bursting entities along with underlying global and local time constraints. Reference [15] proposed the use of

RelGrams which are combinations of Subject-Object-Predicates and statistical reasoning to identify event-schemas from large corpora. Reference [16] developed a probabilistic solution for template generation. The approach requires performing joint probability estimation using EM, which is not scalable to large corpora. Reference [17] worked on extracting significant events from webcast text on sports where significance of an event was computed based on people participating in reporting the event. Since an event is covered from multiple perspectives in different News articles, the focus of [18] was to generate the most compact, objective and informative headline for the event.

It may be observed that while there has been considerable work on event extraction and event schema identification, not much has been done in event-based retrieval or computing significance of an event towards analytical tasks. Consequently not much work has been done on categorizing business-critical events, their correlations and possible impact. In [20] a rule based approach was used to identify feature based sentiment and business event phrases from news documents. The relevance of news articles was decided on the basis of presence of event and sentiment polarity. This is closest to our work. However the proposed mechanisms are not rule-based, hence easily adaptable to different domains.

3   Event Extraction and Classification Architecture

In this section we present a brief overview of the proposed mechanism to extract and classify business events from News articles. The proposed system detects events of predefined types that can be used for particular business analysis tasks. The business context defines the type of events that are useful for the domain. The system is accordingly trained to extract business-critical events from News articles. An event occurrence is assumed to be detected from a single sentence within a News article. There may be multiple event occurrences of same or different classes within a single News article. Analysts can query the system for News articles that contain events of specific types associated to specific entities.

Figure 1 shows the proposed system architecture. News articles are collected from a set of pre-defined news websites using RSS feeds. Each news article is then subjected to sentence extraction, word stemming using Porter's algorithm[1] and then Named entity extraction. Named entities denote names of people, organizations, products, money or date values etc. The Stanford core NLP[2] suite has been used for extracting sentences and named-entities from the news content. All news articles along with the metadata element like source, author, date etc. and extracted sentences and named-entities are stored in a local data repository.

Each sentence is then subjected to event detection and classification. A sentence may be categorized into one of the known event classes or as "others" indicating that it does not contain any business-critical information for the given context. Subsequently

---

[1] http://snowball.tartarus.org/algorithms/porter/stemmer.html.

[2] http://nlp.stanford.edu/software/corenlp.shtml.

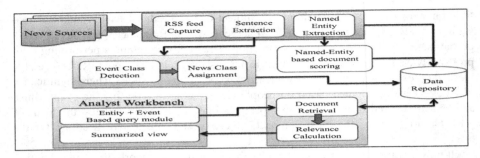

Fig. 1. Proposed event extraction and classification architecture

the event-class distribution within a News article is computed. The analyst's work-bench is a search and analytics module which underlines the use of events for business analysis activities. An analyst can query the News collection for events related to specific entities. The named entities typically represent product, people or company names. The named-entities and event distributions are used for retrieving relevant News articles for a specific entity-event combination. It presents a list of articles to the user ranked by their relevance to the query. Analysts are also presented with sum-marized views of events extracted from the collection in association with pre-specified named entities.

4   Business Event Classification

In this section we present a k-nearest neighbor based classification algorithm for detecting and classifying business events within an article. The novelty of the proposed method lies in sentence representation and similarity computation mechanism within a classical k-NN framework. The system is trained to classify each sentence of an incoming News article based on its similarity to each classified sentence in a training set. The assigned classes of the top k nearest neighbors of the sentence are then used to determine the class of the sentence based on majority voting. Multiple values of k have been used for experiments.

4.1   Word-Vector Based Sentence Representation

We have used the *word2vec*[3] tool to compute continuous distributed representations of words. Each word is represented as a 500 dimensional numerical vector. This tool learns word representations that are not sparse and semantically similar or grammati-cally closer words are placed closer to each other. One billion word benchmark dataset [19] was used to obtain the word vectors. A window size of 10 words was chosen and words with lesser than 10 instances were ignored.

---

[3] https://code.google.com/p/word2vec/.

The proposed system first takes a text corpus as input and produces the word vectors as output. Continuous bag-of-words algorithm is used to construct a vocabulary from the training text data and then learn vector representation of words. The word-vector store is indexed for fast retrieval of a vector for a given word. Using the word-vector representation ensures that use of semantically similar words also does not affect event recognition and classification accuracy.

4.2   k-NN Based Event Classification

The word vectors generated in the earlier step are used for detecting events from News articles, by classifying each sentence into a known event class or a separate class called OTHERS. The vector representation of each word in an incoming sentence is retrieved from the word-vector store. Similarity between a pair of sentences is computed in terms of the word-vectors of the contained words.

For each word in a given sentence, its most similar neighbor in another sentence is located using the cosine similarity of word vectors. This ensures position of words do not affect accuracy of classification.

To find the closest neighbor of a word 'w' in another sentence S_i, the word vector of w is compared with the word vectors of each word in S_i using the cosine similarity measure. The word of S_i which yields the highest similarity with w, is accepted as its neighbor in S_i. It may be noted that the same word of S_i may be returned as neighbor for two different words.

The function $Similarity(S_X, S_i)$ computes similarity of two sentences using the above measure. The function $kNN(S_X)$ assigns the class label for a new sentence S_X. Finally, the two functions are used to compute the event distribution for each News article. Stop-words are ignored while computing similarity. Since words have been already stemmed so different morphological forms of the same word have the same representation and hence the same word vector. This reduces the size of the vocabulary.

Function Similarity(S_X, S_i)

1. Let W_x and W_i denote the sets of words in S_x and S_i respectively
2. For each $\omega_p \in W_x$, find semantically similar word $\omega'_p \in W_i$ as follows:

   (a) Let $\overrightarrow{V_{\omega_p}}$ denote the word-vector for word ω_p and $\overrightarrow{V_{\omega_j}}$ denote the word-vector for word ω_j where $\omega_j \in W_i$

   (b) $\omega'_p = argmax_{\omega_j} \left[cosine\_similarity \left(\overrightarrow{V_{\omega_p}}, \overrightarrow{V_{\omega_j}} \right) \right]$

3. Calculate $Similarity(S_X, S_i) = \dfrac{\sum_{\omega_p} \left[cosine\_similarity \left(\overrightarrow{V_{\omega_p}}, \overrightarrow{V_{\omega'_p}} \right) \right]}{|W_x|}$

Function kNN(S_X)

1. Let $T = \{$Set of labeled sentences where the label denotes an event class$\}$
2. For each sentence S_i in T
   (a) Calculate $Similarity(S_X, S_i)$

3. Select $T' \subset T$ such that T' contains k most similar sentences for S_X based on $[\text{Similarity}(S_X, S_i)]$
4. Obtain labels L_k from sentences in T'
5. Count occurrences for each label l_i in L_k
6. Assign class l_i to S_X if count $(l_i) > k/2$ else assign class "Others"

Event distribution for each News articles is computed as follows.

1. For each News article D construct an event vector $\overrightarrow{E_D}$ which is initially NULL.
   (a) For each sentence $S_i \in D$. Obtain event class C_i using Function $kNN(S_i)$
   (b) For each event class C_i
      (i) Calculate $P(C_i/D) = \left(\frac{\text{Frequency of } C_i \text{ in D}}{\text{total no of sentences in D}} \right)$
   (c) $Event\ vector\ \overrightarrow{E_D}[i] = P(C_i/D)$

It may be noted that the similarity measure used here is not commutative. Since the goal of the computation is to identify for a given sentence the maximally aligned sentence from the training class and thereby its assigned label, one-way matching of the given sentences to training data was needed.

5   Event-Based News Retrieval

As stated in Sect. 3, one of the uses of event classification can be searching of a news collection using implicit or explicit event queries. An event query is of the form (E_i, C_j) where E_i denotes a named-entity of interest and C_j denotes an event class. News articles retrieved from the local repository are ranked by relevance with respect to the above query, where the total relevance score is generated by combining the two scores defined below:

Named Entity Based Relevance Score - A document D may contain multiple entities of the same or different types with different frequencies. The relevance score presented here considers the relative importance of given entity E_i within the document. The first step is to identify entity type T of E_i using lookup tables built from the named entities extracted from existing document collection. Post that the entity is represented as E_i^T.

Let E denote any entity. Let E^T denote an entity of type T

$\sum_D E_i^T$ = Total number of occurrences of E_i^T in document D

$\sum_D E$ = Total number of occurrences of entities in document D

$\sum_D E^T$ = Total number of occurrences of entities of type T in document D

$$\text{Score } (D/E_i^T) = \left[\lambda \left(\frac{\sum_D E_i^T}{\sum_D E^T} \right) + (1 - \lambda) \left(\frac{\sum_D E - \sum_D E_i^T}{\sum_D E} \right) \right]$$

The first factor determines the relative importance of the given entity with respect to other entities of the same type. It takes care that a random mention of an entity name in a document does not fetch a high score for it. For example, even when a News article is predominantly about a company's performance, its competitor names are also mentioned in it. This factor makes sure that this document gets high relevance for the primary company rather than its competitors. The second factor determines whether the document is majorly talking about the relevant type of entity. This factor helps in distinguishing among different types of entities that may be related to each other. For example, when the information required is about a product, though its manufacturing company name may occur in an article, it is expected that the product name will be more frequent than its company name. News articles with fewer mentions of the product and more about the company would be ranked lower than the earlier article. We found experimentally that $\lambda = 0.8$ yields the best results.

Event Based Relevance Score – Each news article's relevance to the queried event is computed using the event distribution vector. The score is represented as:

$$\text{Score } (D/C_j) = \overrightarrow{E_D}[j]$$

Total relevance score – The total score is a linear combination of the above scores:

$$\text{Score } (D/(E_i^T, C_j)) = 0.5 * \text{ Score } (D/E_i^T) + 0.5 * \text{ Score } (D/C_j)$$

6 A Use-Case Scenario for Business Event Detection for Supply Chain Analysts

In this section we present a specific application scenario for business event detection in the supply chain industry. The supply chain industry is involved in logistics and planning for each stage of product manufacturing and delivery. Manufacturing itself is preceded by procurement of raw materials or supply of components. The three stages of supply chain are

1. Sourcing of raw material or components for production - termed as Source.
2. Making or manufacturing of products - termed as Make.
3. Delivery of products are termed as source - termed as Deliver.

The supply chain is impacted by multiple factors at different stages of Source, Make or Deliver processes. Following is a detailed description of different types of events that are considered critical to the domain. Table 1 shows examples of how they occur in News articles.

1. **Market News:** Any event related to sales directly influences the Deliver process of the supply chain. Events of this type report upward or downward trends in the focus company, competitors, or the industry as a whole.
2. **Production News:** News related to manufacturing facility expansions or contractions provide insights that could lead to short-term adjustments or long-term plans.

Table 1. Event examples extracted from news articles

Event class	Example	Source
Market news	Hyundai had record U.S. sales of more than 700,000 in 2012	Forbes
Production	Nissan is already building its third plant in Mexico, breaking ground in July on the $2 billion factory in the central state of Aguascalientes	Bloomberg
Supply disruption	'Toyota Motor faced another tough year in fiscal 2011 (ended March 31, 2012), hit by supply disruptions and production cuts following the Great East Japan Earthquake in March 2011'	Reuters
Launch	Hyundai has just launched the third generation of the Santa Fe which has sleeker lines and a very executive look	Irish Independent
People	Ford Motor Company today announced that Michael Boneham, president and managing director, Ford India has elected to retire effective December 31, 2012 after a successful career spanning over 27 years with Ford Motor Company	The Hindu
Product failure or recall	Ford this month recalled 16,000 Fusions for excessive engine temperatures that could lead to fire and an additional 19,000 Fusions for a defect with low-beam headlights	Reuters

These could include new equipment capabilities, labour requirements, lower cost manufacturing locations, and so on.

3. **Supply Disruption News:** Suppliers form the starting point of any supply chain. Any news regarding supply disruption related to a supplier or a region may have a ripple effect on entire business. Suppliers may supply to competing manufacturers, disruption events help manufacturers take tactical corrective action.

4. **Launch News:** Launching a new product has an indirect effect on the Deliver and Make processes. Tracking news about newly launched products will enable a business to be better prepared for positioning requisite manufacturing and warehouse capacity.

5. **People News:** Sometimes news related to people associated with a company has an impact on the demand for that company's products. Executive movements, new responsibilities, retirements, and death, can change customers' perception of a company and its products.

6. **Product Failure or Recall News:** Product failures and recall events have an indirect effect on all aspects of business. They can affect the demand for the recalled products, cast doubts about manufacturing quality, and supplier non-conformance.

Figure 2 illustrates the relationship among the different classes of events and the supply chain stage that they affect. These events are also classified as primary or secondary factors depending on their impact. Primary factors directly influence core supply chain processes, while secondary factors wield an indirect influence.

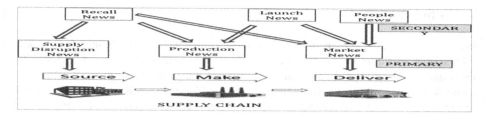

Fig. 2. Impact of events on Supply Chain – an example

7 Results for Supply Chain Use-Case

In this section we first present event detection and classification results for a collection of 30000 news articles published over a period of 15 months from Oct-2012 to Dec-2013 for major Automobile companies that trade in US. Training data for k-NN classification contained 600 manually tagged sentences for the 6 categories mentioned in Sect. 6. Table 2 shows 10-fold cross validation accuracies for different values of k from 3 to 10. It was observed that k = 7 yields best accuracy.

Table 2 10- fold cross validation k-NN classification accuracy for different k

k	Accuracy	k	Accuracy	k	Accuracy	k	Accuracy
3	0.88	5	0.89	7	0.91	9	0.89
4	0.88	6	0.88	8	0.88	10	0.86

The retrieval precision was manually verified for the most relevant article using 120 queries, generated as a combination of four entities across six event classes for five months. The precision yielded was 73.3 %. Figure 3 shows a sample screenshot of the most relevant News articles retrieved for event class 'Market News' for three automobile companies for January, 2013.

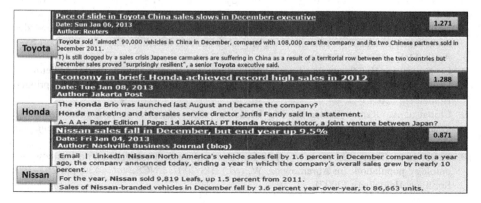

Fig. 3. Most relevant Market news for Honda, Toyota, Nissan and Hyundai (Jan,'13)

8 Conclusion

In this paper, we have presented a method for detecting and classifying business-critical events from Business News articles. It has been shown that use of word vectors can yield quite high accuracies for event classification. This paper also presents an event-based News retrieval mechanism which can be used by analysts. Rigorous experiments have been conducted with a large collection of News articles and examples from a real use case scenario have been shared. This work is being currently extended to generate automated alarms for supply chain analysts on detecting events of interest. Computing impact of events in conjunction with business data is also another area which is being explored.

References

1. Hogenboom, F., Frasincar, F., Kaymak, U., de Jong, F.: An overview of event extraction from text. In: Workshop on Detection, Representation, and Exploitation of Events in the Semantic Web (DeRiVE 2011) (2011)
2. Lei, Z., WU, L.-D., Zhang, Y., Liu, Y.-C.: A system for detecting and tracking internet news event. In: Ho, Y.-S., Kim, H.J. (eds.) PCM 2005. LNCS, vol. 3767, pp. 754–764. Springer, Heidelberg (2005)
3. Tanev, H., Piskorski, J., Atkinson, M.: Real-time news event extraction for global crisis monitoring. In: Kapetanios, E., Sugumaran, V., Spiliopoulou, M. (eds.) NLDB 2008. LNCS, vol. 5039, pp. 207–218. Springer, Heidelberg (2008)
4. Liu, M., Liu, Y., Xiang, L., Chen, X., Yang, Q.: Extracting key entities and significant events from online daily news. In: Fyfe, C., Kim, D., Lee, S.-Y., Yin, H. (eds.) IDEAL 2008. LNCS, vol. 5326, pp. 201–209. Springer, Heidelberg (2008)
5. Okamoto, M., Kikuchi, M.: Discovering volatile events in your neighborhood: local-area topic extraction from blog entries. In: Lee, G.G., Song, D., Lin, C.-Y., Aizawa, A., Kuriyama, K., Yoshioka, M., Sakai, T. (eds.) AIRS 2009. LNCS, vol. 5839, pp. 181–192. Springer, Heidelberg (2009)
6. Nishihara, Y., Sato, K., Sunayama, W.: Event extraction and visualization for obtaining personal experiences from blogs. In: Salvendy, G., Smith, M.J. (eds.) HCH 2009. LNCS, vol. 5618, pp. 315–324. Springer, Heidelberg (2009)
7. Hung, S.-H., Lin, C.-H., Hong, J.-S.: Web mining for event-based commonsense knowledge using lexico-syntactic pattern matching and semantic role labelling. Elsevier J. Expert Syst. Appl. 37(1), 341–347 (2010)
8. Li, F., Sheng, H., Zhang, D.: Event pattern discovery from the stock market bulletin. In: Lange, S., Satoh, K., Smith, C.H. (eds.) DS 2002. LNCS, vol. 2534, pp. 310–3115. Springer, Heidelberg (2002)
9. Chinatsu, A., Ramos Santacruz, M.: REES: a large-scale relation and event extraction system. In: ANLC 2000 Proceedings of the Sixth Conference on Applied Natural Language Processing, pp. 76–83 (2000)
10. Piskorski, J., Tanev, H., Wennerberg, P.O.: Extracting violent events from on-line news for ontology population. In: Abramowicz, W. (ed.) BIS 2007. LNCS, vol. 4439, pp. 287–300. Springer, Heidelberg (2007)

11. Chambers, N., Jurafsky, D.: Unsupervised learning of narrative event chains. In: Proceedings of ACL-08: HLT (2008)
12. Chambers, N., Jurafsky, D.: Unsupervised learning of narrative schemas and their participants. In: Proceedings of ACL (2009)
13. Chambers, N., Jurafsky, D.: A database of narrative schemas. In Proceedings of LREC (2010)
14. Chambers, N., Jurafsky, D.: Template-based information extraction without the templates. In: Proceedings of ACL (2011)
15. Balasubramanian, N., Soderland, S., Mausam., Etzioni, O.: Generating coherent event schemas at scale. In: Proceedings of the 2013 Conference on Empirical Methods in Natural Language Processing, pp. 1721–1731 (2013)
16. Cheung, J., Poon, H., Vandervende, L.: Probabilistic frame induction. In: Proceedings of NAACL, HLT (2013)
17. Chen, C.-M., Chen, L.-H.: A novel approach for semantic event extraction from sports webcast text. In: Multimedia Tools and Applications, December 2012
18. Alfonseca, E., Pighin, D., Garrido, G.: HEADY: news headline abstraction through event pattern clustering. In: ACL 1, pp. 1243–1253. The Association for Computer Linguistics (2013)
19. Chelba, C., Mikolov, T., Schuster, M., Ge, Q., Brants, T., Koehn, P., Robinson, T.: One billion word benchmark for measuring progress in statistical language modeling (2013). arXiv preprint arXiv:1312.3005
20. Drury, B., Almeida, J.J.: Identification of fine grained feature based event and sentiment phrases from business news stories. In: Proceedings of the International Conference on Web Intelligence, Mining and Semantics, p. 27. ACM, May 2011

Author Index